T0318845

Geothermal Power Plants: Principles, Applications, Case Studies and Environmental Impact

Third Edition

Ronald DiPippo, Ph.D.
Chancellor Professor Emeritus
University of Massachusetts Dartmouth
North Dartmouth, Massachusetts

AMSTERDAM • BOSTON • HEIDELBERG • LONDON • NEW YORK • OXFORD
PARIS • SAN DIEGO • SAN FRANCISCO • SINGAPORE • SYDNEY • TOKYO
Butterworth-Heinemann is an imprint of Elsevier

Butterworth-Heinemann is an imprint of Elsevier
The Boulevard, Langford Lane, Kidlington, Oxford, OX5 1GB, UK
225 Wyman Street, Waltham, MA 02451, USA

First edition 2005; Second edition 2008; Third edition 2012

Copyright © 2012 Elsevier Ltd. All rights reserved

No part of this publication may be reproduced or transmitted in any form or by any means, electronic or mechanical, including photocopying, recording, or any information storage and retrieval system, without permission in writing from the publisher. Details on how to seek permission, further information about the Publisher's permissions policies and our arrangements with organizations such as the Copyright Clearance Center and the Copyright Licensing Agency, can be found at our website: www.elsevier.com/permissions.

This book and the individual contributions contained in it are protected under copyright by the Publisher (other than as may be noted herein).

Notices
Neither the author nor the Publisher, nor any of its employees, makes any warranty, express or implied, or assumes any legal liability or responsibility for the accuracy, completeness, or usefulness of any information, apparatus, product, or process disclosed, or represents that its use would not infringe privately owned rights. Reference herein to any specific commercial product, process, or service by trade name, mark, manufacturer, or otherwise, does not necessarily constitute or imply its endorsement, recommendation, or favoring by the author or the Publisher.

Knowledge and best practice in this field are constantly changing. As new research and experience broaden our understanding, changes in research methods, professional practices, or medical treatment may become necessary.

Practitioners and researchers must always rely on their own experience and knowledge in evaluating and using any information, methods, compounds, or experiments described herein. In using such information or methods they should be mindful of their own safety and the safety of others, including parties for whom they have a professional responsibility.

To the fullest extent of the law, neither the Publisher nor the authors, contributors, or editors, assume any liability for any injury and/or damage to persons or property as a matter of products liability, negligence or otherwise, or from any use or operation of any methods, products, instructions, or ideas contained in the material herein.

Library of Congress Cataloging-in-Publication Data
DiPippo, Ronald.
Geothermal power plants : principles, applications, case studies, and environmental impact / Ronald DiPippo. – 3rd ed.
p. cm.
Includes bibliographical references and index.
ISBN 978-0-08-098206-9
1. Geothermal power plants. 2. Geothermal resources. 3. Geothermal engineering. I. Title.

TK1055.D57 2012
621.44–dc23

2011052963

British Library Cataloguing-in-Publication Data
A catalogue record for this book is available from the British Library

For information on all Butterworth-Heinemann publications
visit our website at www.elsevierdirect.com

Printed in the United States of America
12 13 14 15 10 9 8 7 6 5 4 3 2 1

Working together to grow
libraries in developing countries

www.elsevier.com | www.bookaid.org | www.sabre.org

ELSEVIER BOOK AID International Sabre Foundation

Dedication

For my wife
Joan Marjorie DiPippo

for her enduring patience and support
throughout the writing of this book
from the very beginning.

Contents

Foreword to the Third Edition

The third publication and expanded edition of Ron DiPippo's *Geothermal Power Plants* is timely considering the renewed interest in geothermal energy development across the globe.

The potential of geothermal heat energy is enormous; it's estimated that, over the course of a year, the equivalent of more than 100 million GWh of heat energy is conducted from the Earth's interior to the surface. But geothermal energy tends to be relatively diffuse, a phenomenon making it difficult to tap. The Earth itself concentrates geothermal heat in certain regions – typically, regions associated with the boundaries of tectonic plates in the planet's crust – and this makes geothermal energy useful as a heat source for direct use or electricity production using today's technology.

Geothermal resources have not received a great deal of focus in the overall energy picture. This is despite the fact that direct use of geothermal energy is commonplace in some nations. Iceland uses geothermal a great deal for heating and electricity purposes, as does Eastern and Western Europe, where geothermal heating has been used for centuries, and where the first geothermal power generation occurred in Italy in 1904.

Geothermal electricity generation has progressed in spurts over the last 50 years, with the first modern facilities built during the late 1950s and the 1960s, and development expanding significantly following the oil crises of the 1970s and early 1980s. Development activity waned, particularly in the U.S. during the 1990s, but has picked up significantly over the past decade with rising oil prices and global focus on carbon reduction.

In addition to a growing U.S. market, new projects are under construction and development worldwide, from Africa to Central and South America, to the Caribbean islands, to Asia, Oceania, and Europe.

This increased development activity has triggered investment in nonconventional technologies such as Engineered Geothermal Systems (EGS) and coproduction with oil and gas wells. Further investment has occurred in Iceland and Italy where developers are drilling into supercritical reservoirs with hopes to produce 50 MW from single wells.

EGS is important, in particular to improve the performance of marginal geothermal wells that are now considered "sub-commercial" or "dry." It can contribute significantly with new techniques currently being field-tested that may apply to existing fields or be used to create entirely new fields that were once considered lacking in permeability.

As seen with other alternative energy technologies such as wind and solar investment and innovation spur growth dramatically once commercialization is achieved. Whether it be larger wind turbines or more efficient solar cells, the future of geothermal energy relies on similar improvements, particular in drilling and resource exploration to reduce costs and improve efficiency in development activities. Today, the geothermal industry is at a turning point. With nations looking beyond oil and coal and even nuclear power in the case of Germany and Japan, serious efforts are being launched to create a thriving geothermal industry.

In geothermal energy development the multidisciplinary approach is a must. The disciplines range from geology, hydrology, geophysics, drilling, and well design to power plant engineering.

Today's revival of activity in the geothermal industry corresponds with a new generation of young professionals entering the job market in a time of high oil prices and concerns about climate change. This new generation is passionate about alternative energy and their potential to lead the transition from fossil-fuel powered electricity to innovative technologies providing cleaner forms of energy.

This Third Edition fills the space of the previous version and draws from both theory and practical examples of operating power plants. It covers the complete life cycle of development from resource exploration to construction of the power plant to sustaining the hydrothermal reservoir. This edition has added potential to further promote geothermal energy development as it includes elements not previously covered. This edition also emphasizes the concept of exergy-based utilization efficiency, which is crucial for geothermal power plants.

This volume proves beneficial for engineers and designers of technology, as well as for developers, financiers, and agencies lacking a basic knowledge of geothermal energy. While geothermal development has its challenges, the industry has tapped just a fraction of the vast potential of the global geothermal resource. With nearly 11 GW-electric installed at midyear 2011, the most obvious and easiest to access resources have been developed. Had oil exploration ceased after only the most observable fields had been discovered, supply would have expired many decades ago. Thus the value and importance of this book is to illuminate the challenges and opportunities in order to move forward to development of innovations that will further unlock the potential of this resource.

As a comprehensive guide to geothermal technology and development, this book serves as a critical reference for students, engineers, and researchers to help in their particular area of study, as well as gain a better understanding of the industry as a whole. This book is intended for mechanical, electrical, chemical, industrial, and power engineers; system designers; power plant technicians; geoscientific researchers; project developers and managers; and advanced students. It contains both theoretical background material and examples that can be used for training of design engineers. Most engineers are trained more generally and, as with any energy technology, specialization and mastery of the nuances of the technology are required for further innovation. This book familiarizes the reader with the progress in research and development, and charts a course for future improvements so that those mastering the material can "hit the ground running."

DiPippo's work has practical application for today's geothermal industry. Leading geothermal development companies like Ormat have used the material in previous editions to train engineers and other employees. At Ormat, we have also used this material during development of different configurations of geothermal power plants to maximize efficiency for each resource and to manage specific reservoirs to produce the most energy possible without compromising the sustainability of the geothermal field.

As the Chairman and Chief Technical Officer at Ormat, I intend to use this publication to guide our engineers and scientists during development activities and encourage others to do the same.

Lucien Y. Bronicki
Ormat Technologies, Inc.
Reno, Nevada
September 1, 2011

Preface and Acknowledgements o the Third Edition

his edition represents a major revision and extension of the original edition published n 2005. In the roughly six years since the First Edition appeared, the inexorable novement toward cleaner and more sustainable energy resources has continued. Despite growing evidence of global changes to our environment that will have huge onsequences for future generations, the world has not moved as rapidly toward less polluting and more sustainable sources of primary energy as might be expected in the ace of the coming crisis.

Nevertheless geothermal energy can become a significant source of renewable, sus-ainable, and clean energy to the world, both for direct heat and electric power genera-ion. In the latter case – the subject of this book – it is worthwhile to repeat the words of my mentor Professor Joseph Kestin, who back in 1979, wrote: "*Geothermal lectricity, unlike fossil or nuclear, cannot be ordered: it must be developed, for there is nothing more hazardous than a premature order for conversion equipment.*" Apparently this simple lesson needs to be relearned by each generation and especially by those who aim o develop new sites, new resources, or new technologies.

It was in fact this unique aspect of geothermal energy that intrigued me when I first became involved with it back in the 1970s. As an academic and a researcher, I looked for he basic principles that would guide the exploitation of geothermal energy. Beyond the undamental geology, chemistry, and physics, much of what was needed turned out to be highly site-specific. Back then there were few geothermal plants in operation and nearly all of them were dry-steam plants. Even so there were significant differences among the esources that necessitated specific engineering designs to cope with the variations.

Now that a very wide spectrum of resources is being tapped to generate power, the need o study and thoroughly characterize each site and then craft the power plant to match he resource is even more important. We may understand the basic science that governs he processes, but the applications will require innovative solutions to problems encoun-ered all the way from exploration through drilling and plant design, construction, opera-ion, and maintenance. Therefore, this edition includes more case studies in Part Three.

Chapter 19 focuses on two new and growing power plants in Iceland, the Nesjavellir and Hellisheidi plants. Iceland has experienced a spectacular increase in geothermal power generation since 2007, growing its installed capacity from 422.4 MW to 715.4 MW, a 69% increase over four years. And this has been accomplished without retiring any of its older plants. Iceland now has 31 operating units compared to 24 in 2007.

Chapter 20 tells the story of Raft River in Idaho, one of the first fields to be developed for geothermal power back in the 1970s and early 1980s. The original pilot plant was soon dismantled after a short demonstration run but after 20 years of inactivity, the site has been brought back to life. The lessons learned from the original plant and the new one are presented in this chapter.

Chapter 21 highlights another country making great strides in geothermal development, thanks in large measure to legislative reforms that opened geothermal energy to private companies. Turkey now is host to eight geothermal units, three of them modern gleaming examples of highly efficient units. Turkey went from having two units with an installed capacity of 27.8 MW in 2007 to eight units at 95 MW in mid-2011, more than tripling its geothermal generating capacity in four years. And more plants are expected to be built in the near future.

Chapter 22 deals with the future hope for geothermal power, Enhanced Geothermal Systems or EGS. In the Second Edition, EGS was covered as part of the presentation of advanced energy conversion systems (Chapter 9) but so much has happened in the last four years that even a chapter devoted solely to EGS is not adequate to address the subject. It is a tale full of hope yet disappointments. It is not yet certain what role EGS will play in the development of geothermal power, but it remains a most important technology that must be further refined before it can be truly considered commercial. The new chapter details many research and development efforts and the lessons that should be learned from them.

The other new addition is an appendix that explains how a convenient new software program, REFPROP, developed by and available from the U.S. National Institute of Standards and Technology (NIST), can be used to implement spreadsheet systems simulators. Appendix H includes a tutorial on the use of the program and worked examples showing how to use it in conjunction with Excel to model a geothermal binary plant.

All the data tables on power plants around the world have been completely updated to mid-2011 (Appendix A). Updates and revisions have been made to several chapters from the Second Edition, with a new section on solar-geothermal hybrid plants (Chapter 9) and an extension of exergy analysis to pumps and to production well performance (Chapter 10). A couple of typos from the Second Edition have been corrected and hopefully not too many new ones have crept into this one.

I wish to thank the following people for various forms of assistance during the writing of the Third Edition. First of all, I extend my deepest appreciation to Lucien Bronicki, the Chairman and CTO of Ormat, for writing the Foreword. The previous editions lacked a Foreword and to have one of the most successful geothermal entrepreneurs write one for this edition is very special and meaningful to me. Several people provided valuable information on the plants in Turkey: Umran Serpen (Instanbul Technical University); Marshall Ralph and Bill Harvey (Power Engineers); Henry Veizades (Veizades & Associates); and Riza Kaderli (Guris). Lucien Bronicki and Zvi Kreiger (Ormat) generously provided me with flow diagrams and performance data for several of their recent binary installations, and Lucien reviewed some new sections for accuracy. Einar Gunnlaugsson (Orkuveita Reykjavíkur) clarified the historical timeline of development for the Nesjavellir and Hellisheidi plants. Ernst Huenges and Stephanie Frick (Deutsches GeoForschungsZentrum-GFZ) were kind enough to send me a copy of heir excellent book, *Geothermal Energy Systems*, that was edited by Ernst and which as extensive research material related to EGS. Tiffany Gasbarrini, my editor at lsevier, encouraged me to tackle this new edition and steered it through the proposal, pproval, and implementation process in a professional and efficient manner. The lsevier production team, particularly Lisa Lamenzo and Charles Roumeliotis, did an utstanding job of keeping the book on time and as free of errors as is humanly possi-le. I finally thank all those who have helped me grasp the nuances and subtleties of eothermal power plants in various ways through planned and random conversations; ny errors of commission or omission are solely mine.

The color figures are designated with [WWW] at the end of the caption. All of the gures used in this book may be viewed at elsevierdirect.com/companions/9780080982069.

onald DiPippo
artmouth, Massachusetts
ugust 2011

Preface and Acknowledgements to the Second Edition

In the roughly three years since the First Edition of this book appeared, the world has become increasingly alarmed about the consequences of global climate change. The debate over the causes of the documented warming of the planet has essentially come to an end. Attention is focused on collective actions that must be taken now to slow, or reverse, the rising temperatures and to stabilize the situation as soon as possible.

So-called "greenhouse gases" such as carbon dioxide have been identified as the main contributors to global warming. Since the dawn of the Industrial Age some two hundred years ago, humans have been mining and converting carbon, in the form of hydrocarbons stored for millions of years underground, into carbon dioxide. The combustion of fossil fuels is largely responsible for the quality of life that many have come to take for granted and which appears threatened.

Given the diversity of economies and cultures among the world's countries, it may be futile to hope for a coherent, comprehensive, and effective global energy policy. Such a policy must begin by acknowledging the critical role of energy, particularly electricity, in improving the standard of living for all peoples. It must then find ways to minimize the environmental impact of converting energy resources to serve a myriad of applications. As fossil fuels, especially petroleum, become scarcer, it is likely that the world will rely more heavily on electricity for many daily needs, including vehicular transportation. Thus, the use of renewable, non-polluting sources of electric power will be in great demand.

Geothermal energy has been just such an environmentally-friendly supplier of electricity for over 100 years. The First Edition of this book had several sections devoted to the environmental effects of geothermal power generation, but lacked a thorough, self-contained treatment of the subject. Chapter 19 of this edition answers that deficiency. The presentation is objective, and covers the environmental strengths and weaknesses of geothermal power technology. The strengths, relative to other means of generating power, stem from the inherent nature of geothermal energy and how it is used to generate electricity. The potentially detrimental aspects are site-specific, vary in severity depending on the characteristics of a given resource, and present challenges to the developers of geothermal plants. Fortunately these potential problems have been and continue to be addressed with effective mitigation techniques. The result is that geothermal energy has an excellent reputation as one of the most environmentally benign sources of electricity.

Despite its advantages in a world increasingly looking for clean sources of energy, geothermal power has experienced only slow, albeit steady growth over the last 10–15 years (see Appendix A, Fig. A.1). The strong growth experienced in the aftermath of the oil shocks of the 1970s has not been sustained. The availability and price of oil continues to be an important influence on the growth of geothermal power. Also the lack of new discoveries of sizeable hydrothermal resources has certainly been a contributing factor in this slowdown.

The next major growth period may result from the perfection of techniques being developed to create Enhanced or Engineered Geothermal Systems or EGS (see Sect. 9.8). With research starting in 1973 in the United States and continuing in several countries, EGS holds out the promise of affording any region the opportunity to generate power from the hot rocks that lie beneath the surface at great depths. Encouraging work is being done in Europe and Australia in areas lacking typical hydrothermal systems that could provide a template for other regions. A 2006 MIT study by a panel of experts from a variety of disciplines concluded that, with appropriate funding for research, development, and demonstration, as much as 100,000 MW could be competitively on-line by the year 2050 in the United States alone (see Ref. [1] in Chap. 19).

This edition includes updated statistics on the state of geothermal power plants around the world (Appendix A), as well as 43 new practice problems, along with the answers to many of them, augmenting the ones in the First Edition (Appendices F and G).

One of the objectives of the new edition was to produce a less expensive volume. To achieve this we eliminated color illustrations. However, over 70 color photographs and illustrations are available to readers via the Internet. Considering that there were only 15 color illustrations in the First Edition, this constitutes a significant enhancement for the reader.

I owe a debt of thanks to Jonathan Simpson (Elsevier) for all his help in making this new edition a reality, and to the reviewers of the prospectus who convinced him that this was a worthwhile venture. José Antonio Rodriguez (LaGeo S.A. de C.V.) generously gave permission for the use of the Berlín power plant photograph that graces the cover, and for the aerial photograph of the Ahuachapán plant (Fig. 19.12); both plants are in El Salvador. Curt Robinson (Geothermal Resources Council) likewise granted permission to reproduce the photograph of the Zunil landslide (Fig. 19.9). I am grateful to Ian Thain (Geothermal & Energy Technical Services Ltd.) who shared some of his historical recollections from Wairakei, and to Richard Glover (Glover Geothermal Geochemistry) for clarifications regarding the Wairakei Geyser Valley and his review of Sect. 19.5.1. Dan Schochet (Ormat) and Larry Green (Geothermal Development Associates) provided the latest information on their companies' installations. Lastly, I thank my wife, Joan, for her meticulous proofreading of the new material. Her pages of corrections reminded me that somewhere in my life I could have benefited from a course in typing.

Ronald DiPippo
Dartmouth, Massachusetts
May 2007

reface and Acknowledgments o the First Edition

t was the Fourth of July in the year 2004. Brilliant fireworks streamed across the vening sky. Had this been any of a thousand cities and towns in the United States, he event would not be unusual. But why would a tiny village in the Tuscany region f Italy be celebrating on America's Independence Day?

One hundred years ago on that very day in the village of Larderello, geothermal nergy was born as a source of electric power generation. Over the following century, eothermal energy has grown to be the most reliable, efficient, and environmentally-enign renewable source of electric power. Indeed, more than twenty countries across he world where geothermal energy is now used for power production could have een celebrating this historic event.

Geothermal energy is the residual thermal energy in the earth left over from the lanet's origins. The temperature of the core of the earth is estimated to be 6,650°C 12,000°F). The molten rock that flows inexorably beneath the relatively thin surface rust attains temperatures of about 2,200°C (4,000°F). The fact that living creatures can valk on the surface of the earth without feeling the effect of this terribly hot fluid testifies o the insulating nature of the solid rock that separates the surface from the molten rock.

If one were to drill a very deep well from the surface into the earth and were to neasure the temperature as a function of depth, one would find a fairly constant ncrease of about 3°C for each 100 m of depth (1.6°F per 100 ft). This rate is typical in ormal areas, but there are anomalous regions associated with volcanic or tectonic ctivity where the temperature gradients far exceed these normal values. For example, t Larderello, Italy where the first engine driven by geothermal steam was built, the radient is 10–30 times higher than normal. This means that temperatures over 00°C (575°F) can be found at a depth of 1 km (3,300 ft), easily reachable with oday's drilling technology. If fluids exist in reservoirs at such elevated temperatures nd can be brought to the surface through wells, they can serve as the working fluids or electric generating stations. Such fluids are complex mixtures of high-pressure, nineral-laden water, with gases dissolved into solution as a result of the contact of the ery hot water with various types of rocks in the reservoir.

It is the challenge of geothermal scientists and engineers to locate these reservoirs f hot fluids, design means to bring them to the surface in an economical and reliable ashion, process them in a suitable power plant to generate electricity, and then to dis-ose of the spent fluid in an environmentally acceptable manner, usually by returning he fluid to the reservoir through injection wells. The system must be so designed to erform satisfactorily for at least 25–30 years to be deemed economically viable.

The engineering challenge is further heightened by the variable nature of geother-nal reservoirs and fluids. While it is possible to categorize reservoirs and fluids into several general types, the detailed characteristics of each reservoir vary to the exten that each site must be studied and well understood before the resource can be properl developed. The development process includes the determination of the location, depth orientation, number, and type of wells that are needed; the type and size of powe plant to be built; the method of disposal of the spent geothermal fluid; and the type o abatement systems that may be needed to conform to local environmental regulations All of these must be addressed in order to perform a preliminary economic analysis t decide on the project's feasibility.

It is not uncommon for the preliminary work to take several years, and in the case o particularly difficult fluids, a decade or more, before a power plant can be put into opera tion. The geothermal field near the Miravalles volcano in Costa Rica was discovered i 1976 but the first power unit did not come on-line until 1994. The time devoted t geoscientific studies is necessary to avoid unwise investment in risky ventures. Once field is properly understood, however, developers can often proceed with additional unit in short order. At Miravalles, Unit 1 (55 MW) was followed by a small wellhead uni (5 MW) in 1995, Unit 2 (55 MW) in 1998, Unit 3 (27.5 MW) in 2000, and Unit (15.5 MW) in 2004. A geothermal reservoir adjacent to the Salton Sea in souther California was discovered accidentally in the 1950s, but the extremely aggressive geo thermal brines resisted commercial use until the 1980s. It took a massive research an development effort to come up with technological methods to tame these very hot highly corrosive and scaling fluids. These technical advances led to the construction o six power plants with a total installed capacity of over 650 MW as of 2004.

The uses to which geothermal energy can be put cover a wide spectrum fro low-temperature applications, such as greenhouse heating and aquaculture, to high temperature applications, including absorption chillers for refrigeration and air condition ing and power generation. Normal earth conditions can be exploited by earth-coupled hea pumps for year-round climate control in buildings, so-called Geoexchange® systems. Th spectrum of applications is given by the well-known Lindal diagram shown in Figure I.1.

This book focuses on the higher-temperature end of the spectrum where geotherma energy can provide for the generation of electricity – generally regarded as the highes grade and most useful form of energy known to humankind.

The first part of this book deals with the geological nature and origins of geotherma energy, as well as the exploration efforts that are needed in the early stages of th development process. The means of acquiring the geothermal fluids via productio wells conclude the first part.

The designer of geothermal power plants nowadays has available a wide assortmen of plant designs to cope with the wide diversity of geothermal resources. This was no always the case, as was mentioned earlier for example with the Salton Sea resource Through research and painful trial and error, there is now no resource that cannot b exploited by means of a suitably designed system.

In the second part, we describe the many types of energy conversion systems tha are being used to generate electricity from geothermal energy. Both basic principle and practical applications will be covered. Included are some advanced systems tha may one day reach the commercial stage.

The third and last part of the book presents several case studies of power plant selected from six different countries around the world. Examples include plants o

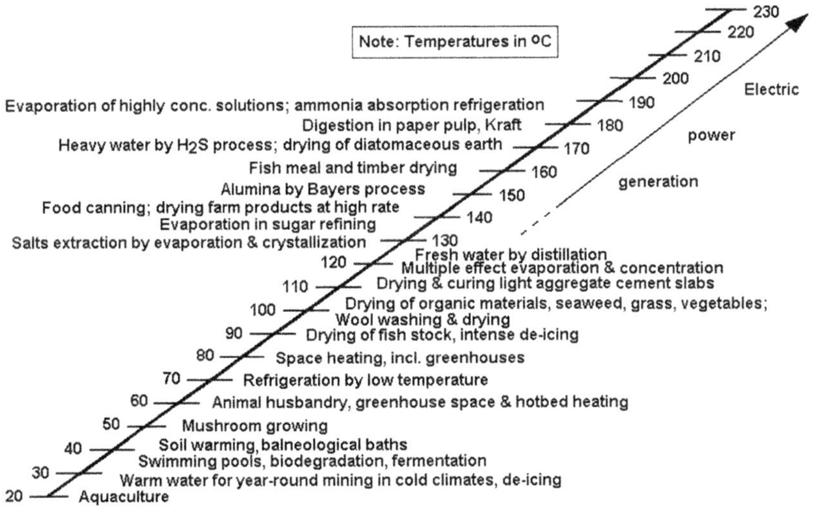

ig. I.1 Modified Lindal diagram showing applications for geothermal fluids.

lifferent types and sizes, using fluids with various characteristics, in areas having videly differing environments. Each case typically includes a description of the geol- gy, the resource, the environment, the wells and piping system, the energy conver- ion system(s), the plant performance, and operating experience. Since geothermal esources are dynamic, it is more than likely that some of the power plants or the geo- luid gathering systems will be different as of the reading of this book relative to the vriting of it (mid 2004). Geothermal engineers and scientists must be prepared to cope vith changes in the properties of the reservoir and the wells, and to modify the energy onversion systems accordingly, over the life of the plant.

In most chapters, I have included quotations, some from long ago and some recent, ome from famous people and some from folks who are known to relatively few. These vords provide insight into the nature of scientific and geothermal phenomena, and emind us of the challenges we face in turning Nature's bounty to the benefit of umankind.

I published my first book on geothermal energy in 1980: Geothermal Energy as a ource of Electricity. That effort was funded by the U.S. Department of Energy. I owe a lebt of thanks to the late Joseph Kestin, the project leader at Brown University, for giv- ng me the opportunity to work on his exciting project that also led to the publication of the reference book, Sourcebook on the Production of Electricity from Geothermal nergy. I am also indebted to Clifton B. McFarland, the DOE program manager, for his ar-sighted leadership and encouragement. During the 1970s and 1980s when I was a esearch professor at Brown University, I benefited greatly from discussions and colla- orations with H. Ezzat Khalifa, now an engineering professor at Syracuse University, vith whom I published several papers on hybrid fossil-geothermal power plant sys- ems. I also wish to thank Gustavo Calderon, now retired from the Interamerican Development Bank, for giving me the opportunity, starting in the early 1980s, to broaden my geothermal experience internationally through service on geothermal advisory panels in Costa Rica, Guatemala, El Salvador and Kenya.

I acknowledge with deep appreciation Gordon Bloomquist (Washington State U.) and John Lund (Oregon Institute of Technology) for their reviews of the prospectus for this book and for their thoughtful comments and suggestions. Ted Clutter (Executive Director of the Geothermal Resources Council) was always ready to help me locate or identify a photo or reference. He also provided me with his photograph of the geothermal power plants at Mammoth, California, which graces the cover of the book.

My editor at Elsevier Advanced Technology, Geoff Smaldon, has been supportive and encouraging throughout the writing and publication process. Thanks to the power of electronic communication, he and I have remained in close contact even though he is in Oxford, England, and I in Dartmouth, Massachusetts. I compliment Geoff and the publication staff at Elsevier for their efficiency in producing this volume in a remarkably short time.

During the writing of this book I received the cooperation of many other individuals in the geothermal community and elsewhere who generously granted permission for me to use their previously published works, to reprint their photographs, who helped track down elusive references, and who offered suggestions that I believe improved the content and presentation of the material.

At the risk of inadvertently offending an overlooked colleague, I am indebted to the following individuals: Stefano Bellani and Marnell Dickson (IGG/CNR), Guido Cappetti, Adolfo Fiordelisi and Iris Perticone (Enel GreenPower), Jamie Claire and Tsuyoshi Yamada (Toshiba), Julie Gonzalez and Christine Hopf-Lovette (EPRI), Steve Cole (Memphis, TN), Paige Gibbs (UMass Dartmouth), Luiz Gutiérrez-Negrín (CFE), Shaun Hardy (Carnegie Inst.), Tom Haynes (Hughes Christensen), Susan Hodgson (CA DOGGR, ret.), Roland Horne (Stanford U.), Donald Hudson (Reno, Nevada), Eduardo Iglesias (IGA), John Jacobson (Kuster), Marcelo Lippmann (LBNL), Kenneth McCombs (McGraw-Hill), Ed McCrae, Shunji Nakamura and Ricky Takada (MHIA), Dick Meeuwig (Nevada BMG), David Michetti (Calpine), Paul Moya (ICE), Oleg Povarov (Assn. Geothermal Energy Soc.), Andrea Rossetto (Torreglia, Italy), Paolo Santini (GE-Florence), Subir Sanyal (GeothermEx), Dan Schochet and Graciela Sapiro-Goldman (Ormat), Marlene Vogelsang (PG&E), and François-D. Vuataz (U. Neuchâtel).

Lastly, I thank my wife, Joan, for her patience, understanding, and support throughout the many months of my self-imposed exile as this work took shape. She was always eager to read drafts and could be counted on to offer valuable suggestions for a clearer exposition of the subject matter.

Ronald DiPippo
Dartmouth, Massachusetts
March 2005

Part 1

Resource Identification and Development

- Geology of Geothermal Regions
- Exploration Strategies and Techniques
- Geothermal Well Drilling
- Reservoir Engineering

"A man must stand in fear of just those things that truly have the power to do us harm, of nothing else, for nothing else is fearsome."

Dante Alighieri, *The Divine Comedy: The Inferno* – 1306–1321

The first part of the book deals with the geological aspects of geothermal resources – how the forces of nature shaped the earth in a way to create reservoirs capable o supplying energy for geothermal power plants. We discuss the means to identify anc characterize geothermal prospects, and the techniques for drilling wells into geotherma formations to extract the hot fluids for use in power stations. The last part of this sectior of the book examines the physical principles of fluid flow through the porous rock that constitute the reservoir, and the modern computer simulation methods tha are used to model the behavior of the reservoirs. The color figures are designated witl [WWW] at the end of the caption. All of the figures used in this book may be viewed a elsevierdirect.com/companions/9780080982069.

(Left) Volcan Rincon de la Vieja, Guanacaste province, Costa Rica
Location: 10.8N, 85.3W
Elevation: 6,286 feet (1,916 meters)
Photo by Federico Chavarria Kopper, published by Smithsonian Inst.
Global Volcanism Program website: http://www.volcano.si.edu/world/volcano.cfm?vnum=1405-02= [WWW].

(Above) Volcan Pacaya, Guatemala (foreground)
Location: 14.38N, 90.60W
Elevation: 8,371 feet (2,552 meters)
Volcan Agua (background)
Location: 14.5N, 90.7W
Elevation: 12,333 feet (3,760 meters)
Ref: Volcano World, U. of N. Dakota:
http://volcano.und.nodak.edu/vwdocs/volcimages/south_america/guat/pacaya.html [WWW].

Chapter 1

Geology of Geothermal Regions

"Birth and death. Like us, geothermal features begin and end, moving through cycles of their own. We draw towards them, lured by change, beauty, and an unusual cast of the familiar – water, rocks, and heat. We search them for answers to mysteries in our own lives, like birth and death."

Susan F. Hodgson – 1995

1.1 Introduction

Geothermal energy – Earth heat – can be found anywhere in the world. But the high-temperature energy that is needed to drive electric generation stations is found in relatively few places. The purpose of this opening chapter is to provide the geologic framework within which high-temperature geothermal resources can be understood, both with regard to their occurrence and their nature.

Readers who are unfamiliar with the rudiments of Earth science may wish to consult any of the standard texts on the subject, e.g., Refs. [1–4]. Those interested in the history of geologic thought, dramatic geological events, and ancient geothermal energy usage will find fascinating reading in Refs. [5–8]. W.A. Duffield provides an excellent, brief introduction to modern geologic theory of volcanoes in a beautifully illustrated book [9]. In selecting general texts on geology, one must be aware that any book written before 1970 will not include the most recent thinking on the structure of the Earth and the dynamic mechanisms that give it its life. We refer to the theory of

Geothermal Power Plants: Principles, Applications, Case Studies and Environmental Impact, Third Edition.

© 2012 Elsevier Ltd. All rights reserved.

plate tectonics, now universally accepted, which provides us with the basic tools to understand the origins of high-temperature geothermal resources.

1.2 The Earth and its atmosphere

In 1915 A.L. Wegener (1880–1930) put forth a highly controversial theory of continental drift in the first edition of his book *The Origin of Continents and Oceans* [10]. Although he elaborated on it in later editions of his book in 1920, 1922 and 1929, the controversy persisted. His theory was motivated by the observation that the continents, particularly South America and Africa, seemed to be pieces of a global jig-saw puzzle that had somehow been pulled apart. He reasoned that all land masses were once connected in a gigantic supercontinent he named "Pangaea." He posited that the now separated continents floated and drifted through a highly viscous sea floor. This part of his theory was later proved incorrect but the basic notion of drifting continents was right. Wegener's problem was in identifying correctly the forces that ripped apart the pieces and in fact keeps them moving.

Studies that began in the 1950s and continued into the 1960s matched the ages of rocks found along the northeastern coast of South America and the northwestern coast of Africa [11]. The correlation of rock ages ran from Recife in Brazil to Trinidad off the coast of Venezuela on the South American side, and from Luanda to Sierra Leone on the African side. Oceanic research also showed that new land was being created on either side of the mid-Atlantic ridge, the so-called "sea-floor spreading" phenomenon [12]. By dating these deposits, Earth scientists were able to confirm the movement of the vast plates that constitute the crust of the Earth. Continents are part of the crust and have been in constant motion since the beginning of the Earth some 4.5 billion years ago.

An excellent animation of this motion starting about 740 million years ago can be viewed at the web site of the University of California at Berkeley's Museum of Paleontology [13]. From this animation it is clear that Pangaea existed as a supercontinent for only a blink of geological time, around 200 million years ago, having itself been formed from the collision of several land masses beginning in the Precambrian era.

While there is no controversy today over the theory of plate tectonics, there remains much uncertainty about the detailed structure of the inner Earth. A great deal of research has gone into exploring and characterizing the Earth's atmosphere but only one or two projects have aimed at probing the depths of the Earth. One of them, Project Moho, intended to drill through the thinnest part of the oceanic crust (about 5 km thickness) to enter the mantle. In 1909 Croatian scientist A. Mohorovičic (1857–1936) had observed, at a certain depth, a discontinuity in the velocity of seismic waves caused by earthquakes. He deduced that this represented a boundary between the generally solid crust and the generally molten mantle. This interface has become known as the Mohorovičic Discontinuity (or simply the Moho) in his honor [1]. However, Project Moho was halted in 1966 apparently for lack of funds and produced no results.

Another deep drilling effort, the Salton Sea Deep Drilling Program, ran from 1984–1988 with funding from the U.S. Dept. of Energy but failed to achieve much [14,15]. One well was drilled to a total depth of 10,564 ft but suffered a collapsed liner at 6380 ft. Although this was later repaired, the deepest measurements were taken at 5822 ft and indicated a temperature of roughly 290°C. Neither the depth nor the able 1.1 **Data on the Earth and its atmosphere from various sources; distances are not shown to cale.**

Region	Distance from surface km	Temperature °C	Density g/cm3
	300	1125	3.6×10^{-14}
Thermosphere			
	85	−95	2×10^{-8}
Mesosphere			
	50	0	1×10^{-6}
Stratosphere			
	12	−60	3×10^{-4}
Troposphere			
Surface	0	10	2.7 continental 3.0 oceanic
Crust			
	35	1100	3.3
Mantle			
	2900	3700 to 4500	5.7 10.2
Liquid (iron) core			
	5100	4300 to 6000	11.5
Solid inner (iron) core			
(center)	6350	4500 to 6600	11.5

emperature was particularly remarkable given the state of geothermal drilling at the ime. At the conclusion of this effort, the following problems were cited as serious bar-iers to any future deep drilling program (to say, 50,000 ft): extremely high tempera-ures in the well, loss of control of the orientation of the well, lost circulation of drilling luids, and fishing for equipment lost downhole.

Currently there is an international consortium of eleven countries called the nternational Continental Scientific Drilling Program [16] that funds projects to give nsight into Earth processes and to test geologic models. So far the deepest proposed vell-drilling project received by the ICDP is for a 5000 m well in China; that well was eported to be at a depth of 3666 m on October 23, 2003 [17].

Thus our knowledge of the planet Earth beyond a depth of a few kilometers is based n indirect evidence. What we accept as the model for the Earth's inner structure is urdened with uncertainty, particularly the temperature as a function of depth. Table 1.1 summarizes the model of the Earth and its atmosphere; shown are the dis-ances from the surface of the Earth to each significant layer, the temperature thought o exist there, and the density. The crustal thickness is for continental areas; oceanic rusts are much thinner, about 7–10 km on average. The wide spread in the tempera-ures at the deepest levels reflects the speculative nature of these estimates.

These layers are usually depicted as concentric spheres, much like the inside of a golf ball, in ultra-simplified schematics. However, the interfaces are likely so irregular nd the boundaries so fuzzy that such a representation is misleading.

Table 1.2 Radioactive elements in common rocks in the Earth's crust.

Rock	Concentration			Heat generation, 10^{-6} cal/g · yr		
	U, ppm	Th, ppm	K, %	U	Th	K
Granite	4.7	20	3.4	3.4	4.0	0.9
Basalt	0.6	2.7	0.8	0.44	0.54	0.23
Peridotite	0.016	0.004	0.0012	0.012	0.001	0.0003

Sometimes the analogy is drawn between the Earth and a chicken's egg, with the Earth's crust compared to the shell of an egg. Relating the thickness of the Earth's crust, 35 km for continental regions, to its diameter, roughly 12,700 km, we get a ratio of 35/12,700 or 0.00276. If we apply the same ratio to an egg with a diameter of say, 50 mm, we would find a shell thickness of 0.138 mm or 0.0054 in. In fact the shell of an egg is about 1/64 in or about 0.016 in. Thus an egg's shell is about three times thicker proportionally than the crust of the Earth. Put in other words, if the Earth's crust were in proportion to the shell of an egg, it would be about 100 km thick instead of 35 km.

Since the temperature at the base of the crust is about 1100°C, the temperature gradient between the surface (assuming a surface at 10°C) and the bottom of the crust is 31.1°C/km or about 3.1°C/100 m. This is usually taken as the normal conductive temperature gradient. Good geothermal prospects occur where the thermal gradient is several times greater than normal. The rate of natural heat flow per unit area is called the normal heat flux; it is roughly 1.2×10^{-6} cal/cm^2 s, in non-thermal areas of the Earth.

The Earth's crust is composed of various types of rock which contain some radioactive isotopes, in particular, uranium (U-235, U-238), thorium (Th-232) and potassium (K-40). The heat released by these nuclear reactions is thought to be responsible for the natural heat that reaches the surface. Table 1.2 lists three rock types and their radioactive constituents.

These basic ideas are enough for us to move on to explore how the motion of the tectonic plates creates the conditions favorable for the exploitation of geothermal energy.

1.3 Active geothermal regions

The relative motion of plates, of any size, gives rise to several possible interactions. These are shown in Fig. 1.1.

When a plate comes under compression, it can relieve the stress by folding, by cracking and thrusting one piece upon the next, by cracking and trenching beneath the next, and by thickening. Trenching or subduction is one of the most important mechanisms that give rise to high-temperature geothermal regions. When a plate is subjected to tension, it can relieve the stress by cracking open and rifting, by cracking in several places leading to down-dropping, and by thinning. All of these responses to tension lead to anomalous geothermal regions that may be conducive to exploitation.

Two plates may also slide past each other along what is called a transform fault, perhaps the most famous of which is the San Andreas fault running along much of the

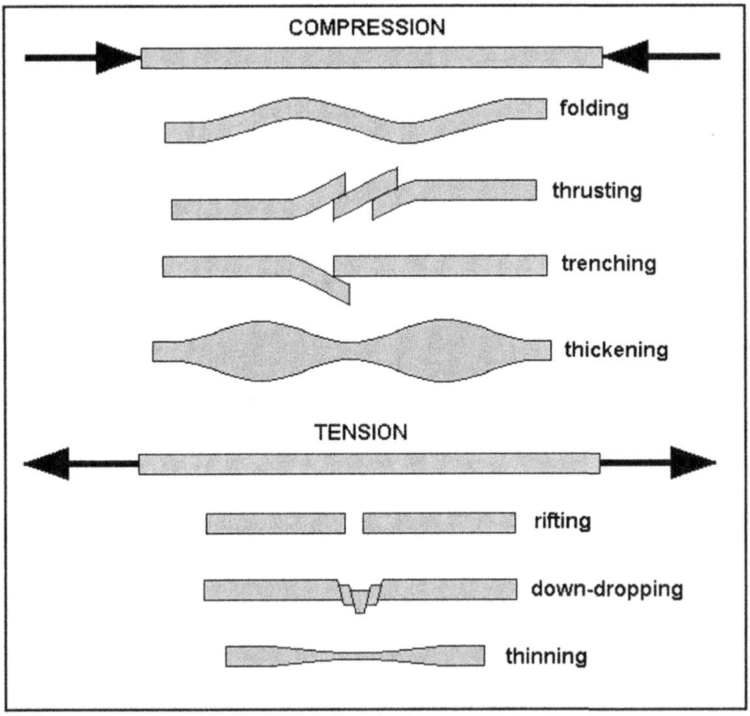

ig. 1.1 Response of plates to compression and tension.

ength of California in the United States [18]. While this fault, and others related to it, ave caused immeasurable harm and financial loss from numerous earthquakes, it also as given rise to several commercial geothermal resources that have been beneficial.

The coincidence of earthquake zones and geothermal regions has been depicted in NOAA map (see Fig. 1.2) showing the collocation of these two phenomena [19]. he loci of earthquakes correspond very closely with the boundaries of the massive rustal plates that comprise the tectonic model of the Earth's surface structure.

The plate boundaries experience various kinds of interactions as depicted in Fig. 1.1. rom the viewpoint of geothermal exploitation, the most important of these occur long the edges of the gigantic Pacific plate, the so-called "Pacific Ring of Fire." If we nclude the two adjacent eastern plates, the Cocos and the Nazca plates, as well as the vestern one, the Philippine plate, then the following countries (in clockwise order) are ffected: United States, Mexico, Guatemala, El Salvador, Honduras, Nicaragua, Costa Rica, Panama, Colombia, Ecuador, Peru, Bolivia, Chile, New Zealand, Micronesia, Papua New Guinea, Indonesia, Philippines, China, Japan, and Russia. All 21 of these ountries have exploitable geothermal resources and 13 of them have geothermal ower plants in operation as of mid-2007. Generally speaking, subduction zones exist eneath all land masses in contact with the Pacific, Cocos, and Nazca plates, except the ontiguous United States and Mexico where transform boundaries exist. The Alaskan

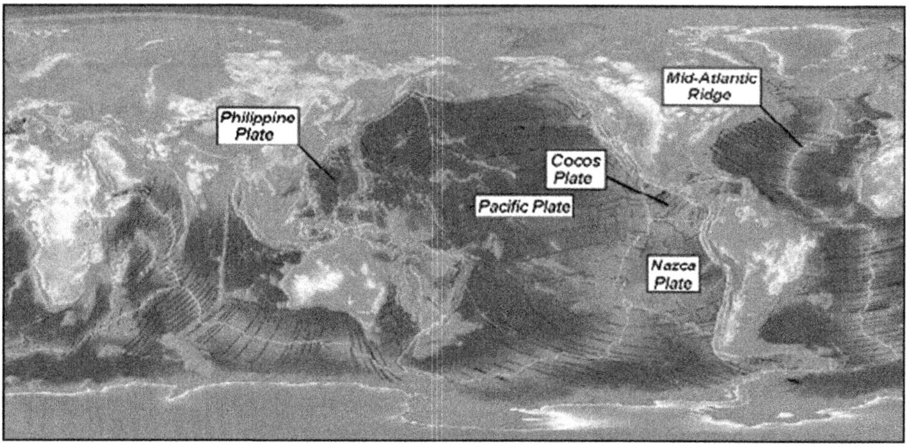

Fig. 1.2 Collocation of earthquakes, tectonic plate boundaries, and geothermal regions; after [19] [WWW].

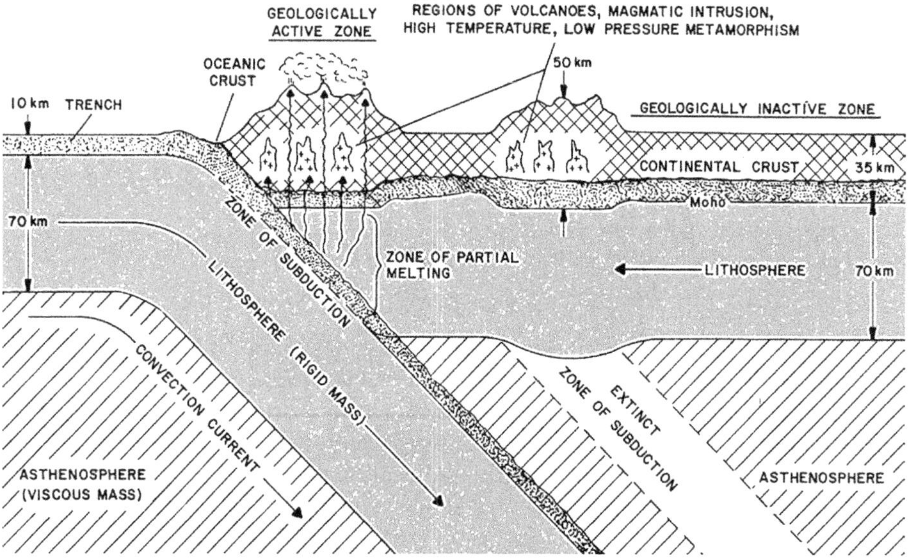

Fig. 1.3 Schematic model of a trench or subduction zone; after [20].

Aleutian Islands lie in a subduction zone and Hawaii lies over a localized hot spot in the middle of the Pacific plate.

The mid-Atlantic ridge, a rift zone, is home to Iceland and the Azores, two volcanic islands that have put their geothermal resources to practical use for many years. Other regions such as the Mediterranean and Himalayan belts and the East African Rift zone also are being exploited for geothermal power.

Figure 1.3 shows a conceptual diagram of a subduction zone typical of what is found along the Central and South American coasts [20]. Oceanic crust is sliding from left to right and diving beneath the continental land mass. The shaded slab labeled lithosphere s the very upper layer of the mantle which moves as a nearly rigid body. Below it lies the asthenosphere which is the molten but highly viscous part of the mantle. The trenching crust carries with it sea-floor sediments as it rubs against the edge of the continent. This relative motion is not smooth but marked by abrupt jolts whenever the stress built up in the interface exceeds the frictional resistance. This creates very deep-seated earthquakes that can have devastating effects. Nearly all the major cities of Central America, for example, have felt the power of such earthquakes. Partial melting of the lithosphere at the interface is believed to give rise to plutons of molten rock that tend to move upwards owing to their lower density and higher temperature than the native rock. It is the emergence of these magma bodies, sometimes violently, that account for the nearly unbroken string of active volcanoes from Guatemala to the southern tip of Chile. These magmatic zones offer ample heat sources to accompany high precipitation that occurs over the region to create many potential hydrothermal geothermal systems.

An example of a non-magmatic hydrothermal system is the geothermal province known as the Basin and Range in Nevada, U.S. This geologic region was created by extensional forces that have given rise to alternating ranges of mountains separated by valleys (basins). Steep faults associated with the edges of the mountain ranges are targets for geothermal development and currently support 50 geothermal power units, totaling 185 MW.

1.4 Model of a hydrothermal geothermal resource

There appear to be five features that are essential to making a hydrothermal (i.e., hot water) geothermal resource commercially viable. They are:

- A large heat source
- A permeable reservoir
- A supply of water
- An overlying layer of impervious rock
- A reliable recharge mechanism.

A highly schematic depiction of such a system is shown in Fig. 1.4, first presented by D.E. White [21].

Cold recharge water is seen arriving as rain (point A) and percolating through faults and fractures deep into the formation where it comes in contact with heated rocks. The permeable layer offers a path of lower resistance (point B) and as the liquid heats it becomes less dense and tends to rise within the formation. If it encounters a major fault (point C) it will ascend toward the surface, losing pressure as it rises until it reaches the boiling point for its temperature (point D). There it flashes into steam which emerges as a fumarole, a hot spring, a mud pot, or a steam-heated pool (point E). The boiling curve is the locus of saturation temperatures that correspond to the local fluid hydrostatic pressure. We will have more to say about this in Chapter 4.

The intent of a geothermal development project is to locate such systems and produce them by means of strategically drilled wells. As might be presumed, most (but not all) hydrothermal systems give away their general location through surface thermal manifestations such as the ones described above.

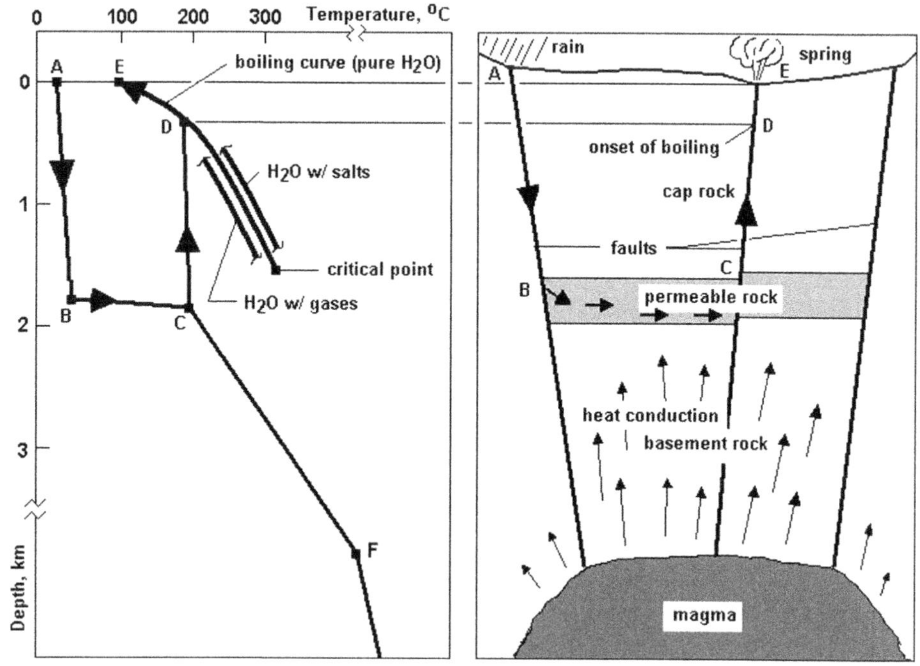

Fig. 1.4 Schematic model of a hydrothermal geothermal system; after [21].

If any one of the five features listed as needed for a viable hydrothermal resource is lacking, the field generally will not be worth exploiting. For example, without a large heat source geofluid temperatures will be relatively low, i.e., the thermal energy of the system will be insufficient to support exploitation long enough to make it economic. Without sufficient permeability in the formation, the fluid will not be able to move readily through it, i.e., it will not be able to remove much of the stored thermal energy in the rock. Furthermore, low permeability will cause poor well flow or, even worse, may prevent any production from the reservoir. Without fluid in the system there is no heat transfer medium and the thermal energy of the formation will remain in the reservoir. Without an impermeable cap rock, the geofluids will easily escape to the surface appearing as numerous thermal manifestations and the pressure in the formation will quickly dissipate. And lastly, without a reliable and ample recharge to the reservoir, the geofluid will eventually become depleted when it supplies a power plant.

With the exception of requirements (a) and (d), deficiencies in the others have been addressed through research and field practice. Insufficient permeability can sometimes be remedied by artificial means such as hydraulic fracturing (called "hydrofracking") in which high-pressure liquid is injected from the surface through wells to open fractures by means of stress cracking. However, unless the newly created widened fractures are held open with "proppants" they will re-close when the injection ceases. If little water is present in the formation or recharge is meager, all unused geofluid from the plant can be reinjected. Furthermore, external fluids can be brought to the site by some means and injected into the formation. In Chapter 12 we discuss such a process

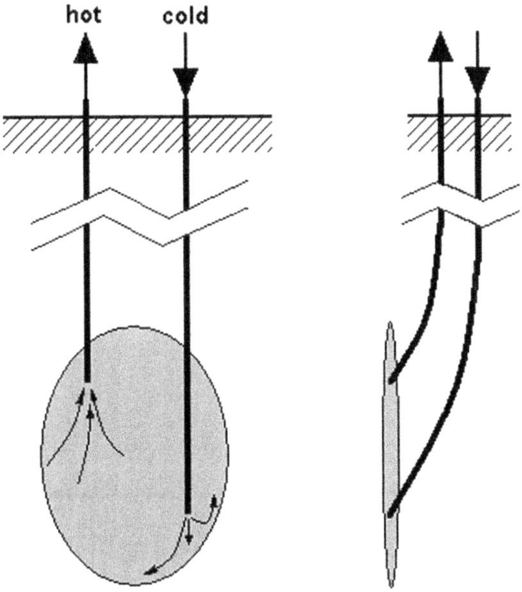

ig. 1.5 Ideal Hot Dry Rock production scheme.

t The Geysers field in Northern California in the United States in which treated nunicipal waste water from nearby communities is sent to the field via pipeline to ssist in the maintenance of an inventory of fluid in the reservoir.

.5 Other types of geothermal resources

ls of 2007, hydrothermal resources are the only geothermal systems that have been leveloped commercially for electric power generation. However, there are four other orms of geothermal energy that someday may reach the commercial stage. They are: Hot Dry Rock, HDR (or Enhanced Geothermal Systems, EGS), geopressured, magma nergy, and deep hydrothermal. We will briefly describe each of these.

.5.1 Hot Dry Rock, HDR

There are many geothermal prospects that have high temperature but are lacking fluid n the formation or the permeability is too low to support commercial development. These systems can be "enhanced" by engineering the reservoirs through hydraulic fracuring. An injection well is drilled into the hot formation to a depth corresponding to the promising zone. Cold water is injected under high pressure to open existing fractures or reate new ones. Once the formation reaches a state of sufficient volume and permeabilty, another well (or wells) is drilled to intercept the newly formed "reservoir." Ideally, a losed loop is thus created whereby cold water is pumped down the injection well nd returned to the surface through the production well after passing through the hot, rtificially-fractured formation [22]. The ideal HDR concept is illustrated in Fig. 1.5.

Table 1.3 Some HDR (EGS) projects worldwide [23].

Country	Location	Dates
United States	Fenton Hill, New Mexico	1973–1996
	Newberry Volcano, Oregon	2010–present
United Kingdom	Rosemanowes	1977–1991
Germany	Bad Urach	1977–1990
Japan	Hijiori	1985–2003
	Ogachi	1986–2008
France	Soultz	1987–present
Switzerland	Basel	1996–2009
Australia	Hunter Valley	2001–present
	Cooper Basin	2002–present

Considerable research has gone into development of the HDR concept and a good deal continues today. Table 1.3 summarizes some of these projects [23]. These and many other HDR (EGS) projects will be described in detail later in the book in Chapter 23, devoted entirely to this subject.

There are many practical problems in developing a HDR system. It is difficult to control very deep, directional, geothermal wells. Drilling techniques in the oil industry now permit wells to be turned 90° while being drilled, allowing the well to drain several vertical pockets of petroleum. However, oil wells tend to be shallower than the ones envisioned for HDR, the temperatures encountered are far lower, and the rocks are not as hard as those found in geothermal regions. Furthermore, the HDR wells must be precisely aimed to hit the deep target in order to form a closed fluid circuit. Lastly, if some of the engineered fractures are not connected to the production well, injected fluid may be lost to the formation. This would require continuous makeup water to maintain the power plant in operation. Some of these difficulties appear to have been at least partially solved in the on-going research, particularly at the Japanese sites; see Chap. 23.

1.5.2 Geopressure

Along the western and northern coastline of the Gulf of Mexico, there is a potent energy resource called "geopressure." During the drilling for oil and gas in the sedimentary coastal areas of Texas and Louisiana, fluids have been encountered with pressures greater than hydrostatic and approaching lithostatic. Hydrostatic pressure increases with depth in proportion to the weight of water, i.e., at about 0.465 lbf/in^2 per ft. However in formations where the fluid plays a supportive role in maintaining the structure of the reservoir, the weight of the solid overburden roughly doubles the gradient to approach the lithostatic value of 1.0 lbf/in^2 per ft.

Geopressured reservoirs were formed along the Gulf Coast through the steady deposition of sediments that created an overburden on the underlying strata. Figure 1.6 is a simplified cross-section through a geopressured reservoir. Periodically, subsidence occurred causing compaction of the rock layers. Subsidence also resulted in steeply dipping faults that can isolate elements of the formation. With the heavy overburden and no way to dissipate the load, the pressure within these lenses of sand grows to levels in excess of hydrostatic.

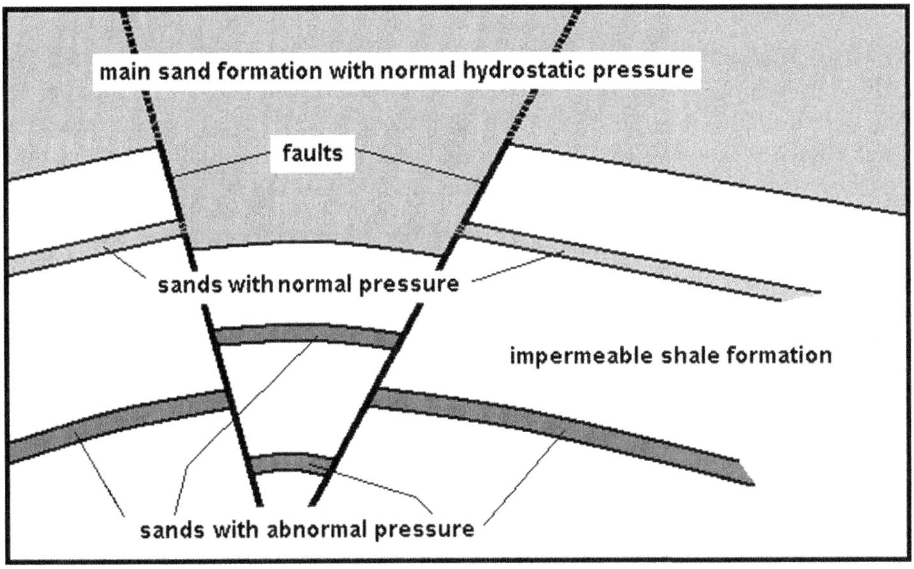

ig. 1.6 *Cross-section schematic of geopressured reservoir.*

In the geopressured reservoirs of the Gulf Coast, the pressures were sufficiently high o prevent drilling for oil and gas. With improved understanding of these zones and etter drilling techniques, these reservoirs can now be safely drilled.

Geopressured reservoirs are characterized by three important properties that make hem potentially attractive for geothermal exploitation: (1) very high pressure, (2) high emperature, and (3) dissolved methane.

The first property allows the use of a hydraulic turbine to extract the mechanical nergy stored in the form of high pressure; the second property allows the use of a eat engine of some kind to extract the thermal energy; and the last property allows or either the combustion of the gas on site for power generation or for sale to enhance he economics of a development project.

However, there are six criteria that must be satisfied before geopressured reservoirs an be commercially developed; these are:

- Is the fluid hot enough, say >230°C?
- Is there sufficient methane dissolved in the fluid?
- Is the high-pressure sand sufficiently permeable?
- Is the high-pressure sand sufficiently thick?
- Is the sand formation fault-bounded but not too fractured?
- Can we guarantee that no subsidence will occur?

he economic viability of a geopressured geothermal project requires a "yes" answer o all of these questions. In Sect. 9.6.3 we discuss a pilot plant that attempted to prove he concept of a geopressured geothermal plant, but this resource has yet to achieve ommercial status.

1.5.3 Magma energy

The next geothermal resource is the one that goes directly to the source of the heat, namely, a magma body relatively close to the surface of the Earth. The concept is to drill a well into the magma, insert an injection pipe, and pump cold water down the well under great pressure. The cold fluid will solidify the molten magma into a glassy substance that should crack under the thermal stress imposed on it. If the water can be made to return to the surface by passing upward through the cracked, extremely hot glassy material, it would reach the surface hot and ready for use in a Rankine-type power plant.

As simple as it is to describe the concept, it is not as easy to carry out such a plan. The U.S. Dept. of Energy conducted two research projects aimed at understanding the magma environment in the 1970s and 1980s. The first one was carried out at the lava lake within the crater of Kilauea Iki on the island of Hawaii [24]. This effort succeeded in drilling through the solidified crust of the lake into the still-molten lava that had a temperature of about 1000°C (1830°F). In fact 105 m of core were obtained from the melt zone and several experiments were run to understand the mechanism of energy extraction from a lava body.

The second research program, the Magma Energy Program, was directed at obtaining a better scientific understanding of the existence and behavior of large magma bodies within calderas. The one selected for study in the mid-1980s was the Long Valley caldera in central California and the research was performed by the Sandia National Laboratory of Albuquerque, New Mexico [25]. The caldera is an oval-shaped region about 18 × 32 km with a prominent resurgent dome. At the time, the dome had risen some 235 mm over the period from 1980–1985, making it both scientifically interesting and practically important to gain a clearer understanding of the phenomenon.

The original goals of the program were to:

- Demonstrate the existence of crustal magma bodies at depths less than 25,000 ft
- Develop and test new drilling technology for hostile environments
- Better understand the creation and evolution of the Long Valley caldera
- Better define the hydrothermal system related to the caldera.

An ambitious exploration well was planned, targeted for a final depth of 20,000 ft (6000 m) [26]. The conceptual design of the well is shown in Fig. 1.7 (to scale in vertical direction). Since an existing 40-in diameter mud riser was in place to a depth of 39 ft from an earlier aborted well, this was used instead of the planned 40-in surface casing. The well was to be drilled in four phases: Phase I – to 2500 ft, Phase II – to 7500 ft, Phase III – to 14,000 ft or 300°C (600°F), whichever came first, and Phase IV – to a total depth of 20,000 ft and 500°C (900°F). In 1989, Phase I was successful in reaching 2568 ft with the 20-in casing, after encountering massive lost circulation at the shallowest depths. Phase II was completed to 7588 ft in November 1991 [27]. Core samples were taken at the 2568 ft and the 7588 ft points by drilling ahead some 100–200 ft. The well was not continued beyond Phase II owing to a shift in DOE policy away from fundamental research and more toward applied research. In 1996 the well was handed over to the U.S. Geological Survey for use as a monitoring well [28].

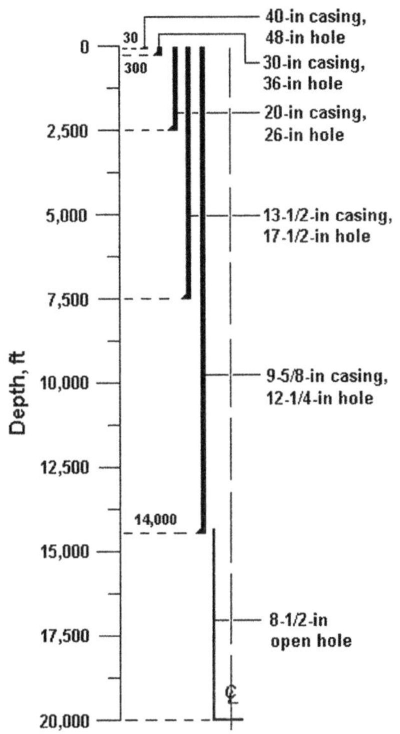

ig. 1.7 *Conceptual design of Long Valley magma energy exploratory well; after [26]. Vertical dimensions are* *hown to scale.*

Since the well only reached depths that were routinely achieved at other geother-nal fields, it failed to produce much new drilling technology. For example, it had been lanned to develop insulated drill pipe to maintain the drilling muds at reasonable emperatures in extremely high temperature formations; this was not done. It did pro-uce some scientific information that led to a better understanding of the nature of the ong Valley caldera, but no further projects have appeared to try to tap the vast mount of thermal energy contained in near-surface magma bodies.

.5.4 Deep hydrothermal

he last of the geothermal resources is one that has only recently been accessed and devel-ped. Deep hydrothermal resources are those that lie at depths of 2500 to 4000 m and dee-er. They may lie in areas marked by normal geothermal temperature gradients, and as uch may yield fluids at only low to moderate temperatures. For example in a place where he gradient is say 30°C/km, fluids found at 4000 m might range from 120–140°C.

Deep drilling in Europe and Australia as a part of HDR (EGS) efforts has discovered hat reservoirs of fluid exist at these depths and that the formations possess some per-neability. In fact, deep sedimentary layers can be exploited even without hydrofractur-ng the formation. These wells produce geofluids that can be used in energy conversion systems specifically designed for lower temperature fluids. The waste discharge fluid from the power plant may often be further utilized for direct heating o buildings and homes before being reinjected. The cost of such deep wells is significantly higher than the usual shallower geothermal wells, but with sufficient financial incentives offered by governments, private developers are able to successfully exploit wha were previously thought to be uneconomic geothermal resources. Chapter 22 covers this subject in more detail.

References

[1] Leet, L.D., S. Judson and M.E. Kauffman, *Physical Geology, 5th Ed.*, Prentice-Hall, Englewood Cliffs, NJ 1978.

[2] Plummer, C.C. and D. McGeary, *Physical Geology, 4th Ed.*, Wm. C. Brown Publishers, Dubuque, IA 1988.

[3] Press, F. and R. Siever, *Earth, 2nd Ed.*, W.H. Freeman and Company, San Francisco, 1978.

[4] Tarbuck, E.J. and F.K. Lutgens, *Earth Science, 8th Ed.*, Prentice-Hall, Englewood Cliffs, NJ, 1997.

[5] Cataldi, R., S.F. Hodgson and J.W. Lund, Eds., *Stories from a Heated Earth: Our Geothermal Heritage* Geothermal Resources Council and International Geothermal Association, Sacramento, CA, 1999.

[6] Sigurdsson, H., *Melting the Earth: The History of Ideas on Volcanic Eruptions*, Oxford University Press New York, 1999.

[7] Krafft, M., *Volcanoes: Fire from the Earth*, Discoveries, Harry N. Abrams, Inc., Publishers, 1993.

[8] Winchester, S., *Krakatoa: The Day the World Exploded: August 27, 1883*, HarperCollins, New York, 2003.

[9] Duffield, W.A., *Volcanoes of Northern Arizona: Sleeping Giants of the Grand Canyon Region*, Grand Canyon Association, Grand Canyon, AZ, 1997.

[10] Waggoner, B.M., "Alfred Wegener (1880–1930)," U. of California Berkeley Museum of Paleontology http://www.ucmp.berkeley.edu/history/wegener.html, 1996.

[11] Hurley, P.M. and J.R. Rand, "Review of Age Data in West Africa and South America," in *The History of the Earth's Crust*, R.A. Phinney, Ed., Princeton University Press, Princeton, NJ, 1968.

[12] Wilson, J.T., "Continental Drift," Offprint No. 868, *Scientific American*, W.H. Freeman and Company San Francisco, April, 1963.

[13] U. of California Berkeley Museum of Paleontology, http://www.ucmp.berkeley.edu/geology/anim1 html, 1999.

[14] Anon., *Salton Sea Scientific Drilling Program Monitor*, U.S. Dept. of Energy, Issues 1–5, Oct. 1995–Apr. 1996.

[15] Anon., *Salton Sea Scientific Drilling Program Quarterly Reports No. 1–11*, Oct. 1984–Jan. 1988.

[16] http://www.ICDP-Online.org.

[17] http://www.ccsd.org.cn/English/index.htm.

[18] Anderson, D.L., "The San Andreas Fault," Offprint No. 896, *Scientific American*, W.H. Freeman and Company, San Francisco, V. 225, No. 5, November 1971, pp. 52–66.

[19] http://www.ngdc.noaa.gov/mgg/global/relief/SLIDES/JPEGfull/Slide18.jpg. Reprinted and modified from National Oceanic and Atmospheric Administration, National Geophysical Data Center, Boulder CO, May 10, 2004.

[20] DiPippo, R., *Geothermal Energy as a Source of Electricity: A Worldwide Survey of the Design and Operation of Geothermal Power Plants*, U.S. Dept. of Energy, DOE/RA/28320-1, U.S. Gov. Printing Office Washington, DC, 1980.

[21] White, D.E., "Characteristics of Geothermal Resources," Chap. 4 in *Geothermal Energy: Resources Production, Stimulation*, P. Kruger and C. Otte, Eds., Stanford University Press, Stanford, CA, 1973.

[22] Smith, M.C., "The Hot Dry Rock Program," *Los Alamos Science*, Winter/Spring, 1983, pp. 86–87.

[23] http://www.nedo.go.jp/chinetsu/iea/ann3e.htm, "Development of a Hot Dry Rock Power Generation System," New Energy and Industrial Technology Development Organization, Kanagawa, Japan, 2004

[24] Hardee, H.C., J.C. Dunn, R.G. Hills and R.W. Ward, "Probing the Melt Zone of Kilauea Iki Lava Lake Kilauea Volcano, Hawaii," *Geophysical Rev. Letters*, V. 8, No. 12, 1981, pp. 1211–1214.

[25] Chu, T.Y., J.C. Dunn, J.T. Finger, J.B. Rundle and H.R. Westrich, "The Magma Energy Program," *Geothermal Resources Council BULLETIN*, V. 19, No. 2, 1990, pp. 42–52.
[26] Finger, J.T and J.C. Eichelberger, "The Magma Energy Exploration Well," *Geothermal Resources Council BULLETIN*, V. 19, No. 2, 1990, pp. 36–41.
[27] Anon., "Phase II Drilling at Long Valley Exploratory Well Completed," *Geothermal Progress Monitor*, Rep. No. 14, DOE/CE-0394, U.S. Dept. of Energy, 1992, p. 6.
[28] Anon., "Transfer of Long Valley Well on Schedule," *Geothermal Progress Monitor*, Rep. No. 18, DOE/EE-0121, U.S. Dept. of Energy, 1996, p. 8.

Problems

1.1 It has been suggested that the radioactive decay of certain elements in the Earth's crust can account for the observed normal heat flux at the surface. Assume the crust is made only of granite containing certain amounts of uranium, thorium and potassium. The average heat flux at the surface of the Earth is about 1.5×10^{-6} cal/cm$^2 \cdot$s. Calculate the thickness of crust needed to give this heat flux using data given in Table 1.2. Compare your calculated thickness with the actual thickness, and discuss the reasonableness of this theory.

1.2 In Problem 1.1, we discussed the possibility that the normal heat flux through the Earth's crust is caused by radioactive decay of certain elements. Let us consider a different possibility, namely, that heat conduction from the hot molten mantle through the crust might give rise to the normal (not anomalous) heat flux. Use Fourier's law of one-dimensional heat conduction (i.e., that the heat flux is proportional to the temperature gradient times the thermal conductivity), choose a reasonable value for the thermal conductivity of the rock comprising the crust, and calculate the required temperature difference to produce the observed normal heat flux. What temperature do you find, therefore, at the bottom of the crust? Interpret your finding.

1.3 In continental areas, the average geothermal heat flux is about 1.2×10^{-6} cal/cm$^2 \cdot$s. Assume that three-fourths of this is attributed to the crust. Using an average thermal conductivity for rock, calculate the temperature (above the average surface temperature) at the Moho.

1.4 In the center of the East Mesa geothermal field in Southern California, U.S. there is a high heat flux anomaly, reaching about 7 times higher than normal. Perhaps this is caused by a magma intrusion inside the crust. Assuming the magma has a temperature of about 1000°C, estimate the depth of the magma intrusion. The surface temperature is about 35°C on average.

1.5 It is generally held that geothermal fluids must exceed 150°C in order to be practical for generating electricity. Allowing a surface temperature of 25°C, how deep must one drill to hit this temperature at a location having a normal gradient?

Chapter 2

ːxploration Strategies ınd Techniques

"For a successful technology, reality must take precedence over public relations, for Nature cannot be fooled."

Richard P. Feynman – 1988

ː.1 Introduction

ɫ has been said that there are no commercial geothermal fields that could not have ›een discovered by an intelligent layperson. If this were true, then the strategies and ophisticated techniques we will discuss in this chapter would be unnecessary.

While it is certainly true that nearly all successful geothermal projects are at sites vhere thermal manifestations existed, it is nevertheless a fact that many false steps ıave been taken in developing these fields – before the introduction of systematic xploration programs. In the early days of the geothermal industry, it was common to lrill wells close to thermal manifestations such as fumaroles, hot springs, mud pots, ınd even geysers. The result was usually the extermination of the manifestations, ›erhaps some short-term production from the well, but often the true source of the ;eothermal energy was missed.

eothermal Power Plants: Principles, Applications, Case Studies and Environmental Impact, Third Edition.
› 2012 Elsevier Ltd. All rights reserved.

The scientific tools that are now available and routinely used at geothermal prospects allow for better characterization of the resource before the costly phase of deep well drilling. By defining the subsurface nature of the field, a more reliable determination of drilling sites can be made, increasing the probability of a successful discovery well and of a successful field development campaign.

2.2 Objectives of an exploration program

There are five things a geothermal exploration program should accomplish:

- Locate areas underlain by hot rock
- Estimate the volume of the reservoir, the temperature of the fluid in it, and the permeability of the formation
- Predict whether the produced fluid will be dry steam, liquid or a two-phase mixture
- Define the chemical nature of the geofluid
- Forecast the electric power potential for a minimum of 20 years.

As mentioned most geothermal fields are marked by thermal features that give away the fact that there is a heat source somewhere in the vicinity. Even without gushing outlets of hot fluids, the ground itself may be hotter than normal. Without the heat supply there will be no geothermal resource.

Unless a significant volume of permeable rock exists, any production from the reservoir will be small and short-lived, and unless a reasonable minimum fluid temperature can be assured, the energy production will be too small to make the project commercially viable.

The physical and chemical properties of the geofluid under production conditions are important in deciding on the feasibility of a power project. It may be difficult to obtain this information without drilling wells, but a good exploration program should give a reasonable estimate for these properties. The ultimate exploration tool is the drilling of deep wells based on the findings of the scientific surveys.

The last bulleted item is the main outcome of the exploration phase, namely, can the field be expected to support a geothermal power plant of a certain capacity for a sufficient period of time to convince investors that the project will be worthwhile.

2.3 Phases of an exploration program

In this section we describe the usual steps in a full exploration program. As mentioned, the site may be fairly promising based on prior observations, but these steps are necessary to fully characterize the resource. In typical chronological order, the phases are:

1. Literature survey
2. Airborne survey
3. Geologic survey
4. Hydrologic survey
5. Geochemical survey
6. Geophysical survey.

Ve could have included drilling as the final phase of the exploration program, but we ıclude here only the geoscientific studies that lead up to the drilling phase. Some of ıe scientific phases may be combined to save time and to create synergy among the arious investigations. It is clear from the wide range of disciplines cited in the listing, ıat a team of individuals from one or more companies with appropriate expertise is eeded to carry out the exploration of a geothermal prospect. Let us examine each of ıese phases in some detail.

.3.1 Literature survey

is likely that someone at some time has taken a look at whatever site is now under con-ideration. With the use of the Internet, it is an easy matter to search the existing literature or prior studies. There are now extensive databases online for most geothermal prospects. or example, Ref. [1] provides information on hot springs in the United States and several ther countries. Ref. [2] contains data on all wells and springs with temperatures greater nan 20°C for the following states in the U.S.: Alaska, Arizona, California, Colorado, daho, Montana, Nebraska, Nevada, New Mexico, North Dakota, Oregon, South Dakota, exas, Utah, Washington, and Wyoming. A total of 11,775 sites are included. There is lso a compilation of chemical information for the fluids at each of these springs and wells.

The United States Geological Survey (USGS) maintains an excellent database for vells drilled in the U.S. [3]. For example, one can find well log data for many wells rilled in Nevada [4]; Table 2.1 and Fig. 2.1 illustrate the type of information that is vailable there [5]. A large body of data can also be found at the USGS web site [6]. or sites that may not have made their way onto the World Wide Web, a trip to a ocal library or town hall will often turn up interesting reports on regional sites. With his type of information in hand, the remaining phases can be planned more ffectively.

Table 2.1 Well data for Fish Lake Valley well FPL-1 [5].

Well ID:	FLP-1	**Log Date:**	12/4/92	**County, State:**	Esmeralda, NV
Latitude:	37° 55.3′	**Longitude:**	117° 59.4′	**Elevation (ft):**	5300
Company:	Magma Power	**Prospect:**	Fish Lake Valley		

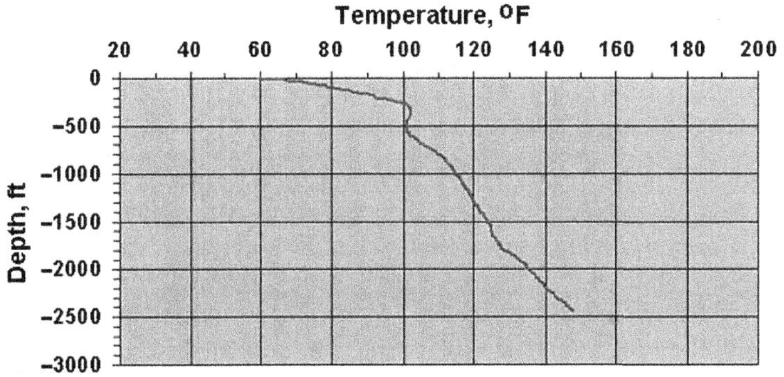

ig. 2.1 *Temperature profile for Fish Lake Valley well FLP-1 [5].*

2.3.2 Airborne survey

Aerial photography of the prospective site can yield the following information:

- Structural nature of the field
- Locations of thermal anomalies and manifestations (via infrared imaging, IR)
- Aeromagnetic data
- Geographic benchmarks to guide later ground surveys
- Geologic mapping (in conjunction with geologic survey).

The use of stereographic images can reveal the surface expression of faults that are very important in defining the possible avenues of fluid flow through the formation. Warm ground will show up in IR photos; if the area receives snowfall, the same information is clearly visible from patterns of snow melt. IR surveys can consist of color photographs in the 1–2 μm wavelength range, false-color photos, or IR scanning using either 3–5 μm or 8–14 μm wavelengths [7]. To be useful, an IR survey ought to be able to detect very small temperature anomalies, of the order of 0.05–0.5°C over areas of up to hundreds of square kilometers. A rise of 0.5°C over the normal background would be the result of a heat flux about 50 times greater than the normal heat flux. Also a reliable IR survey must be able to distinguish a true geothermal effect from one caused by weather, hydrology, terrain or topography [8,9].

Aeromagnetic measurements are thought to reveal areas of hydrothermally-altered rock because the process changes magnetic rock to non-magnetic rock. Thus a magnetic low is interpreted as evidence of possible hot geothermal fluids. However, there are enough cases where this method has not been definitive in outlining a hydrothermal area that a magnetic low by itself is not considered sufficient; it should be integrated with the results of other methods. Reliable aeromagnetic data may be used to help validate and constrain numerical models of the reservoir at a later stage of development; see Chap. 4.

2.3.3 Geologic survey

A geologic survey is usually the first work to be conducted "on the ground" and aims at detecting and characterizing the following:

- Tectonic and stratigraphic setting
- Recent faulting
- Distribution and age of young volcanic rocks
- Location and nature of thermal manifestations
- Hydrothermally-altered ground and rocks.

This phase is typically conducted by a geovolcanologist, a geologist with special training and experience in volcanic systems. The history of the creation and evolution of the site over geologic time is of more than scientific interest; it can offer insights into the present, unseen state of the deep formation. Many geothermal fields owe their existence to past volcanism, sometimes fairly recent. The eruption history, including the timing of events, the nature and volume of the material ejected, and the extent of the debris field, all contribute to the full picture. In a sense, the geologist is in the odd position of having to foretell the past to explain the present.

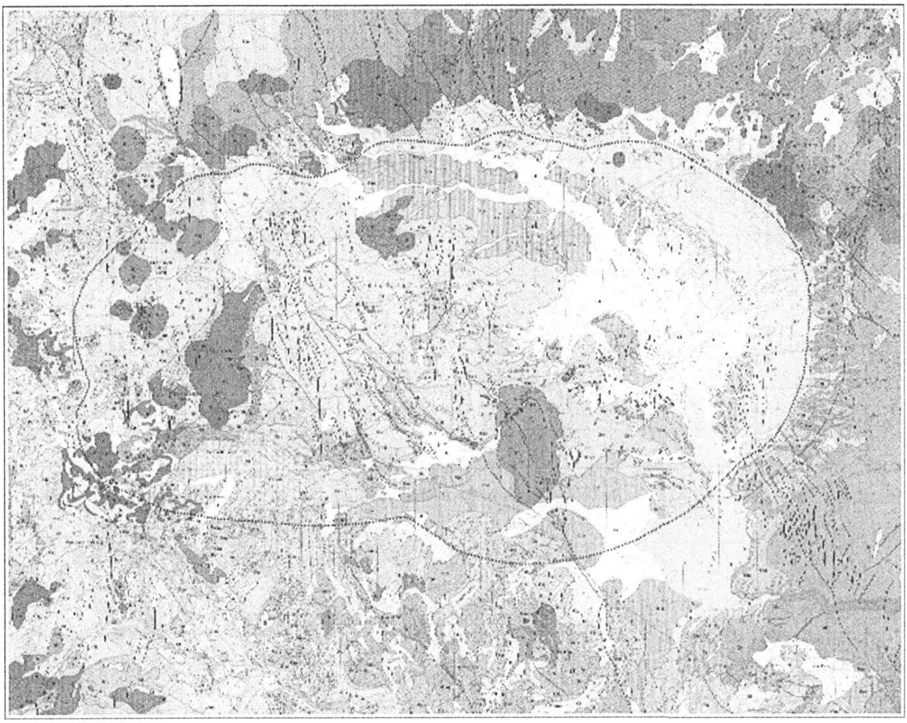

ig. 2.2 Excerpt of geologic map of the Long Valley caldera, California [10] [WWW].

Two products of the geologist's work are a geologic map of the area and the first raft of conceptual model for the hydrothermal system. The geologic map is a large-cale, multi-color rendering of the types of rocks that appear at the surface. An examle of a geologic map is given in Fig. 2.2. This is an excerpt from the full map for the ong Valley caldera in California, provided by the USGS [10]; the reader may wish to xplore the full map in more legible detail by accessing the USGS web site at Ref. [10] r this book's companion web site.

Figure 2.3 is such a detail view for the northwest section of the caldera. Note that nterstate Highway Rt. 395 runs across the middle of the map in a NW-SE direction. ach area represents a different kind of rock unit (type and age) at the surface, lines re drawn to show the type and location of contacts and faults, and strike and ip symbols show, respectively, the planar orientation of the faults and their angle of teepness [11]. Geologists name the various units based on the location where that articular unit was first discovered or best displayed. For example, the unit designated Qqm" in Fig. 2.3 is called "quartz latite of Mammoth Mountain." Other more comnon units are simply given their geologic names and described in detail regarding heir existence at this particular site. For example, the unit "Qp" in Fig. 2.3 is named pyroclastic fall deposits" from the Holocene and Pleistocene periods, and identified as oming from eruptions of the Mono-Inyo Craters chain [10].

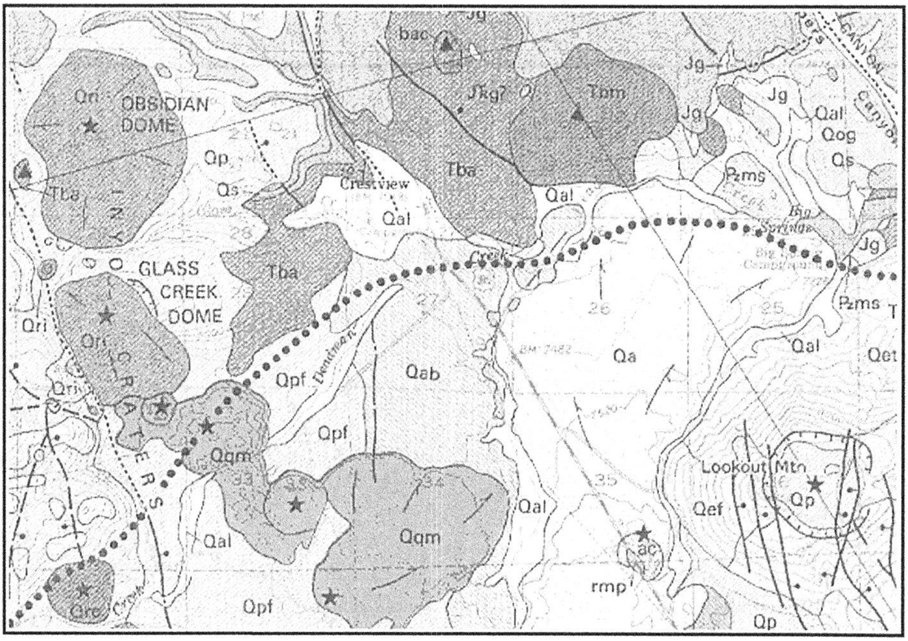

Fig. 2.3 Detail of Long Valley geologic map: northwest section of the caldera (dotted line) [10] [WWW].

The conceptual model of the geothermal system depicts graphically the major subsurface features of the system, including the locations of rock units, faults, and fluid flow paths. The job of creating this map begins with the geologic survey and continues as more data comes in from the other surveys. It is not uncommon for the conceptual model to evolve even after the field goes into production, as reservoir information from production and injection wells becomes available. A sample conceptual model of the Long Valley caldera, adapted from Ref. [12], is shown in Fig. 2.4. This is a cross-section through the same field shown in the geologic map of Fig. 2.2. As the focus of the exploration program narrows to the region of highest geothermal potential, the conceptual model also becomes more limited in areal extent but finer in detail. The development of the conceptual model usually involves reconciling conflicting opinions of geoscientists who may interpret the survey data differently.

2.3.4 Hydrologic survey

Since one of the vital prerequisites for a commercial geothermal field is the presence of water in the formation in ample quantities, the hydrologic survey is an important part of the exploration program. It is conducted by a hydrologist, usually a civil engineer with expertise in groundwater behavior. Sometimes the person or group conducting the geologic survey also performs the hydrologic survey. The objective is to learn as much as possible about the fluids in the system, including their age, chemical and physical properties, abundance, flow paths, and recharge modes.

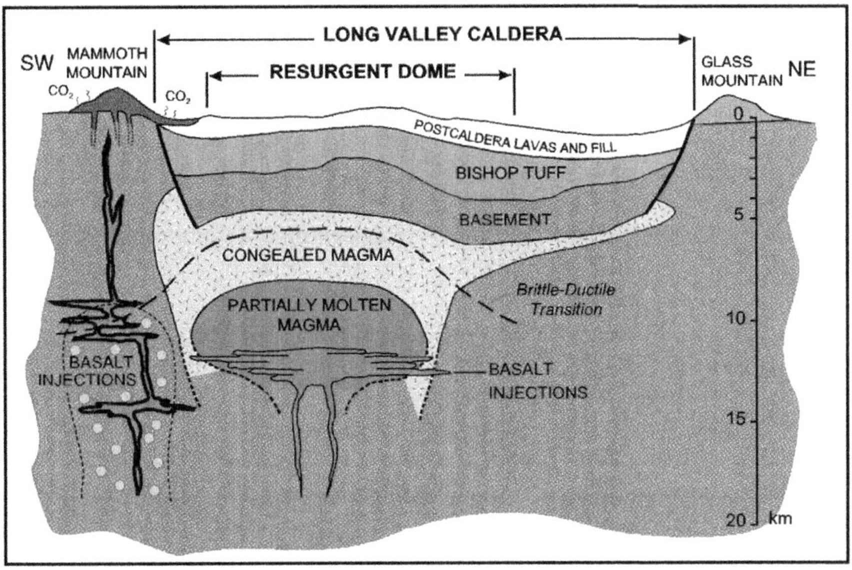

ig. 2.4 *Conceptual model of Long Valley caldera showing main structural features; after [12] [WWW].*

The hydrologic survey will typically involve study of the following:

- Meteorological data
- Temperature and flow rates of both hot and cold springs
- Chemical analysis of springs
- Water table in existing water wells
- Water movements both on the surface and in the reservoir.

y performing a mass balance on all the observed streams, the hydrologist can make n estimate of the volume of fluid in the formation and the rates of throughput, critical nformation for estimating the production capacity of the reservoir. In the area of hemical analysis of the springs, there is a natural overlap with the next phase of the rogram, the geochemical survey.

.3.5 *Geochemical survey*

he geochemist has several important responsibilities, namely:

- Identify whether the resource is vapor- or liquid-dominated
- Estimate the minimum temperature of the geofluid
- Determine the chemical properties of the fluid both in the reservoir and in the produced state
- Characterize the recharge water, including its nature and sources.

here are several well-established methods for carrying out this task. The difficulty ften comes in the interpretation of the results of the various diagnostic tests. The uest for clear and accurate answers to vital questions such as geofluid temperature is ften frustrated by complicating factors that are only vaguely understood at the early stages of exploration. Let us focus on the determination of the geofluid temperature in the reservoir to illustrate this point.

High temperature liquids moving through porous rock quickly reach equilibrium with various elements found in the rocks. The most important of these for estimating the fluid temperature are quartz (silica, SiO_2), sodium (Na), potassium (K), and calcium (Ca).

The concentration of silica in the geofluid produced from a reservoir can be correlated with the geofluid temperature using the solubility of quartz as a function of temperature, $Q = Q(t)$. This function has been determined by laboratory tests in pure water and in solutions with compositions typical of fluids in a geothermal environment. The solubility of quartz depends not only on the fluid temperature but on its salinity and pH. At the high temperatures found in a hydrothermal reservoir, the solubility of silica is controlled by quartz, a crystalline form of silica, but at the lower temperatures encountered in the waste fluids after their use in a power plant, the solubility is controlled by the amorphous form of silica.

The functional relationship between the quartz concentration and the fluid temperature for pure water is given by eqs. (2.1) and (2.2):

$$t(Q) = a_0 + a_1 Q + a_2 Q^2 + a_3 Q^3 + a_4 \log_{10} Q \tag{2.1}$$

where the coefficients are:

$a_0 = -42.198$
$a_1 = 0.28831$
$a_2 = -3.6686$ E-04
$a_3 = 3.1665$ E-07
$a_4 = 77.034$.

The units of the concentration, Q, are mg/kg or ppm, the units of the temperature, t are °C, and the limits of the validity of the correlation are from $t = 20-330$°C [13].

It is sometimes necessary to calculate the quartz concentration for a known temperature; then the following correlation may be used:

$$Q(t) = b_0 + b_1 t + b_2 t^2 + b_3 t^3 + b_4 t^4 + b_5 t^5 + b_6 t^6 + b_7 t^7 \tag{2.2}$$

where the coefficients are:

$b_0 = 7.0281$
$b_1 = -0.228748$
$b_2 = 1.20057$ E-02
$b_3 = -1.50598$ E-04
$b_4 = 1.42552$ E-06
$b_5 = -6.0991$ E-09
$b_6 = 1.34828$ E-11
$b_7 = -1.29355$ E-14.

The units of Q and t, are the same as in eq. (2.1), and the limits of the validity of the correlation are from $t = 20-340$°C.

The corresponding relationship for amorphous silica is given in eq. (2.3):

$$\log_{10} S(t) = c_0 + c_1 t + c_2 t^2 + c_3 t^3 \tag{2.3}$$

where the coefficients are:

$c_0 = -1.34959$
$c_1 = 1.625$ E-02
$c_2 = -1.758$ E-05
$c_3 = 5.257$ E-09.

The units of the amorphous concentration, S, are mg/kg or ppm, the units of the emperature, t, are °C, and the limits of the validity of the correlation are from = 90–340°C [14].

When a geofluid reaches the surface through a hot spring, a fluid sample can be aken and the amount of silica can be measured. If, on the one hand, the fluid emained in a liquid state from the reservoir to the sample with no flashing of steam, hen the silica concentration will be the same in the reservoir as in the sample. This ase leads to what is called the "no steam loss" silica geothermometer: the thermometic function for this case is [15]:

$$t(S) = \frac{1309}{5.19 - \log_{10} S} - 273.15 \tag{2.4}$$

lternatively, one could calculate the temperature using eq. (2.1), inserting the meaured silica concentration for Q.

If, on the other hand, the liquid partially flashed to steam (which is quite likely iven that the spring is at atmospheric pressure), then the silica concentration in the eservoir fluid will be less than what is observed in the sample owing to the concenrating effect of having all the silica in a smaller amount of liquid. The two concentraion values are related as follows:

$$Q(t_{res}) = S(t_{atm}) \times (1 - x_{atm}) \tag{2.5}$$

where x_{atm} is the quality (or dryness fraction) of the liquid-vapor mixture at atmospheric conditions. It is a function of the reservoir fluid enthalpy (which depends on he fluid temperature) and the fluid enthalpy at atmospheric conditions:

$$x_{atm} = \frac{h_f(t_{res}) \quad h_f(t_{atm})}{h_{fg}(t_{atm})} \tag{2.6}$$

where the numerator is the difference between the enthalpy of a saturated liquid at he reservoir temperature and one at atmospheric conditions, and the denominator is he latent heat of evaporation at atmospheric conditions. For standard atmospheric ressure, we can use *Steam Tables* [16] to find h_f (100°C) = 419.04 kJ/kg and h_{fg}(100°C) = 2257.0 kJ/kg.

The problem then is to find a reservoir temperature such that eqs. (2.1) or (2.2), 2.5), and (2.6) give a consistent set of results for the measured value of $S(t_{atm})$. When this is worked out, the result is called the "maximum steam loss" silica geothermometer: the thermometric correlation function for this case is:

$$t(S) = 75.272 + 0.72077S - 1.5807 \times 10^{-3} S^2 + 1.9924 \times 10^{-6} S^3 - 9.6957 \times 10^{-10} S^4 \tag{2.7}$$

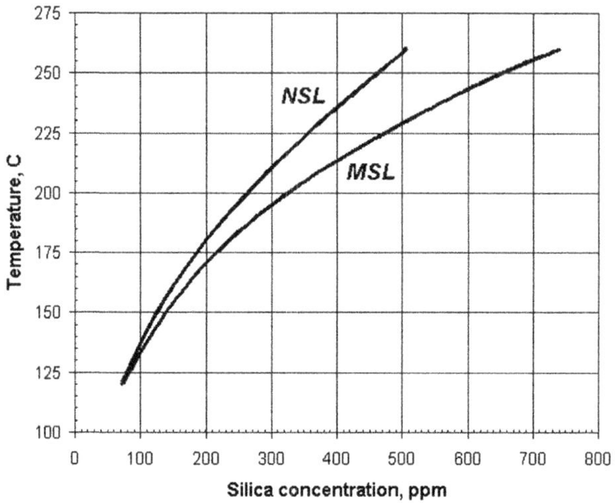

Fig. 2.5 Silica geothermometers: NSL = no steam loss; MSL = maximum steam loss.

Both of these geothermometers assume the geofluid is pure water at neutral pH. Figure 2.5 gives the inferred geofluid temperature for these two cases.

Thus we see that a decision must be made regarding the process followed by the geofluid from the reservoir to the surface before selecting which silica geothermometer to use to estimate the reservoir temperature. For example, a measurement of 500 ppm silica could mean either a reservoir temperature of 260°C (NSL) or 230°C (MSL), a significant difference. Also if any silica precipitates from the geofluid before the sample is taken, the results will be invalid, and the geothermometer will underestimate the reservoir temperature. Another possibility is that the reservoir silica equilibrium might be controlled by a form of silica other than quartz, such as chalcedony or cristobalite, thereby rendering the analysis developed above inapplicable. The molality and pH of the reservoir fluid will also influence the accuracy of silica geothermometry. All of these issues require an intelligent judgment on the part of the geochemist.

Besides using silica equilibrium, one can apply the sodium/potassium, Na/K, ratio as a measure of reservoir temperature. The thermometric function is:

$$t = \frac{1217}{\log_{10}\dfrac{\{Na\}}{\{K\}} + 1.483} - 273.15 \qquad (2.8)$$

where {Na} and {K} are the concentrations of sodium and potassium in the geofluid. The equation applies for temperatures greater than 150°C [15]. This formula is a correlation of field data with considerable scatter, particularly below 150°C. It seems to be reliable in the 180–200°C range, and where the groundwater that could dilute the geofluid is low in Na and K [15].

The last geothermometer based on dissolved minerals in liquid geofluid we will mention is the Na/K/Ca-geothermometer. The following thermometric equation is useful for calcium-rich geofluids:

$$t = \frac{1647}{\log_{10}\dfrac{\{Na\}}{\{K\}} + \beta\left[\log_{10}\dfrac{\sqrt{\{Ca\}}}{\{Na\}} + 2.06\right] + 2.47} - 273.15 \tag{2.9}$$

where $\beta = 4/3$ for $t < 100°C$ and $\beta = 1/3$ for $t > 100°C$ [15].

There is an entirely different suite of geothermometers based on gas geochemistry. These involve equilibrium reactions among carbon dioxide (CO_2), carbon monoxide (CO), hydrogen (H_2), hydrogen sulfide (H_2S), and methane (CH_4) in the presence of water vapor (H_2O). For example, the following three thermometric functions are correlations to worldwide field data that relate the concentrations (in mols/kg) of three gases to the reservoir temperature [17]; note these temperatures, denoted by T, are in kelvins:

$$\log_{10}\{CO_2\} = -1.09 - \frac{3894.55}{T} + 2.532 \log_{10} T \tag{2.10}$$

$$\log_{10}\{H_2S\} = -11.80 - 0.06035\,T - \frac{17,691.09}{T} + 27.163 \log_{10} T \tag{2.11}$$

$$\log_{10}\{H_2\} = -3.04 - \frac{10,763.54}{T} + 7.003 \log_{10} T \tag{2.12}$$

Each of these formulas may be solved numerically for temperature for a given gas concentration. By and large, the method of gas geochemistry leads to sets of simultaneous reaction equations that must be solved iteratively or graphically to find the temperature that satisfies all the governing gas equations [18,19].

Lastly, geochemical studies may be able to identify the general patterns of fluid movement through the formation. A search for radon emissions can often identify active faults and fractures that may be conduits for fluid flow.

2.3.6 Geophysical survey

Geophysics is the application of principles of mechanics, thermal science, and electrical science to the delineation and characterization of geothermal systems. The geophysical phase of exploration typically provides the final piece of the puzzle and should lead directly to the identification of the best locations to drill the first deep wells. The geophysicist usually has available the data from the previously described surveys and uses this information to decide which tests to perform. Some of the most useful techniques include:

- Heat flux measurements
- Temperature gradient surveys
- Electrical resistivity surveys
- Seismic methods, both active and passive
- Gravity surveys.

The properties that are measured during the geophysical phase include:

- Temperature
- Electrical conductivity or resistivity
- Density
- Velocity of propagating waves in solid material
- Magnetic susceptibility
- Local gravitational acceleration.

The geophysicist is aided in determining heat flux and thermal gradients by having data from shallow wells (about 100–200 m deep). Recall from Chapter 1 that in non-thermal areas the normal temperature gradient is about 3.1°C/100 m, and the normal heat flux is roughly 1.2×10^{-6} cal/cm$^2 \cdot$s. By convention, 1×10^{-6} cal/cm$^2 \cdot$s or 1 μcal/cm$^2 \cdot$s is called 1 heat flow unit or 1 HFU. When the mode of heat transfer is pure conduction, the phenomenon conforms to Fourier's Law which, in 1-dimensional form, is:

$$\dot{Q} = -kA\frac{dT}{dx} \tag{2.13}$$

or

$$q = -k\frac{dT}{dx} \tag{2.14}$$

where $\dot{Q}$ is the thermal power or heat flow per unit time (in J/s), k is the thermal conductivity of the material (in J/m·s·°C or W/m·°C), A is the area through which the heat flows (in m^2), and dT/dx is the temperature gradient driving the heat flow (in °C/m). When eq. (2.13) is divided by A, one obtains the equation for the heat flux, q, eq. (2.14). The negative sign indicates that the heat flows in the direction of decreasing temperature, in accordance with the Second Law of thermodynamics; see Appendix D. Approximate values of thermal conductivity for typical rocks are:

- Granite 1.73–3.98 W/m·°C
- Limestone 1.26–1.33 W/m·°C
- Marble 2.07–2.94 W/m·°C
- Sandstone 1.60–2.10 W/m·°C.

Considering the near-surface formation, conduction will be the predominant means of heat transfer, absent fluid circulation. The thermal conductivity will tend to increase with depth as the formation compacts under the increasing weight of the overburden. The heat flux will be more or less constant in steady-state conditions. Therefore, from eq. (2.14) we expect that the temperature gradient will decrease with depth in a conductive environment. This is important in the interpretation of shallow gradients because extrapolation to deeper depths will generally lead to erroneously high values, as can be seen in Fig. 2.6.

Where convection plays a role, as one would hope in a permeable geothermal formation, this tends to create a more or less isothermal zone: the lower end of the A-Cond + Conv curve in Fig. 2.6.

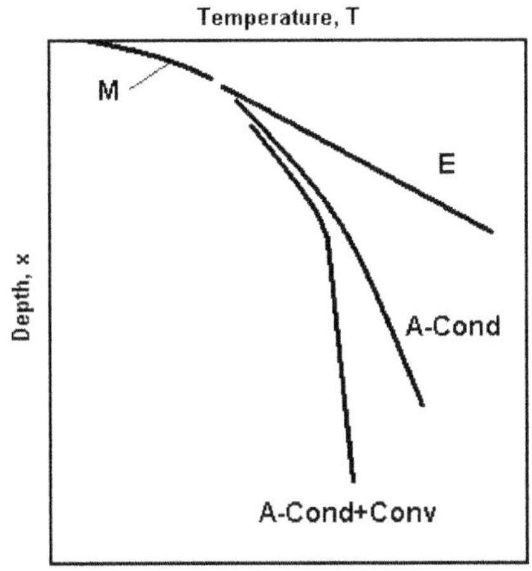

ig. 2.6 Temperature gradients. M = measured; E = linear extrapolation; A-Cond = actual with pure conduction; -Cond + Conv = actual with a deep convective zone.

High heat flow is a good indicator of a geothermal system. High temperatures close ɔ the surface will result in high heat flux but can be misleading as to the hottest sub-urface zone. There are several possible causes of a high heat flow:

- Presence of hot water close to the surface
- Exothermic reactions in the formation
- High radioactive content of the country rocks
- Friction along faults.

ieophysical surveys are geared to detect the first of these as the primary indicator of ne thermal anomaly. But because hot fluids can move laterally as well as vertically in porous medium, the surface expression of the highest heat flux may not coincide vith the hottest reservoir region and may divert attention away from the source of the eofluid. An example is shown in Fig. 2.7, drawn roughly from the case of the Dunes eld in the Imperial Valley, California.

In this case, the impressive heat flow anomaly at point A is associated with a side-ow, laterally displaced from the main source of the geofluid beneath point B. The emperature reversal at a relatively shallow depth is the key indicator of this effect. his phenomenon has been seen in many geothermal fields and is a warning to incor-orate several types of measurements into the conceptual model of a field. Any single ndicator can often be misleading.

Electrical measurements are another important element in the geophysical explora-ion phase. The electrical resistivity, ρ, is one of the most commonly measured quanti-ies; it is measured in units of ohm · m. Low resistivity can be caused by the presence f hot water with dissolved minerals in a porous medium. Generally rocks are poor onductors of electricity and exhibit high resistivity, but when an electrolyte fills the ore spaces, the average resistivity will be reduced. Also rocks that have been hydro-hermally altered to clayey material are better conductors than the native country

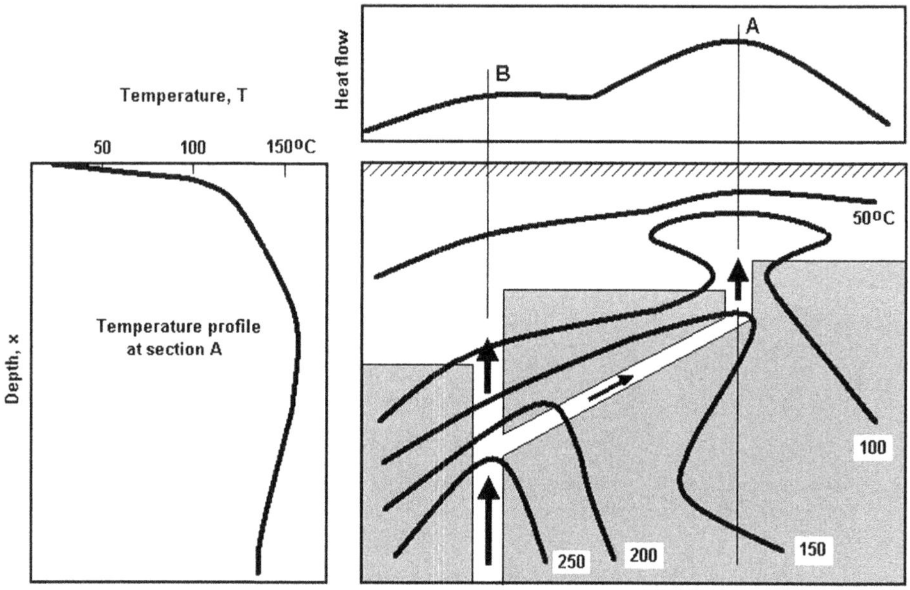

Fig. 2.7 Heat flow and temperature gradient with lateral flow in reservoir.

rock. Therefore whenever low resistivities are found, a high-temperature hydrotherma reservoir may be present.

Dakhnov's Law gives the relation between resistivity and temperature [20]:

$$\rho_0 = \frac{\rho_{18}}{1 + 0.025(t - 18)} \tag{2.15}$$

where ρ_0 is the resistivity of a rock with its pore space filled with a fluid having a resistivity ρ_{18} at 18°C. Archie's Law expresses the formation resistivity ρ_0 relative to that of the fluid filling the pores ρ_w [21]:

$$\rho_0 = \frac{\rho_w}{\phi^m} \tag{2.16}$$

where ϕ is the porosity of the rock (expressed as a fraction) and m is the so-called cementation factor. For loosely consolidated sediments, $m \approx 1.2-1.5$; for well-cemented or crystalline rock, $m \approx 1.9-2.2$ [21]. Meidav [21] presents graphs and nomograms that facilitate the use of these equations.

Because the resistivity depends on the salinity of the pore water as well as its temperature, care must be taken in interpreting resistivity measurements. For example, sea water having about 32,000 ppm of NaCl at 18°C has the same resistivity as chloride water with 10,000 ppm NaCl at 100°C. Also since gases are poor electrical conductors, a dry steam reservoir will appear as a resistivity high in the midst of a relative low. Resistivities of 5–10 ohm · m or lower are usually taken as a good indicator of a hot hydrothermal resource when confirmed by other indicators, such as a high temperature gradient.

Areas of active faults and fissures can be detected by monitoring the field for seismic ctivity. Swarms of micro-earthquakes indicate possible fluid conduits and can lead to ood sites for well drilling.

Measurements of the variation in the local gravitational acceleration across the field an reveal different masses beneath the surface. Sensitive instrumentation is needed, apable of detecting changes of the order of 0.01 mGal, where 1 Gal = 1 cm/s^2. he standard gravitational acceleration at the surface of the Earth is 981 Gal or 81,000 mGal. So a gravimeter with a sensitivity of 1 part in 98 million is needed. he reading essentially gives the density or the specific gravity of the formation. Since 1e reading is affected by the presence of hills or valleys in the neighborhood of the 1strument, and since the data must be referenced to a standard elevation (typically ea level), the raw data must be corrected. To do this accurately, high-precision topo-raphic information must be available for the field. Once the corrected value is found, is compared to what the normal gravity should be for the location of the instrument; ny difference is called the Bouguer gravity anomaly, *BA*, given by

$$BA = g_0 - (\delta_0 - 0.3086h + 2\pi\rho Gh - \tau) \tag{2.17}$$

here g_0 is the standard gravity at the location of the instrument, δ_0 is the theoreti-al gravity at the latitude of the instrument but at sea level, $0.3086h$ gives the hange in gravity as a function of elevation h above sea level, $2\pi\rho Gh$ is the infinite-lab correction to account for the attraction of a slab of material with a density ρ xtending in all horizontal directions and lying between the instrument and sea vel, and τ is the terrain correction to account for elevation changes near the 1strument [22].

Low-density volcanic rocks or partially molten magmatic plutons will create gravity ows. Very sensitive measurements may be able to track the movement of liquids rough porous rock and thus be useful for monitoring the effects of production and ijection of geofluids during the operation of a power plant. As with all geoscientific urveys, gravity data should be viewed in light of other indicators before drawing onclusions.

Although it is impossible to set down hard and fast interpretations of any geother-nal exploration results, rough inferences may be drawn from Fig. 2.8 where electrical esistivity is cross-plotted against the temperature gradient. The shaded region is learly the most desirable. These inferences should be seen as tentative unless con-rmed by other studies. This type of diagram in which more than one indicator is used vill provide a segue to the last section on exploration.

.4 Synthesis and interpretation

Vhen the geoscientific surveys have been completed, it is necessary to synthesize all he acquired data and to draw conclusions based on the totality of the evidence. As we ave seen, any one survey can give a misleading indication. The best way to visualize he comprehensive picture of the geothermal prospect is to construct, on a workable cale, a composite map, or synthesis map, of the site. The main purpose of the synthe-is map is to identify the best targets for the first few deep wells.

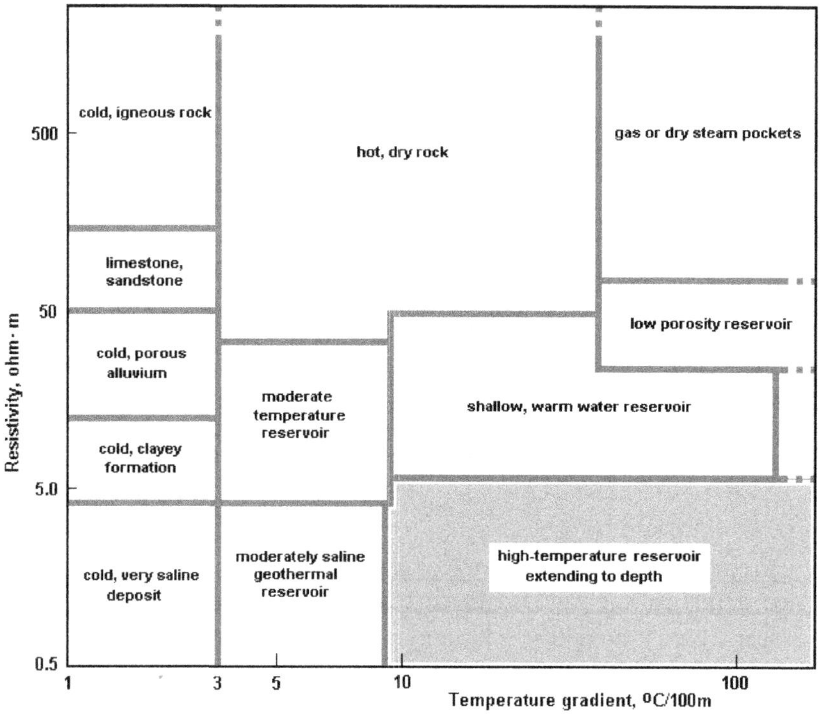

Fig. 2.8 Map of resistivity versus temperature gradient with possible interpretations.

The base map should be a simplified geologic map with at least 50-m topographic contours, and it should contain, at minimum, the following information:

- Roadways, buildings and dwellings
- Major geologic structures such as caldera borders, principal faults both observed and inferred, recent lava flows, basement outcrops, and outlines of suspected buried units of low permeability
- All thermal manifestations, identified by type, temperature, and chemical nature
- Radon concentration contours (if performed)
- Seismic activity
- Gravity contours
- Electrical resistivity contours.

Besides this map, there should be several elevation cross-sections taken through the field, preferably orthogonal, showing the extent of various geologic units down to and including the upper basement, faults, and resistivity layers. These elevation sections usually correspond to the lines along which the resistivity surveys were performed.

With the availability of modern 3-dimensional computer graphics, the synthesis can be presented in a convenient, user-friendly, visually appealing format.

As the field is developed, and more wells are drilled, this vital set of drawings or computer representation must be periodically updated to serve as an ongoing guide

) the next steps. Lithologic logs from deep wells will allow a better definition of the ubsurface and should improve the probabilities for successful wells. Lastly, additional eoscientific work may be required during field development to refine the understand- ng of the field and to guide step-out drilling.

.5 The next step: Drilling

fter all the scientific exploration has been done, the results synthesized and inter- reted, a decision will be made regarding the next phase, namely the drilling of everal, typically three, deep exploratory wells. If the synthesis map and its cross- ections accurately represent the geothermal system, the task of siting the wells will be traightforward. The drilling phase will be more costly by far than the exploration hase and should be undertaken only after a rigorous program of geoscientific studies as been carried out, with positive indications. The next chapter will deal with eothermal drilling.

eferences

[1] http://www.hotspringsenthusiast.com/USsprings.asp, "Hot Springs Enthusiast."

[2] http://geoheat.oit.edu/database.htm, "Western States Geothermal Databases CD," Geo-Heat Center, Oregon Institute of Technology, Klamath Falls, OR.

[3] http://ngmdb.usgs.gov/, "National Geologic Map Database," U.S. Geologic Survey, Menlo Park, CA.

[4] http://geopubs.wr.usgs.gov/open-file/of99–425/webmaps/GB%20map.html, "Great Basin Borehole Temperature Logs," U.S. Geologic Survey, Menlo Park, CA.

[5] http://geopubs.wr.usgs.gov/open-file/of99–425/frameset_FLP-1_ENG.html, "Well log for Fish Lake Valley, Nevada, Well FLP-1, U.S. Geologic Survey, Menlo Park. CA.

[6] http://water.usgs.gov/wid/index-state.html#CA, "USGS Fact Sheets," U.S. Geologic Survey, Menlo Park, CA.

[7] Duprat, A. and G. Omnes, "The Costs of Geophysical Programs in Geothermal Exploration," *Proc. Second U.N. Symposium on the Development and Use of Geothermal Resources*, V. 2, San Francisco, Lawrence Berkeley Laboratory, 1975, pp. 963–970.

[8] Del Grande, N.K., "An Advanced Airborne Infrared Method for Evaluating Geothermal Resources," *Proc. Second U.N. Symposium on the Development and Use of Geothermal Resources.* V. 2, San Francisco, Lawrence Berkeley Laboratory, 1975, pp. 947–953.

[9] Dickinson, D.J., "An Airborne Infrared Survey of the Tauhara Geothermal Field, New Zealand," *Proc. Second U.N. Symposium on the Development and Use of Geothermal Resources*, V. 2, San Francisco, Lawrence Berkeley Laboratory, 1975, pp. 955–961.

10] http://geopubs.wr.usgs.gov/dds/dds-81/GeologicalMaps/ScannedMap/sheet1.pdf, "Geologic Map of Long Valley Caldera, Mono-Inyo Craters Volcanic Chain, and Vicinity, Eastern California," R.A. Bailey, U.S. Geological Survey, Menlo Park, CA, 1989.

11] http://www.nature.nps.gov/geology/usgsnps/gmap/gmap1.html, "Geologic Maps," U.S. Geological Survey, National Park Service, Menlo Park, CA, 2000.

12] http://www.icdp-online.org/sites/longvalley/wellsite/well.html, "Long Valley Coring Project," International Continental Scientific Drilling Program, Potsdam, Germany, 2003.

13] Fournier, R.O. and R.W. Potter II, "A Revised and Expanded Silica (Quartz) Geothermometer," *Geothermal Resources Council BULLETIN*, V. 11, No. 10, 1982, pp. 3–12.

14] Fournier, R.O. and W.L. Marshall, "Calculations of Amorphous Silica Solubilities at 25 and 300°C and Apparent Cation Hydration Numbers in Aqueous Salt Solutions Using the Concept of Effective Density of Water," *Geochim. Cosmochim. Acta.*, V. 47, 1983, pp. 587–596.

[15] Fournier, R.O., "Application of Water Geochemistry to Geothermal Exploration and Reservoi Engineering," Chap. 4 in *Geothermal Systems: Principles and Case Histories*, L. Rybach and L.J.P Muffler, Eds., John Wiley & Sons, New York, 1981.

[16] Keenan, J.H., F.G Keyes, P.G. Hill and J.G. Moore, *Steam Tables: Thermodynamic Properties of Wate Including Vapor, Liquid, and Solid Phases (International Edition – Metric Units)*, John Wiley & Sons, Inc. New York, 1969.

[17] Arnorsson, S. and E. Gunnlaugsson, "New Gas Geothermometers for Geothermal Exploration – Calibration and Application," *Geochim. Cosmochim. Acta.*, V. 49, 1985, pp. 1307–1325.

[18] D'Amore, F. and A.H. Truesdell, "Calculation of Geothermal Reservoir Temperatures and Steam Fraction from Gas Compositions," *Geothermal Resources Council TRANS.*, V. 9-Part I, 1985, pp. 305–310.

[19] D'Amore, F., M. Mussi, S. Grassi and R. Alaimo, "Pantelleria Island (Sicily, Italy): A Gas Geochemica Survey," *Proc. World Geothermal Congress*, V. 2, International Geothermal Association, Auckland, NZ 1995, pp. 1007–1012.

[20] Dakhnov, V.N., Trans. by G.V. Keller, *Geophysical Well Logging, Quart. Colo. Sch. Mines*, V. 57, No. 2 1962, pp. 443.

[21] Meidav, T., "Application of Electrical Resistivity and Gravimetry in Deep Geothermal Exploration," *Geothermics – Special Issue 2, U.N. Symposium on the Development and Utilization of Geotherma Resources*, Pisa, V. 2, Pt. 1, 1970, pp. 303–310.

[22] Diment, W.H., "Resource Characteristics: Exploration, Evaluation, and Development," Chap. 2 in *Sourcebook on the Production of Electricity from Geothermal Energy*, J. Kestin, Ed. in Chief, R. DiPippo H.E. Khalifa and D.J. Ryley, Eds., U.S. Dept. of Energy, DOE/RA/4051-1, U.S. Gov. Printing Office Washington, DC, 1980.

[23] DiPippo, R., *Geothermal Energy as a Source of Electricity: A Worldwide Survey of the Design and Operation of Geothermal Power Plants*, U.S. Dept. of Energy, DOE/RA/28320-1, U.S. Gov. Printing Office Washington, DC, 1980.

Problems

2.1 With reference to Fig. 2.1, calculate the (a) shallow and (b) deep temperature gradients, in °F/100 ft. Compare these to the normal temperature gradient Discuss possible reasons for the relatively isothermal section between the shallow and deep portions.

2.2 For each of the geofluid samples listed below, apply the following geothermometers to obtain estimates of the reservoir fluid temperature: (a) SiO_2 (maximum steam loss), (b) SiO_2 (no steam loss), (c) Na/K, and (d) Na/K/Ca. Geofluid samples: (1) Ahuachapán wells AH-1, AH-5, and AH-26; (2) Hatchobaru wells H-4 and H-6; (3) Cerro Prieto wells (average). For data on wells, see DiPippo [23], Tables 3.4, 6.17, and 7.2.

2.3 Suppose a sample from a well has a Na/K = 6.97.

(a) Estimate the reservoir temperature using the Na/K geothermometer equation.

(b) Suppose further that the sample has a $Ca^{1/2}/Na = 1.5$. Estimate the temperature using the Na/K/Ca geothermometer.

2.4 An alternative formulation for the Na/K geothermometer has been offered [15]:

$$t = \frac{855.6}{\log_{10}\dfrac{\{Na\}}{\{K\}} + 0.8573} - 273.15$$

Compare this formulation to the one given in eq. (2.8). Calculate and plot the results from 150–350°C. On one graph, plot *t* (°C) on the ordinate and {Na}/{K} on the abscissa showing the curves for both formulations. Discuss the disagreement between these curves, the limits of applicability, and the relative reliability.

.5 Use the silica geothermometer to estimate the reservoir fluid temperature if the two-phase fluid from a test well is separated at the surface into its liquid and vapor phases at 100°C and the concentration of silica in the liquid is: (a) 300, (b) 400, (c) 500, (d) 600, (e) 700 ppm.

.6 Compare the results of using eqs. (2.1) and (2.4) to determine the geofluid temperature for the "no steam loss" case over a range of silica concentrations from 200 to 700 ppm. Discuss the possible causes of any differences you find.

Chapter 3

Geothermal Well Drilling

"The well blew up like a volcano, everything came flying up – mud, tools, rocks, and steam. After things settled down, there was just clean steam. But the noise was loud enough to hear all over the valley."

Glen Truitt, on the drilling of well No. 1 at The Geysers – 1922

3.1 Introduction

'he confirmation of the optimistic outcome of the exploration phase comes with the uccessful drilling of the first exploratory well. Figures 3.1 and 3.2 show wells that ave struck productive formations [1].

Generally three wells are drilled as the first step in the field confirmation stage. The vells are sited at the most promising locations as determined by the synthesis of all he exploration studies. If possible they should form a triangle in the hope of defining a roductive area of the field. Since these wells must be viewed as part of the exploration rogram, as must information as possible should be collected during their drilling. This ncludes the taking of core samples from at least some portions of the well to help nderstand the lithology of the formation. This data must be incorporated into the onceptual model of the field prior to undertaking the next step – the drilling of the evelopmental wells.

In this chapter we will briefly describe the drilling process; for a more detailed and uantitative presentation, the reader is advised to consult Edwards *et al* [2] who devote everal chapters of their book to drilling.

3.2 Site preparation and drilling equipment

Geothermal fields are often remote, far from developed areas, and require significant ite preparation before drilling can begin. During the scientific exploration phase,

eothermal Power Plants: Principles, Applications, Case Studies and Environmental Impact, Third Edition.

© 2012 Elsevier Ltd. All rights reserved.

Fig. 3.1 Discharge from a well at the Ahuachapán field, El Salvador [1] [WWW].

access to the site had to be sufficient only to allow a few people to reach and sample thermal springs and other manifestations, to drive small vehicles, run electrical equipment across the terrain, etc. But for this stage, roads will need to be constructed that can carry heavy equipment to the designated drilling targets. In volcanic environments, this can be challenging owing to rugged terrain. To minimize the number of drill pads, several wells can be drilling from a single pad, the wellheads being only tens of meters apart. Directional drilling, however, allows the productive sections of the wells to be widely separated, perhaps by as much as 500–1000 m, to avoid interference. Generally, the first or "discovery" well is a deep vertical well (say, 2500 m) to allow as much information as possible to be gathered about the formation.

ig. 3.2 *Initial blowing of an Indonesian well at the Dieng geothermal field [1] [WWW].*

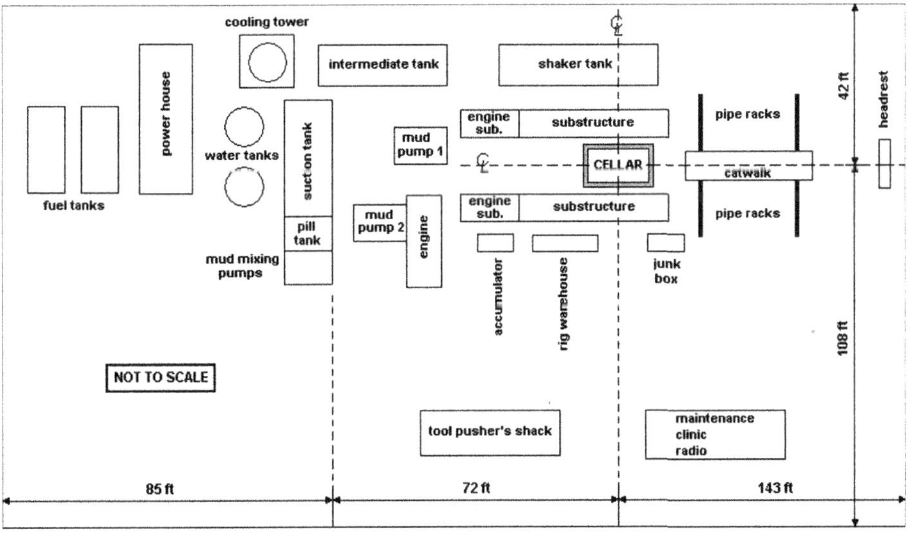

ig. 3.3 *Drill pad to support a 3000 m rig; after [3].*

A typical well pad with all equipment to support a 3000 m drill rig is shown in ig. 3.3. An area of roughly 90 × 45 m (300 × 150 ft) is required, including buffer pace around the equipment and structures.

Figure 3.4 shows the major pieces of equipment used at the drill rig. Note the elative thicknesses of the drill pipe and the drill collar which holds the bit.

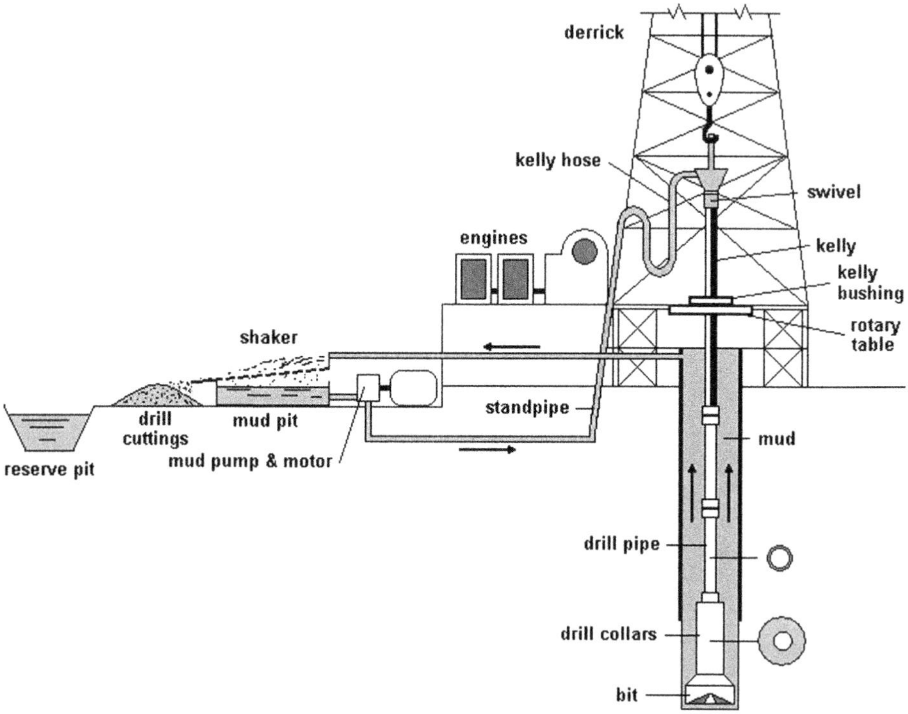

Fig. 3.4 Typical drill rig setup; not to scale.

3.3 Drilling operations

The creation of the hole relies on the compressive forces exerted on the rock by the multi-toothed drill bit. Rotary drilling is the standard method in geothermal drilling whereby a string of drill pipe is hung from a derrick and turned by an engine, typically a diesel. The top section of the pipe, called the "Kelly," is square in cross-section to allow it to be rotated by the action of a rotary table through which it passes. The bit is a tri-cone roller bit, based on the 1909 invention of Howard R. Hughes, Sr. (1870–1924), that applies very concentrated loads on the rock face causing it to crack and spall; see Fig. 3.5. The rock fragments or chips must be removed for the bit to proceed.

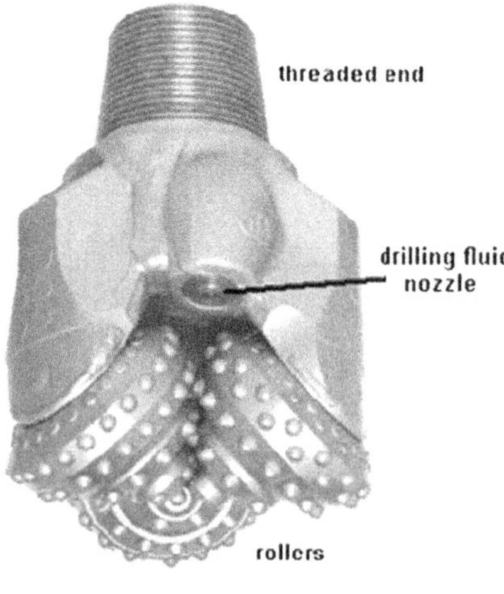

Fig. 3.5 A tri-cone roller bit designed for hard rock formations, manufactured by Hughes Christensen [4].

The drilling fluid, or "mud," is a critical element in the operation; it serves four vital unctions:

- Removes the rock chips
- Cools the drill bit and drill string
- Lubricates the drill string
- Prevents the collapse of the well wall during drilling.

'he mud is pumped down the center of the drill pipe to the bit where it passes through ozzles that accelerate and direct it onto the rock beneath the bit. It sweeps up the hips as it flows under the bit and carries them with it as it returns to the surface hrough the annulus between the well wall and the drill string. When it gets to the urface it flows over a screen and shaker to dislodge the chips which are preserved nd examined by the geologist. The mud then is returned to the bottom by the mud ump. When the formation is hot, it is necessary to cool the mud before returning it o the well. A cooling tower is shown in the plan layout in Fig. 3.3, but has been omited from Fig. 3.4. Owing to unplanned deviations of the bit during drilling, the drill ipe may come in contact with the wall over some length; the mud lubricates the conact surface helping to reduce the friction. Finally, by exerting pressure against the vall, the mud can keep the formation intact until a permanent casing can be put in lace.

The drilling mud is an engineered fluid, designed to perform its specific tasks [5]. It ; basically water with numerous additives that can alter the viscosity and density. 'he density can range from 62.4 lbm/ft^3 (pure water) to 150 lbm/ft^3. Clays are lended into solution to achieve desired properties. A severe disadvantage to using nud as the drilling fluid arises when the formation is tight, with little permeability xcept for a few narrow fractures bearing the geofluid. The mud can clog the fractures nd destroy what little permeability was there. In such cases, mud can be used until he productive zone is approached, but beyond that point, the drilling fluid is changed o air or aerated mud. Whereas the ability of air to lift rock chips from a large diameter ole is poor, at the lower end of the well in the production zone, the well diameter is maller and air can do the job effectively.

If the bit encounters a highly permeable zone, the drilling mud may be absorbed by he formation and the return upflow will be less than the downflow. This is called a lost circulation" problem. Sometimes the loss is great enough to prevent any returns t all – a "total lost circulation." This is a highly desirable outcome when the well is n the production zone because it signals excellent permeability and a potential big roducer. But if this happens in the low-temperature shallow zone, it is very troubleome because drilling cannot continue without drilling mud. Much effort has gone nto finding special materials to close these "thief zones" [6].

It is always a judgment call by the drilling engineer, in consultation with the geocientists, as to whether a lost circulation zone within the upper part of the reservoir hould be cemented. This is a permanent solution to the lost circulation problem, but he permeable zone will be lost for any possible production should the rest of the well urn out to be unproductive. This dilemma tends to occur in the early stages of field levelopment before a clear picture of the reservoir has emerged.

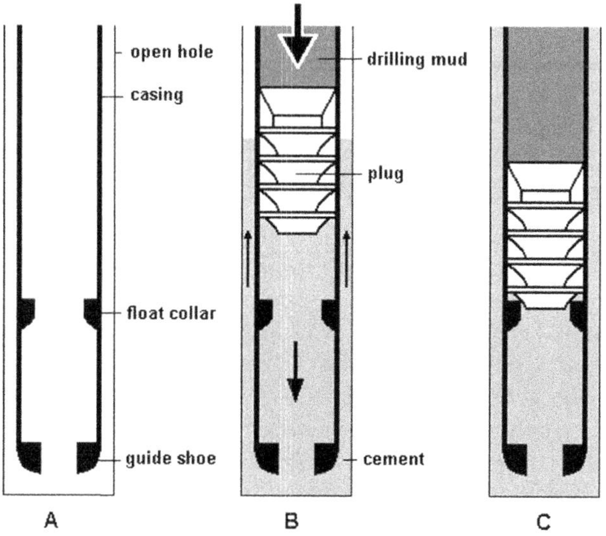

Fig. 3.6 Cementing a casing in three steps.

The running and cementing of the casings is a critical task [7]. A standard way to perform this operation is depicted in Fig. 3.6. When a hole has been dug to the desired depth, the casing is lowered into the hole (Fig. 3.6A). The cement is mixed and a volume somewhat greater than the annular volume between the inside of the hole and the outside of the casing is forced down the inside of the casing with a plug (Fig. 3.6B) using drilling mud as the pusher. Direct displacement of the cement should lead to the complete filling of the annulus with cement by the time the plug comes to rest on the float collar (Fig. 3.6C). The return of the excess cement to the surface signals a successful job.

One of the main problems encountered involves the loss of cement to the formation through permeable zones, the same lost circulation zones mentioned earlier. Geothermal formations are notorious for having large porous zones that can be caused by any number of reasons. For example, interfaces between different rock layers can be highly permeable, or a particularly porous volcanic deposit may be encountered. In fact, cavern-like openings have been found in many fields that cause the drill string to drop precipitously for several meters. If such a zone exists, it provides an easy outlet for the cement and the cement may not return to the surface by the time the plug is on the bottom. This leaves an uncemented length of casing that can no longer be cemented by the displacement method. The casing cannot be left in this unsecured condition. The uncemented annulus is probably filled with liquid. If this section becomes heated during later operation of the well, the trapped liquid will heat up, try to expand, and its pressure will rise dramatically, most likely leading to the collapse of the casing. This certainly would mean the loss of the well and could even cause a blowout if the weak section is fairly close to the surface.

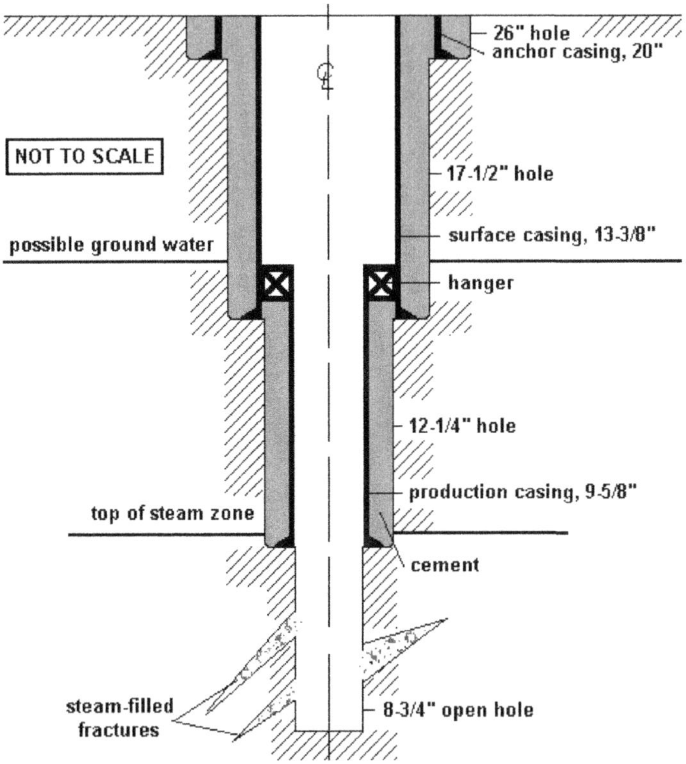

'ig. 3.7 *Typical well completion for a dry steam reservoir.*

The only way to repair this is to perform a so-called "squeeze job" whereby cement f forced directly into the annulus from the surface. This, however, is fraught with risk ecause some drilling mud could still remain in the annulus as the cement falls into lace via rivulets, not as a contiguous volume. One way to mitigate this is first to heat he casing by sending steam or hot water down a central pipe, until all the liquid has vaporated. While this will eliminate the possibility of liquid getting trapped in the nnulus, there is still a chance that air pockets could exist when the job is done. But ir is compressible, unlike a liquid, so that if it is heated, its pressure will not increase s greatly as with a liquid and is therefore less likely to cause a collapse of the casing. nother approach is to add perlite or gels with a bridging agent such as gilsonite o the cement during the displacement method in the hope that these will enter the ermeable zones, swell up, and fill in the voids [6].

As is customary with any kind of deep well, the drilling is done in stages. A shallow nchor casing is cemented into a large hole, followed by the surface casing. The production casing is the longest one and terminates just short of the main production one; see Fig. 3.7. The well may be drilled deeper and left as an open hole or fitted vith a slotted liner, set so that the slots align with the production zone. In some wells, he production casing is carried all the way to the surface, but in others it may be ung from the surface casing.

3.4 Safety precautions

There is always the risk of a blowout when drilling a geothermal well. This occurs when an unexpected, high-pressure permeable zone is encountered. Enough of these dramatic and dangerous events have happened so that there are now strict regulations in most countries on the proper safety precautions to be followed during drilling [8].

The use of blowout preventers is standard practice nowadays; see Fig. 3.8. These are a set of fast-acting ram-type valves attached to the surface casing, and through which the drill pipe rotates. In the event of a "kick" from the well, these valves are slammed tight around the drill string, effectively closing off the well. Another valve attached to the wellhead just above the casing allows for controlled venting of the well to a silencer until the well is brought under control, usually by quenching the well with cold water.

The presence of toxic gases such as hydrogen sulfide, H_2S, can lead to severe injuries and even fatalities unless proper procedures are followed [9]. The cellar of a well is

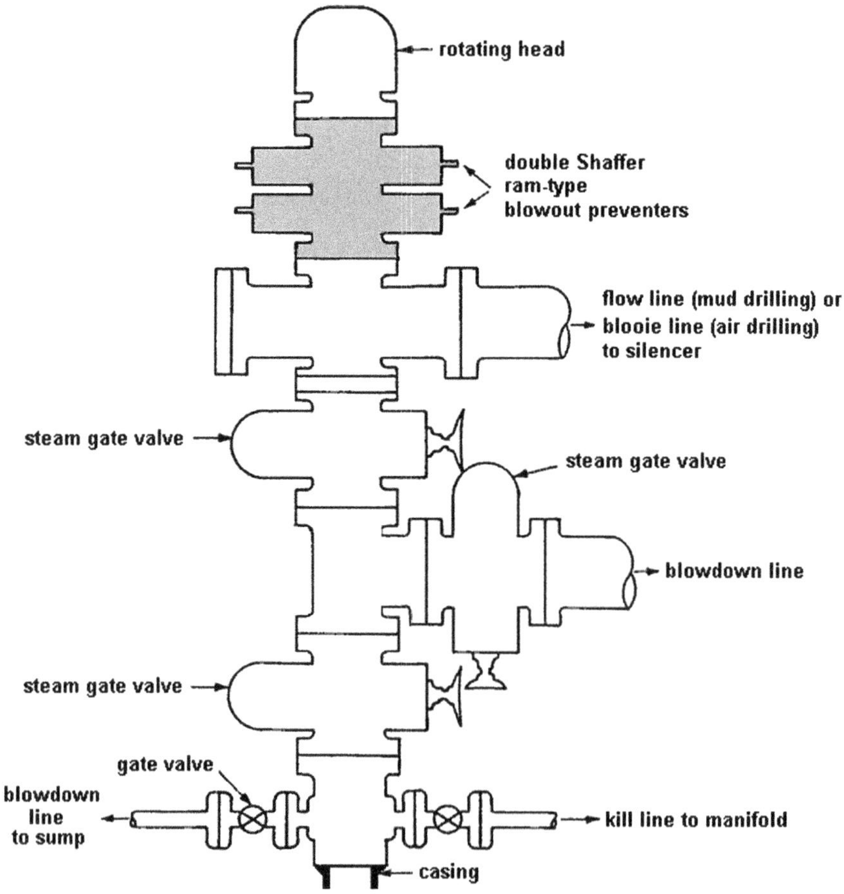

Fig. 3.8 Typical blowout preventer system at The Geysers, California.

particularly dangerous place since both H_2S and carbon dioxide, CO_2, are heavier ıan air. If there are any leaks from the well or casing, these gases can accumulate in ıe cellar. Inexpensive H_2S and CO_2 sensors should be installed wherever there is the ossibility of high concentrations of these gases.

These gases and others can be extremely corrosive and can lead to casing failures nless the proper materials are used. In a new field before the fluids are thoroughly nalyzed and the corrosion potential is assessed, this can lead to blowouts even after ıe well has been completed. If a break occurs in the casing string the pressurized geouids can escape outside the wellbore and cause severe damage to surface equipment nd injuries to personnel who may be nearby. Blowout preventers cannot stop such ilures should they occur during drilling, and of course once a well is completed the lowout preventers are no longer in place and the well is controlled by the set of wellead valves.

eferences

[1] "Foramines: The Geothermal Drilling Contractor," Brochure 606.88.26, Paris, France, undated.

[2] Edwards, L.M., G.V. Chilingar, H.H. Rieke III and W.H. Fertl, Eds., *Handbook of Geothermal Energy*, Chaps. 5–8, Gulf Publishing Company, Houston, TX, 1982.

[3] PNOC Energy Drilling, Inc., PNOC Rig No. 4, National 610, Metro Manila, Philippines, undated.

[4] http://www.bakerhughes.com/hcc/literature/products/catalog.pdf, *Hughes Christensen Drill Bit Catalog*, Hughes Christensen Company, The Woodlands, TX, 2001.

[5] Hutchison, S.O. and G.W. Anderson, "What to Consider When Selecting Drilling Fluids," *World Oil*, October 1974, reprint.

[6] Shyrock, S.H., "Geothermal Cementing," Chap. 7 in *Handbook of Geothermal Energy*, L.M. Edwards et al, Gulf Publishing Company, Houston, TX, 1982.

[7] Shyrock, S.H. and D.K. Smith, "Geothermal Cementing – The State of the Art," Rep. No. C-1274, Halliburton Services, Duncan, OK, undated.

[8] Anon., *Drilling and Operating Geothermal Wells in California, 4th. Ed.*, Pub. No. PR7S, California Dept. of Conservation, Div. of Oil and Gas, Sacramento, CA, 1986.

[9] Dosch, M.W. and S.F. Hodgson, *Drilling and Operating Oil, Gas, and Geothermal Wells in an H_2S Environment*, Pub. No. M10, California Dept. of Conservation, Div. of Oil and Gas, Sacramento, CA, 1986.

Chapter 4

Reservoir Engineering

"I venture to predict that great engineering work will be done in the future in conducting water to Plutonic regions, for the more abundant abstraction and upbringing of their heat energy."

Frank A. Perret – November 12, 1925

Geothermal Power Plants: Principles, Applications, Case Studies and Environmental Impact, Third Edition.
© 2012 Elsevier Ltd. All rights reserved.

4.1 Introduction

Reservoir engineering is the art of describing quantitatively the behavior of fluids in a porous or fractured rock formation and using that description to effectively manage the production and injection of those fluids. This task is complicated by the fact that no one can be sure of the underground patterns of fractures or porosity that give rise to the permeable channels through which the fluids may move. Furthermore, even if that were possible for one moment, the fracture patterns very likely will change over time since the geothermal environment is a geologically dynamic one.

The product of the reservoir engineer is the reservoir working model, built on the conceptual model produced from the earlier scientific studies and the results of exploratory drilling, but incorporating a computer simulation of the reservoir fluid properties. Such a model can be used to forecast the evolution of the field under different scenarios of production and injection. This is an extremely important tool in the development of a geothermal resource because it tells us the long-range anticipated capability of the field to support a given amount of electrical generation.

The flow of fluids in a porous medium and in wells has been thoroughly studied from both a theoretical and an experimental viewpoint [1–9]. Most of the literature pertains to oil and gas applications. However, the vast amount of accumulated data from tests at numerous geothermal fields constitutes a wealth of information to serve the needs of reservoir researchers.

In the rest of this chapter, we will present some of the theoretical findings for specific, idealized conditions, describe some of the tests that are performed on wells, and show how this information is used to generate the reservoir working model.

4.2 Reservoir and well flow

The flow in a geothermal well cannot be viewed in isolation; it must be coupled to the flow of fluid in the reservoir. The only exception to this rule would be when the reservoir offers no resistance to fluid flow, a physical impossibility. While it is not easy to describe the flow of a geofluid passing through a more-or-less vertical well owing to possible changes in phase and flow pattern, at least the geometry of the conduit is known. In the reservoir, it is far more difficult, if not impossible, to describe the flow analytically because the path of the fluid is unknowable. As such, the analytical approach relies on a "lumped parameter" method in which the details are smoothed out and averaged values of important properties are used instead.

4.2.1 Darcy's Law

One analytical approach that can be solved exactly involves flow through an idealized porous medium consisting of a bed of packed granules, the so-called "Darcy flow" problem, the solution to which is known as Darcy's Law, after H.P.G. Darcy (1803–1858) [10]. Here the medium is seen as one of homogeneous permeability, K, the fluid is characterized by an absolute viscosity, μ, and moves very slowly ("slug flow"), in one direction, horizontally, under a pressure gradient, dP/dx.

Darcy's Law for this case yields the following equation for the fluid velocity, $\mathcal{V}$:

$$\mathcal{V} = -\frac{K}{\mu}\frac{dP}{dx} \tag{4.1}$$

where the negative sign indicates that the fluid moves in the direction of decreasing pressure.

In honor of Darcy, the unit of permeability is called the darcy, and is defined as follows: 1 darcy corresponds to the slow flow of a single-phase fluid having a viscosity of 1 centipoise (cP) and a volumetric flow rate of 1 cm^3/s through an area with a cross-section of 1 cm^2 under the driving force of a pressure gradient of 1 atm/cm. For geothermal reservoirs, permeabilities are much smaller than a darcy and a more appropriate unit is 1/1000 of a darcy, the millidarcy, mD. A typical permeability for a geothermal reservoir would be 10–70 mD.

4.2.2 Reservoir-well model: Ideal case

The simplest model of the reservoir-well system is depicted in Fig. 4.1.

By continuity of mass, the flow through the well must equal the flow through the reservoir under steady conditions. If the fluid in the system is liquid (i.e., essentially incompressible), then the volumetric flows must also be equal. Using Darcy's Law and continuity, it can be shown that the pressure distribution in the reservoir as a function of the radial distance from the centerline of the well is given as:

$$P(r) = \frac{\mu \dot{V}_W}{2\pi K L_R} \ln r \tag{4.2}$$

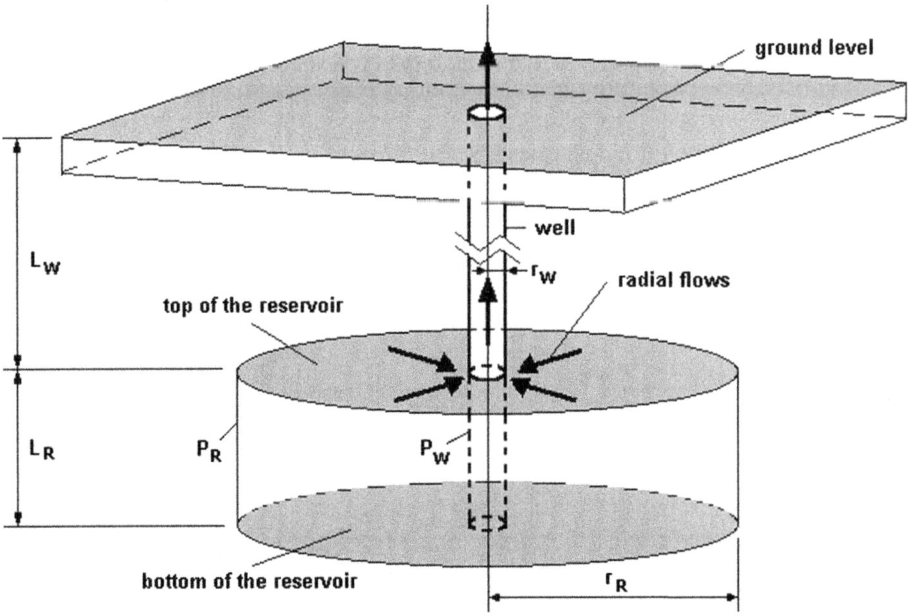

Fig. 4.1 Schematic of simplified reservoir-well system.

where $\dot{V}_W$ is the volumetric flow rate into the well through the interface between the well and the reservoir, and L_R is the reservoir thickness. The pressure increases with increasing distance from the well, approaching the undisturbed reservoir pressure logarithmically. The difference between the far-field reservoir pressure and the pressure at the well face is known as the drawdown ΔP:

$$\Delta P \equiv P(r = r_R) - P(r = r_W) = \frac{\mu \dot{V}_W}{2\pi K L_R} \ln \frac{r_R}{r_W} \tag{4.3}$$

Thus, for this highly idealized case, the volumetric flow rate entering the well is given by

$$\dot{V}_W = \frac{2\pi K L_R}{\mu \ln \dfrac{r_R}{r_W}} \Delta P \tag{4.4}$$

If the fluid as it flows up the well undergoes a phase change (e.g., flashes into vapor), then the volumetric flow rate at the surface will be different from eq. (4.4) owing to compressibility effects. The mass flow rates however will be the same. This case will be studied in Sect. 4.2.6.

4.2.3 Reservoir-well model: Basic principles

The approach described here is based on a model by Ryley [4,11] which uses the fundamental equations of fluid mechanics but replaces the problematic parameters with averaged empirical quantities. This leads to working equations that produce reasonable results when compared to field measurements.

The reservoir pressure may be assumed to be caused by a column of cold water distant from the well, as shown in Fig. 4.2. The density of the cold water exceeds that of the hot geofluid in the reservoir-well system, thereby creating a natural siphon that will cause the well in the figure to spontaneously begin flowing as soon as the wellhead valve is opened. The exact mechanism responsible for the far-field reservoir pressure is unimportant to the argument we make here.

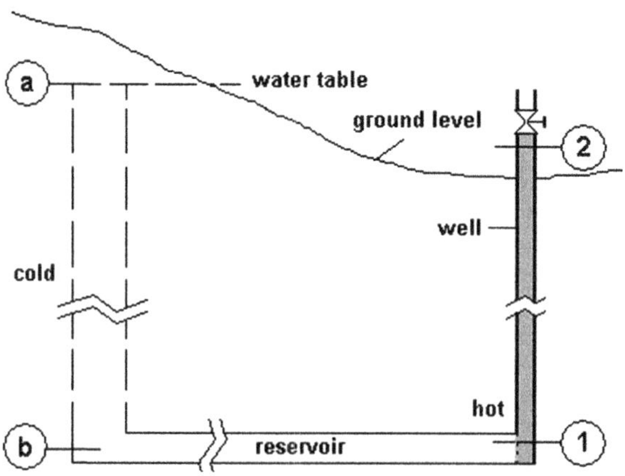

Fig. 4.2 Possible pressurizing mechanism for geothermal well.

Considering first only the well between states 1 and 2, the First Law of thermodynamics for an open system in steady flow may be written as

$$\dot{Q} - \dot{W} = \dot{m}\left[(h_2 - h_1) + \frac{1}{2}(\mathcal{V}_2^2 - \mathcal{V}_1^2) + g(z_2 - z_1)\right] \tag{4.5}$$

where the left-hand terms are the heat and work flow rates between the well and its surroundings, and the right-hand terms are, in order, the enthalpy, kinetic, and potential energy differences from the top of the well to the bottom. Unless there is a pump in the well, the work term will vanish. Given the low thermal conductivity of rock and cement used to secure the well casings) and the rapidity with which the geofluid moves up the well, we can ignore the heat term to a first approximation. When one takes into account the units of the right-hand terms and the realistic values for these terms, it will turn out that the kinetic energy difference is relatively small. Thus we find

$$h_2 \approx h_1 - gL_W \tag{4.6}$$

Unless the well is very deep and low in temperature, the gL_w-term will be negligible compared to the geofluid enthalpy and may be ignored. In this way we find that the well flow can be approximated fairly accurately as isenthalpic, i.e., one of constant enthalpy.

Now we will apply the momentum equation of fluid mechanics to the well flow. This is essentially an application of Newton's Second Law of Motion written per unit of well area, namely:

$$-dP - \frac{dF}{A} - \rho g\, dz = \rho \mathcal{V} d\mathcal{V} \tag{4.7}$$

where the terms on the left-hand side represent all the forces (per unit area) acting on an elemental fluid body of length dz passing up the well, and the right-hand side is the mass times acceleration (per unit area). The forces are, in order, the pressure forces acting on the ends of the fluid column, the friction forces caused by contact with the well casing, and the hydrostatic body force due to weight of the fluid above the elemental volume. Note that all the force terms are negative and oppose the motion of the fluid up the well, and that consequently the acceleration will also be negative.

We are interested in knowing the pressure as a function of height in the well, $P = P(z)$, and eventually in knowing the relationship between the mass flow rate and the wellhead pressure, $\dot{m} = \dot{m}(P_2)$. This requires us to integrate the momentum equation from the bottom of the well to the top. Before we attempt this, we can express the elemental friction force in terms of the friction factor, f, as follows:

$$dF = \frac{1}{2}\rho\mathcal{V}^2 \times f \times C\, dz \tag{4.8}$$

where $C = 2\pi r_W = \pi D$ is the circumference of the well interior. Now we can express the difference between the downhole pressure and that at the wellhead as follows:

$$P_1 - P_2 = \int_{\mathcal{V}_1}^{\mathcal{V}_2} \rho(z)\, \mathcal{V}(z)\, d\mathcal{V} + \frac{2}{D}\int_{z_1}^{z_2} f\rho(z)\, \mathcal{V}^2(z)\, dz + g\int_{z_1}^{z_2} \rho(z)\, dz \tag{4.9}$$

The formal integration of the three right-hand terms cannot be carried out unless the friction factor is known, along with the dependency of the density and velocity on the distance along the well. When flashing occurs in the well, these terms are not obvious.

4.2.4 Liquid-only flow

A particularly simple special case involves only the flow of liquid in the system. Here there is no acceleration term, and the velocity and density can be taken as constants. Thus the pressure difference along the well is given by

$$P_1 - P_2 = \frac{2f\,\rho\,\mathcal{V}^2 L_W}{D} + g\rho L_W \tag{4.10}$$

where the friction factor may be found from the Swamee-Jain equation [12],

$$f = \frac{0.25}{\left\{\log_{10}\left[\frac{\varepsilon/D}{3.7} + \frac{5.74}{\mathrm{Re}^{0.9}}\right]\right\}^2} \tag{4.11}$$

with ε being the absolute roughness of the well casing and Re being the Reynolds number

$$\mathrm{Re} = \frac{\rho \mathcal{V} D}{\mu} = \frac{\dot{m} D}{\mu A} = \frac{4\dot{m}}{\mu \pi D} \tag{4.12}$$

4.2.5 Location of the flash horizon

Another case that can be handled easily involves the determination of the location of the flash horizon in the well, i.e., the elevation where the liquid flashes into vapor. Here we will couple the reservoir to the well and follow the flow of liquid through the system.

With reference to eq. (4.4), if we recognize that the volumetric flow rate is equal to the mass flow rate divided by the density, i.e.,

$$\dot{V} = \dot{m}/\rho \tag{4.13}$$

we can arrive at the following expression for the drawdown pressure difference:

$$\Delta P = P_R - P_1 = \frac{\mu \ln(r_R/r_W)}{2\pi K L_R \rho_R}\dot{m} \equiv C_D \dot{m} \tag{4.14}$$

where we have introduced C_D, the drawdown coefficient. Thus the pressure at the bottom of the well under flowing conditions can be found from the reservoir pressure, the mass flow rate, and the drawdown coefficient:

$$P_1 = P_R - C_D \dot{m} \tag{4.15}$$

The drawdown coefficient may be found as follows. The reservoir pressure, P_R, can be found by measuring the pressure in the well opposite the inflow zone under static conditions, i.e., when $\dot{m} = 0$. If the pressure is measured at the same location under flowing conditions, along with the mass flow rate using surface equipment, then the drawdown coefficient can be calculated using eq. (4.15). Note that the inverse of C_D is called the productivity index, J_I.

To locate the flash horizon, we need to consider the flow from the reservoir into the well and up the well to the place where the geofluid pressure falls to the saturation pressure corresponding to the geofluid temperature. That is, at the flash point

$$P_F = P_{sat}(T_R) \tag{4.16}$$

If we combine eqs. (4.10) and (4.14) and eliminate the velocity using the continuity equation, we end up with

$$P_2 = P_R - \left[\frac{\mu}{\rho_R}\right]\left[\frac{1}{KL_R}\right]\left[\frac{\ln(r_R/r_W)}{2\pi}\right]\dot{m} - \left[\frac{1}{\rho}\right]\left[\frac{32}{\pi^2}\right]\left[\frac{fL_W}{D^5}\right]\dot{m}^2 - \rho g L_W \tag{4.17}$$

This equation is written from the reservoir to the top of the well, but we only want to apply it to the flash point. Therefore, the working equation to solve for the flash horizon is

$$P_{sat}(T_R) = P_R - \left[\frac{\mu}{\rho_R}\right]\left[\frac{1}{KL_R}\right]\left[\frac{\ln(r_R/r_W)}{2\pi}\right]\dot{m} - \left[\frac{1}{\rho}\right]\left[\frac{32}{\pi^2}\right]\left[\frac{fL_F}{D^5}\right]\dot{m}^2 - \rho g L_F \tag{4.18}$$

where L_F is the distance up the well from the feed zone up to the flash horizon. The depth from the surface to the feed zone is L_W, so that the depth to the flash horizon is simply $L_W - L_F$. We can simplify eq. (4.18) using the drawdown coefficient and solve for the flash horizon:

$$L_F = \frac{P_R - P_{sat}(T_R) - C_D\dot{m}}{\rho g + C_2\dot{m}^2} \tag{4.19}$$

where a new term C_2 has been defined as

$$C_2 \equiv \left[\frac{1}{\rho}\right]\left[\frac{32}{\pi^2}\right]\left[\frac{f}{D^5}\right] \tag{4.20}$$

The observant reader will detect that the friction factor f in eq. (4.20) is itself a function of the mass flow rate through the Reynolds number; see eqs. (4.11) and (4.12). This dependency is weak for fully developed, high Reynolds number, turbulent flow, but is not negligible for many geothermal applications.

From eq. (4.19), one can see that the flash horizon moves down the well (i.e., L_F gets smaller) as the mass flow rate increases. The flash point will reach the feed zone of the well for sufficiently high flow rates and eventually will move into the reservoir. Once two-phase flow occurs in the formation, our linear drawdown equation, eq. (4.15), will no longer be valid, and the process becomes highly complex.

The velocities of the liquid and vapor phases will differ as they flow through the formation since the resistance offered by the porous rock to the vapor will be less than for the liquid. This will cause the vapor to be produced at a higher rate than the liquid, and the enthalpy of the two-phase mixture to exceed the value expected for isenthalpic flow. Thus, the appearance of "excess enthalpy" at the wellhead is an indication of flashing in the formation. Likewise, the wellhead dryness fraction will be greater than for isenthalpic flow.

Numerical example: Flash horizon Before moving on to the two-phase flow region above the flash point, it is instructive to consider a numerical example of eq. (4.19). This will also provide a feeling for the order of magnitude of the terms in the equation.

Given data:

$$P_R = 20\text{ MPa},\ \dot{m} = 75\text{ kg/s},\ L_W = 2000\text{ m},\ T_R = 250°\text{C},$$
$$D = 0.244\text{ m (or 9–5/8 in)},\ g = 9.81\text{ m/s}^2,\ P_1 = 17\text{ MPa @ }\dot{m} = 75\text{ kg/s}$$

From *Steam Tables* [13], we can find the appropriate properties for the geofluid (assumed to be pure water): $P_{sat} = 3.973$ MPa, $\rho = 799.2$ kg/m^3 and $\mu = 0.000107$ kg/m·s.

The drawdown is found from the given data:

$$C_D = \frac{P_R - P_1}{\dot{m}} = \frac{20 - 17}{75} = 0.040\frac{MPa}{kg/s}$$

The Reynolds number and the friction factor are found from eqs. (4.12) and (4.11):

$$\text{Re} = \frac{4\dot{m}}{\mu\pi D} = \frac{4\times 75}{0.000107\times\pi\times 0.244} = 3{,}657{,}613$$

$$f = \frac{0.25}{\left\{\log_{10}\left[\frac{\varepsilon/D}{3.7} + \frac{5.74}{\text{Re}^{0.9}}\right]\right\}^2} = \frac{0.25}{\left\{\log_{10}\left[\frac{0.0457/244}{3.7} + \frac{5.74}{3{,}657{,}613^{0.9}}\right]\right\}^2} = 0.0139$$

We have used a surface roughness of 0.0457 mm, assuming commercial steel casing.

The factor C_2 may now be calculated:

$$C_2 = \left[\frac{1}{\rho}\right]\left[\frac{32}{\pi^2}\right]\left[\frac{f}{D^5}\right] = \left[\frac{1}{799.2}\right]\left[\frac{32}{\pi^2}\right]\left[\frac{0.01392}{0.244^5}\right] = 0.06528\text{ kg}^{-1}\text{ m}^{-2}$$

And finally we can calculate the height of the flash horizon above the feed zone (assumed to be at the bottom of the well):

$$L_F = \frac{P_R - P_{sat}(T_R) - C_D\dot{m}}{\rho g + C_2\dot{m}^2} = \frac{20 - 3.973 - 0.040\times 75}{799.2\times 9.81 + 0.06528\times 75^2}$$
$$= \frac{13.027}{7840.15 + 367.2} = 1587\text{ m}$$

Thus, the fluid flashes from liquid to a two-phase mixture starting at a depth of 413 m below the wellhead.

Without going into the numbers, if this well could flow at 150 kg/s, then the flash depth would be 921 m below the wellhead, still within the wellbore but some 508 m deeper.

heoretically, a mass flow rate of 400.67 kg/s would place the flash point at the entrance the well since the drawdown (0.040 × 400.67 = 16.027 MPa) results in the saturation ressure being reached as the fluid arrives at the well (20 – 16.027 = 3.973 MPa). Since his is a very high flow rate for a single well, for the conditions considered in this example, is likely that flashing would always occur in the wellbore. Over time however, should he reservoir pressure fall, as can be expected during exploitation, the drawdown may ause the fluid to flash in the formation. For example, in this case, if the reservoir pressure vere to drop to 10 MPa, then a mass flow greater than 151 kg/s will lead to flashing in he formation.

.2.6 Two-phase flow in the well

Ve now return to the more difficult task of modeling the two-phase flow that takes lace from the flash horizon upward to the wellhead. We may begin with eq. (4.9) ewritten to apply from the flash point, state F, to the wellhead, state 2:

$$P_F - P_2 = \int_{\mathcal{V}_1}^{\mathcal{V}_2} \rho_{LV}(z)\,\mathcal{V}_{LV}(z)\,d\mathcal{V} + \frac{2}{D}\int_{z_F}^{z_2} f_{LV}\,\rho_{LV}(z)\,\mathcal{V}_{LV}{}^2(z)\,dz + g\int_{z_F}^{z_2} \rho_{LV}(z)\,dz \tag{4.21}$$

In order to integrate eq. (4.21), we are at once confronted with the problem f deciding what to use for the various two-phase properties denoted by the ubscript LV. These include the density, velocity, and friction factor. This general roblem has been the subject of many research efforts and continues to this day. n addition to Refs. [1–9], the reader may wish to consult Refs. [14–17] for more etails.

We will adopt the lumped-parameter approach presented by Ryley [11] in which nean effective values are used for these two-phase terms. These may then be viewed s constants and the equation readily integrates to:

$$P_F - P_2 = \frac{\bar{\rho}_{LV}}{2}(\mathcal{V}_2^2 - \mathcal{V}_1^2) + \frac{2}{D}\bar{f}_{LV}\,\bar{\rho}_{LV}\,\overline{\mathcal{V}}_{LV}{}^2(z_2 - z_F) + g\bar{\rho}_{LV}(z_2 - z_F) \tag{4.22}$$

Jote that the velocity at the flash point is taken as equal to that at the well feed zone ince the liquid is incompressible up to that point. We now must give quantitative neaning to each of the mean effective terms.

- $\mathcal{V}_2$: The exit velocity is that velocity which will generate the same kinetic energy carried by the two-phase mixture. That is,

$$\frac{\mathcal{V}_2^2}{2} = x_2\frac{\mathcal{V}_{g2}^2}{2} + (1 - x_2)\frac{\mathcal{V}_{f2}^2}{2} \tag{4.23}$$

where x_2 is the mass fraction of vapor (called the quality) of the exiting two-phase mixture, defined as [18]:

$$x_2 = \frac{\rho_{g2}A_{g2}\mathcal{V}_{g2}}{\rho_{f2}A_{f2}\mathcal{V}_{f2} + \rho_{g2}A_{g2}\mathcal{V}_{g2}} \tag{4.24}$$

The area terms account for the portions of the exit area occupied by the two phases. The slip ratio, k_2, relates to the velocity of the two phases:

$$k_2 \equiv \frac{V_{g2}}{V_{f2}} \tag{4.25}$$

Thus the exit velocity can be expressed as

$$V_2 = V_{g2}\left[x_2 + \frac{1-x_2}{k_2^2}\right]^{1/2} \tag{4.26}$$

- $\overline{\rho}_{LV}$: The average density over the two-phase region is taken simply as the average of the values at the flash horizon and at the exit:

$$\overline{\rho}_{LV} = \frac{1}{2}\left[\rho_R + x_2\rho_{g2} + (1-x_2)\rho_{f2}\right] \tag{4.27}$$

- $\overline{V}_{LV}$: Recognizing that the mass flow rate is equal to the product of the density, the cross-sectional area, and the velocity, we apply this notion to get an expression for the mean effective two-phase velocity:

$$\overline{V}_{LV} = \frac{\dot{m}}{\overline{\rho}_{LV}A} \tag{4.28}$$

- $\overline{f}_{LV}$: The two-phase friction factor cannot be expressed in terms of any other property as we have done for the other mean effective quantities. The only thing we can do is to recognize that it will be larger than the liquid friction factor, f, but how much larger, we cannot say. Multipliers in the range of 2–3 are probably acceptable.

4.2.7 Complete model: Reservoir to wellhead with wellbore flashing

We may now assemble the pieces of the model to obtain the governing equation for the pressure behavior from the undisturbed reservoir to the wellhead including flashing within the well. The pressure drop from the far reservoir to the well entrance point is given by eq. (4.15); from the entry point to the flash horizon by eq. (4.10); and from the flash horizon to the wellhead by eq. (4.22). Thus we obtain:

$$P_2 = P_R - C_D\dot{m} - g(\rho_R L_F + \overline{\rho}_{LV}d_F) - \frac{2}{D}\overline{f}_{LV}\overline{\rho}_{LV}\overline{V}_{LV}^2 d_F$$
$$- \frac{32\, fL_F}{\pi^2 D^5 \rho_R}\dot{m}^2 - \frac{\overline{\rho}_{LV}}{2}\left[C_3^2\left(x_2 + \frac{1-x_2}{k_2^2}\right) - \frac{16}{\pi^2 D^4 \rho_R^2}\right]\dot{m}^2 \tag{4.29}$$

where d_F is the depth to the flash horizon from the wellhead, and the factor C_3 is defined as

$$C_3 = \frac{x_2\rho_{f2} + (1-x_2)\rho_{g2}k_2}{\rho_{f2}\rho_{g2}A} \tag{4.30}$$

he wellhead dryness fraction, x_2, can be computed from the following equation using n isenthalpic process from the reservoir to the surface:

$$x_2 = \frac{h_R - h_{f2}}{h_{g2} - h_{f2}} \tag{4.31}$$

ith h_R evaluated for a saturated liquid at the reservoir temperature.

By the time the fluid has made its way from the far reaches of the reservoir to the ellhead, the pressure has been reduced from P_R to P_2 by the following effects, starting ith the second term on the right-hand side of eq. (4.29):

- Drawdown in the reservoir
- Hydrostatic pressure drop in the well, liquid section, and two-phase section
- Frictional pressure drop in the two-phase section
- Frictional pressure drop in the liquid section
- Accelerational pressure drop in the two-phase section.

Our objective of finding the dependency of the mass flow rate on the wellhead ressure, $\dot{m} = \dot{m}(P_2)$, is implicit in eq. (4.29). All properties of the liquid may be found om *Steam Tables* at the reservoir temperature, using saturated liquid values. The roperties of the two phases at the wellhead, saturated liquid and saturated vapor, can lso be taken from the *Steam Tables* once the wellhead pressure is specified. The draw-own coefficient may be assumed from earlier measurements. The two-phase friction actor may either be treated as a variable parameter or set equal to some multiple, say , or 3, of the liquid friction factor. Ryley [11] showed that a wellhead slip ratio of bout 5 gave the correct behavior.

The algorithm for the calculation of the mass flow rate as a function of the wellhead ressure begins by positing a mass flow rate and then determining the flash horizon s described in Sect. 4.2.5 for a well of known diameter. All values that depend on the nass flow rate in eq. (4.29) are then found. Then one assumes a value for P_2 and valuates all the terms that depend on it. If eq. (4.29) is satisfied when all these values re inserted in the equation, then that wellhead pressure corresponds to the selected nass flow rate. If not, then an iterative procedure is followed until the equation is atisfied. By choosing a suite of values for mass flow rate and carrying out the proce-ure just described, the so-called productivity curve can be determined, i.e., the mass ow rate m. versus the wellhead pressure P_2.

Several numerical examples are presented in Ref. [11] which demonstrate that he method is capable of giving both qualitative and quantitative results that compare ery favorably with actual well behavior. Typical productivity curves are shown chematically in Fig. 4.3. The general shape of the curves is not affected by the values f the friction factors – liquid and two-phase – but as the friction factors increase, the ow rate decreases for a given wellhead pressure.

As a practical matter, at a certain wellhead pressure, further lowering of the ressure does not result in an increase in the flow rate, and we say the flow is choked. Note that the numerical solution of eq. (4.29) shows that the mass flow rate reaches maximum as the wellhead pressure is reduced, and tends to decrease as P_2 is further educed. Thus, one simply cuts off the calculation at the maximum point and keeps he flow rate constant for lower pressures.

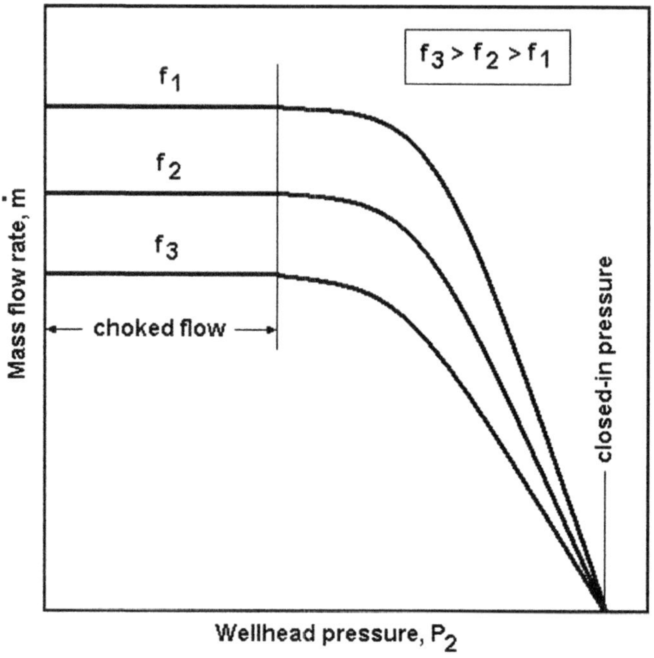

Fig. 4.3 Typical productivity curves from eq. (4.29).

We may sum up the phenomenon of reservoir-well flow with reference to Fig. 4.4 where the mass flow rate is plotted against the pressure. The productivity curve is the heavy line at the left. The line labeled P_1 is the pressure at the inlet to the well from the feed zone. The reservoir pressure is P_R and the pressure corresponding to the lowest stable mass flow rate is the maximum discharge pressure, P_{MD}. The shaded region at bottom left is unstable and is not modeled by the method developed in this section. Also shown is the saturation pressure P_{sat} corresponding to the reservoir temperature.

Three mass flows are highlighted. For $\dot{m}_1$, starting at the right and moving to the left we first see the drawdown within the reservoir, followed by the frictional and hydrostatic pressure drop in the liquid, and finally the two-phase pressure drop. The flashing begins within the well. For $\dot{m}_2$, the flashing begins just at the point where the fluid enters the well, state α. For $\dot{m}_3$, the flashing begins somewhere in the formation, state β, and the pressure drop from there to the wellhead is caused solely by two-phase flow effects.

4.3 Well testing

The testing methods for geothermal wells have been adapted from the oil and gas industry [3,5]. The objective of a well testing program is to determine as much as possible about the reservoir and the geofluid, and in particular the ability of the reservoir-well combination to produce an acceptable flow rate under thermodynamic conditions suitable to generate electricity. It would be very useful to be able also to forecast the

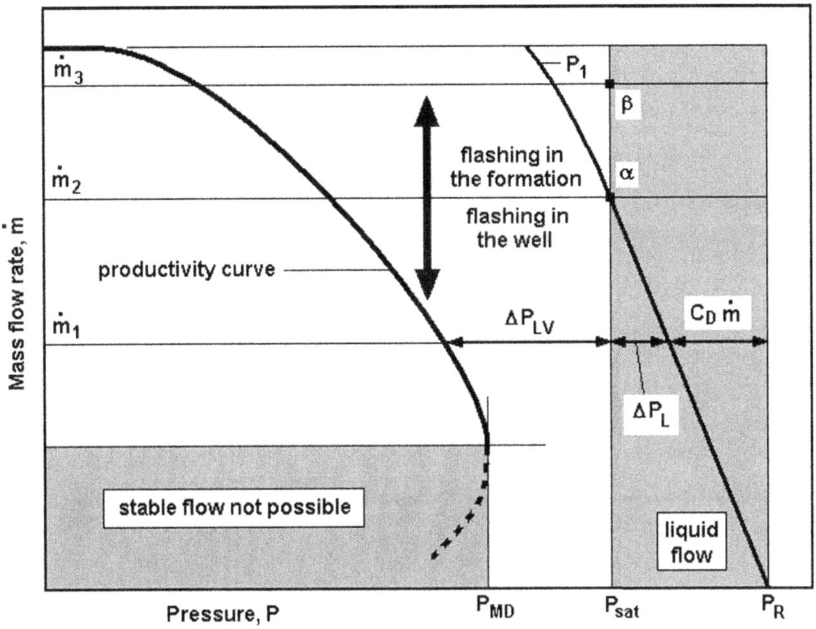

ig. 4.4 *Flow domains from reservoir to wellhead.*

fetime over which the well can maintain this level of production, but this is more ifficult, if not impossible, at the early well-testing stage.

.3.1 *Desired information*

Iere is what we want to know at the outset of the testing program:

- Pressure in the well as a function of depth, including at feed zones
- Temperature in the well as a function of depth, including at feed zones
- Pressure in the reservoir, under static and flowing conditions
- Reservoir permeability or permeability-thickness product
- Mass flow rates of both liquid and vapor phases
- Chemical composition of both liquid and vapor phases, downhole and at the surface
- Condition of the well.

The last item refers to possible damage to the permeable zones in the formation, or example, by drilling mud caking or cement invasion. The condition of the casings s also important.

.3.2 *Pressure and temperature instrumentation*

ressures measured on the surface can be made using any standard means such as Bourdon gauges, a Fortin barometer, differential manometers, etc. Pressure gauges ttached to pipes or vessels give readings of pressure relative to atmospheric pressure so-called "gauge" pressure). The absolute pressure at the site must be added to the gauge pressure to obtain absolute pressure; property tables require absolute pressure. The atmospheric pressure varies with altitude, a, as will the boiling temperature of water. The following two equations may be used to estimate these two effects up to an altitude, a, of 2600 m above sea level:

$$P_{atm}(a) = 101.3 - 0.0120\, a \quad (4.32)$$

where the pressure is measured in kPa and altitude is in m;

$$T_{sat}(P_{atm}) = 27.14 \ln P_{atm} - 25.397 \quad (4.33)$$

where the temperature is measured in °C.

Pressures in the geothermal well can be accurately measured with piezoelectric pressure transducers. One device that is commercially available can measure pressures up to 34,000 kPa with an accuracy of 8 kPa. The same device also incorporates a resistance thermometer with an accuracy of 0.15% (full scale). The whole package can operate in a hot geothermal well at up to 300°C (570°F) for as long as 9 hours [19]. The data is recorded with the aid of a USB port on a computer running Windows 98 or higher. As can be seen in Fig. 4.5, the assembly is about 5 ft long and has a

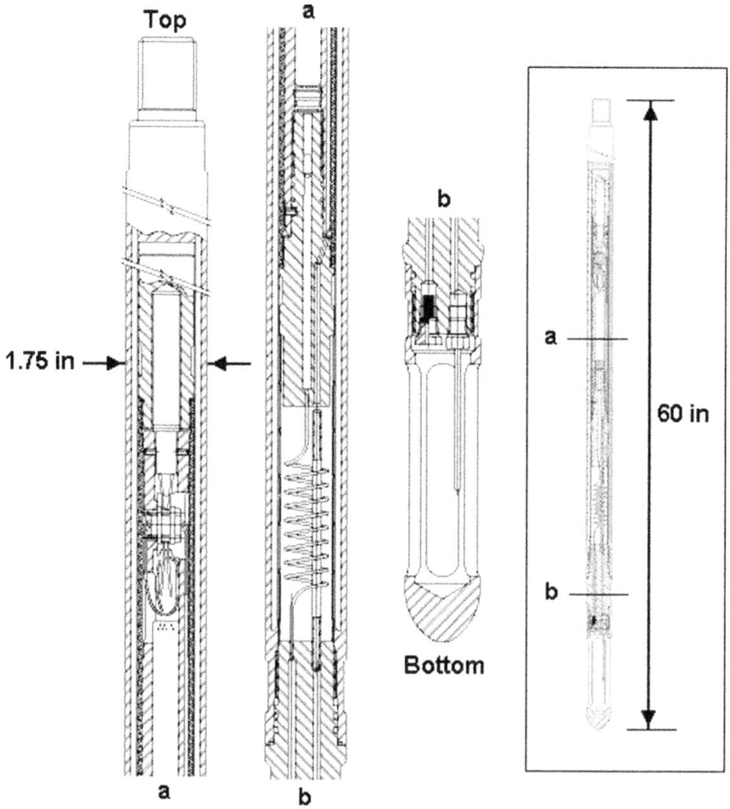

Fig. 4.5 Kuster K10 geothermal tool for pressure and temperature measurements [19].

iameter of less than 2 in and weighs about 21 lbs (9.5 kg). It is lowered on a wireline hrough a gland seal and set at any depth to measure the pressure and temperature imultaneously.

.3.3 *Direct mass flow rate measurements*

Iass flow rates of two-phase geofluids are best measured directly by first separating the wo phases, and measuring the flow rates of the liquid and vapor individually by means f venturi meters or calibrated orifices. Venturis are very accurate for determining olumetric flow rates; see Fig. 4.6. If the fluid density is known, then the mass flow rate an easily be calculated. The volumetric flow rate can be found from

$$\dot{V} = C_V A_2 \sqrt{\frac{2g(P_1 - P_2)/\gamma}{1 - (A_2/A_1)^2}} \tag{4.34}$$

vhere C_V is the venturi calibration constant, γ is the specific gravity of the fluid in he U-tube manometer, and the pressure difference is related to the reading of the nanometer:

$$\frac{P_1 - P_2}{\gamma} = \gamma H_D - H_D = H_D(\gamma - 1) \quad \text{[liquid flow]} \tag{4.35}$$

vhere the units of both sides of eq. (4.35) are equivalent head of water. For steam low, the density of steam above the manometer is negligible and eq. (4.35) becomes imply

$$\frac{P_1 - P_2}{\gamma} = \gamma H_D \quad \text{[steam flow]} \tag{4.36}$$

Care must be taken to ensure that the liquid from the separator is cooled some 0–30°C below saturation and is free of entrained steam. Even a small amount on he order of 0.5% can introduce unacceptably large errors in the liquid flow measurenents [20].

A sharp-edged orifice plate (see Fig. 4.7) may also be used for these measurements hey are much less expensive than venturi meters and are more suitable for field

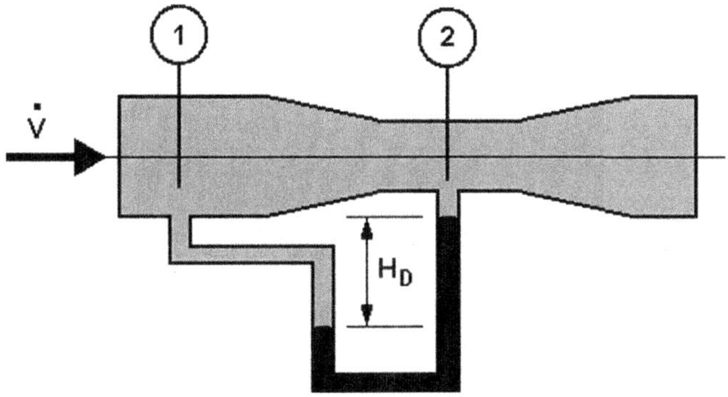

ig. 4.6 Venturi flow meter.

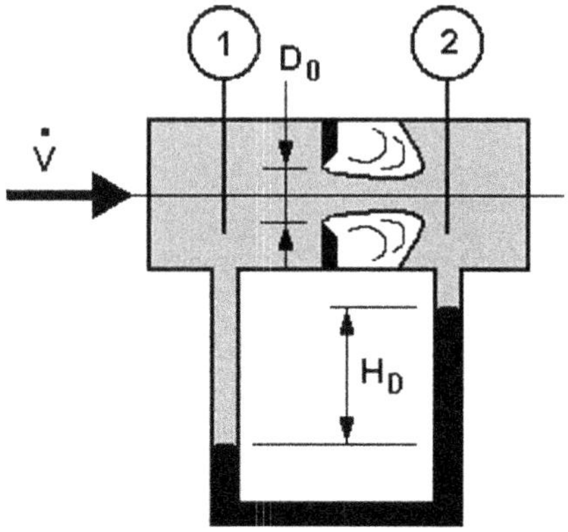

Fig. 4.7 Sharp-edged orifice flow meter.

measurements. Venturi meters are more appropriate when accurate measurements are needed for steam flow rates being delivered to a power station.

The theoretical advantage of a sharp-edged orifice meter (compared with a rounded-edged one) is that the reading is not sensitive to the fluid viscosity and therefore should not be sensitive to the fluid temperature. However it is difficult to make a true sharp-edged orifice, and abrasive fluids will soon wear down the knife-edge anyway, so that orifices need to be calibrated. The volumetric flow rate is found from

$$\dot{V} = C_0 A_0 \sqrt{2g(P_1 - P_2)/\gamma}\left[1 + \frac{1}{2} C_0^2 \left(\frac{D_0}{D_1}\right)^4\right] \tag{4.37}$$

where C_0, the orifice coefficient, is determined by calibration.

The mass flow rate is calculated from the volumetric flow rate using

$$\dot{m} = \rho\dot{V} = \dot{V}/v \tag{4.38}$$

where ρ is the fluid density and v is the specific volume.

4.3.4 Indirect mass flow rate measurements

One means to determine the mass flow rates of the liquid and vapor phases involves the somewhat cumbersome procedure of feeding the total well flow (or a known fraction thereof) into a large insulated vessel partially filled with a known amount of cold water. This technique requires fast-acting valves to allow an accurate timing of the flow into the vessel; see Fig. 4.8.

At time t_1 the liquid in the vessel resides at state 1; during a period of time Δt, the vessel is filled to state 2, receiving two-phase fluid at state *WH*. The First Law of thermodynamics applied to the vessel over the time interval Δt (open, unsteady system) yields the working equation for the enthalpy of the two-phase mixture:

$$h_{WH} = \frac{m_2 u_2 - m_1 u_1}{m_2 - m_1} \tag{4.39}$$

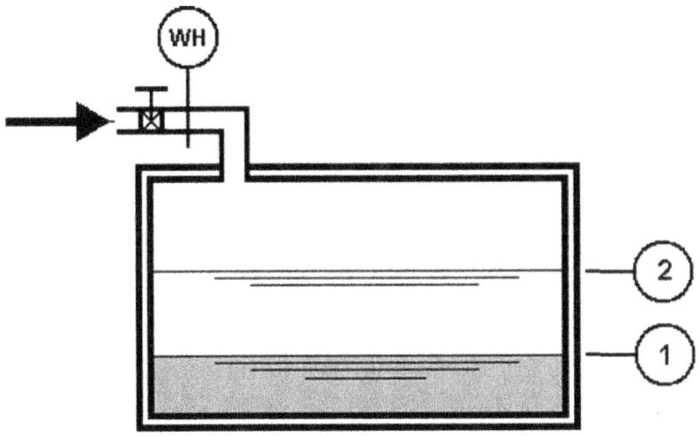

ig. 4.8 Calorimeter method for wellhead flow measurement.

here u stands for the internal energy of the geofluid. If the wellhead pressure, P_{WH}, known, this allows one to calculate the quality at the wellhead and the flow rates f the two phases.

Another more commonly used method is the James lip pressure method [20]. ames conducted experiments using discharge pipes of 3, 6, and 8 in diameter and rrived at an empirical correlation that can be used to determine the total, two-phase ell flow:

$$\dot{m} = 8,953.5 \frac{DP^{0.96}}{h^{1.102}} \quad (4.40)$$

here $\dot{m}$ is the total mass flow rate (in t/h or 10^3 kg/h), D is the internal diameter of e pipe (in mm), h is the two-phase enthalpy (in kcal/kg), and P is the pressure (in g/cm^2, abs) measured by a pressure gauge located on a 1/4-in tap, 1/4-in back from e lip of the pipe discharging freely into the atmosphere under choked conditions. he enthalpy cannot be measured directly and must be found using a silencer in hich the liquid and vapor are separated at atmospheric pressure. A 90°-notched weir easures the flow of liquid, and the enthalpy can be found by solving the equation

$$h^{1.102} + C_W h = C_W h_g \quad (4.41)$$

which h_g is the enthalpy of saturated vapor at atmospheric pressure (from *Steam ables*) and the weir factor C_W is given by

$$C_W = \frac{1.052P^{0.96}D^2}{\dot{m}_f h_{fg}} \quad (4.42)$$

ith h_{fg} being the enthalpy of evaporation at atmospheric pressure. The liquid mass ow rate can be found from the standard weir formula

$$\dot{m}_f \approx 2.6\, L_W^{2.5} \quad (4.43)$$

where L_w is the height of the water passing over the weir measured from the bottom of the notch ($L_w > 1$ ft).

4.3.5 Transient pressure measurements and analysis

A series of time-dependent pressure measurements made after changes in the mass flow rate from a well can be used to determine several properties of the reservoir [5]. A common test is the *pressure buildup test*. The procedure is to conduct a steady flow test at a certain volumetric flow rate $\dot{V}_1$ for a period of time, t_1, during which the downhole pressure opposite the feed zone is monitored. The well is then closed in and the pressure is monitored for an additional time duration, Δt. This procedure is shown schematically in Fig. 4.9.

The governing equation for the shut-in, downhole pressure behavior is [5]

$$P_{dh,s} = P_i - \frac{\dot{V}\mu}{4\pi K L_R} \ln\left[\frac{t_1 + \Delta t}{\Delta t}\right] \tag{4.44}$$

where P_i is the static reservoir pressure. Since the pressure difference depends on the log of the time, a plot of pressure versus log of the dimensionless time should yield a straight line. This behavior is shown in Fig. 4.10, a semi-log graph known as a Horner plot, for some hypothetical data.

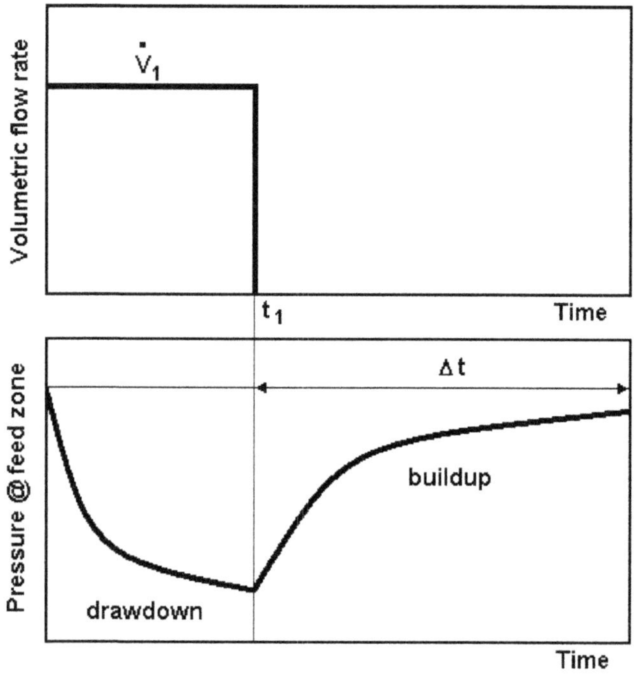

Fig. 4.9 Pressure buildup test.

The data may be extrapolated to infinite time, i.e., $\Delta t \rightarrow \infty$ or $(t_1 + \Delta t)/\Delta t \rightarrow 1$, to btain the initial static pressure, P_i. From eq. (4. 44) it can be seen that the magnitude f the slope of the line, m, is given by

$$m = 2.3026 \frac{\dot{V}\mu}{4\pi K L_R} \tag{4.45}$$

r

$$K L_R = 2.3026 \frac{\dot{V}\mu}{4\pi m} \tag{4.46}$$

Thus the product of the formation permeability and the reservoir thickness can be .educed from a pressure buildup test.

It is also possible to infer whether or not there is a "skin effect," i.e., an added resis- ance to flow caused by well damage. In such a case, a "skin factor" s is defined as hown below to account for an additional pressure drop between the far reservoir and he interior of the well:

$$\Delta P_{skin} \equiv s \frac{\dot{V}\mu}{2\pi K L_R} \tag{4.47}$$

Many other kinds of transient flow tests may be carried out. These include draw- own tests with multiple steps and interference tests among several wells. In the latter, chemical tracer is injected into one well while other wells are producing. The produ- ers are monitored for the return of the tracer, thereby revealing the connectivity mong the various wells. The reader is referred to Refs. [2, 9] for more details.

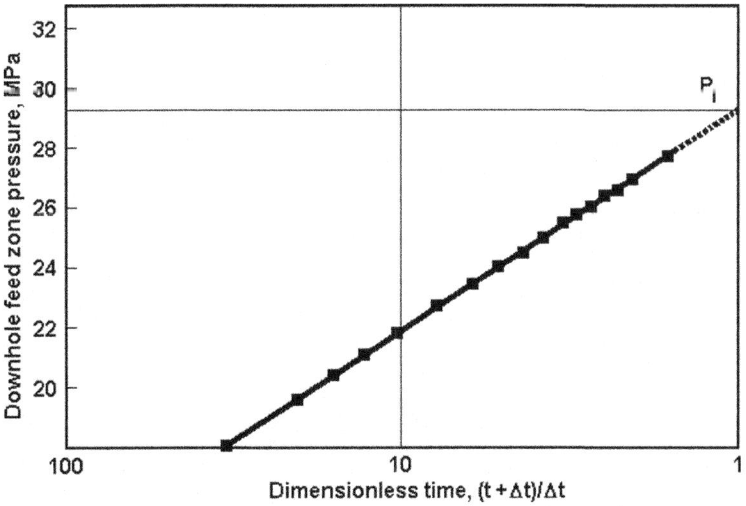

ig. 4.10 Horner plot for pressure buildup test.

4.4 Calcite scaling in well casings

One of the common problems seen in geothermal wells is the deposition of calcium carbonate or calcite, $CaCO_3$, in the well casing, starting just above the flash horizon. It is not uncommon for high-temperature geofluids to be close to saturation with respect to calcite as they flow through the formation. The solubility of calcite, however, varies inversely with temperature, so that it cannot precipitate from the geofluid merely because of a decrease in temperature, other factors being constant. The other properties of the geofluid that influence the solubility are:

- Partial pressure of carbon dioxide CO_2
- pH
- Salinity
- Calcium ion concentration.

The first two factors are interrelated; when the geofluid flashes in the well, the steam that is released carries with it most of the CO_2. This in turn causes the liquid pH to rise dramatically, and results in the geofluid becoming supersaturated with respect to calcite. Precipitation occurs immediately and can lead to severe narrowing of the wellbore for several meters just above the flash horizon.

The chemical equilibrium reactions that control the process are [21]:

$$CO_2(aq) \Leftrightarrow CO_2(vap) \tag{4.48}$$

$$HCO_3^- + H^+ \Leftrightarrow CO_2 + H_2O \tag{4.49}$$

$$Ca^{++} + 2HCO_3^- \Leftrightarrow CaCO_3\downarrow + H_2O + CO_2\uparrow \tag{4.50}$$

The loss of carbon dioxide ($CO_2\uparrow$) upon flashing can be seen to cause the precipitation of calcite ($CaCO_3\downarrow$) from eq. (4.50). The fraction of CO_2, relative to the original amount dissolved in the brine, that leaves the brine with the flashed steam can be found from [22]:

$$X_{CO_2} = \frac{1/A}{[(1/x)-1]+(1/A)} \tag{4.51}$$

where A is a separation coefficient [23] and x is the dryness fraction of the two-phase, steam-liquid mixture. The separation coefficient is a function of temperature and the salinity of the brine, as shown in Figs. 4.11 and 4.12. The dryness fraction can be calculated from eq. (4.31). For example, if a 1 M brine at 200°C undergoes a 12% flash, then 98.5% of the CO_2 will end up in the vapor phase.

4.5 Reservoir modeling and simulation

A reservoir simulator is a computer code that embodies all of the essential properties of the formation, the geofluid, and the environment together with the physical and chemical laws controlling their interactions to allow the extrapolation of the

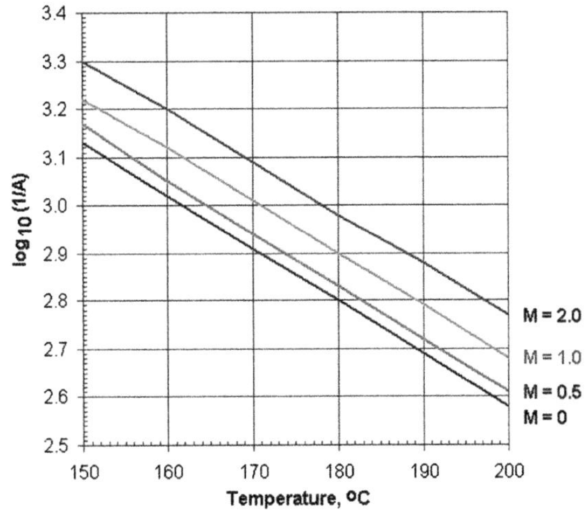

ig. 4.11 *Log_{10} (1/A) as a function of temperature for various molalities.*

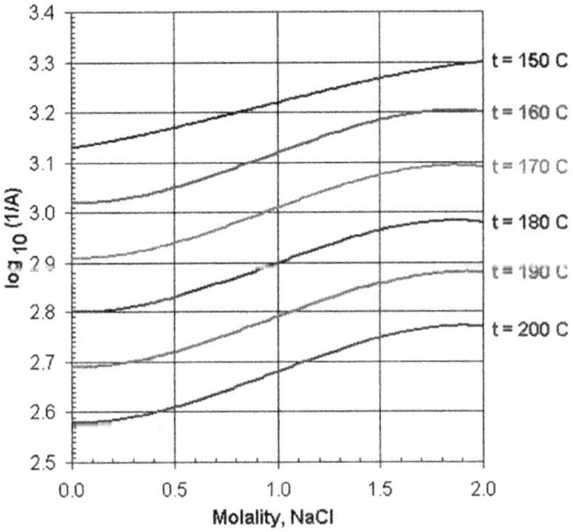

ig. 4.12 *Log_{10} (1/A) as a function of molality for various temperatures.*

erformance of the reservoir from the present time to a future time, usually several ecades away.

The information derived from such a code is valuable to those interested in developng and financing a power plant. Often a reservoir model is the only predictive tool vailable at the early stages of development to assure potential investors that their noney will be wisely spent. It is also a requirement for most feasibility studies when onsidering a new project or the expansion of an existing one.

Nowadays, with the widespread use of personal computers with powerful capabilities once reserved for mainframe or mini-computers, it is far easier to conduct reservoir modeling studies. However, the relative ease with which such studies can be performed does not mean that the results can be viewed as any more reliable. Reservoir simulations require a solid understanding of the geothermal system, based on the geoscientific exploration results including the results of the first few deep wells. Unless this information has been crafted into a model truly representative of the geothermal system, one that may extend for tens of square kilometers on the surface and several thousand meters below the surface, and unless suitable methods of solution for complex equations have been devised, then the results of the simulation will have little predictive value.

Reservoir modeling and simulation is the subject of annual symposia, and numerous technical articles. The reader is referred to the following sources for detailed presentations: Refs. [1, 2, 24–38]. Several universities, companies, and governmental laboratories have developed their own versions of reservoir simulators, usually identified by unique acronyms. The private ones are usually proprietary and all computer programs periodically evolve to newer versions. In Sect. 4.5.6, we list some of the more commonly used geothermal reservoir simulators, briefly highlight their features and include references for details on each one.

4.5.1 Input

A reservoir model must be built on a proper conceptual model of the field. If the conceptual model has flaws, as often happens in the early stages of field development the reservoir model will reflect these errors. For this reason, it is recognized that the reservoir model must be updated as the conceptual model improves.

The input includes as much of the following information as possible:

- Geology: rock types and ages, outcrops, surface topology, faults, calderas, craters, eruptive centers, discontinuities
- Hydrology: rainfall patterns, surface water flows
- Geophysics: surveys of electrical resistivity, gravity, magnetics, temperature gradients, heat flux
- Geochemistry: hot and cold spring compositions, flow rates and temperatures, estimated reservoir temperatures
- Well data: locations, depths, casing design, pressure and temperature profiles both static and flowing, productivity curves.

All of this data permits the construction of the conceptual model which should indicate the areal extent of the reservoir, the areas of natural fluid discharge and recharge, the natural heat flow from the reservoir by conduction through rock and by convection through thermal manifestations, and an estimate of the vertical extent of the geothermal system.

4.5.2 Architecture

Since it is virtually impossible to write down analytical equations to describe all the physical and chemical processes going on in a geothermal reservoir and to solve them nalytically in a closed form, all reservoir models rely on some type of finite difference r finite element numerical technique. The field is divided into discrete cells, both horiontally and vertically, and the governing equations of mass and energy conservation re written in finite difference form for each element of the matrix. The conditions n the boundaries of each cell provide the connections among the cells and permit the olution to extend over the entire reservoir.

The choices of the cell boundaries are arbitrary and may be selected for computational convenience, but usually they coincide with major structural features uch as faults or permeability barriers. The size of the cells is also arbitrary and ie grid usually is made finer in the neighborhood of well-defined features, such s wells, where data is plentiful and reliable. Outlying areas can be represented by rge blocks having average properties without greatly influencing the results of ie model. The geometric shape of the cells is also arbitrary, ranging from simple quares or rectangles to circles or irregular polygons. It is often convenient to ituate each well within its own element.

The vertical extent of the reservoir is modeled by layering the matrix into several istinct horizons. For young fields with little production history, it is not uncommon o have six or more layers covering, for example, the surface and the atmosphere, ie impermeable cap rock, a liquid-dominated zone, a particular rock formation, a deep ermeable zone, and the basement rock. In more developed fields with a long production history, models are much finer in structure and can have dozens of layers [39].

.5.3 Calibration and validation

Once the governing spatial- and time-dependent physical and chemical equations have een written, and the code created along with the method of solution, it is then necesary to calibrate and validate the model. This crucial step is needed to guarantee that he model (and its solution technique) is reliable, stable, and correctly represents the tate of the reservoir going back in time perhaps as much as 250,000 years – from he unexploited, natural state of the reservoir to the present. In this way, the model an be relied upon to accurately model the ancient natural state of the system, as well s the present, unexploited state of the system, and to offer the modeler confidence in naking predictions about the future state of the system.

This is a very tedious computational process. Imagine a mathematical model involving strongly-coupled, nonlinear, partial differential equations represented by a computer code with hundreds of input parameters, many of them ill-defined owing to nsufficient data. The set of equations is said to be mathematically under-constrained, having many more parameters than can be determined or inferred from field measurenents. A solution in such a case will not be unique, there being many possible conditions that will satisfy this set of equations. Judgment is needed by the user of the code o make good choices for the adjustable parameters to arrive at reasonable results.

In what is called the "manual" method, the parameters first are adjusted by the eservoir engineer to give the best fit possible to the current state of the system. The two parameters most often used are the permeabilities of the various rock fornations and the boundary conditions [40]. Then the code is run from the natural tate forward in time and the physical characteristics of reservoir are computed at large time steps. The evolution of the field is thus depicted. Stability of the code requires that eventually the system should reach a steady state that conforms to the observed present state. If the results of this computation are unsatisfactory for any reason, then the modeler must adjust the parameters and rerun the simulation until satisfactory results are obtained.

This process can be partly automated in a sense by having the reservoir properties adjusted by a programmed subroutine within the code to minimize the difference between the code output and selected observed data. The automatic method is called "inverse" modeling, but still requires the intervention of the modeler in choosing the properties that will serve as criteria of the goodness of the fit and the magnitude of the errors that are acceptable. Inverse modeling is more demanding on computer speed and memory than manual modeling, and may require parallel processing using several personal computers.

4.5.4 History matching

The first use of the model is to match the data obtained from the early wells. The model should be able to reproduce the observed changes in the reservoir over the period of production and injection. When this is done during the feasibility stage of a new field, such data may be scanty, but the model still must be able to match it. This usually means another adjustment of the parameters in the model, and a recalculation of the natural state model, followed by more iterations to bring the model into agreement with all the data. As more production and injection data become available, the model should be revised accordingly.

Here again, this process of history matching can be done either manually or automatically. This cycle of periodically updating the reservoir simulator is tedious, time-consuming, and expensive, but absolutely essential if one is to have any confidence in projecting the behavior of the field over many years of operation.

4.5.5 Use of the model

A validated reservoir model may be used to forecast the behavior of the system when it is subjected to various exploitation scenarios. This involves running the model for future times after having placed projected future production and injection wells at specific locations in the field, i.e., mathematically inserting fluid sinks (production wells) and sources (injection wells) into the reservoir. The desired outputs from the model include the evolution of pressure and temperature, the formation of vapor zones, the movement of chemical and thermal fronts through the reservoir, the onset of chemical and thermal breakthroughs from injection wells to production wells, the level of electrical generation that can be supported for a specified number of years, typically 20–30 years, and the number of replacement wells that must be drilled over the years to maintain the electrical output.

Given the complexity of the rock-fluid interaction in an imprecisely known geometric structure, it is easy to see how such a modeling process is subject to considerable uncertainty. Furthermore, since geothermal regions are dynamic and subject to geologic forces that can change the structure of the formation, opening new fissures or closing old ones, moving fault blocks laterally as well as vertically, the assumed onstant geometric structure may be altered by natural forces. This could play a role hen the simulator is used over the millennia of the natural state computation, during hich the reservoir is assumed to have a fixed configuration. Indeed, the field may be ltered by geologic forces during the anticipated life of the power plant, as has been e case at some geothermal power plants around the world.

Notwithstanding these difficulties, reservoir models, after they have been validated nd calibrated against whatever data is available and periodically updated, have roved to be reasonably reliable tools for simulating the state of geothermal systems. n excellent review of over 125 field simulations based on questionnaires filled out by ompanies and individuals can be found in Ref. [40].

.5.6 Examples of reservoir simulators

able 4.1 lists some of the reservoir models in use today. The models fall into one f three categories: finite difference (FDM), finite element (FEM), or integral finite ifference (IFDM). Some of the features are cited but the reader should consult he references for the particulars on each model.

able 4.1 Selected geothermal reservoir simulators.

ode	Type	Features	Reference
ETRAD	FDM	Simulates 3-D multiphase ground-water flow and heat transport; uses regular rectangular mesh.	Vinsome & Shook[28]
YDRO-THERM	FDM	Simulates 3-D multiphase ground-water flow and heat transport, handles temperatures from 0–1,200°C; uses pressure and enthalpy as independent variables.	Hayba & Ingebritsen [29]
IGHTS	FDM	Handles natural state calibration for liquid-dominated reservoirs.	Pritchett [30]
TAR	FDM	5th generation code for multiphase, multicomponent heat and mass transfer in 3-D geometric structures.	Pritchett [31]
QUA 3D	FEM	3-D groundwater flow and contamination transport.	Vatnaskil [32]
EHMN	FEM	Handles nonisothermal multiphase, multicomponent flow in porous media; permeability and porosity of the medium depend on pressure and temperature.	Zyvoloski *et al.* [33]
HAFT	IFDM	Uses integral FDM; original LBL reservoir.	Pruess [34]
IULKOM	IFDM	Multidimensional model, simulates coupled transport of water, vapor, noncondensable gas, and heat in porous and fractured media.	Pruess [34]
OUGH	IFDM	Handles transport of unsaturated groundwater and heat.	Pruess [34]
OUGH2	IFDM	Able to handle different fluid mixtures; water + water with tracer; water + CO_2; water + air; water + air, with vapor pressure lowering; and water + H_2	Pruess [35]
TOUGH2	IFDM	Inverse modeling; automatic calibration, history matching, and error analysis.	Finsterle [36,37]

References

[1] {General bibliographic reference} *Proc. Stanford Workshops on Geothermal Reservoir Engineering*, Vols 1–37, 1975–2012, Stanford University, Stanford, CA.

[2] Samaniego, V.F. and H. Cinco-Ley, "Reservoir Engineering Concepts," Chap. 9 in *Handbook o Geothermal Energy*, Edwards, L.M., G.V. Chilingar, H.H. Rieke III and W.H. Fertl, Eds., Gulf Publishin Company, Houston, TX, 1982.

[3] Golan, M. and C.H. Whitson, *Well Performance, 2nd Ed.*, Prentice-Hall, Englewood Cliffs, NJ, 1991.

[4] Ryley, D.J., "Analysis of the Flow in the Reservoir-Well System," Sect. 2.6 in *Sourcebook on the Productio of Electricity from Geothermal Energy*, Kestin, J., Ed. in Chief, R. DiPippo, H.E. Khalifa and D.J. Ryley, Eds. U.S. Dept. of Energy, DOE/RA/4051–1, U.S. Gov. Printing Office, Washington, DC, 1980.

[5] Matthews, C.S. and D.G. Russell, *Pressure Buildup and Flow Tests in Wells*, Society of Petroleum Engineers, American Institute of Mining, Metallurgical, and Petroleum Engineers, New York, 1967.

[6] Cheremisinoff, N.P., Ed., *Encyclopedia of Fluid Mechanics, V. 3, Gas-Liquid Flows*, Gulf Publishing Company, Houston, TX, 1986.

[7] Dake, L.P., *Fundamentals of Reservoir Engineering*, Elsevier Scientific, 2001.

[8] Ahmed, T., *Reservoir Engineering Handbook, 2nd Ed.*, Butterworth-Heinemann, 2002.

[9] Smith, C., G.W. Tracy and R.L. Farrar, *Applied Reservoir Engineering, Vols. 1 and 2*, Oil & Ga Consulting International, Tulsa, OK, 1999.

[10] Darcy, H., "The Public Fountains of the City of Dijon: Experience and Application, Principles to Follow and Formulas to be Used in the Question of the Distribution of Water," (Trans. by G. Brown and B. Cateni, 1999), Original published in 1856.

[11] Ryley, D.J., "The Mass Discharge of a Geofluid from a Geothermal Reservoir-Well System with Flashing Flow in the Bore," *Geothermics*, V. 9, 1980, pp. 221–235.

[12] Swamee, P.K. and A.K. Jain, "Explicit Equations for Pipe Flow Problems," *Journal of Hydraulic Engineering*, V. 102, ASCE, 1976, p. 657.

[13] Keenan, J.H., F.G. Keyes, P.G. Hill and J.G. Moore, *Steam Tables: Thermodynamic Properties of Water Including Vapor, Liquid, and Solid Phases (International Edition – Metric Units)*, John Wiley & Sons, Inc. New York, 1969.

[14] Delhaye, J.M., M. Giot and M.L. Riethmuller, Eds., *Thermohydraulics of Two-Phase Systems for Industrial Design and Nuclear Engineering*, Hemisphere Publishing Corp, McGraw-Hill, New York, 1981.

[15] Ginoux, J.J., Ed., *Two-Phase Flows and Heat Transfer with Application to Nuclear Reactor Design Problems* Hemisphere Publishing Corp, McGraw-Hill, New York, 1978.

[16] Hsu, Y.-Y. and R.W. Graham, *Transport Processes in Boiling and Two-Phase Systems Including Near-Critical Fluids*, Hemisphere Publishing Corp, McGraw-Hill, New York, 1976.

[17] Wallis, G.B., *One-dimensional Two-phase Flow*, McGraw-Hill, New York, 1969.

[18] Ryley, D.J., "Property Definition in Equilibrium Wet Steam," *Int. J. Mech. Sci.*, V. 6, 1964, pp. 445–454.

[19] http://www.kusterco.com/manuals.htm, Kuster Company, Long Beach, CA, 2003.

[20] James, R., "Factors Controlling Borehole Performance," *Geothermics – Special Issue 2, U.N. Symposium on the Development and Utilization of Geothermal Resources*, Pisa, V. 2, Pt. 2, 1970, pp. 1502–1515.

[21] Ellis, A.J. and W.A.J. Mahon, *Chemistry and Geothermal Systems*, Academic Press, New York, 1977.

[22] Eskesen, J.H., A. Whitehead and A.W. Brunot, "Cycle Thermodynamics," Sect. 4.1.2 in *Sourcebook on the Production of Electricity from Geothermal Energy*, Kestin, J., Ed. in Chief, R. DiPippo, H.E. Khalifa and D.J. Ryley, Eds., U.S. Dept. of Energy, DOE/RA/4051-1, U.S. Gov. Printing Office, Washington, DC, 1980.

[23] Ellis, A.J. and R.M. Golding, "Solubility of Carbon Dioxide above 100°C in Water and Sodium Cloride Solutions," *Amer. J. Sci.*, V. 261, 1963, p. 47.

[24] Lippmann, M.J., "Numerical Modeling of Injection," Chap. 3 in *A Course on Injection Technology*, J. Rivera, Convenor, Int'l. School of Geothermics, Pisa, Italy, 1995, pp. 113–145.

[25] Bullivant, D.P., M.J. O'Sullivan and M.R. Blakeley, "A Graphical Interface for a Geothermal Reservoir Simulator," *Proc. World Geothermal Congress*, V. 4, Florence, Italy, 1995, pp. 2971–2976.

[26] White, S.P., W.M. Kissling and M.J. McGuinness, "Models of the Kawareu Geothermal Reservoir," *Geothermal Resources Council TRANSACTIONS*, V. 21, 1997, pp. 33–39.

[27] Williamson, K.H., "Development of a Reservoir Model for The Geysers Geothermal Field," *Monograph on The Geysers Geothermal Field*, Special Report No. 17, Geothermal Resources Council, Davis, CA, 1991, pp. 179–187.

[28] Vinsome, P.K.W. and G.M. Shook, "Multipurpose Simulation," *J. Pet. Sci. and Eng.*, V. 9, 1993, pp. 29–38.

[29] Hayba, D.O. and S.E. Ingebritsen, "The Computer Model HYDROTHERM, a Three-Dimensional Finite-Difference Model to Simulate Ground-Water Flow and Heat Transfer in the Temperature Range of 0 to 1,200°C," Rep. 94–4045, U.S. Geological Survey Water Resources Investigations, Menlo Park, CA, 1994.

[30] Pritchett, J.W., "NIGHTS: A Single-Phase Geothermal Reservoir Simulator," *Proc. World Geothermal Congress*, V. 4. Florence, Italy, 1995, pp. 2955–2958.

[31] Pritchett, J.W., "STAR: A Geothermal Reservoir Simulation System," *Proc. World Geothermal Congress*, V. 4, Florence, Italy, 1995, pp. 2959–2963.

[32] Vatnaskil Consulting Engineers, *AQUA Lecture Notes and User's Manual*, Reykjavik, Iceland, 1993.

[33] Zyvoloski, G.A., Z. Dash and K. Kelkar. "FEHMN 1.0: Finite Element Heat and Mass Transfer Code," Rep. LANL-12062, Los Alamos National Laboratory, Los Alamos, NM, 1991.

[34] Pruess, K., "SHAFT, MULKOM, TOUGH: A Set of Numerical Simulators for Multiphase Fluid and Heat Flow," *Geotermia, Revista Mexicana de Geoenergía*, V. 4, No. 1, 1988, pp. 185–202.

[35] Pruess, K., "TOUGH2 – A General-Purpose Numerical Simulator for Multiphase Fluid and Heat Flow," Rep. LBL-29400. Lawrence Berkeley Laboratory, Berkeley, CA, 1991.

[36] Finsterle, S., "Multiphase Inverse Modeling: Review and iTOUGH2 Applications," *Vadose Zone Journal*, V. 3, 2004, pp. 747–762.

[37] http://esd.lbl.gov/iTOUGH2, Finsterle, S., "Multiphase Inverse Modeling Lecture Notes," Lawrence Berkeley National Laboratory, Berkeley, CA, 2000.

[38] O'Sullivan, M.J., K. Pruess and M.J. Lippmann, "Geothermal Reservoir Simulation: The State-of-Practice and Emerging Trends," *Proc. World Geothermal Congress*, Kyushu-Tohoku, Japan, 2000, pp. 4065–4070.

[39] Mannington, W., M. O'Sullivan and D. Bullivant, "Computer Modelling of the Wairakei-Tauhara Geothermal System, New Zealand," *Geothermics*, V. 33, 2004, pp. 401–419.

[40] O'Sullivan, M.J., K. Pruess and M.J. Lippmann, "State of the Art of Geothermal Reservoir Simulation," *Geothermics*, V. 30, 2001, pp. 395–429.

Problems

4.1 Consider a deep well filled with pure water from the surface to its total depth. Calculate the so-called "boiling point curve," i.e., calculate the temperature required to just begin to boil the water at any given depth. Your answer will be a table and a graph showing the boiling temperature versus depth. Do the calculation in two ways:

(a) assume that the water has a uniform constant standard density throughout the entire column, and

(b) account for the variability of density with depth (or equivalently with temperature) by dividing the column into a series of finite steps over which an average density may be used for each step.

When you have completed the calculations, discuss any difference between these two approaches and compare the results with the simple formula given by James, namely, $t/°C = 69.56\ (z/m)^{0.2085}$. At what depth is the critical point reached for both of your calculations and for James' formulation?

4.2 A hot water reservoir is discovered at a depth of 3000 ft. Downhole instrumentation indicates a pressure of 1500 lbf/in^2 and a temperature of 500°F. A control valve at the surface maintains a constant mass flow rate up the well and keeps the fluid pressure at the wellhead equal to 120 lbf/in^2. Assume adiabatic conditions along the well and neglect the effects of friction. Calculate the state of the geofluid at the wellhead under these conditions. Sketch the process from the well bottom to the wellhead on a temperature-entropy diagram. If flashing occurs in the wellbore, estimate the depth below the surface at which it takes place.

4.3 A production well exists in a hypothetical hydrothermal reservoir with a permeability $K = 100$ mD and a thickness (in the production zone) $L_R = 100$ m. The reservoir fluid has a viscosity $\mu = 1$ cp, and the well has a diameter of 9–5/8 in.
   (a) Assuming a volumetric well flow rate $\dot{V} = 100$ liters/s, calculate the drawdown pressure difference ΔP in atm as a function of the size of the reservoir (i.e., as a function of the reservoir radius r_r). Express r_r in m, and plot your results out to $r_r = 1000$ m.
   (b) Assuming a reservoir radius $r_r = 200$ m, find the drawdown as a function of flow rate $\dot{V}$, and plot your results up to $\dot{V} = 200$ liters/s.

4.4 A vertical geothermal well is 1400 m deep and has a 9–5/8" O.D. (8.921" I.D.) commercial steel casing from top to bottom. During flow tests, it is observed that the bottom-hole pressure is 13.75 MPa when the flow rate is 50 kg/s, and that it is 12.5 MPa when the flow rate is 100 kg/s. The reservoir temperature is 240°C.
   (a) Calculate and plot the depth from the surface to the flash horizon as a function of well flow rate for the range 0–150 kg/s, assuming the geofluid is pure water.
   (b) Write a computer program or spreadsheet to analyze the wellflow problem described in Part (a). The program should receive as input: reservoir temperature, well depth and casing diameter, and two pairs of data on flow rate versus bottom-hole pressure. The output should include the depth to the flash horizon plus anything else you think is important.

4.5 Consider a vertical, constant diameter, geothermal well. The well has a total depth L_T, a diameter $D = 7$–5/8", and a wall friction factor f, and carries a flow rate $\dot{m}$. The reservoir is characterized by a pressure P_R, a temperature T_R, and a drawdown coefficient C_D. The flash horizon is at an elevation L_F above the reservoir (or above the well bottom).
   (a) Determine the flash horizon as a function of wall friction factor (over a range of 0–0.020) for the following conditions:
   $T_R = 500$°F, $P_R = 800$ lbf/in^2, $\dot{m} = 100{,}000$ lbm/h, $C_D = 5$ (lbf/in^2)/(lbm/s).
   (b) Determine the flash horizon as a function of mass flow rate (over a range of 50,000–500,000 lbm/h) for the following conditions:
   $T_R = 450$°F, $P_R = 840$ lbf/in^2, $f = 0.008$, $C_D = 5$ (lbf/in^2)/(lbm/s).
   (c) Determine the flash horizon as a function of drawdown coefficient (over a range of 3–8 (lbf/in^2)/(lbm/s)) for the following conditions:
   $T_R = 400$°F, $P_R = 2400$ lbf/in^2, $f = 0.008$, $\dot{m} = 100{,}000$ lbm/h.
   (d) Determine the flash horizon as a function of reservoir temperature (over a range of 300–600°F) for the following conditions:
   $P_R = P_{sat}(T_R) + 500$, lbf/in^2, $\dot{m} = 100{,}000$ lbm/h, $f = 0.008$, $C_D = 5$(lbf/in^2)/(lbm/s).
   (e) Suppose, for each part, the total well depth is $L_T = 6000$ ft. Discuss the ramifications of the flash horizon calculations. In particular, focus on the

feasibility of placing a downhole pump in the well to increase the flow rate, i.e., the productivity of the well, and/or to prevent flashing anywhere in the well. When you consider a pumped well, you may assume that a pump will produce a flow rate 2.5 times greater than the self-flowing rate. Assuming that the pump requires a net positive suction head (NPSH) of 50 ft of water to avoid cavitation at the pump inlet, identify an example from the above calculations where it is possible to locate a pump in a well at a depth (from the surface) of less than 1000 ft.

Part 2

Geothermal Power Generating Systems

- Single-Flash Steam Power Plants
- Double-Flash Steam Power Plants
- Dry-Steam Power Plants
- Binary Cycle Power Plants
- Advanced Geothermal Energy Conversion Systems
- Exergy Analysis Applied to Geothermal Power Systems

"The First Law of thermodynamics says you can't get something for nothing; the Second Law says you can't even break even."

Anonymous

The second part of the book covers the energy conversion systems that take the geothermal fluids from the production wells, process them for use, produce electricity in a power plant, and finally dispose of the fluids in an effective and environmentally benign manner. We present the thermodynamic principles governing the design and operation of the power plants, and illustrate them with several practical examples. The traditional types of plant include flash-steam and direct dry-steam plants, as well as binary plants. Advanced systems are covered, some of which are already at the commercial stage while others require more development. We conclude this section with a general presentation of the use of the Second Law of thermodynamics for power plant analysis. This so-called exergy analysis is an effective tool for the design of efficient power plants.

Hatchobaru power station, Units 1 & 2, Kyushu, Japan.
Photo: Kyushu Electric Power Co., Inc., Fukuoka, Japan [WWW].

Chapter 5

Single-Flash Steam Power Plants

Geothermal Power Plants: Principles, Applications, Case Studies and Environmental Impact, Third Edition.

© 2012 Elsevier Ltd. All rights reserved.

"In theory there is no difference between theory and practice. In practice there is."
Yogi Berra

5.1 Introduction

The single-flash steam plant is the mainstay of the geothermal power industry. It is often the first power plant installed at a newly-developed liquid-dominated geothermal field. As of August 2011, there were 169 units of this kind in operation in 16 countries around the world. Single-flash plants account for about 29% of all geothermal plants. They constitute nearly 43% of the total installed geothermal power capacity in the world. The unit power capacity ranges from 3 to 117 MW, and the average power rating is 27 MW per unit. Full details can be found in Appendix A.

5.2 Gathering system design considerations

When the geothermal wells produce a mixture of steam and liquid, the single-flash plant is a relatively simple way to convert the geothermal energy into electricity. First the mixture is separated into distinct steam and liquid phases with a minimum loss of pressure. This is done in a cylindrical cyclonic pressure vessel, usually oriented with its axis vertical, where the two phases disengage owing to their inherently large density difference. The siting of the separators is part of the general design of the plant and there are several possible arrangements.

A typical 30 MW single-flash power plant needs 5–6 production wells and 2–3 injection wells. These may be drilled at sites distributed across the field or several may be drilled from a single pad using directional drilling to intercept a wide zone of the reservoir. In either case, a piping system is needed to gather the geofluids from the production wells and transport them to the powerhouse and to the points of disposal. Often the initial piping system is modified if new power units are added later on.

5.2.1 Piping layouts

The separators can be located (1) at the powerhouse, (2) at satellite stations in the field, or (3) at the wellheads.

Figure 5.1 shows five production wells feeding two-phase fluid to a large cyclone separator at the powerhouse. The separated steam enters the turbine via short pipelines and the separated liquid is sent to two injection wells.

Figure 5.2 depicts an arrangement in which the production wells are connected to two satellite separator stations located in the field. Steam from the separators flows to the powerhouse in two pipelines and meet at the steam collector. The separated liquid streams from the satellites flow to the two injection wells. An arrangement of this type is used at the Miravalles power station in Costa Rica; Figure 5.3 is a photograph of one of several separator stations at the Miravalles plant [1].

Figure 5.4 shows wellhead separators at each production well. This design requires individual steam lines from each separator to the steam collector at the powerhouse. Water lines run from the separators to the injection wells and may be combined. This system was installed, for example, at the Ahuachapán power station in El Salvador; the Ahuachapán layout for the first two units is shown in Fig. 5.5 [2].

.2.2 Pressure losses

ne of the main concerns in the design of the gathering system is the pressure loss in ne steam lines from the wellhead to the powerhouse. The steam pressure drop is a unction of the diameter, length, and configuration of the steam piping, as well as the ensity and mass flow rate of the steam. Of these the most critical variable is the pipe iameter. Equation 5.1 is a correlation for steam pressure drop caused by friction:

$$\Delta P_f = 0.8\frac{L\dot{m}^{1.85}}{\rho D^{4.97}} \quad (5.1)$$

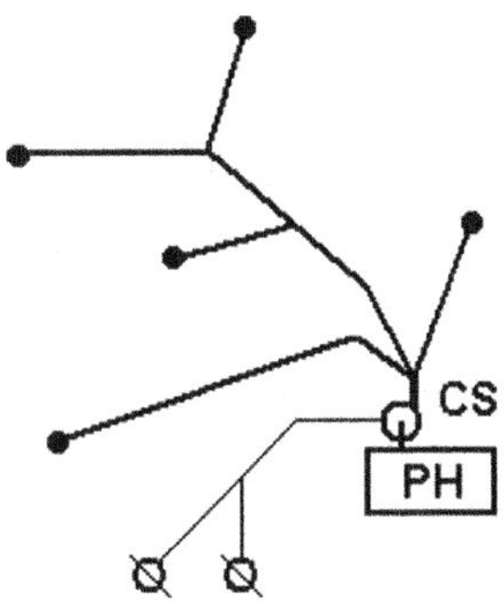

ig. 5.1 Two-phase gathering system: cyclone eparator (CS) at the powerhouse (PH). Filled rcles = production wells; open circles = ijection wells.

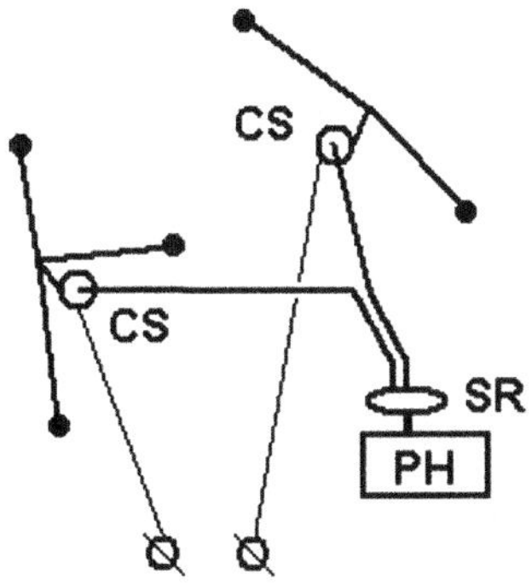

Fig. 5.2 Gathering system with satellite separator stations: steam pipelines to a steam receiver (SR) at the powerhouse.

ig. 5.3 Miravalles separator station. Photo by P. Moya [1] [WWW].

where L is the length of the pipe (ft), $\dot{m}$ is the mass flow rate (lbm/s), ρ is the density (lbm/ft^3), and D is the inside diameter of the pipe (in). The pressure drop is given in lbf/in^2. Note that eq. (5.1) is not dimensionally homogeneous; the 0.8 factor accounts for the particular units chosen.

Since the density of steam is relatively low, the change in pressure due to changes in pipe elevation are much smaller than the friction term given by eq. (5.1). The diameter plays a huge role in the pressure drop since it is inversely proportional to the diameter raised to essentially the fifth power. By installing pipes of a larger diameter the pressure loss can be drastically reduced but the extra cost of the larger pipes may be unacceptable economically. A thermodynamic-economic optimization study will lead to the optimum pipe size.

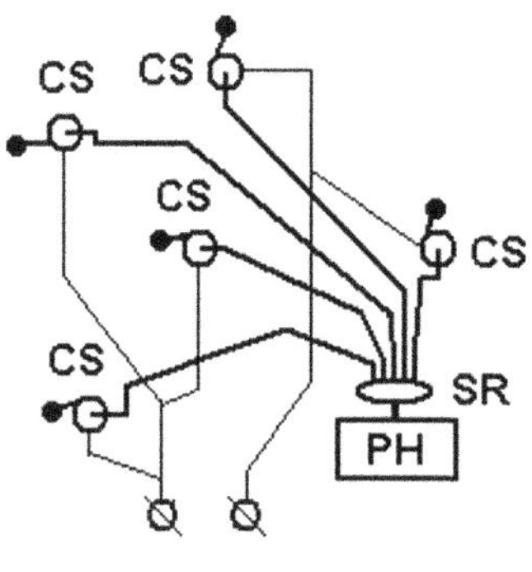

Fig. 5.4 Gathering system with individual wellhead separators.

The pressure drop in the liquid lines is less of a concern since the liquid is going to be disposed of by injection, but unnecessarily high pressure losses might require pumps to maintain sufficient reinjection pressure. The frictional pressure drop in the liquid pipes depends on the same variables as in steam pipes plus the friction factor, which in turn is a function of the pipe diameter, internal roughness,

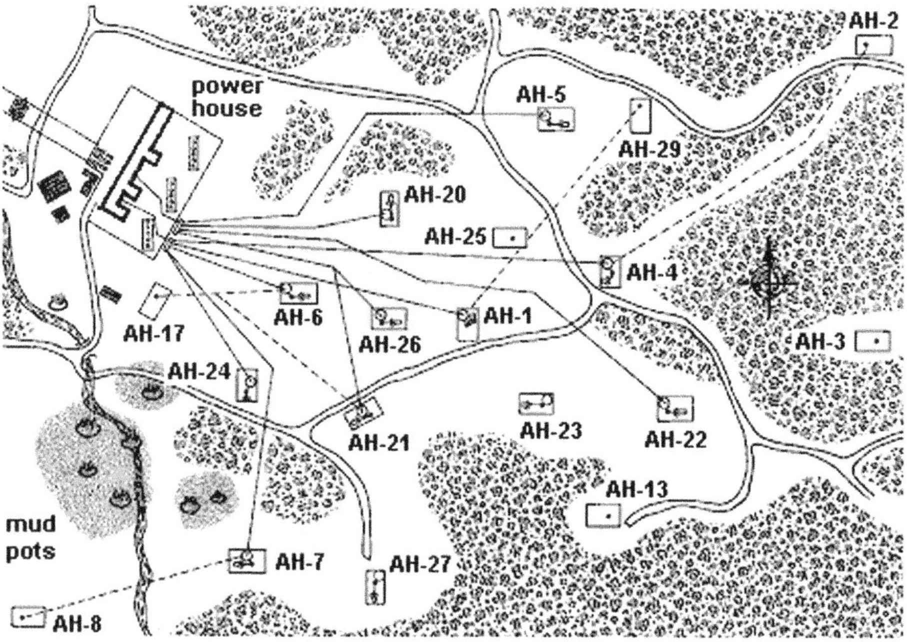

Fig. 5.5 Ahuachapán pipelines for Units 1 and 2; after [2]. This was modified when Unit 3 was added and when new wells were completed.

nd the viscosity of the liquid. Equation 5.2 gives the pressure loss (lbf/in^2) in a hori-ontal liquid pipeline:

$$\Delta P_f = 1.75 \times 10^{-4} \frac{fL\dot{m}^2}{\rho D^5} \tag{5.2}$$

here f is the friction factor given by the Swamee-Jain equation

$$f = \frac{0.25}{\left\{\log_{10}\left[\frac{\varepsilon/D}{3.7} + \frac{5.74}{\mathrm{Re}^{0.9}}\right]\right\}^2} \tag{5.3}$$

here ε is the pipe internal roughness (e.g., 1.48×10^{-4} ft for commercial steel pipe) nd Re is the Reynolds number

$$\mathrm{Re} = \frac{4}{\pi} \frac{\dot{m}}{\mu D} \tag{5.4}$$

here μ is the absolute viscosity (e.g., 1.223×10^{-4} lbm/ft·s for water at 145 lbf/in^2, bs and 300°F). If there is a change in the elevation of the pipe, the gravity head con-ribution must be included:

$$\Delta P_g = \rho g\, \Delta H \tag{5.5}$$

here g is the local gravitational acceleration (e.g., $g = 32.2$ ft/s^2 at sea level) and ΔH s the change in elevation (ft). The gravity-head term is positive for downcomers and egative for risers.

The pressure loss in a two-phase, steam-liquid pipeline is far more complex and less eliably predicted analytically [3]. Correlations may be used to estimate the pressure drop ut often field tests are conducted to determine the losses experimentally. The situation is omplicated by the fact that the two phases may flow in any of several different patterns epending on the pipe orientation and the relative amounts of the phases present.

For upward flow in a vertical pipe, starting from all liquid flow at the bottom, the ollowing flow patterns are encountered in sequence as the fluid moves up the pipe:

- Bubbly flow (bottom of pipe)
- Bubbly-Slug flow
- Slug flow
- Slug-Annular flow
- Annular flow
- Annular-Mist flow
- Mist flow
- All vapor flow, eventually for a sufficiently long pipe.

For flow in a horizontal pipe, the following flow patterns are encountered under cer-ain circumstances:

- Bubbly flow
- Stratified flow
- Wavy flow
- Plug flow

- Slug flow
- Annular flow.

For each flow pattern, the mechanism for the pressure loss is different and requires the use of empirical correlations. Many of these have been developed and the subject is still being researched.

In general, the pressure loss in a two-phase pipeline consists of three terms: (1) the frictional pressure drop (viscous effects), (2) the gravitational pressure drop (body force effects), and (3) the accelerational pressure drop (inertial effects). The viscous term may be found, for example, from a correlation by Lockhart and Martinelli [4] that gives the ratio of the two-phase pressure drop to the single-phase, steam-only pressure drop. The gravity term enters when the pipeline changes its elevation and requires knowledge of the average density of the two-phase fluid over a length of pipe, and the variation of the density along the pipeline. This is usually expressed in terms of the void fraction, i.e., the fraction occupied by the vapor of the cross-section of a short length of the pipe. Lockhart and Martinelli also provided correlations for void fraction [4]. The acceleration term results from the application of the momentum equation and plays a significant role at and just beyond the point where the liquid initially flashes from a liquid to a vapor when the pressure falls to the saturation pressure corresponding to the local fluid temperature. This is important in a production well, for example if the flash point is encountered between the fluid entry point and the wellhead, as we saw in Sect. 4.2. This term is often small for horizontal or gentle-sloped two-phase pipelines and may be neglected for these cases.

James [5] has offered a very simple formulation for the two-phase pressure drop as simply the steam-phase pressure drop divided by the square root of the local dryness fraction. His formula reduces to the steam pressure drop if the dryness fraction is equal to 1, but it fails to give the correct result at the other extreme for all liquid flow, i.e., when the dryness fraction becomes zero. Thus, the simple James rule should be viewed as a rough approximation that should only be used when the dryness fraction exceeds 0.5. In any case, its use for a long two-phase pipeline requires a step-by-step, iterative calculation with adjustments being made to the dryness fraction as the pressure falls and the dryness fraction increases.

The important conclusion from these considerations is that two-phase pipelines can be designed as elements of a geothermal gathering system but proper account must be taken of the pressure drop since it can be larger than that in single-phase steam lines. The presence of unsteady flow patterns such as slug flow can cause excessive vibrations and should be avoided by proper selection of pipe diameters. So-called flow pattern "maps" [4] can guide the designer to safe regimes.

Another important aspect concerns the flow of liquid that is removed from the cyclone separators. That fluid is in a saturated state and any loss in pressure can cause it to flash into vapor. For example, if the liquid is conveyed upward immediately after leaving the separator, the reduction in the gravity head will lower the pressure and one can expect flashing and vibrations that could damage equipment. Furthermore, if the fluid is conveyed horizontally over a significant distance, the frictional pressure drop may lead to flashing in the pipeline before the fluid reaches the injection well. This will create a vapor barrier and inhibit the flow of the fluid down the well. In such a case it may be necessary to bleed the vapor from the wellhead or ɔ install a booster pump upstream of the vapor breakout point. It is preferable to .ave the injection piping run downhill. Any drop in temperature of the liquid will mit- ;ate the problem of flashing but will exacerbate the potential problem of chemical recipitation; see Sect. 6.6.

.3 Energy conversion system

'he terminology single-flash system indicates that the geofluid has undergone a single ashing process, i.e., a process of transitioning from a pressurized liquid to a mixture f liquid and vapor, as a result of lowering the geofluid pressure below the saturation ressure corresponding to the fluid temperature.

The flash process may occur in a number of places: (1) in the reservoir as the fluid ows through the permeable formation with an accompanying pressure drop; (2) in he production well anywhere from the entry point to the wellhead as a result of the ɔss of pressure due to friction and the gravity head; or (3) in the inlet to the cyclone eparator as a result of a throttling process induced by a control valve or an orifice late. It is often the case in a newly developed field that the flashing occurs in the well- ore initially, but with time as the field undergoes exploitation and the reservoir pres- ure declines, the flash point may move down the well and even enter the formation. ometimes the term "separated steam" is used for this type of plant owing to the nanner in which the steam is obtained for use in the turbine.

While the actual location of the flash point can be important in the operation of a ower plant, from the point of view of understanding the thermodynamics of the energy onversion process, it is irrelevant. We will assume that the geofluid starts off as a com- ressed liquid somewhere in the reservoir, that it experiences a flashing process some- vhere, that the two-phases are separated, and that the steam is then used to drive a urbine which in turn drives the electric generator. A simple schematic of this operation s given in Fig. 5.6 [6], where the main components of a single-flash plant are shown.

At each production well, PW, there is a assemblage of equipment to control and mon- tor the flow of the geofluid from the well to the plant. This equipment includes: several alves, WV, a silencer, S, (simple cyclone separator for emergency venting), piping, and nstrumentation (pressure and temperature gauges). If wellhead separators are used, he cyclone separator, CS, will be located close to the wellhead on the same pad.

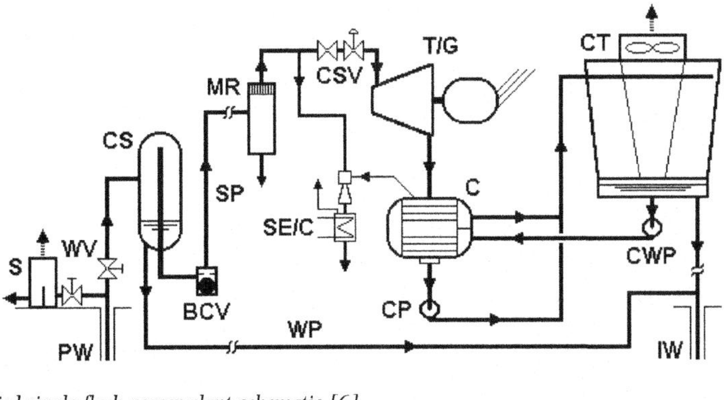

'ig. 5.6 *Simplified single-flash power plant schematic [6].*

Fig. 5.7 Wellhead separator system at Ahuachapán, El Salvador. Photo by author [2] [WWW].

A classic example of a wellhead arrangement showing the separator and other equipment is given in Fig. 5.7 [2]. Wellhead valves are seen above the cellar at right. Two-phase flow passes through the long-radius bend and enters the cyclone separator tangentially. Steam is collected from the center of the vessel by a standpipe, seen emerging from the bottom. The steam then passes through a ball check valve, BCV, at the left, emerges from the top and flows to the powerhouse. The separated liquid flows into a holding tank (small vertical vessel to the right of the bottom of the separator), from which it can go either to the silencer seen to right rear or to injection wells.

It is important to separate the two phases efficiently prior to the steam being admitted to the turbine. Liquid entrained in the steam can cause scaling and/or erosion of piping and turbine components. Generally, the quality of the steam entering the turbine should be at least 99.995% dry. Although there are a few designs in use for the cyclone separators, the industry has generally settled on the simple Webre-type separator, which is depicted in Fig. 5.7. Lazalde-Crabtree [7] published an approach to designing such vessels. He presented two variations: one for a primary 2-phase separator and one for a moisture remover. The designs were based on a combination of theory and empirical correlations. His recommended geometry of the two vessels is given in Fig. 5.8.

To achieve a very high level of steam quality, Lazalde-Crabtree recommends the guidelines given in Table 5.1.

A variation on these designs is used at several plants in Iceland. Originally vertical separators were deployed but since about 1995 these have been replaced with horizontal ones of the design shown in Fig. 5.8a. The main principle of separation here is gravity, augmented by a set of vane baffle plates fitted to the bottom of the vessel and a horizontal perforated droplet removal plate at the entrance to the steam exit chamber.

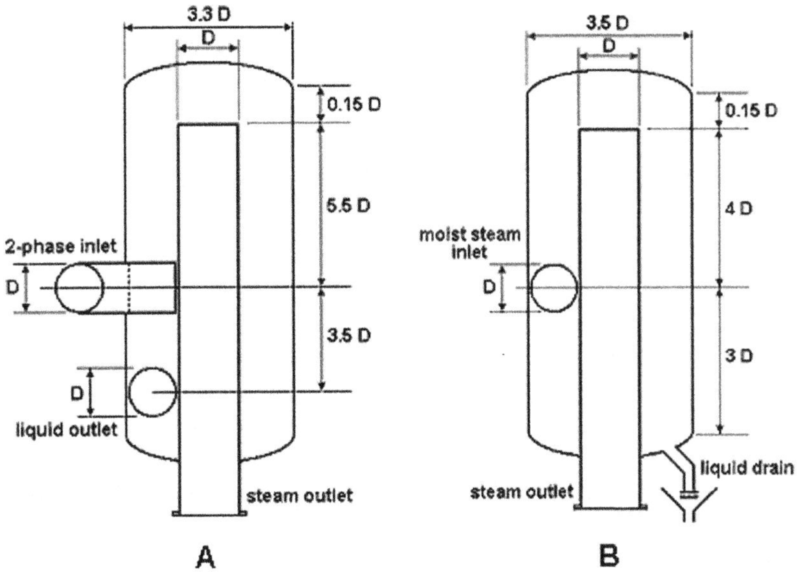

'ig. 5.8 General design specifications for cyclone separator (A) and cyclone moisture remover (B) [7].

'able 5.1 Separator and moisture remover design guidelines [7].

'arameter	Separator	Moisture remover
'laximum steam velocity at the 2-phase inlet pipe	45 m/s (150 ft/s)	60 m/s (195 ft/s)
.ecommended range of steam velocity at the 2-phase inlet pipe	25–40 m/s (80–130 ft/s)	35–50 m/s (115–160 ft/s)
'laximum upward annular steam velocity inside cyclone	4.5 m/s (14.5 ft/s)	6.0 m/s (20 ft/s)
.ecommended range of upward annular steam velocity inside cyclone	2.5–4.0 m/s (8–13 ft/s)	1.2–4.0 m/s (4–13 ft/s)

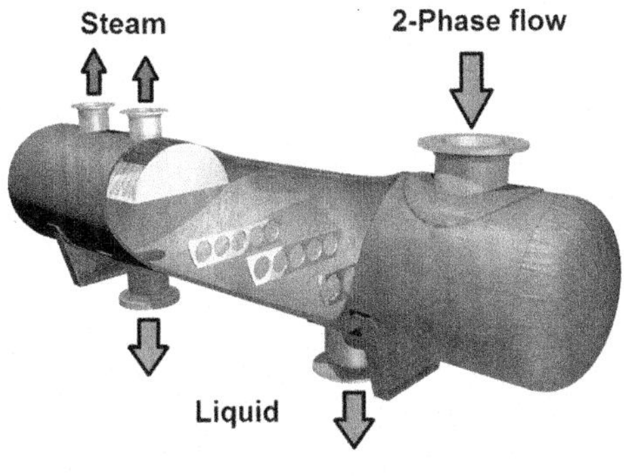

'igure 5.8a Sectioned view of a horizontal separator as used at Icelandic plants.

There are advantages and disadvantages to each type of separator [7a]. The advantages to the vertical design include cleaner steam, a sharper cut-off, wider pressure range, and easier maintenance; the disadvantages include size limitations and the height of construction. For the horizontal design the advantages include no size constraints and greater throughput per vessel; the disadvantages include the need for horizontal mist eliminators for high quality steam and greater maintenance. Of the two separators, generally horizontal ones are less expensive to build and install.

Where the separators are situated at a distance from the powerhouse, the steam transmission pipelines are fitted with traps to capture and remove moisture that may form from condensation within the pipes. Prior to being admitted to the turbine, the steam may be scrubbed to remove any fine moisture droplets that may have formed in the transmission pipelines and escaped the steam traps. The moisture remover, MR in Fig. 5.6, is usually located directly outside the powerhouse.

The turbines used in geothermal applications must be made of corrosion-resistant materials owing to the presence of gases such as hydrogen sulfide that can attack ordinary steel. Various alloys have been successfully used for turbine steam-path elements e.g., nozzles, blades, diaphragms, etc. Generally, 12% chromium steel is used for steam-path components: AISI Type 403/410 or AISI Type 405 alloy steels [8].

The design of the steam path through the blades is similar to that for a nuclear generating station because the steam enters the turbine essentially as a saturated vapor at a moderate to low pressure. This is markedly different from the steam conditions normally found at a coal, oil, or gas-fired power plant where highly superheated steam is used. Typical geothermal turbine inlet steam conditions are saturated with pressures that range from 5–10 bar (80–140 lbf/in^2). As a result, significant amounts of moisture appear in the steam path of geothermal turbines, particularly in the lowest pressure stages. These relatively slow moving droplets strike the back of the leading edge of the blades causing erosion unless this area is reinforced. Cobalt-rich alloy strips, such as Stellite©, are inlaid at these critical areas to protect them from damage.

Since the corrosive effects of geothermal fluids depend on the chemical composition of the geofluids, it is often considered wise to conduct *in situ* materials testing before deciding on the selection of materials for the plant. In these field tests, various samples (coupons) of alternative materials are subjected to long-term exposure to the geothermal liquids and vapors under conditions closely matching those expected during plant operation. The following properties are observed: corrosion, corrosion fatigue, stress corrosion cracking, erosion, and tensile strength [8,9]. Table 5.2 shows typical materials used for critical turbine elements [10].

Turbines for single-flash units are typically rated at 25–55 MW and consist of 4–5 stages of impulse-reaction blades. Both single-flow and double-flow designs are in use. Overall isentropic efficiencies in the high 80 percent range have been obtained.

The steam from the turbine is condensed by means of either a surface-type condenser, C, as shown in Fig. 5.6 or in a direct-contact condenser of either the barometric or low-level type. Most plants now employ surface condensers in which the geothermal steam passes through the shell side and cooling water passes through the tube side. This maintains physical and chemical separation between the geothermal steam and the cooling water, and allows more effective removal and treatment of noncondensable gases. Gases such as carbon dioxide and hydrogen sulfide exist with the natural steam and do not condense at the temperatures in the condenser. Therefore

able 5.2 Typical turbine element materials [10].

omponent	Material
iping	ASTM A106, Gr B; ASTM A335, Gr P11 or P22
.P. casings	ASTM A356, Gr 1, 6, 9, or 10
P. casings	ASTM A285 or A515
alve bodies	ASTM A216 or A217
asteners	ASTM A193 and A194
otors	ASTM A470
lades	AISI 403
ozzle blades	AISI 403
ands	AISI 405

nless they are removed they will increase the overall pressure in the condenser and ower the turbine power output. Steam jet ejectors with aftercondensers, SE/C (in ig. 5.6), and/or vacuum pumps are used for this purpose.

The cooling water is usually obtained from a cooling tower that recirculates a poron of the condensed steam after it has been cooled by partial evaporation in the presnce of a moving air stream (items CT and CWP in Fig. 5.6). This means that eothermal flash-steam plants do not need a significant supply of cooling water, a najor advantage in areas that are arid. A small amount of fresh water is needed, howver, to provide for replacement of tower blowdown.

.4 Thermodynamics of the conversion process

'he analysis presented here is based on fundamental thermodynamic principles, namely he principle of energy conservation (i.e., the First Law of thermodynamics) and the rinciple of mass conservation. An exposition of the subject may be found in any stanard text such as those by Moran and Shapiro [11] and Çengel and Boles [12], to menon only two. The general equation of the First Law is given in Sect. 10.2.

.4.1 *Temperature-entropy process diagram*

'he processes undergone by the geofluid are best viewed in a thermodynamic state iagram in which the fluid temperature is plotted on the ordinate and the fluid specific ntropy is plotted on the abscissa. A temperature-entropy diagram for the single-flash lant is shown in Fig. 5.9.

.4.2 *Flashing process*

'he sequence of processes begins with geofluid under pressure at state 1, close to the aturation curve. The flashing process is modeled as one at constant enthalpy, i.e., an senthalpic process, because it occurs steadily, spontaneously, essentially adiabatically, nd with no work involvement. We also neglect any change in the kinetic or potential nergy of the fluid as it undergoes the flash. Thus we may write

$$h_1 = h_2 \tag{5.6}$$

This was discussed in Sect. 4.2.3 when we examined the flow of a geofluid from the eservoir to the wellhead.

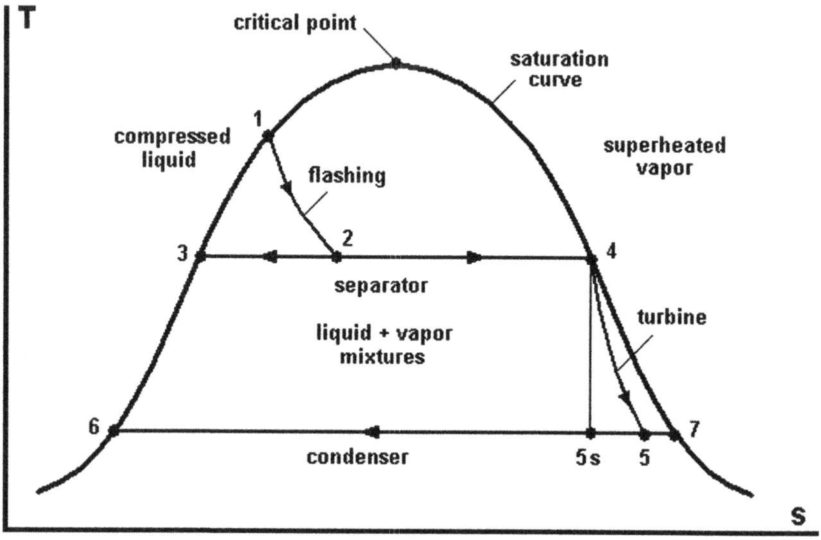

Fig. 5.9 Temperature-entropy state diagram for single-flash plants.

5.4.3 Separation process

The separation process is modeled as one at constant pressure, i.e., an isobaric process once the flash has taken place. The quality or dryness fraction, x, of the mixture that forms after the flash, state 2, can be found from

$$x_2 = \frac{h_2 - h_3}{h_4 - h_3} \tag{5.7}$$

by using the so-called lever rule from thermodynamics. This gives the steam mass fraction of the mixture and is the amount of steam that goes to the turbine per unit total mass flow into the separator.

5.4.4 Turbine expansion process

The work produced by the turbine per unit mass of steam flowing through it is given by

$$w_t = h_4 - h_5 \tag{5.8}$$

assuming no heat loss from the turbine and neglecting the changes in kinetic and potential energy of the fluid entering and leaving the turbine. The maximum possible work would be generated if the turbine operated adiabatically and reversibly, i.e., at constant entropy or isentropically. The process shown in Fig. 5.9 from 4–5s is the ideal process. We define the isentropic turbine efficiency, η_t, as the ratio of the actual work to the isentropic work, namely,

$$\eta_t = \frac{h_4 - h_5}{h_4 - h_{5s}} \tag{5.9}$$

The power developed by the turbine is given by

$$\dot{W}_t = \dot{m}_s w_t = x_2 \dot{m}_{total} w_t \tag{5.10}$$

his represents the gross mechanical power developed by the turbine. The gross elec-ical power will be equal to the turbine power times the generator efficiency:

$$\dot{W}_e = \eta_g \dot{W}_t \tag{5.11}$$

ll auxiliary power requirements for the plant must be subtracted from this to obtain e net, salable power. These so-called parasitic loads include all pumping power, cool-g tower fan power, and station lighting.

Before eq. (5.9) can be used computationally, it must be recognized that the isentro-ic efficiency of a turbine is affected by the amount of moisture that is present during e expansion process; the higher the moisture, the lower the efficiency. This effect an be quantified by using the so-called Baumann rule [13], which says that a 1% verage moisture causes roughly a 1% drop in turbine efficiency. Since geothermal urbines generally operate in the wet region, we must account for the degradation in erformance. Adopting the Baumann rule, we find the isentropic efficiency for a tur-ine operating with wet steam to be given by

$$\eta_{tw} = \eta_{td} \times \left[\frac{x_4 + x_5}{2}\right] \tag{5.12}$$

here the dry turbine efficiency, η_{td}, may be conservatively assumed to be constant at, ay, 85%:

$$\eta_{td} = 0.850 \tag{5.13}$$

rom Fig. 5.9, it is clear that the quality at the turbine outlet, state 5, depends on the urbine efficiency. State 5 is determined by solving eq. (5.9) using the turbine effi-ency and the fluid properties at state 5s, the ideal turbine outlet state, which are eas-y calculated from the known pressure and entropy values at state 5s. The ideal outlet nthalpy is found from

$$h_{5s} = h_6 + [h_7 - h_6] \times \left[\frac{s_4 - s_6}{s_7 - s_6}\right] \tag{5.14}$$

here the entropy term, by itself, gives the fluid outlet dryness fraction for an ideal urbine. When the Baumann rule is incorporated into the calculation, the following orking equation emerges for the enthalpy at the actual turbine outlet state:

$$h_5 = \frac{h_4 - A\left[1 - \dfrac{h_6}{h_7 - h_6}\right]}{1 + \dfrac{A}{h_7 - h_6}} \tag{5.15}$$

here the factor A is defined as

$$A \equiv 0.425(h_4 - h_{5s}) \tag{5.16}$$

hese equations are based on the assumption that the quality at the turbine inlet, x_4, equal to one, i.e., the entering steam is a saturated vapor. If the inlet is wet (as will be the case for the double-flash system to be discussed in the next chapter), then eq. (5.15) must be modified as follows:

$$h_5 = \frac{h_4 - A\left[x_4 - \dfrac{h_6}{h_7 - h_6}\right]}{1 + \dfrac{A}{h_7 - h_6}} \quad \text{(for } x_4 < 1\text{)} \tag{5.17}$$

5.4.5 Condensing process

Turning next to the surface-type condenser shown in Fig. 5.6, the First Law of thermodynamics leads to the following equation that relates the required flow rate of cooling water, $\dot{m}_{cw}$, to the steam flow rate, $x_2\dot{m}_{total}$:

$$\dot{m}_{cw} = x_2\dot{m}_{total}\left[\frac{h_5 - h_6}{\bar{c}\Delta T}\right] \tag{5.18}$$

where $\bar{c}$ is the assumed constant specific heat of the cooling water ($\approx$ 1 Btu/lbm · °F or 4.2 kJ/kg · K) and ΔT is the rise in cooling water temperature as it passes through the condenser.

For a direct-contact condenser (see Fig. 5.10), the appropriate equation is:

$$\dot{m}_{cw} = x_2\dot{m}_{total}\left[\frac{h_5 - h_6}{\bar{c}(T_6 - T_{cw})}\right] \tag{5.19}$$

5.4.6 Cooling tower process

The cooling tower must be designed to accommodate the heat load from the condensing steam. With reference to Fig. 5.11, the steam condensate that has been pumped from the condenser hotwell is sprayed into the tower where it falls through an air stream drawn into the tower by a motor-driven fan at the top of the tower. The ambient air enters with a certain amount of water vapor, determined by its relative

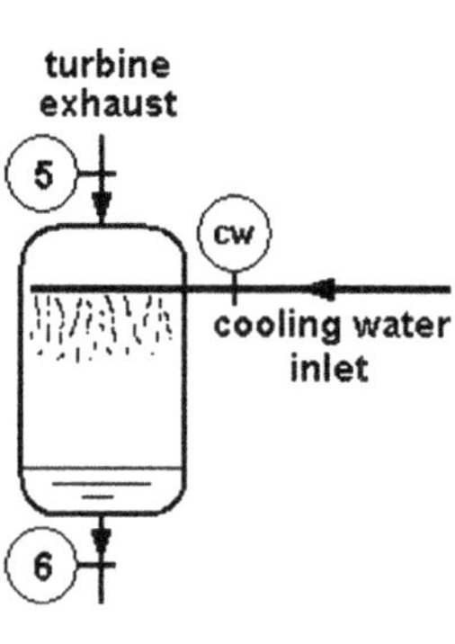

Fig. 5.10 Direct-contact condenser. Vessel is assumed to be perfectly insulated.

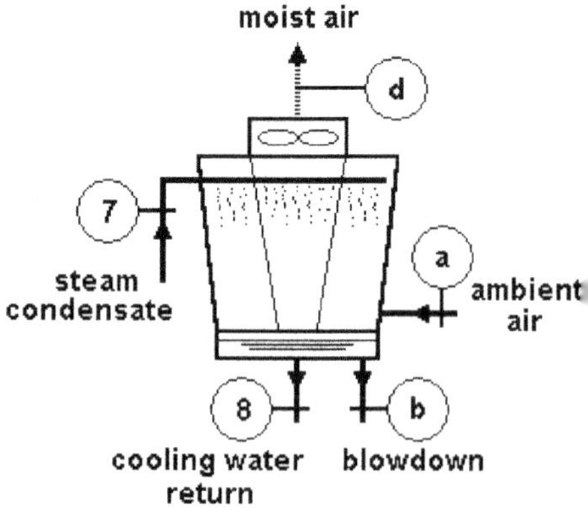

Fig. 5.11 Cooling tower schematic. This type is called a mechanical, induced-draft cooling tower.

umidity, and picks up more water vapor as the condensate partially evaporates. The vaporation process requires heat that comes from the water itself, thereby dropping s temperature.

The internal process involves the exchange of both heat and mass between the air nd the water. The following First Law equation describes the overall operation of the ɔwer, excluding the fan and assuming steady flow and overall adiabatic conditions:

$$\dot{m}_7 h_7 - \dot{m}_8 h_8 = \dot{m}_d h_d - \dot{m}_a h_a + \dot{m}_b h_b \qquad (5.20)$$

There are two other equations needed to analyze the process: mass conservation of vater and mass conservation of air. Recall that both the entering and leaving air treams contain water in the vapor phase (in different percentages). The conservation quations are:

$$\dot{m}_7 + \dot{m}_{wa} = \dot{m}_8 + \dot{m}_b + \dot{m}_{wd} \quad \text{(Conservation of water)} \qquad (5.21)$$

$$\dot{m}_{ad} = \dot{m}_{aa} \quad \text{(Conservation of dry air)} \qquad (5.22)$$

vhere the terms $\dot{m}_{wa}$ and $\dot{m}_{wd}$ represent the water content of the incoming and leaving ir streams, respectively. These can be found from the specific humidity, ω, of the air treams:

$$\dot{m}_{wa} = \omega_a \dot{m}_a \qquad (5.23)$$

nd

$$\dot{m}_{wd} = \omega_d \dot{m}_d \qquad (5.24)$$

'hese five governing equations are used with the properties of steam, water, and moist ir, either in tabular, graphic (psychrometric chart), or electronic form to determine he various flow rates needed for given design conditions.

Cooling towers are also characterized by two other parameters: the range and the pproach. The range is the change in water temperature as it flows through the tower, amely, $T_7 - T_8$, with reference to Fig. 5.11; the approach is the difference between he water outlet temperature and the wet-bulb temperature of the incoming air, amely, $T_8 - T_{wb,a}$. Since the ideal outlet water temperature is the wet-bulb temperaure of the incoming air, the approach is a measure of how closely the tower pproaches ideal performance, i.e., zero approach or $T_8 = T_{wb,a}$.

Cooling towers for geothermal power plants are much larger in cooling capacity han for conventional fossil or nuclear power plants of the same power rating. ince the cooling tower must be sized to reject the heat of condensation of the eothermal steam, we can examine the ratio of that heat, $\dot{Q}_o$, to the power output of he plant, $\dot{W}_e$.

Let us consider first a power plant that operates on a cycle. In Chapter 8 we will over such geothermal plants in detail. Here it is necessary only to understand that for ny cyclic power plant, the First Law of thermodynamics says that the net heat added o the cycle must equal the net work delivered by the cycle, or in terms of thermal and nechanical power,

$$\dot{Q}_{in} - \dot{Q}_o = \dot{W}_e \qquad (5.25)$$

Also for a cyclic plant, we can define the thermal efficiency, η_{th}, as follows:

$$\eta_{th} = \frac{\dot{W}_e}{\dot{Q}_{in}} \tag{5.26}$$

Binary geothermal plants (see Chap. 8) have thermal efficiencies in the range of 10–13%.

It is easy to show [14] that

$$\frac{\dot{Q}_o}{\dot{W}_e} = \frac{1}{\eta_{th}} - 1 \tag{5.27}$$

Fossil-fired, combined steam-and-gas-turbine plants typically have thermal efficiencies of 50–55%, coal-fired plants are about 35–40% efficient, and nuclear plants about 33–35% efficient. Thus, the rate of heat discharged per unit power generated for these three types of plant are, on average, 0.9, 1.7, and 1.9, respectively, whereas a typical geothermal binary plant discharges 7.7 units of waste heat for each unit of useful output. Thus, a 50 MWe geothermal binary plant must have a cooling tower 8.5 times larger in cooling capacity than that for a 50 MWe combined cycle plant.

Although flash-steam plants are not cyclic in operation and the thermal efficiency given in eq. (5.26) is not applicable, the waste heat can nevertheless be calculated using

$$\dot{Q}_o = x_2 \dot{m}_{total}(h_5 - h_6) \tag{5.28}$$

for the single-flash plant described above. This can then be compared to the net power using the equations in Sect. 5.4.4. The general qualitative conclusion regarding the relative size of the cooling systems for flash-steam plants in comparison to conventional plants is the same as for a binary plant, namely, they are larger than cooling systems at conventional plants of the same power rating. One often sees a 5-cell cooling tower used for a 30 MW geothermal flash plant, whereas that same tower could easily accommodate a 250 MW state-of-the-art combined cycle plant.

5.4.7 Utilization efficiency

Lastly, the performance of the entire plant may be assessed using the Second Law of thermodynamics by comparing the actual power output to the maximum theoretical power that could be produced from the given geothermal fluid. This involves determining the rate of exergy carried into the plant with the incoming geofluid. An in-depth presentation of Second Law analysis of geothermal plants is given in Chapter 10.

The specific exergy, e, of a fluid that has a pressure, P, and a temperature, T, in the presence of an ambient pressure, P_0, and an ambient temperature, T_0, is given by

$$e = h(T, P) - h(T_0, P_0) - T_0[s(T, P) - s(T_0, P_0)] \tag{5.29}$$

When this is multiplied by the total incoming geofluid mass flow rate, we obtain the maximum theoretical thermodynamic power or the exergetic power:

$$\dot{E} = \dot{m}_{total}\, e \tag{5.30}$$

he ratio of the actual net power to the exergetic power is defined as the utilization fficiency or the Second Law (exergetic) efficiency of the plant:

$$\eta_u \equiv \frac{\dot{W}_{net}}{\dot{E}} \tag{5.31}$$

All types of power plant can be compared on the basis of the utilization efficiency, o matter the source of the primary energy – be it coal, oil, nuclear, biomass, hydro, olar, wind, or geothermal. Plants can also be designed to maximize η_u when the value f the primary energy (or exergy) is a significant factor in the economics of the peration.

.5 Example: Single-flash optimization

n this section, we will examine the problem of selecting the separator conditions that vill yield the best overall plant performance in terms of the power generation. Two ypothetical cases will be studied.

.5.1 *Choked well flow*

et us consider a geothermal resource of the liquid-dominated type having a known eservoir temperature of 240°C. A single geothermal well is drilled having the follow-ng productivity data, i.e., the total geofluid mass flow rate (liquid and vapor) mea-ured in kg/s as a function of the wellhead pressure measured in bar,a:

P-wh	m-dot total
2	96.0
3	96.0
4	96.0
5	96.0
6	96.0
8	94.0
10	88.0
12	79.5
14	61.5
15	49.0
16	33.0
bar,a	kg/s

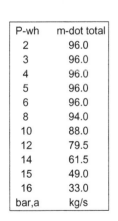

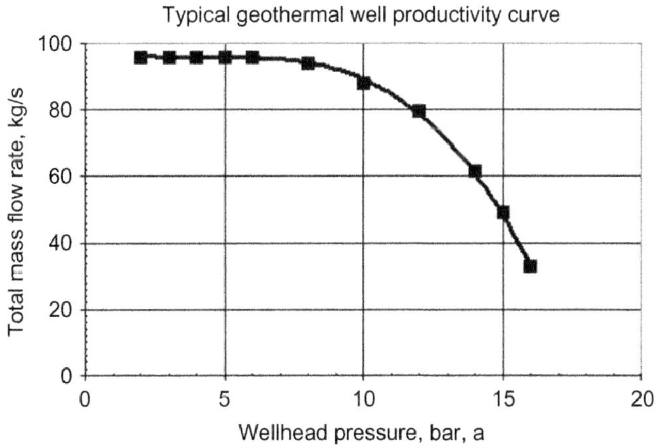

The data are plotted in the figure where it can be seen that the flow increases rap-dly as the well is opened and the pressure is lowered. However, once the pressure eaches about 7–8 bar, the flow rate stabilizes and no further lowering of the pressure an raise the total mass flow rate. Such a well is called "choked."

The question now is posed: What wellhead pressure should be chosen to maximize he power output from a single-flash plant connected to this well?

In solving this problem, we will assume that the well is connected directly to a vellhead separator, allowing us to neglect any pressure loss between the wellhead and the separator. Furthermore, we will neglect any other pressure losses associated with transmitting the separated steam from the separator to the turbine. It is not difficult to accommodate these losses, but the methodology of optimization is more clearly demonstrated by ignoring them here. We will also assume that it is possible to operate the condenser at a pressure of 0.123 bar or a condensing steam temperature of 50°C.

Equations (5.6)–(5.17) are used to analyze the flashing, separation, and turbine expansion processes. The calculations will proceed in two phases: in phase 1 we will determine the specific power output for a range of separator pressures (or equivalently, temperatures), and in phase 2 we will find the total power by factoring in the variation of the total flow rate as a function of the separator pressure. Phase 1 calculations are independent of the well productivity and yield the result in units of MW/(kg/s).

The calculations rely on accurate properties of the geothermal fluid, here assumed to be pure water. Thus, normal *Steam Tables* [15,16] are used to find all thermodynamic properties for the liquid and the vapor.

The results of the phase 1 calculations are given in the table and figure below.

T-sep	P-sep	w (MW/kg/s)
125	2.321	0.08405
130	2.701	0.08515
135	3.120	0.08578
140	3.613	0.08595
145	4.154	0.08568
150	4.758	0.08496
155	5.431	0.08382
160	6.178	0.08225

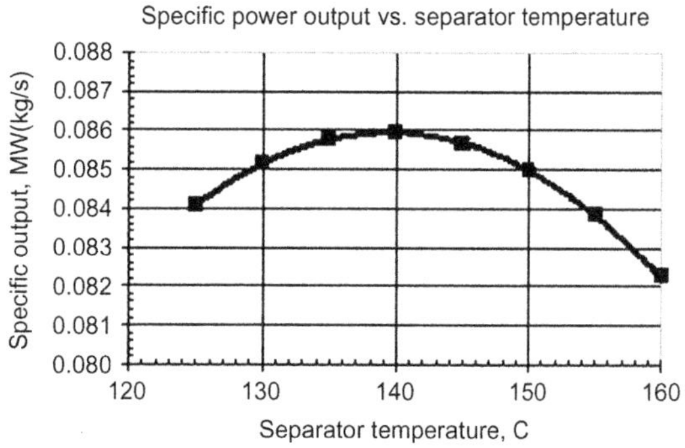

It can be seen that the maximum specific power occurs at a separator temperature of about 140°C or a separator pressure of 3.6 bar,a. The optimum specific power is about 86 kW for each kg/s of total flow from the well. Notice that the optimum separator temperature is close to the average temperature between the reservoir and the condenser, namely, 145°C [(240°C + 50°C)/2 = 145°C]. This simple relationship can be used as a first approximation to determine the best separator temperature. In the following section, we will derive this "rule of thumb."

Next, the actual well flow rate is found from the productivity curve at each assumed wellhead pressure and multiplied by the corresponding specific power to obtain the actual power in MW. For this purpose it is convenient to correlate the productivity curve with the best least-squares fit. An excellent fit is obtained using a cubic equation:

$$\dot{m}_{total} = 99.663 - 2.6287\, P_2 + 0.5802\, P_2^2 - 0.04212\, P_2^3, \tag{5.32}$$

where the pressure is in bar,a and the mass flow rate is in kg/s.

The results of the phase 2 calculations are shown below. The maximum gross power is 8.23 MW, indicative of a very good well. In this case, the optimum power occurs at essentially the same separator conditions as does the optimum specific power. This happens because the well flow curve is choked and the flow rate is nearly constant in the range of pressures around the optimum point. We will next consider an example where the well is not choked.

T-sep	P-sep	m-dot total	W-dot (MW)
125	2.321	96.16	8.082
130	2.701	95.97	8.171
135	3.120	95.83	8.220
140	3.613	95.75	8.230
145	4.154	95.74	8.203
150	4.758	95.75	8.135
155	5.431	95.75	8.026
160	6.178	95.64	7.866

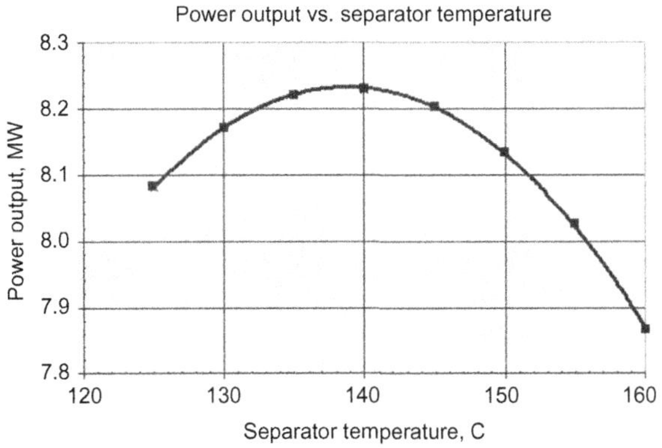

5.5.2 Non-choked well flow

Many wells do not reach their maximum flow rate except at very low wellhead pressures. These very low pressures are not usually appropriate as turbine inlet conditions. This type of productivity curve is often the result of low reservoir permeability or too small a diameter for the well casing. Such wells are not very productive and will lead to a different outcome in the optimization procedure.

Let us now assume the well is characterized by the following production data:

P-wh	m-dot total
2	43.0
3	42.1
4	40.6
5	39.7
6	38.1
7	35.0
8	32.2
9	30.0
10	28.0
11	24.7
12	20.0
13	17.3
bar,a	kg/s

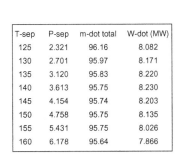

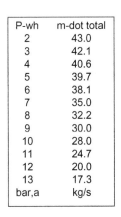

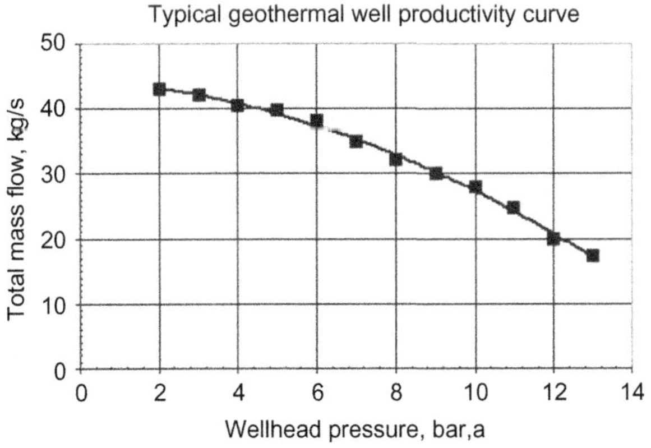

The data can be correlated very well with the second-order equation:

$$\dot{m}_{total} = 44.333 - 0.3363\, P_2 - 0.1357\, P_2^2 \quad (5.33)$$

The results of the phase 1 calculations are the same as for the first case and are not repeated here. The results of the phase 2 calculations are shown on the following page. The optimum power output is about 3.61 MW and occurs at a separator temperature of 130°C and a pressure of 2.7 bar,a. Here we see that the two optima, namely the best specific power and the best total power do not occur at the same values of separator pressure and temperature.

T-sep	P-sep	m-dot total	W-dot (MW)
125	2.321	42.82	3.599
130	2.701	42.43	3.613
135	3.120	41.96	3.600
140	3.613	41.35	3.554
145	4.154	40.59	3.478
150	4.758	39.66	3.370
155	5.431	38.50	3.227
160	6.178	37.07	3.049

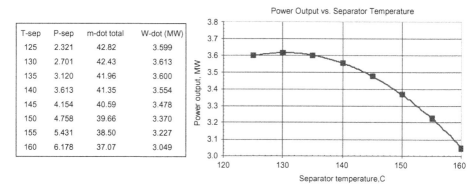

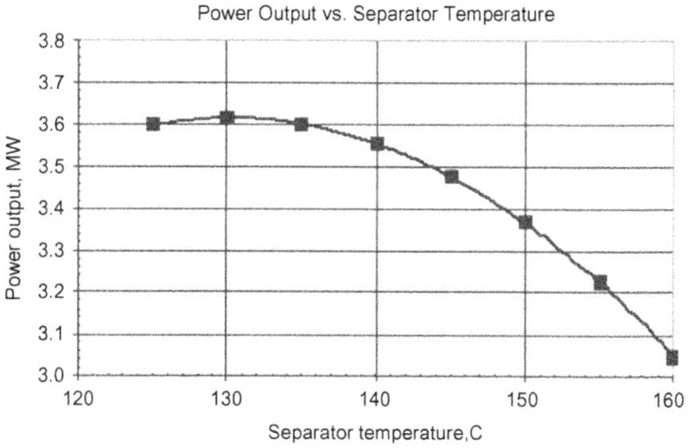

The utilization efficiency, eq. (5.31), can be seen to be equal to the ratio of the specific power output to the specific exergetic power (since the total mass flow rate appears in both the numerator and the denominator, and so cancels out). Thus the maximum utilization efficiency will not be influenced by the total mass flow rate and will coincide with the conditions for the optimum specific power. If we take the dead-state temperature to be 25°C, then the specific exergy for the 240°C liquid-dominated reservoir fluid is 236.52 kJ/kg. Thus, for the two examples shown in this section, the optimum value of the utilization efficiency is the same, namely, 36.3% (=85.95/236.52). For the latter case, the designer must decide on whether it is more economical to design for the highest utilization of the reservoir fluid or for the highest power output.

5.6 Optimum separator temperature: An approximate formulation

This derivation is based on the process diagram shown in Fig. 5.9. The goal is to find the separator temperature, T_2 ($=T_3 = T_4$), that maximizes the specific work output, w, from the plant. We will consider only the turbine work and neglect any pumping or other parasitic work loads.

The specific work per unit mass of geofluid is

$$w = x_2(h_4 - h_5) \tag{5.34}$$

To a first approximation,

$$h_1 - h_6 \approx c(T_1 - T_6) \tag{5.35}$$

and

$$h_3 - h_6 \approx c(T_3 - T_6) \tag{5.36}$$

assuming that c, the average specific heat for saturated liquid, is constant between states 1 and 6. It is convenient for this derivation to arbitrarily set the enthalpy to zero at state 6:

$$h_6 \equiv 0 \quad \text{(reference datum)} \tag{5.37}$$

hus,

$$h_1 = c(T_1 - T_6) = h_2 \tag{5.38}$$

nd

$$h_3 = c(T_3 - T_6) \tag{5.39}$$

rom the definition of the latent heat of evaporation, it follows that

$$h_4 - h_3 = h_{fg} \tag{5.40}$$

where we adopt the *Steam Table* "*fg*" subscript notation for the latent heat. Thus,

$$h_4 = h_3 + h_{fg} = c(T_3 - T_6) + h_{fg} \tag{5.41}$$

Now we invoke a little-known approximation, namely,

$$h_4 - h_3 \approx h_5 - h_6 \tag{5.42}$$

or the range of temperatures typical of geothermal systems, this approximation is ood to about 1.5%. Note that the actual turbine outlet state 5 is used in eq. (5.42), not the ideal isentropic state 5s. Then, it follows that

$$h_{fg} \approx h_5 - h_6 \approx h_5 \tag{5.43}$$

Now we can express the separator quality, x_2, as follows:

$$x_2 = \frac{h_2 - h_3}{h_{fg}} = \frac{c(T_1 - T_6) - c(T_3 - T_6)}{h_{fg}} = \frac{c(T_1 - T_3)}{h_{fg}} \tag{5.44}$$

'he specific work can now be expressed as the following function:

$$w = \frac{c}{h_{fg}}(T_1 - T_3)(h_4 - h_5) = \frac{c^2}{h_{fg}}(T_1 - T_3)(T_3 - T_6) \tag{5.45}$$

t is necessary only to differentiate this expression with respect to T_3, set the result qual to zero, and solve for T_3, a typical application of the calculus of variations [17]. n carrying out the differentiation, we simplify the task by assuming that the multiplier, c^2/h_{fg}, is a constant. This approximation is weaker than the one used in q. (5.42) but for this purpose is acceptable. The differentiation goes as follows:

$$\frac{dw}{dT_3} = (T_1 - T_3)(1) + (T_3 - T_6)(-1) = T_1 - T_3 - T_3 + T_6 = 0 \tag{5.46}$$

Solving for T_3, we find the optimum value of the separator temperature is given approximately as

$$T_{3,opt} = \frac{T_1 + T_6}{2} \tag{5.47}$$

Since this rule indicates that the temperature range between the reservoir and he condenser is divided into two equal segments, this rule is sometimes called the equal-temperature-split" rule. This approximate rule applies to all flash plants egardless of the number of flashes [10]. For a double-flash plant (see Chap. 6), the ule says: (1) the temperature difference between the reservoir and the first flash is equal to (2) the temperature difference between the first flash and the second flash and is also equal to (3) the temperature difference between the second flash and the condenser. Chapter 6 presents a full discussion of this topic.

5.7 Environmental aspects for single-flash plants

5.7.1 General considerations

There are several potential environmental impacts from geothermal plants [10,18,19] Table 5.3 shows a listing that may be applied to all kinds of geothermal power plants.

5.7.2 Considerations pertaining to single-flash plants

There are specific places at a single-flash plant where emissions can occur during normal operation. These include:

- Wellhead and station silencers and mufflers
- Steam traps and drains from pipelines
- Vents from the noncondensable gas ejectors
- Water vapor plume from a wet cooling tower
- Cooling tower blowdown.

Geothermal steam contains noncondensable gases such as hydrogen sulfide, H_2S, carbon dioxide, CO_2, methane, CH_4, and others in very small amounts. Uncontrolled venting of steam releases all these gases to the atmosphere. Under normal conditions, these gases are isolated in the condenser, drawn into the ejectors, and if necessary, treated before release to the atmosphere. There are many reliable, cost-effective means for removing H_2S if the amount that would be vented exceeds regulated limits [10]. The CO_2

Table 5.3 Geothermal power plant environmental concerns.

Possible impact	Details	Abatement techniques
Air pollution	H_2S emissions	Several effective commercial systems in use
Water pollution	Surface discharge of waste brine; groundwater contamination	Reinjection
Noise pollution	Drilling; well testing	Rock mufflers; silencers
Visual pollution	Unsightly pipes and buildings in pristine areas	Use low-level structures; paint equipment in blending colors
Land usage	Well pads, pipe routes, powerhouse, and substation	Much lower impact than conventional plants
Water usage	Cooling tower makeup (for binary plants only)	Use air-cooled condensers
Land subsidence	Liquid removal from subsurface can lead to surface depressions	Rare, most dramatic at Wairakei, New Zealand
Greenhouse gases	CO_2 emissions	Very low emissions relative to conventional fossil plants
Loss of natural wonders	Thermal manifestations may disappear; e.g., the geysers at Beowawe and Steamboat Springs, Nevada	Do not develop resources in or adjacent to national parks

that is released from flash plants is not abated but constitutes a relatively minor source of greenhouse gases: flash plants emit about 0.06 kg/kWh compared with 0.59 kg/kWh for a natural-gas-fired gas turbine and 1.13 kg/kWh for a coal-fired plant [18].

The separated brine contains practically all the dissolved minerals that existed in the geofluid in the reservoir but in higher concentrations. Some of the elements typically found in brines would adversely affect surface or groundwaters if allowed to mix with them. These elements include:

Arsenic, As	Lithium, Li
Boron, B	Magnesium, Mg
Calcium,	Ca Potassium, K
Chloride,	Cl Silicon, Si
Fluoride, F	Sodium, Na.

The main way to prevent water contamination is to reinject the waste brine back into the reservoir. This is now the accepted means of disposing of geothermal waste fluids at plants around the world. Reinjection has the added benefit of restoring some of the fluid that was extracted during production and in providing pressure support in the reservoir. Poor siting of the injection wells, however, can cause premature cooling of the production wells; reservoir simulators can be used to avoid this problem (see Chap. 4).

The noise associated with well drilling and the testing of wells occurs mainly during the early phases of field development and later sporadically as makeup wells are needed during normal operation. There are several effective methods to abate the noise coming from movement of geothermal steam, including cyclone silencers and rock mufflers that reduce the velocity of the steam being vented. An unabated, wide-open geothermal well discharging vertically into the atmosphere produces a noise level of 71–83 dB(A) at a distance of 900 m; by comparison, a noisy urban area has a level of 80–90 dB(A) [18]. Unabated venting is a rare occurrence nowadays but was very much in fashion in the early days of geothermal development to demonstrate the raw power of the natural steam.

A single-flash plant is relatively economical in terms of land required to support the operation compared to other means of generating electricity. A flash plant needs roughly 1,200 m^2/MW, including well pads, pipe routes, power plant, and substation. By comparison, a nuclear plant needs about 10,000 m^2/MW (power station only), a coal-fired plant needs 40,000 m^2/MW (power station plus area to be strip-mined for 30 years), and a solar photovoltaic plant needs 66,000 m^2/MW (power station only) [19].

Whenever large quantities of subsurface water are removed, there is the possibility that the overburden may sink into the space previously occupied by the water. Most geothermal reservoirs, however, reside in competent rock and the geofluid does not materially contribute to the structural strength of the formation. The one exception is Wairakei, New Zealand, where significant subsidence has occurred over the 45 years of exploitation [2]. With the general adoption of reinjection at flash-steam plants, the problem of subsidence has all but disappeared.

There is no question that commercial development of liquid-dominated geothermal reservoirs has destroyed some natural thermal mainfestations, in particular, geysers [20]. The hydrothermal and geological conditions that are necessary to produce a

Fig. 5.12 Geyser at Steamboat Springs in 1986. Photo by D. Hudson [21] [WWW].

geyser are extremely delicate and fragile. Nature itself routinely disrupts these mechanisms through earthquakes, as is evidenced by the ever-changing thermal features at Yellowstone National Park in the United States.

Two other sites in the United States used to have natural geysers, Beowawe and Steamboat Springs, both in Nevada, but they have been extinguished. Each is the site of geothermal power plants. Other kinds of thermal manifestations, such as fumaroles and hot ground, still exist at both sites. A photo of a geyser erupting at Steamboat in 1986 is shown in Fig. 5.12 [21], and one of the same area taken in 1998 is shown in Fig. 5.13 where only steam is seen rising from a deep boiling water table within a narrow fracture. The first power plant opened at Steamboat in 1986 and the site now is home to 17 units having a total rated capacity of 78 MW.

5.8 Equipment list for single-flash plants

The outline below gives the major equipment for a single-flash power plant [6].

5.8.1 Wellhead, brine and steam supply system

- Wellhead valves and controls
    - Blowout preventer (while drilling)
    - Master valves
    - Bleed lines
- Separator vessels
    - Vertical cyclone type
        - Bottom-outlet steam discharge
        - External or integral water collecting tank

Fig. 5.13 *Steam rising from fractures at Steamboat Springs in 1998. Photo by author [WWW].*

- Ball check valves
- Steam piping, insulation and supports
  - Condensate traps
  - Expansion loops or spools
- Steam header
- Final moisture remover
  - Vertical demister
- Atmospheric discharge silencers
  - Rock mufflers or cyclone silencers with weir flow control
- Brine piping, insulation and supports

5.8.2 Turbine-generator and controls

- Steam turbine-generator with accessories
  - Multistage, impulse-reaction turbine
    - Interstage moisture removal (optional)
    - Single-cylinder, single-flow or double-flow
    - Tandom-compound, four-flow
  - Rotor material: stainless steel (typ. 12% Cr, 6% Ni, 1.5% Mo)
  - Blade material: stainless steel (typ. 403, 13% Cr)
  - Stator material: carbon steel
  - Direct coupled, hydrogen or air cooled, two- or four-pole synchronous generator with static excitation
  - Lubricating oil system

- Control system
  - Digital-computer-based distributed control system
  - Continuous date acquisition system
  - Programmable component controller
  - Full automation and remote control (optional)
- Air compressor
  - One or two stage, motor driven units for plant and/or instrument air

5.8.3 Condenser, gas ejection and pollution control (where needed)

- Condenser
  - Direct-contact or surface-type
  - Barometric or low-level jet type
  - Integral gas cooler
  - Material for wetted surfaces: stainless steel (typ. 316 or 316L)
- Condensate pumps and motors
  - Vertical, centrifugal can pumps
  - Stainless-steel wetted surfaces
  - Low-head, high-volume design
  - Two 100 percent capacity units
  - Electric-motor driven
- Gas removal system
  - Steam jet ejectors with inter- and after-coolers
  - Turbocompressors
  - Hybrid ejector/compressor
- NCG treatment system
  - H_2S removal via commercially available methods

5.8.4 Heat rejection system

- Water cooling tower
  - Multi-cell, mechanically-induced-draft, counterflow or crossflow type
  - Natural-draft type (rarely used)
  - Drift eliminator
  - Fire-retardant materials of construction
- Cooling water pumps and motors
  - Vertical, centrifugal, wet-pit type
  - Stainless steel wetted surfaces
  - Low-head, high-volume flow type
  - Four 25 percent or two 50 percent capacity units
  - Electric-motor driven
- Cooling water treatment system
  - Chemical additives to control pH to 6.5–8.0

.8.5 Back-up systems

- Standby power supply
  - Back-feed from grid
  - Diesel generator

.8.6 Noise abatement system (where required)

- Rock mufflers for stacked steam
- Acoustic insulation for noisy fluid handling components

.8.7 Geofluid disposal system

- Injection wells for excess condensate and cooling tower blowdown
- Emergency holding ponds for wells and separators
  - Impermeable lagoons for temporary disposal of waste brine

References

[1] Moya R., P., "Operations and Maintenance of Geothermal Power Plants," World Geothermal Congress, Antalya, Turkey, 2005.

[2] DiPippo, R., *Geothermal Energy as a Source of Electricity: A Worldwide Survey of the Design and Operation of Geothermal Power Plants*, U.S. Dept. of Energy, DOE/RA/28320-1, U.S. Gov. Printing Office, Washington, DC, 1980.

[3] Cheremisinoff, N.P., Ed., *Encyclopedia of Fluid Mechanics, Vol. 3, Gas-Liquid Flows*, Gulf Publishing Company, Houston, 1986.

[4] Wallis, G.B., *One-dimensional Two-phase Flow*, McGraw-Hill, Inc., 1969.

[5] James, R., "Pipeline Transmission of Steam-Water Mixtures for Geothermal Power," *New Zealand Engineering*, V. 23, 1968, pp. 55–61.

[6] DiPippo, R., "Geothermal Power Systems," Sect. 8.2 in *Standard Handbook of Powerplant Engineering, 2nd ed.*, T.C. Elliott, K. Chen and R.C. Swanekamp, Eds., McGraw-Hill, Inc., New York, 1998, pp. 8.27–8.60.

[7] Lazalde-Crabtree, H., "Design Approach of Steam-Water Separators and Steam Dryers for Geothermal Applications," *Geothermal Resources Council BULLETIN*, Sept. 1984, pp. 11–20.

7a] Eliasson, E.T., "Power Generation from High-Enthalpy Geothermal Resources," *Geo-Heat Center BULLETIN*, June 2001, pp. 26–34.

[8] Kozaki, K., "Geothermal Power Plant," *Fuji Electric Review*, V. 26, No. 4, 1980, pp. 110–119.

[9] Anon., *Basic Planning of Geothermal Steam Turbine Plant*, KSI-E1057-1, Toshiba Corp., Tokyo, Japan, 1983.

10] Kestin, J., Ed. in Chief, R. DiPippo, H.E. Khalifa and D.J. Ryley, Eds., *Sourcebook on the Production of Electricity from Geothermal Energy*, U.S. Dept. of Energy, DOE/RA/4051-1, U.S. Gov. Printing Office, Washington, DC, 1980.

11] Moran, M.J. and H.N. Shapiro, *Fundamentals of Engineering Thermodynamics, 5th Ed.*, John Wiley & Sons, Hoboken, NJ, 2004.

12] Çengel, Y.A. and M.A. Boles, *Thermodynamics: An Engineering Approach, 4th Ed.*, McGraw-Hill, New York, 2002.

13] Baumann, K., "Some Recent Developments in Large Steam Turbine Practice," *J. Inst. Elect. Eng.*, V. 59, 1921, p. 565.

14] DiPippo, R., "Geothermal Power Technology," Chap. 18 in *Handbook of Energy Technology & Economics*, R.A. Meyers, Ed., Wiley-Interscience, New York, 1983.

[15] CATT2: Computer-Aided Thermodynamic Tables 2 (Ver. 1.0a), Intellipro, Inc., Wiley College Software, 2001.
[16] Keenan, J.H., F.G Keyes, P.G. Hill and J.G. Moore, *Steam Tables: Thermodynamic Properties of Water Including Vapor, Liquid, and Solid Phases (International Edition – Metric Units)*, John Wiley & Sons, Inc., New York, 1969.
[17] Sokolnikoff, I.S. and R.M. Redheffer, *Mathematics of Physics and Modern Engineering, 2nd Ed.*, McGraw Hill, New York, 1966.
[18] Pasqualetti, M.J., "Geothermal Energy and the Environment: The Global Experience," *Energy*, V. 5 1980, pp. 111–165.
[19] DiPippo, R., "Geothermal Energy: Electricity Generation and Environmental Impact," in *Renewable Energy: Prospects for Implementation*, T. Jackson, Ed., Stockholm Environment Inst., Sweden, 1993 pp. 113–122.
[20] http://www.wyojones.com/destorye.htm, Jones, G.L., "Geysers/Hot Springs Damaged or Destroyed by Man," *Wyo-Jones Geysers Page*, 2002.
[21] http://www.nbmg.unr.edu/geothermal/pix.php?id=geyser-MainTer1986, Nevada Bureau of Mines and Geology, U. of Nevada, Reno, 2004.

Nomenclature for figures in Chapter 5

BCV	Ball check valve
C	Condenser
CP	Condensate pump
CS	Cyclone separator
CSV	Control & stop valves
CT	Cooling tower
CW	Cooling water
CWP	Cooling water pump
IW	Injection well
MR	Moisture remover
PH	Powerhouse
PW	Production well
S	Silencer
SE/C	Steam ejector/condenser
SP	Steam piping
SR	Steam receiver
T/G	Turbine/generator
WP	Water (brine) piping
WV	Wellhead valve

Problems

5.1 Many geothermal power plants use a flash process to generate steam.

(a) Using the *Steam Tables* for pure water substance, calculate and graph the percentage of steam that can be obtained by flashing (i.e, at constant enthalpy) a saturated liquid from a given initial pressure to various final pressures. Show the calculated results both in tabular and graphical form. Initial pressure should range from 60 bar down to 1.0 bar; values of steam percent should be found for the following values of final pressure: 50, 25, 20, 15, 5, 2, and 1 bar. The graph should show steam percent on the ordinate, initial pressure (in bar and lbf/in^2) on the abscissa, and final pressure on a set of parametric curves.

(b) Repeat this problem using computerized property correlations on a computer.

.2 Consider two alternative steam turbine designs for a single-flash geothermal plant. Turbine A has four stages with moisture removal between stages. The moisture is simply extracted after each row of moving blades and throttled to the condenser. Thus, each stage sees saturated vapor at the leading edge. Turbine B has no moisture removal and can be characterized by an overall wet-turbine efficiency, η_{tw} (using the Baumann rule). Inlet conditions to both machines are: 360°F, saturated vapor; condenser temperature is 120°F. The 4-stage turbine A has its stages designed such that the temperature differences between successive stages are the same. Each stage i is described by its own wet-turbine efficiency, $\eta_{tw}(i)$. If the mass flow rate at turbine inlet is 1,000,000 lbm/h, compare the power developed by the two designs. What are the implications of the results?

.3 A single-flash geothermal steam plant receives geofluid from a reservoir having a temperature of 240°C. The condenser temperature is 50°C. Neglect pressure losses in surface pipelines. Use the Baumann rule for the turbine efficiency.

(a) Determine the specific work output (in kJ/kg of geofluid) if the separator operates at 170°C.

(b) Write a computer program or spreadsheet to investigate the effect of separator temperature on the specific work output. From your program, find the optimum value of the separator temperature and pressure, as well as the maximum specific work output. Compare your finding to the prediction of the "equal-temperature-split" rule.

.4 Your task is to analyze a single-flash plant. You must set it up for approximately optimum utilization efficiency. Use the "rule of thumb" given in this chapter to find the optimum separator temperature. The turbine efficiency may be found using the Baumann rule. The geofluid exists in the reservoir as a pressurized liquid at a temperature of $T_1 = 270°C$ and $h_1 = 1185$ kJ/kg. The turbine exhaust temperature is $T_5 = 50°C$. The productivity curve for an average well is given by

$$\dot{m} = 100.23 - 2.339 \times 10^{-2} P + 4.028 \times 10^{-5} P^2 - 1.02 \times 10^{-7} P^3$$

where P is the absolute wellhead pressure (kPa) and $\dot{m}$ is the total well flow (kg/s).

(a) Calculate the power output of the turbine, $\dot{W}_t$ in kW.

(b) Calculate the ratio of the heat rejected from the geofluid in the condenser, $\dot{Q}_0$, to the power output of the turbine, $\dot{W}_t$.

(c) Assuming a dead state at $T_o = 25°C$, calculate the Second Law utilization efficiency, η_u, based on: (i) wellhead conditions; (ii) reservoir conditions.

Chapter 6

Double-Flash Steam Power Plants

"To the engineer, all matter in the universe can be placed into one of two categories: (1) things that need to be fixed, and (2) things that will need to be fixed after you've had a few minutes to play with them. Engineers like to solve problems. If there are no problems handily available, they will create their own problems. Normal people don't understand this concept; they believe that if it ain't broke, don't fix it. Engineers believe that if it ain't broke, it doesn't have enough features yet."

Scott Adams, *The Dilbert Principle* – 2004

Geothermal Power Plants: Principles, Applications, Case Studies and Environmental Impact, Third Edition

© 2012 Elsevier Ltd. All rights reserved.

6.1 Introduction

The double-flash steam plant is an improvement on the single-flash design in that it can produce 15–25% more power output for the same geothermal fluid conditions. The plant is more complex, more costly and requires more maintenance, but the extra power output often justifies the installation of such plants. Double-flash plants are fairly numerous and are in operation in ten countries. As of August 2011, there are 59 units of this kind in operation, 10% of all geothermal plants. The power capacity ranges from 4.7 to 110 MW, and the average power is about 31 MW per unit. See Appendix A for more details.

Since many aspects of a double-flash plant are similar to a single-flash plant, we will generally follow the same format as in Chap. 5 but will focus on the differences between the two concepts. The fundamental new feature is a second flash process imposed on the separated liquid leaving the primary separator in order to generate additional steam, albeit at a lower pressure than the primary steam.

6.2 Gathering system design considerations

For double-flash plants, the addition of the second flash process increases the number of possible arrangements beyond the ones discussed in Sect. 5.2.1. Practically, the alternatives are as follows:

- Separators and flashers at each wellhead, with high- and low-pressure steam lines to the powerhouse, and hot water pipelines from each production well to the injection wells (Fig. 6.1).
- Two-phase pipelines from each well to the powerhouse, with separator(s) and flasher(s) at the powerhouse, short high- and low-pressure steam lines to the turbine(s), and hot water pipelines from the powerhouse to the injection wells (Fig. 6.2).
- Two-phase pipelines from several wells to satellite separator/flasher stations in the field, with high- and low-pressure steam lines from the satellites to the powerhouse, and hot water pipelines from the satellites to the injection wells (Fig. 6.3).

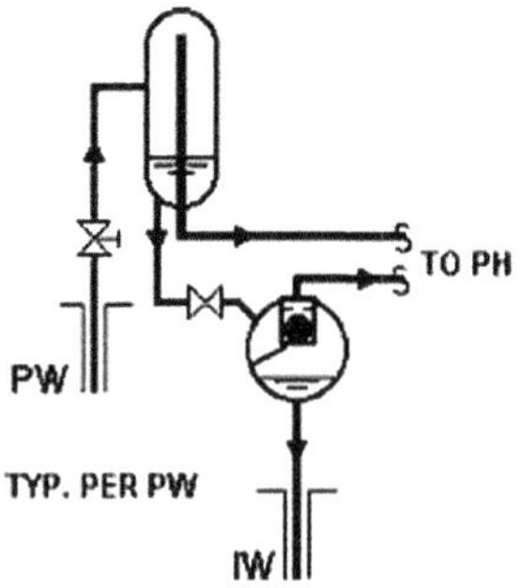

Fig. 6.1 Wellhead separators and flashers. There will be several production wells; injection wells may serve a number of production wells.

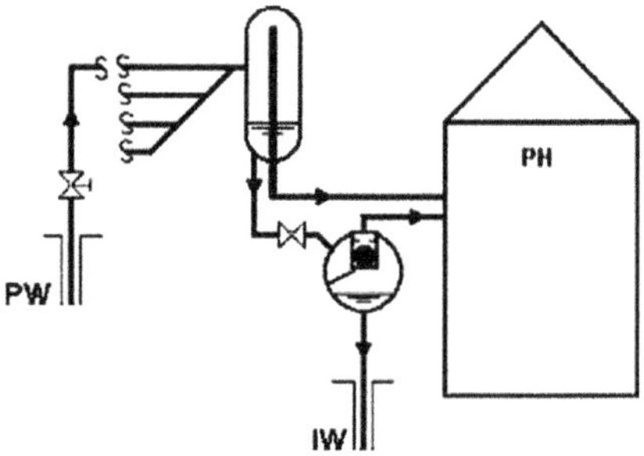

Fig. 6.2 Separators and flashers at the powerhouse. Two-phase flow lines from a number of production wells converge at the separator(s) and flasher(s) at the powerhouse.

- Two-phase pipelines from several wells to satellite separators in the field, with high-pressure steam lines and hot water pipelines to the powerhouse, flasher(s) at the powerhouse, short low-pressure steam lines to the turbine, and hot water pipelines from the powerhouse to the injection wells (Fig. 6.4).

An example of the type of system shown in Fig. 6.2 is presented in Fig. 6.5.

The list of arrangements is not exhaustive. For example, there may be a set of onditions that favor wellhead separators with satellite flashers, or some combination f the above arrangements. The best choice will be determined by thermodynamic and

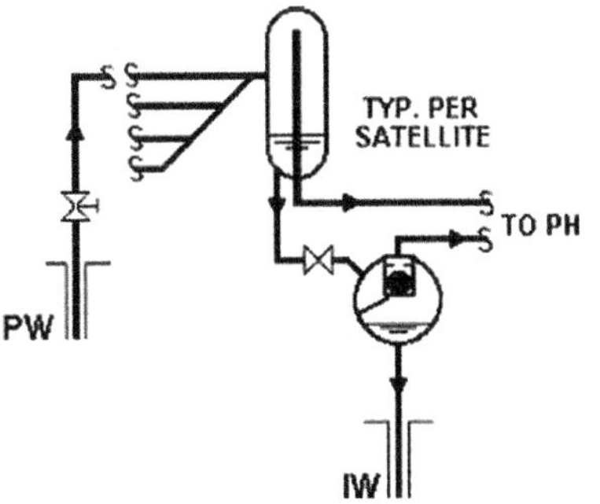

ig. 6.3 Satellite separator-flasher stations. everal sets may be located throughout the field.

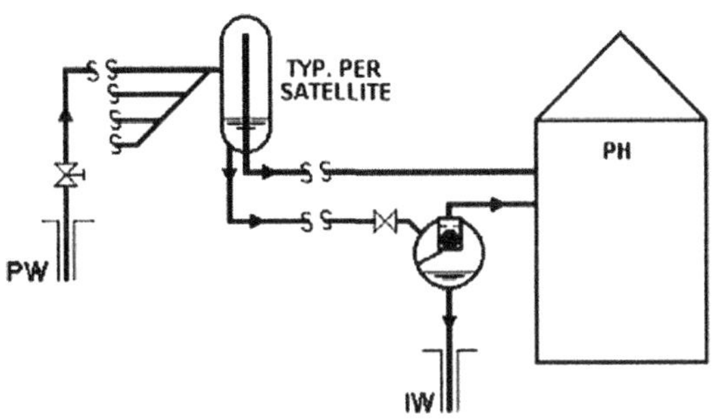

Fig. 6.4 Satellite separator stations and flasher(s) at the powerhouse.

'ig. 6.5 Separator-flasher arrangement at the 17 MW Beowawe double-flash plant, Nevada, U.S.; photo by author. 'wo 2-phase pipelines enter from left rear; the vertical vessel is the cyclone separator and the large horizontal vessel s the flasher. The smaller horizontal vessel in the cellar is the hot water holding tank from which the waste fluid is umped to the injection wells [WWW].

economic analysis, taking into account site-specific conditions including the tempera ture, pressure, and chemical nature of the geofluid, the location of production and injection wells relative to the powerhouse, topography of the site, and method of fluid disposal, including any required scale-control techniques. The latter are now routinely used at many plants and involve both down-well treatment to prevent calcite scaling in the production wells, and/or post-plant treatment to prevent silica deposition in the injection piping and wells. These potential problems are common to both single- and double-flash plants and will be discussed in Sect. 6.6.

The formulations presented in Sect. 5.2.2 for pressure drops need not be repeated since they apply equally well for pipelines at double-flash plants.

6.3 Energy conversion system

The schematic diagram for a double-flash plant is shown in Fig. 6.6 [1]. The design differs from the single-flash plant in Fig. 5.6 in that a flasher F has been added and there is a low-pressure steam line from it to the turbine in addition to the high-pressure one from the separator. The cooling tower that provides the cooling water CW is not shown in Fig. 6.6.

The turbine shown is a dual-admission, single-flow machine where the low-pressure steam is admitted to the steam path at an appropriate stage so as to merge smoothly with the partially expanded high-pressure steam. Other designs are possible; for example, two separate turbines could be used, one for the high-pressure steam and one for the low-pressure steam. In this case, the turbines could exhaust to a common condenser (see Fig. 6.7(A)) or to two separate condensers operating at different levels of vacuum (see Fig. 6.7(B)). For larger power ratings, say, 55 MW or higher, double-flow turbines would be a good choice in order to minimize the height of the last-stage blades. Usually, the last-stage blades in geothermal turbines are at most 25–27 in (635–686 mm) long, but at least one plant, the Darajat Unit 2 in Indonesia, uses 30 in (762 mm) long blades [2].

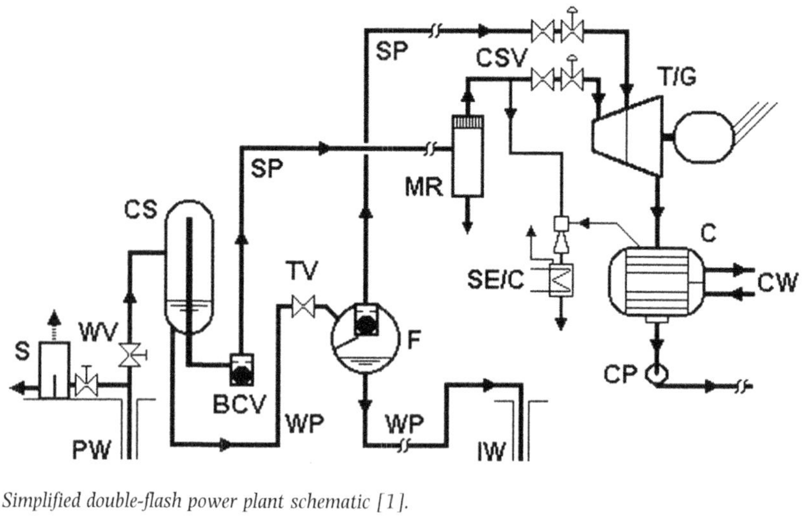

Fig. 6.6 Simplified double-flash power plant schematic [1].

.4 Thermodynamics of the conversion process

.4.1 Temperature-entropy process diagram

'he processes for the double-flash plant are shown in Fig. 6.8, a temperature-entropy iagram.

.4.2 Flash and separation processes

Vith reference to Fig. 6.8, the two flash processes, 1–2 and 3–6, are analyzed in the ame way as the flash process for the single-flash plant in Sect. 5.4.2. Each process

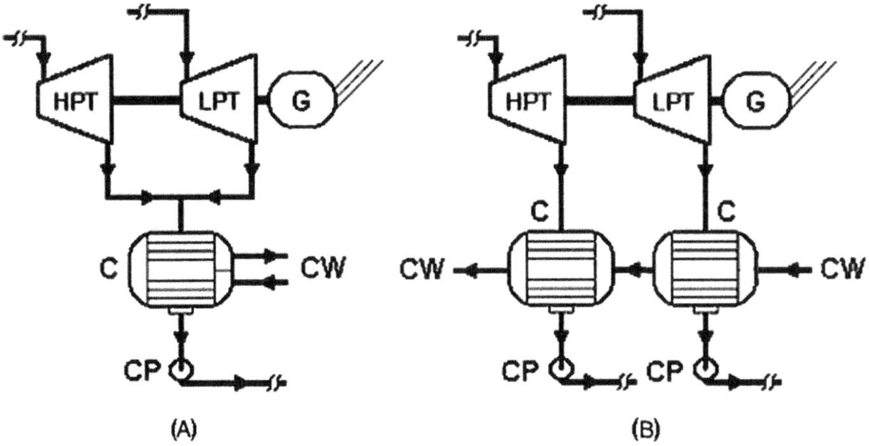

'ig. 6.7 Double-flash plant with separate high- and low-pressure turbines.

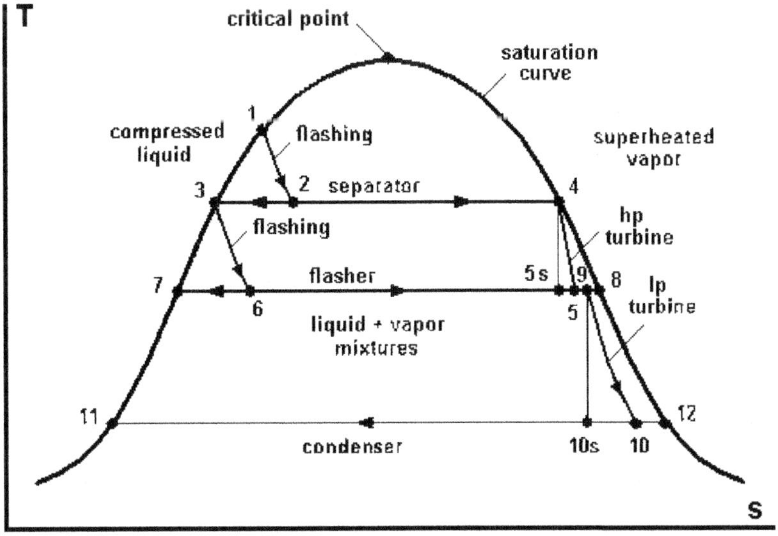

'ig. 6.8 Temperature-entropy process diagram for double-flash plant with a dual-admission turbine.

generates a fractional amount of steam given by the quality, x, of the 2-phase mixture. Each flash is followed by a separation process. The governing equations are as follows:

$$h_1 = h_2 \tag{6.1}$$

$$x_2 = \frac{h_2 - h_3}{h_4 - h_3} \tag{6.2}$$

$$h_3 = h_6 \tag{6.3}$$

$$x_6 = \frac{h_3 - h_7}{h_8 - h_7} \tag{6.4}$$

The mass flow rates of the steam and liquid (brine) for the high- and low-pressure stages are found from:

$$\dot{m}_{hps} = x_2\, \dot{m}_{total} = \dot{m}_4 = \dot{m}_5 \tag{6.5}$$

$$\dot{m}_{hpb} = (1 - x_2)\, \dot{m}_{total} = \dot{m}_3 = \dot{m}_6 \tag{6.6}$$

$$\dot{m}_{lps} = (1 - x_2)\, x_6\, \dot{m}_{total} = \dot{m}_8 \tag{6.7}$$

$$\dot{m}_{lpb} = (1 - x_2)(1 - x_6)\, \dot{m}_{total} = \dot{m}_7 \tag{6.8}$$

These mass flows will be used to calculate the power generated from the two stages of turbine expansion, the amount of waste liquid to be disposed of, and the heat that must be rejected through the condenser and ultimately from the cooling tower.

6.4.3 HP- and LP-turbine expansion processes

Using the same assumptions as in Sect. 5.4.4, we can find the power generated from each of the turbines. We will assume that a dual-admission turbine is used: Fig 6.9 shows a cross-section of a dual-admission, double-flow turbine. If two individual turbines are employed, the analysis for each one follows closely the method used for the turbine in a single-flash plant (see Sect. 5.4.4).

The HP-stages of the turbine may be analyzed according to the methodology used for the single-flash turbine, namely:

$$w_{hpt} = h_4 - h_5 \tag{6.9}$$

$$\eta_{hpt} = \frac{h_4 - h_5}{h_4 - h_{5s}} \tag{6.10}$$

$$\dot{W}_{hpt} = \dot{m}_{hps} w_{hpt} = x_2 \dot{m}_{total} w_{hpt} \tag{6.11}$$

The actual outlet state from the high-pressure section of the turbine must be found using the Baumann rule, exactly as in Sect. 5.4.4:

$$h_5 = \frac{h_4 - A\left[1 - \dfrac{h_7}{h_8 - h_7}\right]}{1 + \dfrac{A}{h_8 - h_7}} \tag{6.12}$$

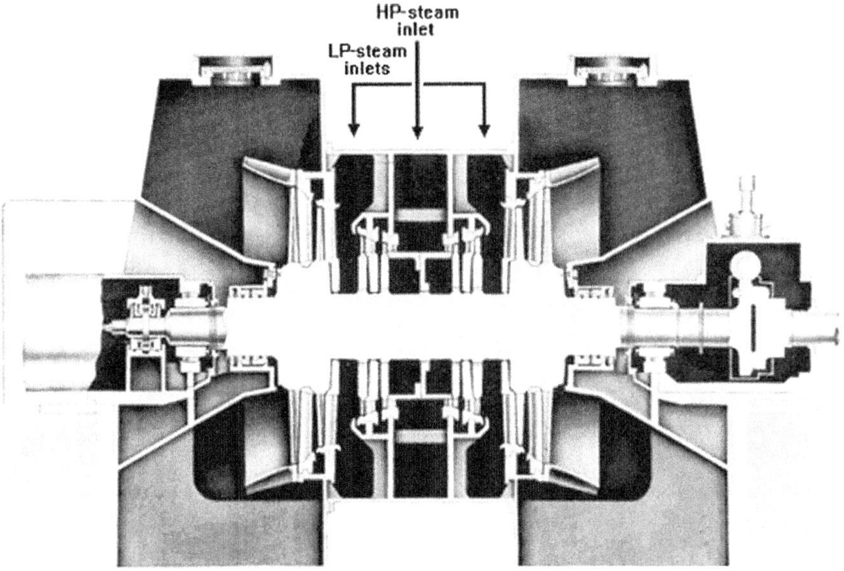

'ig. 6.9 *Cross-section of a dual-admission steam turbine; modified from [3].*

›here the factor A is defined as

$$A = 0.425(h_4 - h_{5s}) \tag{6.13}$$

At this point in the analysis, we must increase the mass flow through the turbine ›ecause the low-pressure steam from the flasher is admitted to the steam path, and oins the partially expanded high-pressure steam at state 5. With reference to Fig. 6.8, he partially expanded steam is at state 5, the low-pressure steam is at state 8 (satu-ated vapor), and the mixed steam, ready to enter the low-pressure turbine stages, is ıt state 9. The First Law of thermodynamics and conservation of mass allow us to find he properties of the mixed state:

$$\dot{m}_5 h_5 + \dot{m}_8 h_8 = (\dot{m}_5 + \dot{m}_8) h_9 \tag{6.14}$$

›r

$$h_9 = \frac{x_2 h_5 + (1 - x_2) x_6 h_8}{x_2 + (1 - x_2) x_6} \tag{6.15}$$

'he low-pressure turbine may now be analyzed as follows:

$$w_{lpt} = h_9 - h_{10} \tag{6.16}$$

$$\dot{W}_{lpt} = \dot{m}_9 (h_9 - h_{10}) = (\dot{m}_5 + \dot{m}_8)(h_9 - h_{10}) \tag{6.17}$$

$$h_{10} = \frac{h_9 - A\left[x_9 - \dfrac{h_{11}}{h_{12} - h_{11}}\right]}{1 + \dfrac{A}{h_{12} - h_{11}}} \tag{6.18}$$

$$A = 0.425(h_9 - h_{10s}) \tag{6.19}$$

$$h_{10s} = h_{11} + [h_{12} - h_{11}] \times \left[\frac{s_9 - s_{11}}{s_{12} - s_{11}}\right] \tag{6.20}$$

$$\eta_{lpt} = \frac{h_9 - h_{10}}{h_9 - h_{10s}} \tag{6.21}$$

The total power generated is the sum of the power from each turbine:

$$\dot{W}_{total} = \dot{W}_{hpt} + \dot{W}_{lpt} \tag{6.22}$$

Finally, the gross electrical power is found from

$$\dot{W}_{e,gross} = \eta_g \dot{W}_{total} \tag{6.23}$$

6.4.4 Condensing and cooling tower processes; utilization efficiency

The analysis presented in Sects. 5.4.5, 5.4.6 and 5.4.7 may be used here, provided the mass flow rate is changed to the sum of the HP- and LP-steam flows. The exergy of the incoming geofluid is found in the same way as before.

6.4.5 Optimization methodology

The optimization process for a double-flash plant is more complicated than for a single-flash plant because of the extra degree of freedom in the choice of operating parameters. For each choice of separator pressure (or temperature), there will be a range of possible flasher pressures (or temperatures), one of which will yield the highest power output. Over the spectrum of separator pressures, there will be corresponding flash pressures that yield the best output. Among this array of results there will be a single overall best optimum point. We can display this concept in a simple diagram, Fig. 6.10.

In the upper portion, P_2 is the separator pressure and T_6/T_2 is the ratio of the flash temperature to the separator temperature. For each point, a, on the upper curve, there will be some best power output $\dot{W}$(point b) that sits atop a curve (shown by the thin line) that comes from varying the flash pressure P_6. The lower heavy curve represents the locus of all such "best" points that come from allowing P_2 to vary over its practical range of values. Point d is the best of the best and defines the optimum plant choices for both separator and flash conditions. Point c shows the optimum ratio of temperatures for the separator and flasher.

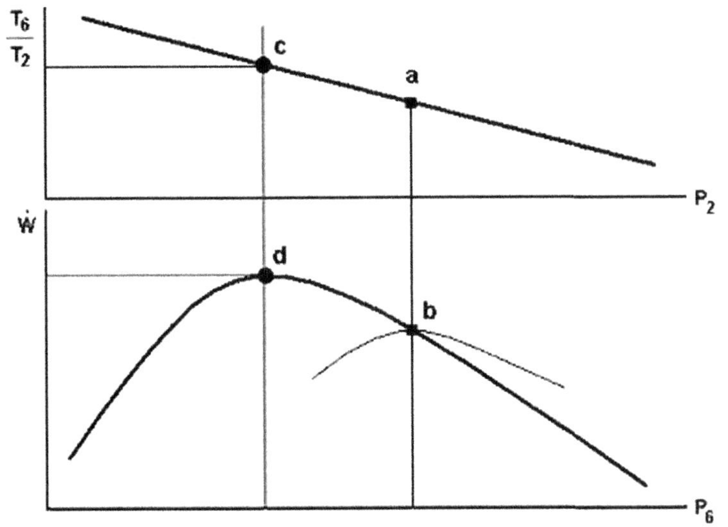

ig. 6.10 Double-flash optimization process.

The computation of all the states needed to accomplish this optimization can be asily programmed and the results plotted for any combination of reservoir and conenser conditions, as was shown in Sect. 5.5. This two-parameter search can begin by sing the "equal-temperature-split" rule to get first approximations to the optimum eparator and flash temperatures, and then the computer can explore combinations of alues on either side of these first guesses.

.5 Example: Double-flash optimization

he following example is taken from a report written by the author for the Electric ower Research Institute [4]. It is worked in U.S. Customary units to offer an alternaive to the S.I. units used in the single-flash example in Chap. 5.

Let the reservoir fluid temperature be 465°F and the well productivity curve be efined by the following second-order function:

$$\dot{m}_{total} = 740 + 0.910P_2 - 0.0123P_2^2 \qquad (6.24)$$

vhere P_2 is the wellhead pressure in lbf/in^2 (all pressures are absolute, unless othervise noted) and $\dot{m}_{total}$ is the total 2-phase mass flow rate in 10^3 lbm/h. The condenser perates at a pressure $P_{10} = 2$ lbf/in^2; the corresponding condenser temperature is 26°F.

The optimization considers two cases: (1) seeking the best power output, $\dot{W}_{total}$, for he given productivity curve, and (2) seeking the best specific output per unit mass of otal well flow, w_{total}. The results are not the same. Table 6.1 shows the results of the wo-parameter search; only the best values for the outputs are shown as functions of he separator temperature.

Table 6.1 Double-flash optimum conditions for $T_1 = 465°F$ and $T_{10} = 126°F$ [4].

T_2 °F	P_2 lbf/in^2	T_6 °F	P_6 lbf/in^2	w Btu/lbm	$\dot{m}_{total}$ 10^3 lbm/h	$\dot{W}_{total}$ MWe
392	225.3	250.7	30.2	44.19	319.58	4.138
374	181.9	242.9	26.2	45.14	497.92	6.585
356	145.3	234.0	22.4	45.67	612.09	8.192
338	114.8	225.7	19.2	45.77	682.13	9.149
320	89.6	217.4	16.3	45.42	722.68	9.617
302	69.0	208.9	13.8	44.56	744.17	9.718
284	52.4	200.7	11.7	43.19	753.89	9.540
266	39.2	192.2	9.8	41.25	756.77	9.146
248	28.8	183.9	8.2	38.70	756.01	8.572

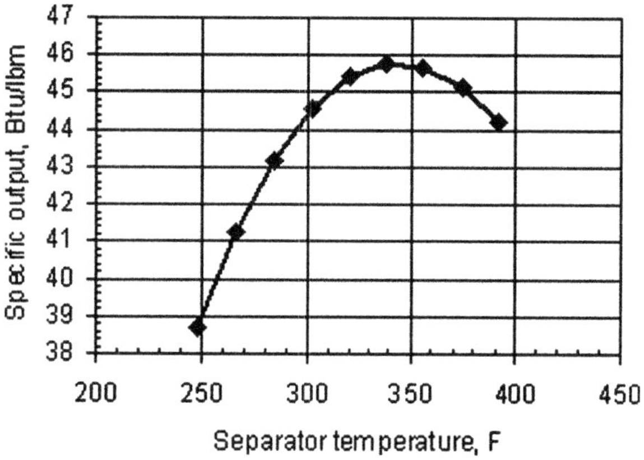

Fig. 6.11 Best specific output as a function of separator temperature.

As can be seen, the specific and total power outputs reach their respective maxima at different points. The variation in these terms can be seen in Figs. 6.11 and 6.12. If one desires the best total power, then the separator should operate at 304°F and the flash temperature should be 210°F; this will yield a power of 9.72 MWe. Alternatively, if one desires the best specific power (i.e., the highest utilization efficiency), then the separator should be at 342°F and the flash temperature should be 227.5°F; this will yield a specific power of 45.8 Btu/lbm. For a dead-state temperature of 65°F, this set of conditions yields a utilization efficiency of 42.1%. This may be compared to a utilization efficiency of 41.1% at the best power point. This is a one percentage point increase or a 2.5% improvement.

If one were to perform the optimization of a single-flash plant for the same given conditions as specified here, the optimum power output would be 7.46 MW, thereby demonstrating that the optimized double-flash plant is capable of generating 31% more gross power than an optimized single-flash plant, all other assumptions being the same [4]. The advantage would be somewhat less in terms of the net electrical power.

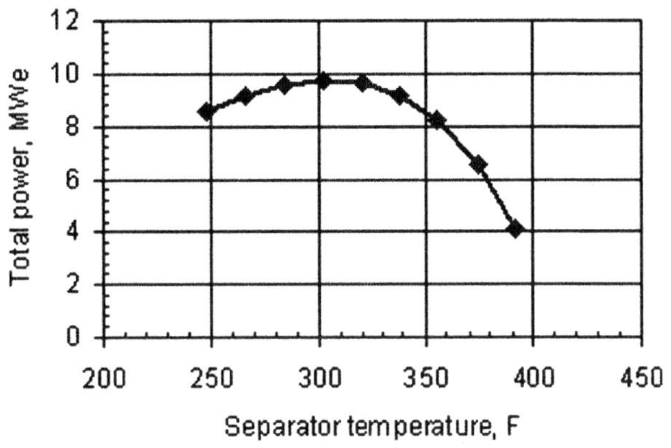

ig. 6.12 Best total power output as a function of separator temperature.

In carrying out these calculations, no attention has been paid to any practical mitations that may exist which would restrict the range of separator and flasher perating conditions. For example, it is not practical to operate the flasher at a presure below atmospheric pressure because the waste brine would have to be pumped ut, thereby increasing the parasitic power load and adding complexity to the overall peration. Another consideration that is particularly important for double-flash plants s the potential of chemical scaling caused by the waste liquid within or leaving the ash vessel. This will be addressed in the next section.

.6 Scale potential in waste brine

Jearly all minerals exhibit higher solubility in water as the temperature of the water ncreases. Conversely, when the water temperature drops, the dissolved minerals ecome less soluble and may precipitate from the solution. One notable exception is alcium carbonate which we have seen in Chap. 4 can become a problem in geothernal production wells. In this section, we will examine the potential problems associted with the dissolution of silica from the waste brines leaving the power plant.

.6.1 *Silica chemistry*

)ne of the minerals that is always found in geofluids is silica, SiO_2. Silica can exist in everal structural forms, from amorphous to highly crystalline, e.g., quartz. Each orm has its own solubility characteristic and all of them show increasing solubility vith increasing temperature, in the range of temperatures normally seen in geothernal reservoirs; see Figs. 6.13 and 6.14. This means that when the temperature of he geofluid decreases as it undergoes the processes in the power plant, the silica hat dissolved into the brine when it was hot and flowing through the fractured ocks of the reservoir will eventually begin to precipitate from the solution as solid ilica.

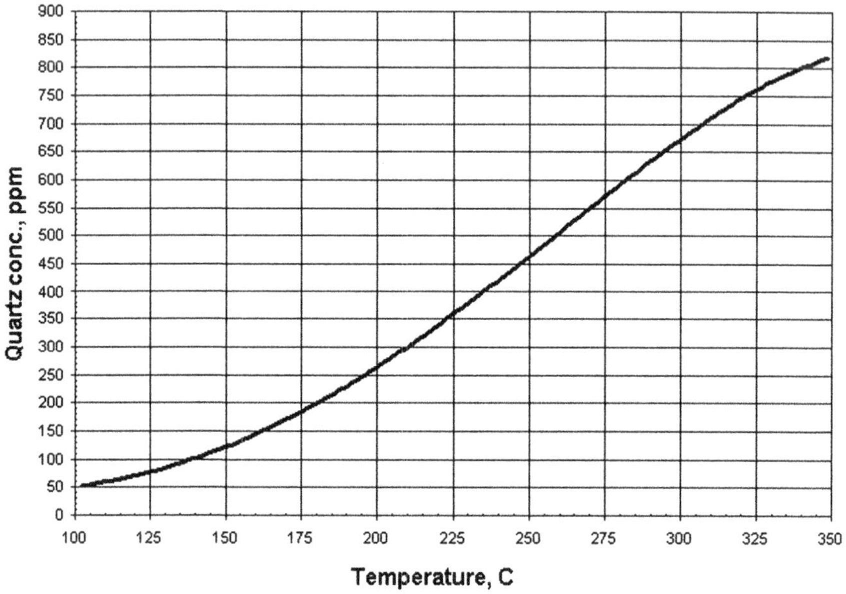

Fig. 6.13 Solubility of quartz as a function of temperature in pure water.

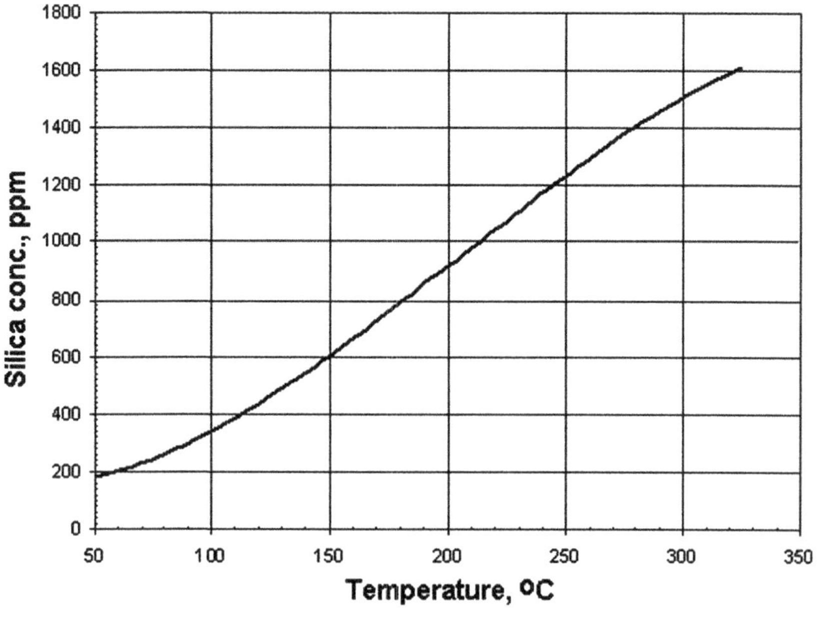

Fig. 6.14 Solubility of amorphous silica as a function of temperature in pure water.

The interesting fact is that the solubility in the hot reservoir fluid is controlled by uartz whereas it is controlled by amorphous silica at the low temperatures typical f waste brines [5]. Fortunately, the latter form has a much higher solubility than he former; compare Figs. 6.13 and 6.14 at the same temperature. By the time a rine has cooled to, say 125°C, it is typically supersaturated with respect to quartz ut undersaturated with respect to amorphous silica. Thus with proper design of he processes, it is possible to avoid the precipitation of the amorphous silica while he fluid is traveling through the plant components. It certainly is possible for recipitation to occur in the injection wells or in the reservoir once the waste uid returns to the formation. This would adversely affect the permeability of the ormation and reduce the injectability of the waste fluid. Of course, any reheating f the waste brine in the formation will reduce the potential for precipitation in the eservoir.

The solubility of silica is a function not only of the fluid temperature but also its alinity and pH. The figures shown above are for pure water. Qualitatively, for a given emperature and pH, the higher the salinity (i.e., higher molality), the lower the olubility of both quartz and amorphous silica in aqueous solutions. For a given emperature and salinity, the solubility of amorphous silica is essentially independent f the pH for low (acidic) values but rises dramatically as pH climbs above neutral, i.e., H > 7. The effect is more pronounced for fluids with high salinity [6].

The kinetics of the precipitation reaction play a critical role in the potential for scal-ng in geothermal plant components. If precipitation can be slowed it may be possible o process the fluid and dispose of it before scaling can occur. Alternatively, if the recipitation can be accelerated, it may be possible to force the fluid to give up its cale-causing minerals in a rapid and controlled manner before it enters the plant roper, thereby allowing the purged fluid to be used without fear of further precipita-ion. Both of these effects have been used at plants near the Salton Sea in the Imperial Valley of the U.S. where highly mineralized, corrosive brines are present. There re five parameters that influence the kinetics of the silica precipitation (essentially a olymerization process) [7]:

1. The initial degree of supersaturation (i.e., actual SiO_2 concentration − equilibrium concentration)
2. Temperature
3. Salinity or molality of the solution
4. pH of the solution
5. Presence (or absence) of colloidal or particulate siliceous material.

The first and second factors can be controlled by proper selection of separator and lash conditions for a given geofluid. The third factor is a fluid characteristic that annot be controlled. The fourth and fifth factors can be adjusted as the fluid moves hrough the plant from the production wells, through pipes and other components, nd eventually into the injection wells. When the brine is acidified, the rate of precipi-ation is very slow and the fluid can be viewed as temporarily stabilized. As the pH is aised, the rate increases dramatically, reaching a maximum at near-neutral pH values, about 6.0–7.5, and then slowing as pH approaches 9.0–9.5. In fact, the rates or pH = 5.3 and pH = 9.0 are roughly the same [5].

The last factor in the list has been exploited successfully for the Imperial Valley plants The geofluid is "seeded" with silica particles in large vessels called flash-crystallizers providing favorable precipitation sites for the supersaturated solution. After two stages of this process, the precipitated silica is removed, dried, compacted, and disposed of The steam that is generated is ready for use in turbines and the waste liquid is clear enough to be reinjected without fear of clogging the reservoir [8].

6.6.2 Silica scaling potential in flash plants

Since double-flash plants are often built on the hotter resources, the dissolved silica is typically higher than for single-flash plants. Furthermore, because of the additional flash process, the waste brine becomes more highly concentrated in silica than in a single-flash plant.

The derivation of the working equations to describe the silica scaling problem is based on Fig. 6.8; the reader should review it before proceeding.

Let us begin at state 1 where the geofluid exists at the reservoir temperature, t_1. It will carry a load of dissolved silica determined by the solubility of quartz. If we assume the water is pure, then we can represent the quartz concentration with the polynomial

$$Q(t) = 41.598 + 0.23932t_1 - 0.011172t_1^2 + 1.1713 \times 10^{-4}t_1^3 - 1.9708 \times 10^{-7}t_1^4 \tag{6.25}$$

where the temperature t is in degrees Celsius and the quartz concentration Q is in mg/kg or parts per million (ppm). As the fluid undergoes various flashing and separation processes, we will assume that the silica remains in the liquid phase. Thus the concentration of the silica in the brine will increase according to the following analysis:

$$S(t_3) = \frac{Q(t_1)}{1 - x_2} \tag{6.26}$$

and

$$S(t_7) = \frac{Q(t_1)}{(1 - x_2)(1 - x_6)} \tag{6.27}$$

where $S(t_3)$ and $S(t_7)$ represent the silica concentrations in the liquid at states 3 and 7, respectively. The symbol S is used to remind us that the controlling silica equilibrium is that of amorphous silica, not quartz. Since the quality after each flash is about 0.12–0.15, the concentration will increase by roughly 15.5% after the first flash (well to separator) and overall by 33.5% (well to waste brine from flash vessel).

Figure 6.15 shows the results of calculations for a single-flash plant at a 225°C resource and a double-flash plant at a 250°C resource. The curve labeled Q is the equilibrium solubility for quartz; the one labeled S is for amorphous silica. The line a-b shows the increase in concentration of silica in the brine as it undergoes a 2-stage flash process from 250°C to 117°C. The latter temperature is the flash temperature found from the "equal-temperature-split" rule for an optimum double-flash plant with a 50°C condenser. The vertical distance from point f to point b above the equilibrium curve indicates that the solution is supersaturated with respect to amorphous silica,

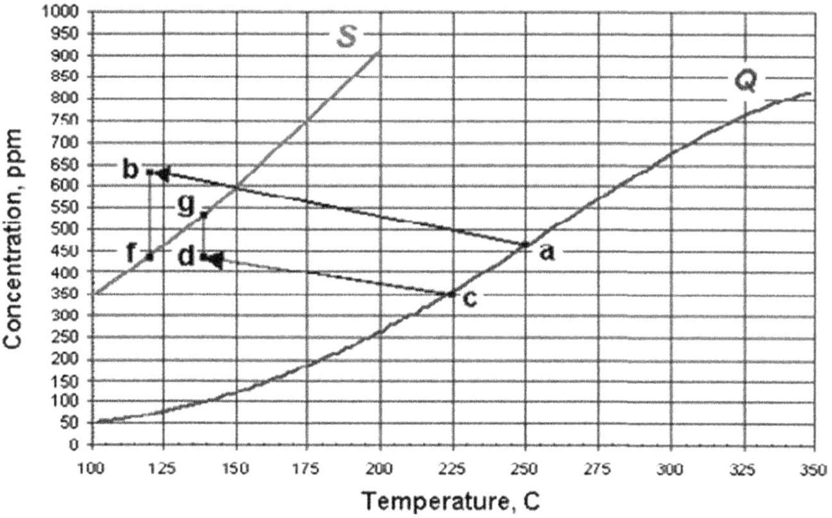

ig. 6.15 Silica condition after flash processes [9].

he amount of supersaturation being 214 ppm. Thus, the fluid is prepared to lose that mount of silica as soon as precipitation commences.

The line c-d shows the increase in concentration of silica in the brine as it ndergoes a 1-stage flash process from 225°C to 137.5°C. The latter temperature s the separation temperature found from the "equal-temperature-split" rule for an ptimum single-flash plant with a 50°C condenser. Since point d lies below point g, he solution is in equilibrium respect to amorphous silica, and no precipitation vould be expected.

In general, the hotter the resource, the higher the silica concentration in the incomng geofluid, the greater the supersaturation in the waste brine, from either type of lash plant, and the greater the likelihood of precipitation in the flash vessel, in the iping leading to the injection wells, in the injection wells, or in the formation. A more horough treatment of this subject can be found in Ref. [9].

One last comment on this phenomenon: The potential for silica precipitation is nitigated to some degree when binary plants are used (see Chap. 8) because the eofluid is not flashed, but only cooled. Thus, there is no increase in the concentration f the silica as the fluid passes through the plant. With reference to Fig. 6.15, the eofluid follows a horizontal line from right to left in a binary plant, which tends o keep the fluid in the safe region below the amorphous silica equilibrium curve. In omparison to a flash plant, this allows the geofluid to be cooled to a lower temperature efore silica precipitation can occur.

.7 Environmental aspects for double-flash plants

Double-flash plants have the same potential environmental impacts as single-flash lant that were described in Sect. 5.7. In the matter of water pollution, the waste brine from a double-flash plant will in general carry more highly concentrated contaminants than single-flash plants. Thus the need for reinjection in this case is even more urgent.

6.8 Equipment list for double-flash plants

Since the equipment used at double-flash plants is very similar to that found at single-flash plants, we will present here a non-annotated list and include annotations only for items unique to double-flash plants [1].

6.8.1 *Wellhead, brine and steam supply system*

- Wellhead valves and controls
- Separator vessels
- Flash vessels
  - Vertical or horizontal low-pressure tanks
  - Inlet valves or orifice plates to control pressure drop
  - Operating pressure greater than atmospheric
- Final moisture remover
  - For high- and low-pressure steam lines
- Ball check valves
- Steam piping, insulation and supports
- Steam header for high- and low-pressure steam
- Final moisture remover
- Atmospheric discharge silencers
- Brine piping, insulation and supports

6.8.2 *Turbine-generator and controls*

- Dual-pressure steam turbine-generator with accessories
  - Single cylinder with dual admission or two single-admission cylinders
- Control system
- Air compressor

6.8.3 *Condenser, gas ejection and pollution control (where needed)*

- Condenser
- Condensate pumps and motors
- Gas removal system
- NCG treatment system

6.8.4 *Heat rejection system*

- Water cooling tower
- Cooling water pumps and motors
- Cooling water treatment system

.8.5 *Back-up systems*

- Standby power supply

.8.6 *Noise abatement system (where required)*

- Rock mufflers for stacked steam
- Acoustic insulation for noisy fluid handling components

.8.7 *Geofluid disposal system*

- Injection wells for excess condensate and cooling tower blowdown
- Emergency holding ponds for wells and separators

‹eferences

1] DiPippo, R., "Geothermal Power Systems," Sect. 8.2 in *Standard Handbook of Powerplant Engineering, 2nd ed.*, T.C. Elliott, K. Chen and R.C. Swanekamp, Eds., pp. 8.27–8.60, McGraw-Hill, Inc., New York, 1998.

2] Anon., "Darajat Geothermal Power Plant: Largest Geothermal Turbine with Single Casing Construction," Mitsubishi Heavy Industries, Ltd., Tokyo, Japan, 1997.

3] Anon., *Geothermal Power Generation by Mitsubishi*, JA-243, Mitsubishi Heavy Industries, LTD., Tokyo, Japan, 1976.

4] DiPippo, R., "Geothermal Power Cycle Selection Guidelines," Part 2 of *Geothermal Information Series*, DCN 90–213–142–02–02, Electric Power Research Institute, Palo Alto, CA, 1990.

5] Kitahara, S., "The Polymerization of Silicic Acid Obtained by the Hydrothermal Treatment of Quartz and the Solubility of Amorphous Silica," *Rev. Phys. Chem. Japan*, V. 30, 1960, pp. 131–137.

6] Kindle, C.H., B.W. Mercer, R.P. Elmore, S.C. Blair and D.A. Myers, "Geothermal Injection Treatment: Process Chemistry, Field Experiences, and Design Options," DE-AC06-76RLO 1830, PNL-4767, Pacific Northwest Laboratory, Sept. 1984.

7] Ellis, A.J. and W.A.J. Mahon, *Chemistry and Geothermal Systems*, Academic Press, New York, 1977.

8] Newell, D.G., O.D. Whitescarver and P.H. Messer, "Salton Sea Unit 3 47.5 MWE Geothermal Power Plant," *Geothermal Resources Council BULLETIN*, V. 18, No. 5, May 1989, pp. 3–5.

9] DiPippo, R., "A Simplified Method for Estimating the Silica Scaling Potential in Geothermal Power Plants," *Geothermal Resources Council BULLETIN*, V. 14, No. 5, May 1985, pp. 3–9.

Nomenclature for figures in Chapter 6

BCV	Ball check valve
C	Condenser
CP	Condensate pump
CS	Cyclone separator
CSV	Control & stop valves
CW	Cooling water
F	Flasher
G	Generator
HPT	High-pressure turbine
IW	Injection well
LPT	Low-pressure turbine
MR	Moisture remover

PH	Powerhouse
PW	Production well
SE/C	Steam ejector/condenser
SP	Steam piping
T/G	Turbine/generator
TV	Throttle valve
WP	Water (brine) piping

Problems

6.1 A double-flash geothermal steam power plant operates from a reservoir at 300°C. The condenser temperature is 50°C. Secondary steam from the flash vessel is mixed with the primary steam at the plenum between the high- and low-pressure sections of the dual-admission turbine. Neglect pressure losses in surface piping and use the Baumann rule for the turbine efficiencies.

(a) Determine the specific work output (in kJ/kg of geofluid) if the separator and flash vessel operate at 230°C and 140°C, respectively.

(b) Write a computer program or spreadsheet to determine the optimum combination of separator and flash vessel temperatures, i.e., that combination which yields the maximum specific work output. Compare your findings to the prediction of the "equal-temperature-split" rule.

6.2 The flow diagram shown on the next page is for the new "El Diablo" geothermal power plant at "Misty Hot Springs," which is being built by the Terra Electric Power Company (TEPCO). The wells produce a mixture of liquid and vapor at state 1. An ideal separator (i.e., no pressure loss) produces saturated vapor (primary steam) at 2 and saturated liquid at 8. The secondary steam is mixed with the primary steam at the pass-in section of the turbine, between the first and second sections of the turbine. A closed-type, shell-and-tube heat exchanger is used in conjunction with a supply of cooling water. The condensed geothermal steam is pumped and mixed with the liquid coming from the flasher, and is returned to the formation through reinjection wells by means of reinjection pumps. You may use the following assumptions:

1. Assume flow from state b to 1 follows a throttling process.
2. Assume turbines operate isentropically.
3. Neglect pump work, i.e., for computational purposes, state 7 is identical to 6, and states 6, 7, 8, and 9 all lie on the saturated liquid line.
4. Assume a wet cooling tower is used at the plant, and that the ambient wet-bulb temperature is 80°F.
5. Apply the "equal-temperature-split" rule for flash-point selection.
6. Assume the geofluid is pure water substance.
7. Wellhead temperature, $T_1 = 340°F$.
8. Wellhead geofluid dryness fraction, $x_1 = 0.133$.
9. Condenser temperature, $T_5 = 124°F$.

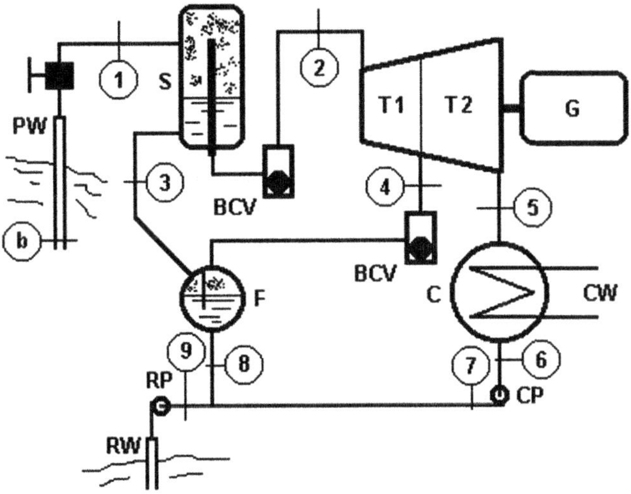

Legend: PW = Production Well (typical); S = Separator; F = Flasher; BCV = Ball Check Valve; T1, T2 = 1st and 2nd Sections of the Turbine; G = Generator; CW = Cooling Water; C = Condenser; CP = Condensate Pump; RP = Reinjection Pump, RW = Reinjection Well.

(a) Determine the bottom-hole temperature, T_b, if flashing takes place just at the bottom of the well.

(b) Sketch the processes involved in the operation of the plant in a temperature-entropy diagram. Label all states in agreement with the given plant schematic. Additional states may be defined as needed.

(c) Make a table giving the specific enthalpy at all 10 labeled state points (including state b), and at any other points that you think are important.

(d) For each 1 lbm that flows from the wells, calculate the mass flowing through each section of the turbine.

(e) Calculate the work output of the plant per unit mass flowing from the wells.

(f) Calculate the geothermal utilization efficiency, η_u.

(g) Assume a *typical* flow rate for a single well, and calculate roughly how many production wells would be needed for a 50 MW unit.

6.3 Consider the double-flash power plant shown in the figure on the following page. The geofluid is produced from a liquid-dominated reservoir (state 1). The separator (S) and the flasher (F) generate high- and low-pressure steam (saturated) for the two turbines T1 and T2. Turbine T1 is equipped with a moisture removal section where moisture which forms during the expansion from state 4 to state 9 is drained away at state 11, and flashed to the condenser. This drain is located at the point where the steam temperature is exactly 50°C below that of the inlet steam.

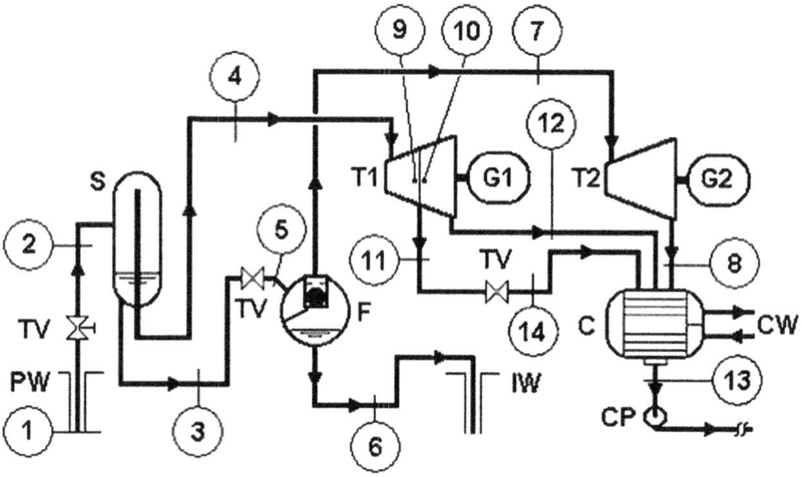

Plant Data: $T_1 = 280°C$; $T_9 = T_{10} = T_4–50°C$; $T_{13} = 40°C$; dead state $T_0 = 25°C$.

Your task is to design and analyze this plant. You may employ the "equal-temperature-split" rule for setting the separator and flasher conditions.

(a) Begin by making a careful sketch of the plant processes in a T-s diagram. Label all states in accordance with the state points shown in the schematic. You may use the Baumann rule in determining the actual efficiency of each turbine, i.e., each of the two sections of T1 and the whole of T2, taking the dry expansion efficiency as 85%.

(b) The analysis will be facilitated if you first find the specific enthalpy h at state points 1 through 12.

(c) Assuming that the total mass flow rate from the well is 1 kg/s, calculate the power output (in kW) of: (i) the high-pressure section of T1; (ii) the low-pressure section of T1; and (iii) turbine T2.

(d) If the actual total well mass flow rate is 1000 kg/s, calculate the total gross power of the plant in MW.

(e) Calculate the Second Law utilization efficiency, η_u, for the plant, based on the condition of the geofluid in the reservoir.

(f) If the original geofluid in the reservoir has a concentration of 595 ppm of silica (as quartz), calculate the concentration of silica in the separated liquid at states 3 and 6.

(g) Using the Fournier and Marshall correlation for amorphous silica solubility, namely,

$$\log_{10} s = -6.116 + 0.01625T - 1.758 \times 10^{-5}T^2 + 5.257 \times 10^{-9}T^3$$

where T is in kelvins and s must be multiplied by 58,400 to obtain ppm, determine whether or not the separated liquid at states 3 and 6 is supersaturated with respect to amorphous silica. What are the implications of your findings?

Chapter 7

Dry-Steam Power Plants

"A few yards further brought us into the midst of puffing geysers, or steam-jets, for I knew not by what other name to call them. Fumes of sulphur here met our nostrils at every step, while the rustling steam, as it spouted from a hundred cavities, completely enveloped us ... The whole of this violent commotion was accompanied by a tremendous noise beneath the Earth's surface ...."

John Russell Bartlett, describing *The Geysers* in California – 1854

7.1 Introduction

Dry-steam plants were the first type of geothermal power plant to achieve commercial status. Their history goes back 100 years to 1904 when Prince Piero Ginori Conti built and operated a tiny steam engine using the natural steam jets that issued from the ground at Larderello in the Tuscany region of Italy; see Fig. 7.1. Since the geofluid consisted solely

Geothermal Power Plants: Principles, Applications, Case Studies and Environmental Impact, Third Edition.
© 2012 Elsevier Ltd. All rights reserved.

Fig. 7.1 Prince Piero Ginori Conti and his 15 kW geothermal steam engine; Enel S.p.A. – Museo della Geotermia Larderello – Archivio Fotografico [1].

of steam, it was fairly easy to hook up a mechanical device to take advantage of the available energy. Although the Prince's engine only generated enough electrical power to illuminate five light bulbs in his factory, it was the springboard for larger plants.

Dry-steam plants tend to be simpler and less expensive than their flash-steam cousins in that there is no geothermal brine to contend with. As we will see, this is a mixed blessing when it comes to maintaining reservoir performance. Although there are only two major dry-steam fields in the world – Larderello and The Geysers, in Northern California, U.S. – there are 71 units of this type in operation in August 2011, about 12% of all geothermal plants. These plants account for 2,893 MW installed or about 27% of the total geothermal worldwide capacity. The average dry-steam unit has a rating of 40.75 MW. See Appendix A for more details.

7.2 Origins and nature of dry-steam resources

Large dry-steam reservoirs have been discovered only in two areas of the world: Larderello and The Geysers. There are limited dry-steam areas in Japan (Matsukawa), Indonesia (Kamojang), New Zealand (Poihipi Road section of Wairakei), and the United States (Cove Fort, Utah). White [2] estimated that only about 5% of all hydrothermal systems with temperatures greater than 200°C are of the dry-steam type. The rare occurrence of geothermal reservoirs producing dry or superheated steam rather than liquid-dominated reservoirs producing liquid-vapor mixtures begs the questions: "Why are these so unique?" and "What gives rise to them?"

The general characteristic of a dry-steam reservoir is that it comprises porous rock featuring fissures or fractures, either occluded or interconnected, that are filled vith steam. Whereas the steam also contains gases such as carbon dioxide, hydrogen ulfide, methane, and others in trace amounts, there is little or no liquid present. The team appears to have either magmatic or meteoric origins. The first possibility nvolves the slow evolution of vapor from magma chambers located at great depth and t very high temperatures close to that of molten rock [3]. The second one involves he percolation of rainwater through faults and fractures to great depth where it ncounters high temperature rock [2].

Measurements of the relative amounts of various isotopes of water, $H_2{}^{16}O$, $H_2{}^{18}O$, $I_2{}^{17}O$, and HDO, in geothermal fluid samples, compared to the natural amounts of he same isotopes, indicate that the most likely source of the fluids in geothermal reservoirs is meteoric waters [4]. This simple conclusion must be tempered by the uncertainty arising from the possible mixing of meteoric fluids that have percolated to great epths with magmatic vapors, making it difficult, if not impossible, to distinguish one rom the other once the fluids have reached the surface.

The mechanism of fluid behavior in a dry-steam reservoir is complex and several nodels have been proposed [2–9]. The emerging consensus is that there are three ources for the steam that is seen in the production wells. The first contribution comes rom steam residing in fissures and fractures in competent rock. The second comes from he vaporization of liquid that formed as condensate from steam that has come n contact with the cooler lateral and upper boundaries of the reservoir. The last one rises from evaporation off the top of a deep brine reservoir over a prolonged period of steam production which causes a decrease in reservoir pressure [8,9].

These mechanisms are depicted schematically in Fig. 7.2. Note that the lateral boundaries of the vapor-dominated reservoir are seen as being highly impermeable. f this were not true, liquid would flood the steam zone from the sides, collapsing the team in the formation. The only liquid in the steam zone comes from the condensate and water trapped in fissures that have only limited vertical extent. A production well provides a preferential flow path allowing the removal of steam and causing a cone of depression in the pressure in the surrounding formation. This pressure reduction leads o further steam generation by causing the trapped liquid to evaporate. In this way he formation may eventually completely dry out, leaving only the condensate and the deep liquid as the sole means to provide additional steam.

The natural recharge enters the system primarily through the lateral boundaries hat may be demarcated by major faults with significant offsets. Surface water hen can reach the depth of the liquid-dominated zone. If the rate of recharge is ess than the rate of production, the deep liquid zone will retreat to even greater depth as the reduced pressure causes more and more evaporation to occur. The nitial presence of the steam zone is thought to be attributable to just such an mbalance between the natural steam loss through the upper layers of the formation relative to the natural recharge. In their natural states prior to exploitation, both the Larderello and The Geysers fields were marked by extensive surface hermal manifestations such as fumaroles, mud volcanos, steam-heated pools, and steaming, acid-altered ground and rocks [10–12]. Over time, the permeability of the near-surface formation decreases as minerals precipitate from the geofluid and seal the fissures and fractures that had originally served as fluid conduits.

Many authors have cited the close correlation between the observed temperatures n dry-steam reservoirs and the temperature that corresponds to the maximum

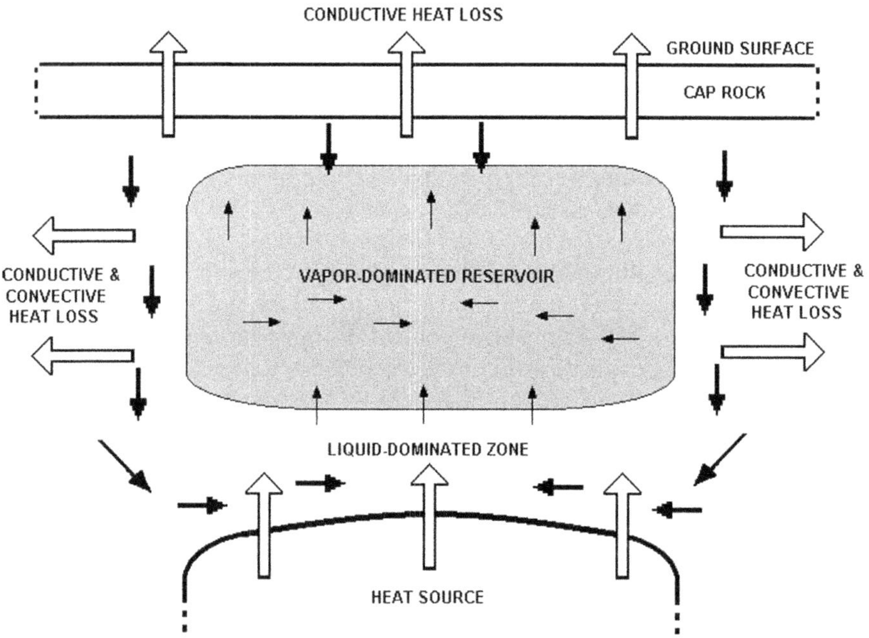

Fig. 7.2 Proposed model for creation and operation of a dry-steam reservoir, after [7]. Open arrows represent heat flow; filled arrows show movement of liquid; small arrows show steam movement.

enthalpy for saturated steam [3–8]. Saturated steam exhibits a maximum enthalpy of $h = 2804.2$ kJ/kg at roughly $T = 235°C$ and $P = 30.6$ bar [13]. The enthalpy-entropy (or Mollier) diagram for water is shown schematically in Fig. 7.3 (to scale, but with an arbitrary numerical entropy scale).

Since the First Law of thermodynamics dictates that steam undergoing an expansion process without heat or work transfer must do so at constant enthalpy, saturated steam at a temperature lower than 235°C experiencing an isenthalpic pressure loss will become superheated (process $3 \rightarrow 4$), whereas one at a temperature higher than 235°C will first form a liquid-vapor mixture (process $1 \rightarrow 2$). It is reasonable to assume that the geofluid in the deepest part of the reservoir will be hottest. It will tend to rise because of its lower density and will undergo a pressure drop as it passes through the restrictive passages in the formation. At each step then, a fraction of liquid will separate out of the steam and tend to flow downward due to its higher density, leaving the now dry steam to continue upward. This brings the steam into a yet lower pressure domain and the process repeats itself [5]. In fact, in whatever direction the steam travels, the pressure will continue to fall due to frictional effects.

This reasoning explains well how the rising steam can reach the maximum enthalpy point but does not explain how the steam stabilizes there as a saturated vapor at 235°C. Further pressure reduction, for instance, would mean the steam would become superheated. James [5] points out that during production, the steam would flow at essentially constant temperature through the hot rock and would tend to follow an isothermal process. He estimated that this could result in as much as 35°C superheat.

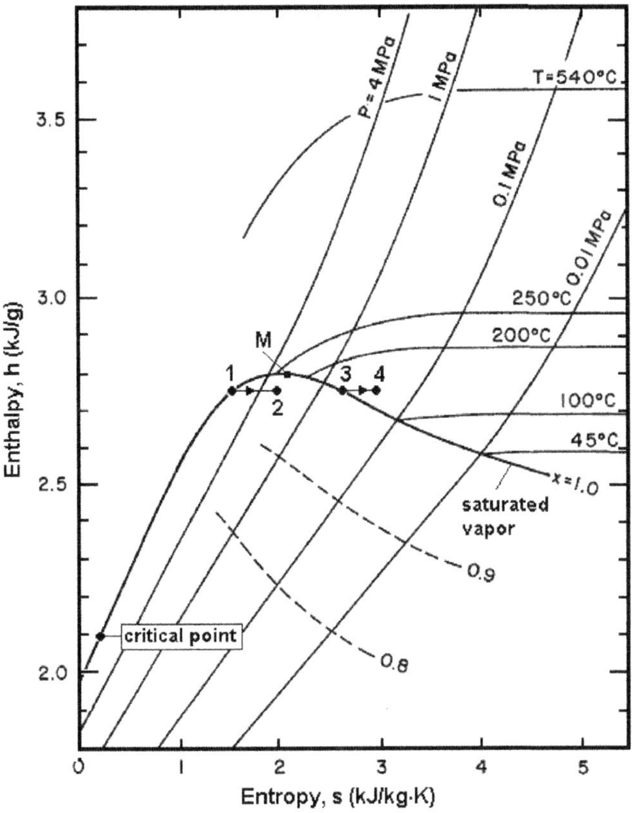

'ig. 7.3 Mollier diagram for water; M = maximum enthalpy point on the saturation line.

McNitt [6] offers a thermodynamic argument to explain this phenomenon based on he amount of heat needed to evaporate a unit of liquid relative to the amount of heat vailable when a steam bubble condenses.

Referring to Fig. 7.4, when a hypothetical bubble forms at level j and begins to nove upward, there are two possibilities: either it will shrink and eventually collapse r it will continue to rise. If it collapses, it would release its heat of condensation to he surrounding liquid. If that amount of heat is sufficient to create another bubble t the higher level, then a continuous steam phase will be created. If not, then the iquid will form the continuous phase. Since this hypothesis calls for the internal xchange of heat between the vapor and liquid phases, the overall process may be iewed as adiabatic, and the process followed by the bubble may be modeled as senthalpic. The lines in the temperature-entropy diagram labeled j-i represent lines f constant enthalpy.

The argument turns on the relative sizes of the following two terms:

$$q_C \equiv h_g(P_j) - h_f(P_i) \tag{7.1}$$

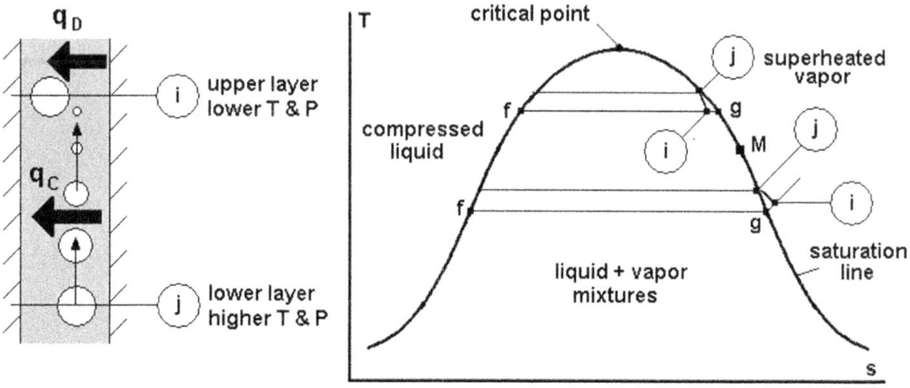

Fig. 7.4 Hypothesis for creation of stable steam zone at 235°C; after [6].

and

$$q_D \equiv h_g(P_i) - h_f(P_i) \tag{7.2}$$

The term defined as q_C is the heat per unit mass delivered by the rising bubble as it condenses in going from state j to state i. The term defined as q_D is the heat per unit mass required to form a bubble at the upper level, at state i. If $q_C > q_D$, then the steam phase will be continuous; if $q_C < q_D$, then the liquid phase will be continuous. But it is obvious that this difference is simply $h_g(P_j) - h_g(P_i)$, which is always negative for initial states with temperatures greater than T_M, the temperature corresponding to the maximum enthalpy point on the saturated vapor line (shown as point M), and which is always positive for initial states with temperatures lower than T_M.

Thus, we see that the steam phase should begin where the reservoir temperature approximates the maximum-enthalpy temperature of 235°C. At depths below this temperature level, we would expect the pressure to vary with depth hydrostatically, but above this level, i.e., throughout the dry-steam reservoir, we would expect roughly constant temperature and pressure. The pressure and temperature gradients through the shallow levels of cap rock including the zone of condensation would be close to the boiling point curve, i.e., a column of water that has a temperature at each point along the column that is equal to the saturation temperature corresponding to its hydrostatic pressure. See Fig. 7.5.

It is clear from Fig. 7.5 that the dry-steam reservoir is characterized by a pressure far below the hydrostatic pressure at the same depth. This pressure deficiency can only be maintained if the permeability of the surrounding formation is very low, effectively isolating the steam field from the influx of liquid from the lateral portions of the greater field. This phenomenon will play a key role when fluid is injected into the formation to prolong the life and productivity of the field. Clearly, the fluid will be readily accepted, in effect, sucked into the reservoir, but the liquid could have the unwanted result of flooding the steam field unless the injection wells are carefully sited and the injection rate is balanced against the rate of steam extraction.

In the first major attempt to restore fluid to a dry-steam field during exploitation, a program of injection was begun in 1989 in the southeast area of The Geysers [14].

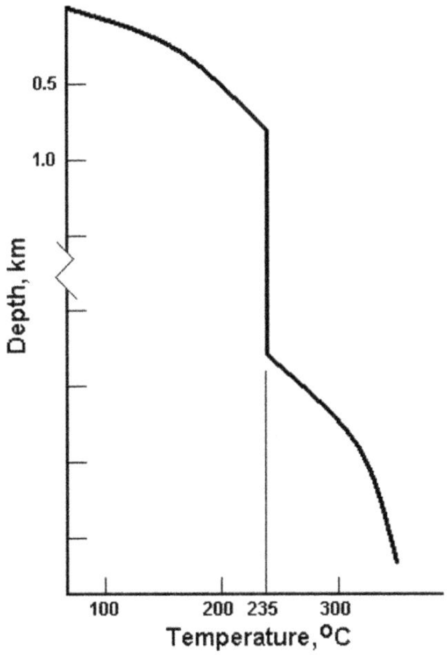

'ig. 7.5 Temperature versus depth schematic for a dry-steam system, after [4].

'he site was selected with the objective of counteracting the observed drying out f the reservoir in a region still marked by high temperature. Significant amounts of uperheat, up to 80°F (44°C), had been seen in the output of wells in the southeast rea of the field but the reservoir temperature had remained constant at about 450°F 232°C). The experiment showed that the injection resulted in an increase in steam low from the production wells, equivalent to about 20 MW of additional power, and et the stage for larger programs of injection involving wastewater from neighboring ommunities in Lake County [15] and Santa Rosa [16]. This will be discussed further n Sect. 12.6.

In summary, a dry-steam reservoir owes its existence to a combination of highly ortuitous circumstances. First there must be heat source relatively close to the surace (c. 5 km depth) to raise the temperature of connate water to the boiling point. 'here must be sufficient permeability above the reservoir to allow steam to escape to he surface over geologically long periods of time, thereby lowering the liquid level lramatically. There must be sufficient interconnectedness of fissures and fractures vithin the reservoir to allow fluid to circulate inside the reservoir. There must be ufficient lateral impermeability between the reservoir and the surrounding rock to revent flooding of the steam reservoir by cooler groundwaters. And finally, the ippermost levels of the formation need to become largely impermeable via the mechinism of self-sealing caused by mineral precipitation. The difference between the 1uge dry-steam reservoirs at Larderello and The Geysers and other dry-steam "caps" ies in the extent to which these conditions are present and how long the processes 1ave been in progress.

7.3 Steam gathering system

The connection between the wells and the powerhouse for a dry-steam plant is relatively simple compared to a flash-steam plant. At the well, there are the usual valves plus a steam purifier. The latter is merely an in-line, axial centrifugal separator designed to remove particulate matter from the steam before it enters the piping system. The steam pipes are covered with insulation and mounted on stanchions, and include expansion loops to accommodate pipe movement from a cold to a hot condition. Steam traps are sited strategically along the pipes to remove condensate which is then conveyed by separate lines to holding ponds and eventual reinjection. See Figs. 7.6 and 7.7.

As the steam line approaches the powerhouse, there is an emergency pressure relief station. This allows for the temporary release of steam in the event of a turbine trip. The steam generally passes through a silencer before entering the atmosphere. It has been found preferable to maintain the wells in a steady open mode rather than cycling the wells through open and closed positions. At the powerhouse one finds a steam header, a final moisture remover, typically a vertical cyclone separator or a baffled demister, and a venturi meter for accurate measurement of the steam flow rate. See Figs. 7.8 and 7.9.

7.4 Energy conversion system

Once the steam reaches the powerhouse, a dry-steam plant is essentially the same as a single-flash steam plant. The turbines are single-pressure units with impulse-reaction blading, either single-flow for smaller units or double-flow for larger units (say,

Fig. 7.6 Wellhead pipelines at The Geysers. Note axial separators to remove particulate matter. Several wells are sited on a single drilling pad to reduce the amount of land needed for well pads. The wells are drilled directionally to intercept a large reservoir volume. Photo courtesy of Calpine Corporation [17] [WWW].

'ig. 7.7 *Steam pipeline and steam trap with condensate drain line at The Geysers. Photo by author [WWW].*

'ig. 7.8 *Emergency steam relief valves at The Geysers. Photo by author [WWW].*

ɔ0 MW or greater). The condensers can be either direct-contact (barometric or ow-level) or surface-type (shell-and-tube). For small units, it is often advantageous o arrange the turbine and condenser side-by-side, rather than the more usual condenser-elow-turbine arrangement seen in most power plants.

Fig. 7.9 Rock mufflers for emergency steam stacking at The Geysers. Photo by author [WWW].

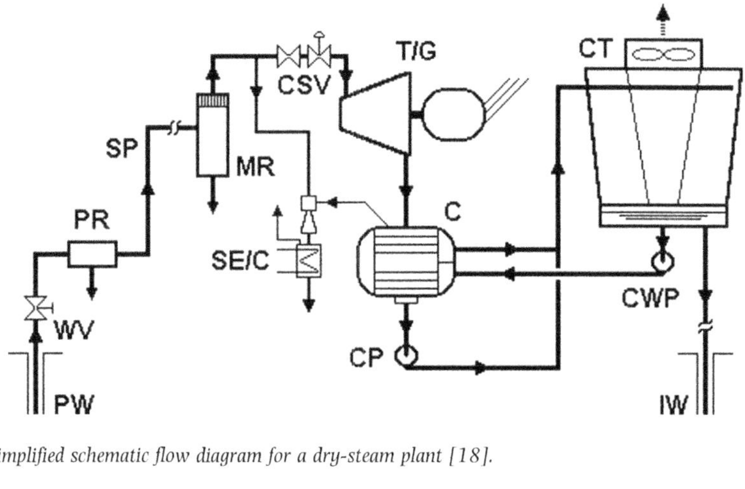

Fig. 7.10 Simplified schematic flow diagram for a dry-steam plant [18].

With the exception of the particulate remover (PR) in place of the cyclone separator (CS), the flow diagram shown in Fig. 7.10 is identical to Fig. 5.6 for a single-flash plant [18].

The processes undergone by the steam are shown in Fig. 7.11. Since the wells produce saturated steam (or slightly superheated steam), the starting point (state 1) is located on the saturated vapor curve. If the steam is superheated, point 1 merely moves slightly to the right. The turbine expansion process 1–2 generates somewhat

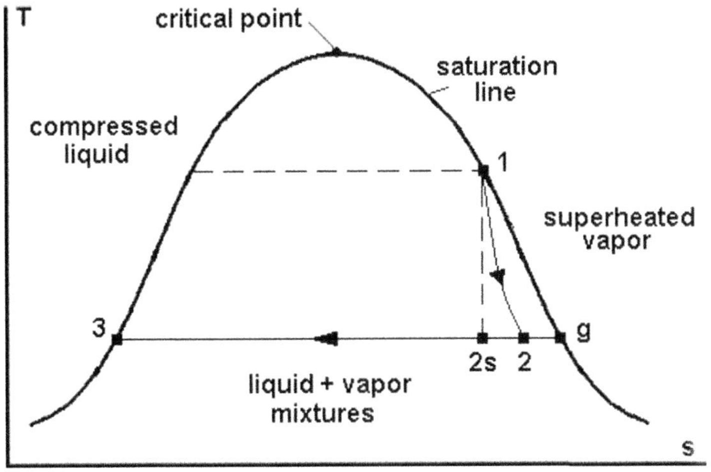

'ig. 7.11 *Temperature-entropy diagram for a dry-steam plant with saturated steam at the turbine inlet.*

ess power output than the ideal, isentropic process 1–2s. Heat is rejected to the urroundings in the condenser via the cooling water in process 2–3. Although the anal-'sis of the operation is the same as for the single-flash plant, we repeat the equations ere for ease of reference.

'.4.1 Turbine expansion process

'he work produced by the turbine per unit mass of steam flowing through it is given by

$$w_t - h_1 - h_2 \tag{7.3}$$

ssuming an adiabatic turbine and neglecting the changes in kinetic and potential nergy of the fluid entering and leaving the turbine. The maximum possible work would e generated if the turbine operated adiabatically and reversibly, i.e., at constant entropy r isentropically.

The isentropic turbine efficiency, η_t, is the ratio of the actual work to the isentropic vork, namely,

$$\eta_t = \frac{h_1 - h_2}{h_1 - h_{2s}} \tag{7.4}$$

'he power developed by the turbine is given by

$$\dot{W}_t = \dot{m}_s w_t = \dot{m}_s(h_1 - h_2) = \dot{m}_s \eta_t (h_1 - h_{2s}) \tag{7.5}$$

'he gross electrical power will be equal to the turbine power times the generator fficiency:

$$\dot{W}_e = \eta_g \dot{W}_t \tag{7.6}$$

'he net power is this amount reduced by all parasitic loads including condensate umping power, cooling tower fan power, and station lighting.

Adopting the Baumann rule to account for the degradation in performance of a wet steam turbine, we find

$$\eta_{tw} = \eta_{td} \times \left[\frac{1 + x_2}{2}\right] \tag{7.7}$$

where the dry turbine efficiency, η_{td}, may conservatively be taken to be constant at 85%:

$$\eta_{td} = 0.850 \tag{7.8}$$

The thermodynamics properties at state 2 are determined by solving eq. (7.4) using the turbine efficiency and the fluid properties at state 2s, the ideal turbine outlet state which are easily calculated from the known pressure and entropy values at state 2s. The ideal enthalpy is found from

$$h_{2s} = h_3 + \left[h_g - h_3\right] \times \left[\frac{s_1 - s_3}{s_g - s_3}\right] \tag{7.9}$$

where the entropy term gives the fluid outlet dryness fraction for an ideal turbine. When the Baumann rule is incorporated into the calculation, the following working equation emerges for the enthalpy at the actual turbine outlet state:

$$h_2 = \frac{h_1 - A\left[1 - \dfrac{h_3}{h_g - h_3}\right]}{1 + \dfrac{A}{h_g - h_3}} \tag{7.10}$$

where the factor A is defined as

$$A \equiv 0.425\,(h_1 - h_{2s}) \tag{7.11}$$

These equations assume that the turbine inlet steam is saturated. If the inlet is superheated (as often happens after a period of operation), then a more complex algorithm must be followed. First, the portion of the expansion process that occurs in the superheated region is analyzed using the dry turbine expansion efficiency of 85%. Then when the steam enters the wet region, the remaining expansion is analyzed using the same equations given above.

The location of the state point where the expansion passes through the saturated vapor curve is found by trial-and-error. The method for doing the calculation is outlined below with the aid of Fig. 7.12. Three isobars are shown: P_1 is the inlet steam pressure, P_2 is the condenser pressure, and P_4 is the pressure at which the expansion enters the wet region (unknown). Since we assume that the dry turbine efficiency is known and constant for the process from 1–4, we may write

$$\eta_d = \frac{h_1 - h_4}{h_1 - h_{4s}} \tag{7.12}$$

Both h_4 and h_{4s} depend on P_4 which is what we are trying to determine. The first step is to guess a value for P_4. Then h_4 is found directly from *Steam Tables*, and h_{4s} may be calculated from

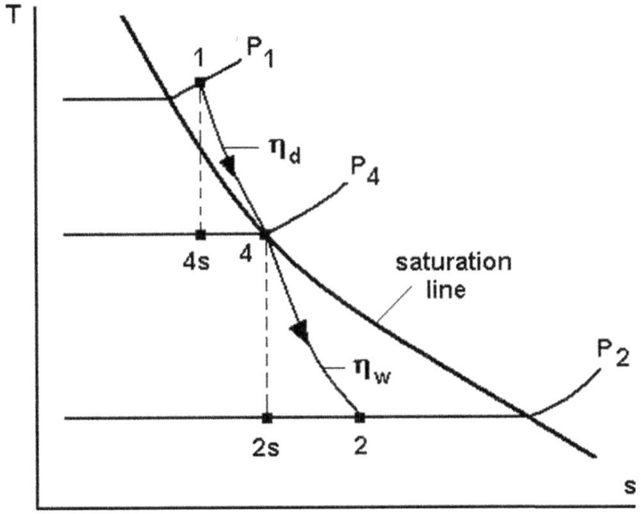

ig. 7.12 *Dry and wet turbine expansion processes for superheated inlet steam.*

$$h_{4s} = h_f + x_{4s}h_{fg} = h_f + \left[\frac{s_1 - s_f}{s_{fg}}\right]h_{fg} \tag{7.13}$$

vhere the enthalpy of saturated liquid and the enthalpy of evaporation at P_4 are given y h_f and h_{fg}, respectively. With h_4 and h_{4s} now both determined for the assumed value f P_4, the turbine efficiency is calculated from eq. (7.12) and compared to 0.85. Unless he calculated efficiency is equal to 0.85, it is necessary to try another pressure and epeat the calculations until agreement is reached. Then the specific power output rom the dry expansion is

$$w_{td} = h_1 - h_4 \tag{7.14}$$

nd the specific power output from the wet expansion is

$$w_{tw} = h_4 - h_2 \tag{7.15}$$

vhere a process similar to that outlined earlier will lead to the value of h_2.

As with all thermodynamic modeling of complex processes, this must be viewed s an approximation. It is known that steam can enter the wet region and remain n the vapor phase, out of thermodynamic equilibrium, for a portion of the full process. The locus of states where the first liquid droplets appear is called the Wilson ine. It is generally accepted that the line of 95% dryness fraction is the limit for team to persist in a state of metastable equilibrium [19]. Given this uncertainty ogether with the fact that the dry expansion in a dry-steam turbine is a small art of the total expansion, it is often acceptable to treat the entire process as taking lace in the wet region and to use the analysis given at the beginning of this section, .e., eqs. (7.3–7.11).

7.4.2 Condensing and cooling tower processes; utilization efficiency

The working equations for the condenser, cooling tower, and the utilization efficiency are the same as for the flash-steam plants already considered and will not be repeated. The reader is referred to Sects. 5.4.5–5.4.7.

7.5 Example: Optimum wellhead pressure

This example will treat the problem of deciding on the optimum wellhead pressure for a dry-steam plant receiving saturated vapor at the wellhead. As with the previous examples, we will ignore pressure losses in pipelines. We will assume that we can control the pressure at the wellhead by means of a throttle valve. The well productivity curve can be approximated as an elliptical equation in terms of the mass flow rate of steam as a function of the wellhead pressure:

$$\left[\frac{\dot{m}}{\dot{m}_{max}}\right]^2 + \left[\frac{P}{P_{ci}}\right]^2 = 1 \tag{7.16}$$

where $\dot{m}_{max}$ is the maximum observed mass flow rate and P_{ci} is the closed-in wellhead pressure. This function is shown schematically in Fig. 7.13. Assuming that values for these two parameters are available from well tests, the mass flow rate at any pressure can be calculated from

$$\dot{m} = \dot{m}_{max}\sqrt{1 - (P/P_{ci})^2} \tag{7.17}$$

Opening the wellhead valve will result in lower pressures, and higher flow rates, but the enthalpy of the steam will remain the same since it is a throttling process.

The effect of this operation can be seen in Fig. 7.14, a Mollier diagram for steam.

The turbine power is proportional to the product of the steam mass flow rate and the enthalpy drop Δh (shown for an ideal isentropic process for simplicity). There are two limits to the wellhead pressure: the closed-in pressure, P_{ci}, for which there is no steam flow, and the condenser pressure, P_c, for which there is no

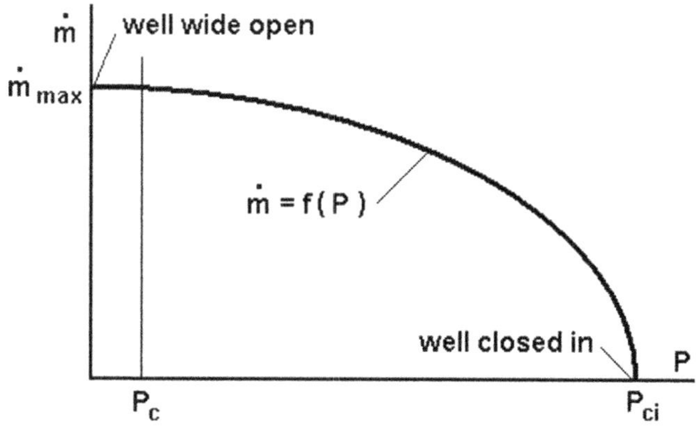

Fig. 7.13 Dry-steam productivity curve.

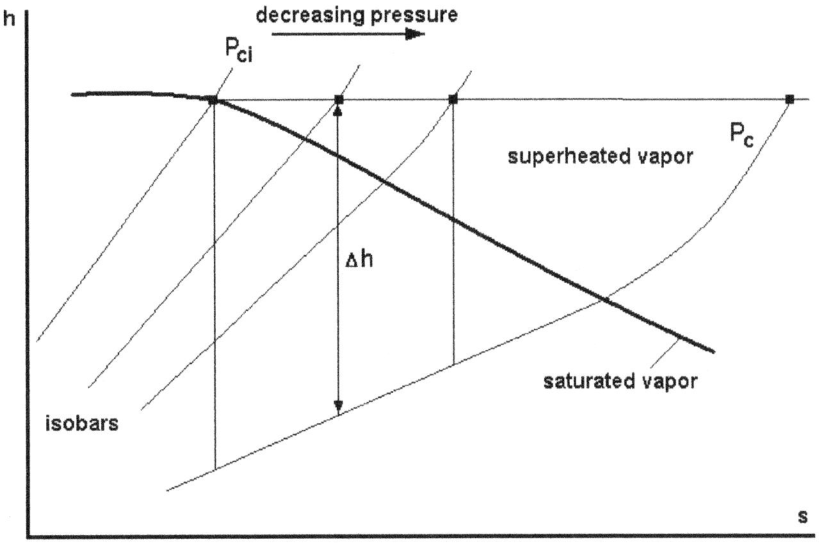

Fig. 7.14 *Ideal turbine expansion processes for various wellhead pressures.*

enthalpy drop. Thus the power vanishes at the two extreme positions for the wellhead pressure, indicating there is some wellhead pressure for which the power will be a maximum.

It is convenient computationally to solve for the power output per maximum steam flow rate:

$$\frac{\dot{W}}{\dot{m}_{max}} = \frac{\dot{W}}{\dot{m}} \times \frac{\dot{m}}{\dot{m}_{max}} = (h_1 - h_2) \times \sqrt{1 - (P/P_{ci})^2} \tag{7.18}$$

where $(h_1 - h_2)$ is the isentropic enthalpy drop across the turbine (shown as Δh in Fig. 7.14). It is not hard to solve this problem for a non-isentropic turbine; see Problem 7.1.

For this illustration, we have selected the following parameters to define the problem (note – all pressures are absolute):

closed-in pressure, $P_{ci} = 300$ lbf/in^2; $P_2 = P_c = 2$ lbf/in^2; $h_1 = 1200.4$ Btu/lbm.

The results of the calculations are shown in Table 7.1. These were obtained graphically using the very large scale Mollier diagram that accompanies the *Steam Tables* [13].

A graph of the power output per maximum steam flow is shown in Fig. 7.15. It can be seen that the optimum operating point occurs at a wellhead pressure of 127 lbf/in^2. At that setting, the plant can produce a gross power of 247.4 Btu/lbm.

This optimum output is equivalent to 72.5 kW per maximum steam flow in 10^3 lbm/h. Thus, for a well capable of a maximum steam flow of 200×10^3 lbm/h, the optimum turbine power generation would be 14.5 MW. Recall that the turbine is assumed to be ideal; for a realistic turbine, this would drop to about 11.5 MW.

Table 7.1 Results of optimization calculations.

P	h_1	h_2	$h_1 - h_2$	p/p_{ci}	$\dot{W}/\dot{m}_{max}$
50	1200.4	985.0	215.4	0.1667	212.39
60	1200.4	973.5	226.9	0.2000	222.32
70	1200.4	964.0	236.4	0.2333	229.87
80	1200.4	956.0	244.4	0.2667	235.55
90	1200.4	949.5	250.9	0.3000	239.34
100	1200.4	942.0	258.4	0.3333	243.62
110	1200.4	936.0	264.4	0.3667	245.99
120	1200.4	930.8	269.6	0.4000	247.09
130	1200.4	926.0	274.4	0.4333	247.30
140	1200.4	922.0	278.4	0.4667	246.23
160	1200.4	913.6	286.8	0.5333	242.61
180	1200.4	906.8	293.6	0.6000	234.88
200	1200.4	900.0	300.4	0.6667	223.90
220	1200.4	895.0	305.4	0.7333	207.63
240	1200.4	890.6	309.8	0.8000	185.88
260	1200.4	885.0	315.4	0.8667	157.35
280	1200.4	881.2	319.2	0.9333	114.60
300	1200.4	877.0	323.4	1.0000	0.00

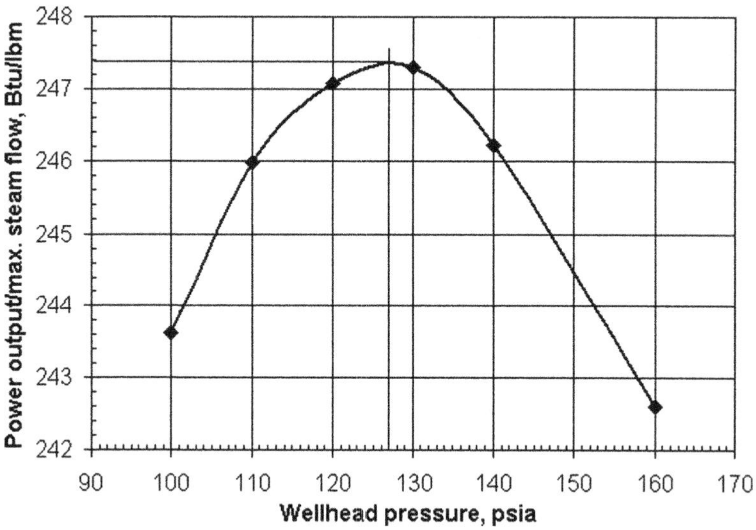

Fig. 7.15 Power output per maximum steam flow rate in the neighborhood of the optimum point.

It is also noteworthy that the curve is relatively flat near the optimum point. Over the range of wellhead pressures from 118–134 lbf/in^2, the power output is within 0.2% of the optimum value. This offers the designer a wide latitude of options without sacrificing much in the way of power generation.

7.6 Environmental aspects of dry-steam plants

Dry-steam geothermal plants have very low potential impact on the environment. The geofluid consists of only steam – no liquid – so there is no mineral-laden brine to dispose of. The noncondensable gases in the steam are isolated in the condenser and removed by means of vacuum pumps or steam-jet ejectors, and they can be treated to remove hydrogen sulfide, if it is present in objectionable levels. The sulfur from certain types of abatement systems is in pure form and may be sold commercially or disposed of in an appropriate landfill [20]. The excess condensate from the cooling tower is reinjected as is any liquid trapped from the steam transmission pipelines.

7.7 Equipment list for dry-steam plants

The major equipment items found at dry-steam plants are quite similar to those used at a single-flash plant. The list below is drawn for dry-steam plants with annotations applied only for items that differ from those at single-flash plants [18].

7.7.1 Steam supply system

- Wellhead valves and controls
- Steam purifier
  Wellhead, axial particulate remover
- Steam piping, insulation and supports
- Steam header
- Final moisture remover

7.7.2 Turbine-generator and controls

- Steam turbine-generator with accessories
- Control system
- Air compressor

7.7.3 Condenser, gas ejection and pollution control (where needed)

- Condenser
- Condensate pumps and motors
- Gas removal system
  Steam jet ejectors with inter- and after-coolers
  Turbocompressors
  Hybrid ejector/compressor
- NCG treatment system

7.7.4 Heat rejection system

- Water cooling tower
- Cooling water pumps and motors
- Cooling water treatment system

7.7.5 Back-up systems

- Standby power supply

7.7.6 Noise abatement system (where required)

- Rock mufflers for stacked steam
- Acoustic insulation for noisy fluid handling components

7.7.7 Condensate disposal system

- Injection wells for excess condensate, and cooling tower blowdown

References

[1] Photograph courtesy of Enel S.p.A. – Museo della Geotermia Larderello – Archivio Fotografico, November 2004.
[2] White, D.E., "Characteristics of Geothermal Resources," Chap. 4 in *Geothermal Energy: Resources, Production, Stimulation*, P. Kruger and C. Otte, Eds., Stanford University Press, Stanford, CA, 1973, pp. 69–94.
[3] Chierici, A., "Planning of a Geothermoelectric Power Plant: Technical and Economic Principles," *Proc. U.N. Conference on New Sources of Energy*, V. 3, Geothermal Energy: II, 1964, pp. 299–311.
[4] Ellis, A.J. and W.A.J. Mahon, *Chemistry and Geothermal Systems*, Academic Press, New York, 1977.
[5] James, R., "Wairakei and Larderello: Geothermal Power Systems Compared," *N. Z. J. Sci.*, V.11, 1968, pp. 706–719.
[6] McNitt, J.R., "Origin of Steam in Geothermal Reservoirs," *Proc. Annual Fall Tech. Conf. and Exhib. Soc. Petroleum Engineers*, AIME, 1977, Paper SPE 6764.
[7] D'Amore, F. and A.H. Truesdell, "Models for Steam Chemistry at Larderello and The Geysers," *Proc. Fifth Workshop Geothermal Reservoir Engineering*, Stanford University, Stanford, CA, 1979, pp. 262–276.
[8] Truesdell, A.H., W.T. Box, Jr., J.R. Haizlip and F. D'Amore, "A Geochemical Overview of the Geysers Geothermal Reservoir," *Monograph on The Geysers Geothermal Field*, Spec. Rep. 17, C. Stone, Ed., Geothermal Resources Council, Davis, CA, 1992, pp. 121–132.
[9] White, D.E., L.J.P. Muffler and A.H. Truesdell, "Vapor-Dominated Hydrothermal Systems Compared with Hot-Water Systems," *Econ. Geol.*, V.66, 1971, pp. 75–97.
[10] Allen, E.T. and A.L. Day, *Steam Wells and Other Thermal Activity at The Geysers, California*, Publ. No. 378, Carnegie Institution of Washington, DC, 1927.
[11] Hodgson, S.F., "A Geysers Album," *Monograph on The Geysers Geothermal Field*, Spec. Rep. 17, C. Stone, Ed., Geothermal Resources Council, Davis, CA, 1992, pp. 19–40.
[12] Burgassi, P.D., "Historical Outline of Geothermal Technology in the Larderello Region to the Middle of the 20th Century," Chap. 13 in *Stories from a Heated Earth: Our Geothermal Heritage*, R. Cataldi, S.F. Hodgson and J.W. Lund, Eds., Geothermal Resources Council and International Geothermal Association, Sacramento, CA, 1999, pp. 195–219.
[13] Keenan, J.H., F.G Keyes, P.G. Hill and J.G. Moore, *Steam Tables: Thermodynamic Properties of Water Including Vapor, Liquid, and Solid Phases (International Edition – Metric Units)*, John Wiley & Sons, Inc., New York, 1969.
[14] Enedy, S.L., K.L. Enedy and J. Maney, "Reservoir Response to Injection in the Southeast Geysers," *Monograph on The Geysers Geothermal Field*, Spec. Rep. 17, C. Stone, Ed., Geothermal Resources Council, Davis, CA, 1992, pp. 211–219.
[15] Dellinger, M., "Lake County Success: Generating Environmental Gains with Geothermal Power," *Geothermal Resources Council BULLETIN*, V. 33, No. 3, 2004, pp. 115–119.
[16] Anon., "Recharging The Geysers: Calpine Corp. and Santa Rosa Celebrate Completion of the World's Largest Wastewater-to-Energy Project," *Geothermal Resources Council BULLETIN*, V. 32, No. 6, 2003, p. 242.

[17] Personal communication, Michetti, D., P.R. Communications Specialist, Calpine Corporation, San Jose, CA, Dec. 27, 2004.

[18] DiPippo, R., "Geothermal Power Systems," Sect. 8.2 in *Standard Handbook of Powerplant Engineering, 2nd ed.*, T.C. Elliott, K. Chen and R.C. Swanekamp, Eds., McGraw-Hill, Inc., New York, 1998. pp. 8.27–8.60

[19] IAPWS, *Guideline on the Tabular Taylor Series Expansion (TTSE) Method for Calculation of Thermodynamic Properties of Water and Steam Applied to IAPWS-95 as an Example*, International Association for the Properties of Water and Steam, Vejle, Denmark, August 2003.

[20] DiPippo, R., *Geothermal Energy as a Source of Electricity: A Worldwide Survey of the Design and Operation of Geothermal Power Plants*, U.S. Dept. of Energy, DOE/RA/28320-1, U.S. Gov. Printing Office, Washington, DC, 1980.

Nomenclature for figures in Chapter 7

C	Condenser
CP	Condensate pump
CSV	Control & stop valves
CT	Cooling tower
CWP	Cooling water pump
	Saturated liquid
g	Saturated vapor
IW	Injection well
M	Point of maximum enthalpy for saturated steam
MR	Moisture remover
PR	Particulate remover
PW	Production well
q_C	Heat of condensation from a steam bubble
q_D	Heat of evaporation to form a steam bubble
SE/C	Steam ejector/condenser
SP	Steam piping
T/G	Turbine/generator
WV	Wellhead valve

Problems

7.1 Consider the dry-steam field at The Geysers in California. The following data are available: closed-in pressure = 225 lbf/in^2; closed-in enthalpy = 1200 Btu/lbm; enthalpy = constant for all settings of the wellhead valve; turbine is *not* isentropic, but is characterized by a wet-turbine efficiency, η_{tw}, given by the Baumann rule; condenser pressure = 4 in Hg ($\approx$2 lbf/in^2).

(a) Find the optimum wellhead pressure for a typical well.

(b) Assuming all the wells behave alike, determine the number of wells needed to power a 110 MW unit if the *maximum* well flow rate is 200,000 lbm/h, and the wells are set at: (i) the optimum wellhead pressure, or (ii) a pressure 30% greater than the optimum pressure.

7.2 Unit No. 6 at The Geysers produces 52 MW of power at the generator from a steam flow of 417,000 kg/h at a pressure of 6.9 bar and a temperature

of 182°C. The reservoir conditions are known to be 32.3 bar and 238°C. The turbine exhaust pressure (condenser pressure) is 0.13 bar.
(a) Sketch the location of the reservoir and turbine-inlet state points on a Mollier chart (h-s); the states should be placed on a *scale-drawn* portion of the h-s diagram. What do you observe about the enthalpy of these two states? What do you notice about these values relative to the maximum enthalpy along the saturated vapor line?
(b) On the same diagram, locate the turbine exhaust state. Calculate the turbine isentropic expansion efficiency, η_t. Calculate the overall geothermal energy utilization efficiency, η_u, based on (i) reservoir and (ii) turbine-inlet conditions. A sink temperature of 48°C is appropriate for Unit No. 6.

7.3 A dry-steam geothermal plant receives saturated steam at a temperature of 180°C. The condenser operates at 50°C and the local dead-state temperature is 20°C. The turbine obeys the Baumann rule; assume 85% for the dry expansion efficiency. Calculate the following:
(a) The actual turbine isentropic efficiency.
(b) The specific work output of the turbine in kJ/kg.
(c) The net utilization efficiency if 5% of the turbine output is needed for all parasitic loads.

7.4 A dry-steam reservoir produces saturated vapor at the maximum enthalpy point. By adjustment of the wellhead valve, the steam is throttled to lower pressures. A pressure $P_1 = 0.800$ MPa is chosen as the wellhead and turbine inlet pressure (ignore pressure losses in steam piping). At this point the well produces 15 kg/s. The turbine exhausts to a condenser at a pressure $P_2 = 10$ kPa.
(a) Find the turbine inlet temperature in °C.
(b) Calculate the specific work output of the turbine in kJ/kg.
(c) Calculate the actual turbine isentropic efficiency.
(d) Calculate the gross power output for the turbine in MW.

7.5 A dry-steam well is characterized by a closed-in pressure of $P_1 = 360$ lbf/in^2, a and saturated vapor. The wellhead valve is set at $P_2 = 150$ lbf/in^2, a at which condition the well flows 200,000 lbm/h. You may assume the turbine is adiabatic and has a constant isentropic efficiency of 75%.
(a) Calculate the power in MW that the turbine will generate under the following two cases: (i) there is no condenser and the turbine exhausts to the atmosphere at 1 atm, and (ii) there is a condenser that has a pressure of 4 in Hg.
(b) Calculate the utilization efficiency in both cases, based on the closed-in condition of the geofluid.

Chapter 8

Binary Cycle Power Plants

Geothermal Power Plants: Principles, Applications, Case Studies and Environmental Impact, Third Edition.
© 2012 Elsevier Ltd. All rights reserved.

8.1 Introduction

Binary cycle geothermal power plants are the closest in thermodynamic principle to conventional fossil or nuclear plants in that the working fluid undergoes an actual closed cycle. The working fluid, chosen for its appropriate thermodynamic properties receives heat from the geofluid, evaporates, expands through a prime-mover, condenses, and is returned to the evaporator by means of a feed pump.

Although it is generally believed that the first geothermal binary power plant was put into operation at Paratunka near the city of Petropavlovsk on Russia's Kamchatka peninsula in 1967 [1], there is evidence that an earlier binary plant existed in the Democratic Republic of the Congo in Africa. A small 200 kW binary unit was installed in 1952 at Kiabukwa about 18 km west of the city of Kamina in southern Katanga province; see Fig. 8.0 [1a]. The site is some 300 km west of the East African Rift System [1b]. Hot water at a temperature of 91°C and a flow rate of 40 kg/s from a geothermal spring in the Upemba graben fed the plant. During its few years of operation, it supplied power to a mining company in the north of Katanga province [1c]. However, owing to lack of maintenance, the plant fell into disuse, was vandalized, and today no trace remains of it [1d]. Other than the brief description and photo in Ref. [1a], there is scant mention of the plant in the literature. There is interest in redeveloping the site.

The Paratunka plant in Russia was rated at 670 kW and served a small village and some farms with both electricity and heat for use in greenhouses. It ran successfully for many years, proving the concept of binary plants as we know them today.

At the birth of the commercial geothermal power age in 1912 at Larderello, Italy, a so-called "indirect cycle" was adopted for a 250 kW plant; this was in effect a "binary"

Fig. 8.0 Geothermal binary plant at Kiabukwa, Democratic Republic of the Congo (best available copy) [1a].

plant. The geothermal steam from wells was too contaminated with dissolved gases and minerals to be sent directly to a steam turbine so it was passed through a heat exchanger where it boiled clean water that then drove the turbine. This allowed the use of standard materials for the turbine components and permitted the minerals to be recovered from the steam condensate [2].

Today binary plants are the most widely used type of geothermal power plant with 235 units in operation in August 2011, generating over 708 MW of power in 15 countries. They constitute 40% of all geothermal units in operation but generate only 6.6% of the total power. Thus, the average power rating per unit is small, only 3 MW/unit, but units with ratings of up to 20–21 MW are coming into use with advanced cycle design using a pair of turbines driving a single generator. Several binary units recently have been added to existing flash-steam plants to recover more power from hot, waste brine. See Appendix A for more statistics.

8.2 Basic binary systems

If one were to plot a histogram of geothermal resources worldwide arranged by temperature, it would be heavily skewed toward low-temperature resources. If the geofluid temperature is 150°C (300°F) or less, it becomes difficult, although not impossible, to build a flash-steam plant that can efficiently and economically put such a resource to use. The lower the resource temperature the worse the problem becomes for flash technology. Indeed at such low temperatures it is unlikely that the wells will flow spontaneously, and if they do, there is a strong likelihood of calcium carbonate scaling in the wells.

One way to prevent the scaling problem is to produce the geofluid as a pressurized liquid by means of down-well pumps. When geofluids are produced this way, it is generally not thermodynamically wise to then flash the fluid in surface vessels and use a flash-steam plant. However, there is one plant that does so, the GEM plant at East Mesa in the Imperial Valley of California in the United States [3]. It is simpler to pass the geofluid as a compressed liquid through heat exchangers and dispose of it in injection wells still in the liquid phase. The thermodynamic irreversibilities associated with the flash process are replaced with irreversibilities from heat transfer across a finite temperature difference. With imaginative design of the heat exchangers, these losses can be minimized, as we will see later.

In its simplest form, a binary plant follows the schematic flow diagram given in Fig. 8.1. The production wells PW are fitted with pumps P that are set below the flash depth determined by the reservoir properties and the desired flow rate. Sand removers SR may be needed to prevent scouring and erosion of the piping and heat exchanger tubes. Typically there are two steps in the heating-boiling process, conducted in the preheater PH where the working fluid is brought to its boiling point and in the evaporator E from which it emerges as a saturated vapor. The geofluid is everywhere kept at a pressure above its flash point for the fluid temperature to prevent the breakout of steam and noncondensable gases that could lead to calcite scaling in the piping. Furthermore, the fluid temperature is not allowed to drop to the point where silica scaling could become an issue in the preheater and in the piping and injection wells

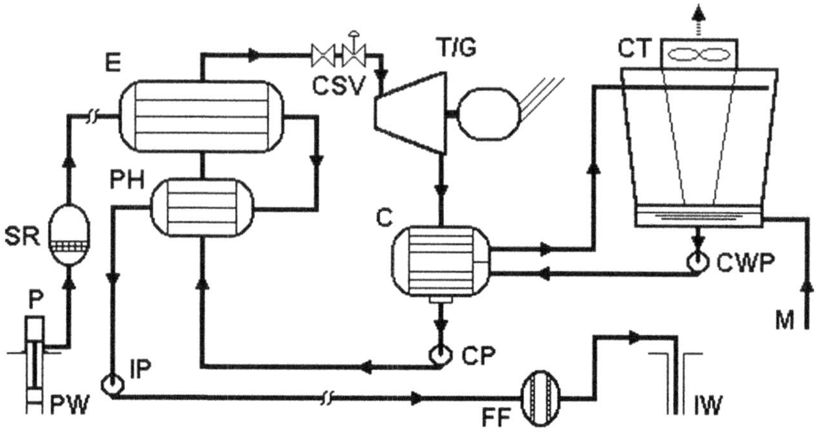

Fig. 8.1 Simplified schematic of a basic binary geothermal power plant [4].

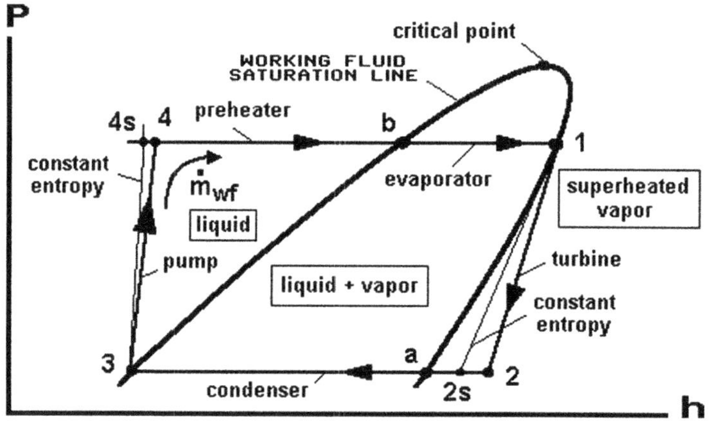

Fig. 8.2 Pressure-enthalpy diagram for a basic binary plant.

downstream of it. Thus the chemical problems described in Chaps. 4 and 6 can be eliminated in principle.

The thermodynamic processes undergone by the working fluid are shown in Fig. 8.2, a pressure-enthalpy, *P-h* diagram. This type of diagram is most often used for refrigeration and air conditioning cycles, but lends itself very well to geothermal binary cycles.

8.2.1 Turbine analysis

The thermodynamic analysis of the cycle is fairly straightforward. Beginning with the binary turbine, we find that the analysis is the same as for steam turbines. It will be useful to select the components from the system flow diagram for easy reference as we present the analysis; see Fig. 8.3.

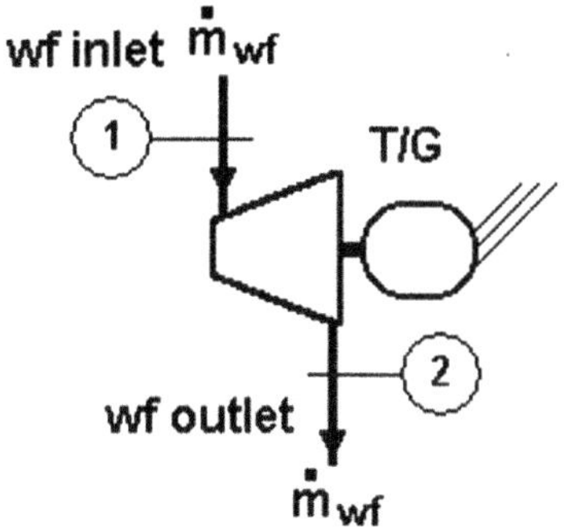

Fig. 8.3 Turbine-generator for binary cycle.

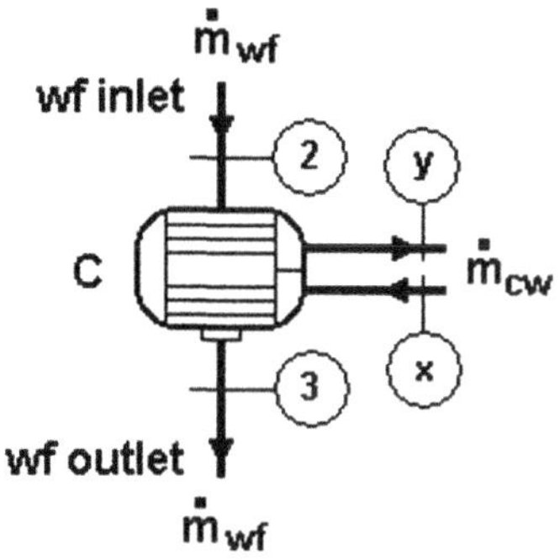

Fig. 8.4 Condenser with cooling water.

With the usual assumptions of negligible potential and kinetic energy terms together with steady, adiabatic operation, the power is found from

$$\dot{W}_t = \dot{m}_{wf}(h_1 - h_2) = \dot{m}_{wf}\eta_t(h_1 - h_{2s}) \tag{8.1}$$

where η_t is the isentropic turbine efficiency, which is a known quantity. For a given working fluid, the thermodynamic properties can easily be found from tables or correlations for whatever design parameters are chosen. The desired turbine power output will then determine the required working fluid mass flow rate.

8.2.2 Condenser analysis

Again the basic working equation is the same as for condensers in flash or dry-steam plants; see Fig. 8.4.

The heat that must be rejected from the working fluid to the cooling medium, be it water (shown here) or air, is given by

$$\dot{Q}_c = \dot{m}_{wf}(h_2 - h_3) \tag{8.2}$$

The relationship between the flow rates of the working fluid and the cooling water is

$$\dot{m}_{cw}(h_y - h_x) = \dot{m}_{wf}(h_2 - h_3) \tag{8.3}$$

or

$$\dot{m}_{cw}\bar{c}(T_y - T_x) = \dot{m}_{wf}(h_2 - h_3) \tag{8.4}$$

since the cooling water may be taken as having a constant specific heat $\bar{c}$ for the small temperature range from inlet to outlet. To dissipate the required amount of waste heat, a cooling tower with a specified range, $T_y - T_x$, will need a mass flow rate determined by eq. (8.4).

8.2.3 Feed pump analysis

Using the same kind of assumptions as for the other components, the power imparted to the working fluid from the feed pump (see Fig. 8.5) is

$$\dot{W}_p = \dot{m}_{wf}(h_4 - h_3) = \dot{m}_{wf}(h_{4s} - h_3)/\eta_p \tag{8.5}$$

where η_p is the isentropic pump efficiency.

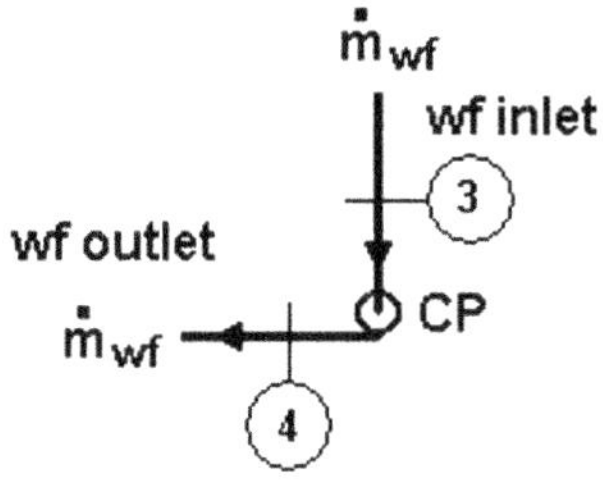

Fig. 8.5 Feed pump for condensate.

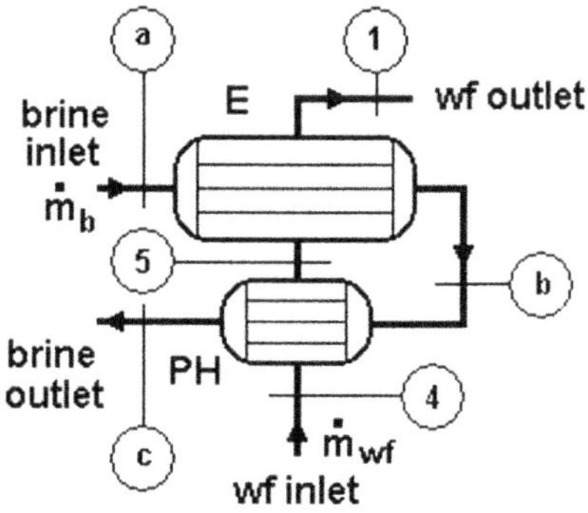

Fig. 8.6 Preheater and evaporator.

8.2.4 Heat exchanger analysis: Preheater and evaporator

The analysis of the heat exchanger where the geothermal brine transfers some of its thermal energy to the working fluid is another straightforward application of the principles of thermodynamics and mass conservation; see Fig. 8.6.

We assume that the heat exchangers are well-insulated so that all the heat transfer is between the brine and the working fluid. In keeping with our overall assumptions, we also assume that the flow is steady, and that the differences in entering and leaving potential energy and kinetic energy are negligible. Considering the entire package as the thermodynamic system, the governing equation is

$$\dot{m}_b(h_a - h_c) = \dot{m}_{wf}(h_1 - h_4) \tag{8.6}$$

If the brine has low dissolved gases and solids, the left-hand side of the equation may be replaced by the average specific heat of the brine $\bar{c}_b$ times the temperature drop:

$$\dot{m}_b\bar{c}_b(T_a - T_c) = \dot{m}_{wf}(h_1 - h_4) \tag{8.7}$$

The following equation may be used to find the required brine flow rate for a given set of cycle design parameters:

$$\dot{m}_b = \dot{m}_{wf}\frac{h_1 - h_4}{\bar{c}_b(T_a - T_c)} \tag{8.8}$$

The design of the individual heat exchangers requires us to examine another thermodynamic diagram: the temperature-heat transfer or *T-q* diagram; see Fig. 8.7. The abscissa represents the total amount of heat that is passed from the brine to the working fluid. It can be shown either in percent or in heat units (say, kJ/kg wf).

The preheater PH provides sensible heat to raise the working fluid to its boiling point, state 5. The evaporation occurs from 5–1 along an isotherm for a pure working fluid. The place in the heat exchanger where the brine and the working fluid experience the minimum temperature difference is called the pinch point, and the value of that difference is designated the pinch-point temperature difference, ΔT_{pp}.

State points 4, 5, and 1 should be known from the cycle specifications: state 4 is a compressed liquid, the outlet from the feed pump; state 5 is a saturated liquid at the

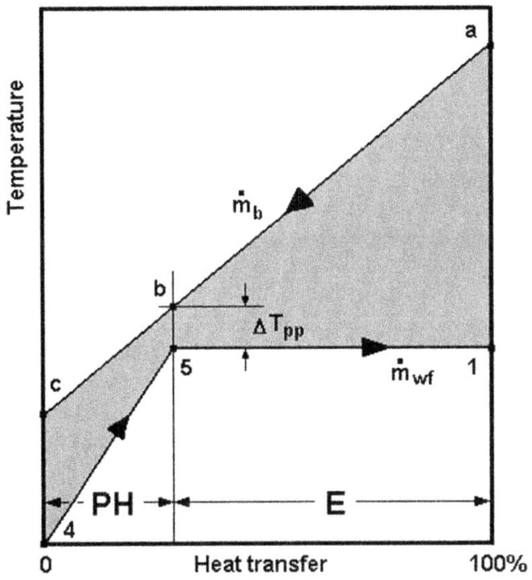

ig. 8.7 *Temperature-heat transfer diagram for preheater and evaporator.*

oiler pressure; and state 1 is a saturated vapor, the same as the turbine inlet condition. hus, the two heat exchangers may be analyzed separately as follows:

Preheater: $\dot{m}_b\bar{c}_b(T_b - T_c) = \dot{m}_{wf}(h_5 - h_4)$ (8.9)

Evaporator: $\dot{m}_b\bar{c}_b(T_a - T_b) = \dot{m}_{wf}(h_1 - h_5)$ (8.10)

The brine inlet temperature T_a is always known. The pinch-point temperature differ-nce is generally known from manufacturer's specifications; this allows T_b to be found rom the known value for T_5. While it is theoretically possible for the pinch point o occur at the cold end of the preheater (for a very steep brine cooling line), this ractically never happens.

The evaporator heat transfer surface area between the two fluids, A_E, can be determined rom the basic heat transfer relationship:

$$\dot{Q}_E = \overline{U}A_E\, LMTD\big|_E \tag{8.11}$$

vhere $\overline{U}$ is the overall heat transfer coefficient, *LMTD* is the log mean temperature lifference, which for the evaporator is found from

$$LMTD\big|_E = \frac{(T_a - T_1) - (T_b - T_5)}{\ln\left[\dfrac{T_a - T_1}{T_b - T_5}\right]} \tag{8.12}$$

nd the evaporation heat transfer rate is given by

$$\dot{Q}_E = \dot{m}_b\bar{c}_b(T_a - T_b) = \dot{m}_{wf}(h_1 - h_5) \tag{8.13}$$

Table 8.1 Approximate values for $\overline{U}$ for several situations [5–8].

Fluids	Overall heat transfer coefficient $\overline{U}$	
	Btu/h · ft^2 · °F	W/m^2 · K
Ammonia (condensing) – Water	150–250	850–1400
Propane or Butane (condensing) – Water	125–135	700–765
Refrigerant (condensing) – Water	80–150	450–850
Refrigerant (evaporating) – Brine	30–150	170–850
Refrigerant (evaporating) – Water	30–150	170–850
Steam – Gases	5–50	30–285
Steam – Water	175–600	1000–3400
Steam (condensing) – Water	175–1050	1000–6000
Water – Air	5–10	25–50
Water – Brine	100–200	570–1135
Water – Water	180–200	1020–1140

The corresponding equations for the preheater are:

$$\dot{Q}_{PH} = \overline{U}A_{PH}\, LMTD|_{PH} \tag{8.14}$$

$$LMTD|_{PH} = \frac{(T_b - T_5) - (T_c - T_4)}{\ln\left[\dfrac{T_b - T_5}{T_c - T_4}\right]} \tag{8.15}$$

$$\dot{Q}_{PH} = \dot{m}_b \bar{c}_b (T_b - T_c) = \dot{m}_{wf}(h_5 - h_4) \tag{8.16}$$

The overall heat transfer coefficient $\overline{U}$ should be determined by experiment with the appropriate fluids to be used in the plant. As a first approximation for preliminary calculations, the values shown in Table 8.1 may be used [4–7]. The uncertainty is large so caution is advised in the use of these values.

Since heat exchangers can be built in a variety of geometrical arrangements (e.g., shell-and-tube, plate, parallel flow, pure counterflow, multiple-pass counterflow, or crossflow), there are correction factors that must be used with the equations given above depending on the configuration, and the reader is referred to any heat transfer book, such as Ref. [8], for more details.

8.2.5 Overall cycle analysis

Having analyzed each of the components of the basic binary plant, we can now sum up by looking at the cycle as a whole. The cycle performance can be assessed by the First Law using the thermal efficiency:

$$\eta_{th} \equiv \frac{\dot{W}_{net}}{\dot{Q}_{PH/E}} \tag{8.17}$$

ince the net power of the cycle is the difference between the thermal power input and ne thermal power rejected, this formula may be rewritten as

$$\eta_{th} = 1 - \frac{\dot{Q}_c}{\dot{Q}_{PH/E}} = 1 - \frac{h_2 - h_3}{h_1 - h_4} \tag{8.18}$$

he heat rejection ratio was derived as eq. (5.27) in Sect. 5.4.6 and is repeated here:

$$\frac{\dot{Q}_c}{\dot{W}_{net}} = \frac{1}{\eta_{th}} - 1 \tag{8.19}$$

These formulas apply to the cycle, not the plant. If the net cycle power is used to upply plant auxiliary power needs such as well pumps, cooling tower fans, station ghting, etc., then all these parasitic loads must be subtracted from the net cycle ower to obtain the net plant power. Since binary cycles tend to have thermal efficienies in the 10–13% range, any further reduction in net power can have a serious npact on plant performance.

Another measure of cycle and plant performance can be obtained using the Second aw in the form of the utilization efficiency, η_u, which is defined as the ratio of the ctual net plant power to the maximum theoretical power obtainable from the geofluid n the reservoir state:

$$\eta_u \equiv \frac{\dot{W}_{net}}{\dot{E}_{res}} = \frac{\dot{W}_{net}}{\dot{m}_b[(h_{res} - h_0) - T_0(s_{res} - s_0)]} \tag{8.20}$$

vhere T_0 is the dead-state temperature (e.g., the local wet-bulb temperature if a water ooling tower is used), and h_0 and s_0 are the enthalpy and entropy values for the eofluid evaluated at the dead-state pressure and temperature (usually approximated s the saturated liquid values at T_0). The subjects of utilization efficiency and Second aw analysis are covered in depth in Chap. 10.

3.3 Working fluid selection

Before we move on to study more complex binary cycles, let us here consider the mportant matter of the selection of the working fluid. This design decision has great mplications for the performance of a binary plant. While there are many choices vailable for working fluids, there are also many constraints on that selection that elate to the thermodynamic properties of the fluids as well as considerations of health, afety, and environmental impact.

3.3.1 Thermodynamic properties

Table 8.2 lists some candidate fluids and their relevant thermodynamic properties; ure water is included for comparison [9]. Clearly all of the candidate fluids have critial temperatures and pressures far lower than water. Furthermore, since the critical ressures are reasonably low, it is feasible to consider supercritical cycles for the ydrocarbons. As we will see shortly, this allows a better match between the brine ooling curve and the working fluid heating-boiling line, reducing the thermodynamic osses in the heat exchangers.

Table 8.2 Thermodynamic properties of some candidate working fluids for binary plants.

Fluid	Formula	T_c °C	T_c °F	P_c MPa	P_c lbf/in^2	P_s @ 300 K MPa	P_s @ 400 K MPa
Propane	C_3H_8	96.95	206.5	4.236	614.4	0.9935	n.a.
i-Butane	i-C_4H_{10}	135.92	276.7	3.685	534.4	0.3727	3.204
n-Butane	C_4H_{10}	150.8	303.4	3.718	539.2	0.2559	2.488
i-Pentane	i-C_5H_{12}	187.8	370.1	3.409	494.4	0.09759	1.238
n-Pentane	C_5H_{12}	193.9	380.9	3.240	469.9	0.07376	1.036
Ammonia	NH_3	133.65	272.57	11.627	1686.3	1.061	10.3
Water	H_2O	374.14	705.45	22.089	3203.6	0.003536	0.24559

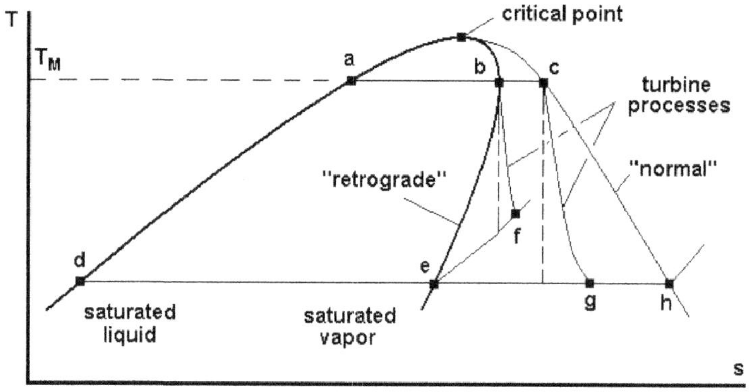

Fig. 8.8 Temperature-entropy diagram contrasting normal and retrograde saturated vapor curves.

Binary mixtures of these fluids have also been studied for use in geothermal binary plants. In particular, the thermodynamic properties of 90% i-C_4H_{10} and 10% i-C_5H_{12} were determined by the National Bureau of Standards (predecessor of NIST) in Washington [10] when it was chosen as the working fluid for the Heber Binary Demonstration plant in the 1980s; see Chap. 18. Mixtures evaporate and condense at variable temperature, unlike pure fluids that change phase at constant temperature. This means that subcritical-pressure boilers for mixed fluids can be better matched to the brine curves, in a manner similar to, but not exactly like, supercritical pure fluids.

Another important characteristic of binary candidate fluids is the shape of the saturated vapor curve as viewed in temperature-entropy coordinates; see Fig. 8.8. This curve for water (shown as the thin line) has a negative slope everywhere, but certain hydrocarbons and refrigerants show a positive slope for portions of the saturation line. That is, there exists a local minimum in the entropy at some low temperature, T_m, and a local maximum in entropy at a higher temperature, T_M. Retrograde fluids include normal butane, isobutane, normal pentane, and isopentane. These fluids exhibit retrograde behavior over the following temperature ranges, $T_m \rightarrow T_M$: C_4H_{10}, −3°C → 127°C; i-C_4H_{10}, −3°C → 117°C; C_5H_{12}, −3°C → 177°C; i-C_5H_{12}, −13°C → 177°C. Since T_m is lower than any temperatures encountered in geothermal binary plants, for practical purposes these fluids can be taken as having saturated vapor lines similar to that shown in Fig. 8.8. This has major implications for Rankine cycles.

On the one hand, normal fluids such as water require considerable superheat, extending the isobar a-b-c upwards, to avoid excessive moisture at the turbine exhaust, state g. On the other hand, retrograde fluids allow expansion from the saturated vapor line into the superheated region, process b-f, avoiding any moisture during the turbine expansion process. It has been shown [11] that it is possible to run a supercritical cycle in which the turbine inlet state lies above the critical point and the expansion line lies inside the wet region for a portion of the process, emerging into the superheated region, without suffering any wetness penalty in efficiency. Apparently, the fluid remains in a metastable vapor state while passing through the wet region by staying on the dry side of the Wilson line [12].

8.3.2 Sonic velocity and turbine size

To a first approximation, the size of the turbine determines its cost, and its size can be estimated from its exit area. The mass flow rate through the turbine exit can be expressed as the product of the working fluid density, the cross-sectional area, and the flow velocity:

$$\dot{m} = \rho A V \tag{8.21}$$

Since the mass flow rate is also given by

$$\dot{m} = \dot{W}_t / w_t = \dot{W} / (h_1 - h_2) \tag{8.22}$$

the exit area can be found from

$$A = \frac{\dot{W}_t}{h_1 - h_2} \frac{v_2}{K a_2} \tag{8.23}$$

where we have replaced the density by its reciprocal, the specific volume v, and the fluid velocity by $K a_2$ where K is a fraction and a_2 is the speed of sound in the fluid at the turbine exit. By definition, the speed of sound in a compressible medium is given by

$$a = \left[\frac{dP}{d\rho} \right]_{s=const}^{1/2} \tag{8.24}$$

The sonic speed can be approximated from property tables and charts using finite differences:

$$a \approx \left[\frac{\Delta P}{\Delta \rho} \right]_{s=const}^{1/2} = \left[\frac{\Delta P}{\Delta (1/v)} \right]_{s=const}^{1/2} \tag{8.25}$$

We may now compare working fluids using the same power outputs and the same fraction of sonic speed at the turbine exhaust to examine their relative cross-sectional areas, and thus the relative sizes of their turbines. It turns out that ammonia has the smallest size turbine for the chosen comparison, and it is convenient to present the results as multiples of the area required for an ammonia turbine.

Table 8.3 gives the results using the approach outlined above. A similar analysis was carried out by Milora and Tester [13] using a different methodology involving the law of corresponding states, and their results are included in the table for comparison purposes. The analysis involves estimates and approximations, exhibited in the

Table 8.3 Turbine size comparisons for several working fluids.(1)

Fluid	Formula	Molar mass	Relative exit area(2)
Ammonia	NH_3	17.03	1.0 (1.0)
Propane	C_3H_8	44.09	2.3 (1.9)
i-Butane	i-C_4H_{10}	58.12	4.1 (4.9)
n-Butane	C_4H_{10}	58.12	5.5 (6.3)
i-Pentane	i-C_5H_{12}	72.15	12.2 (n.a.)
n-Pentane	C_5H_{12}	72.15	14.6 (n.a.)

(1)Turbine inlet temperature = 400 K, saturated, except superheated for NH_3 and C_3H_8; condensing temperature = 320 K. (2)Numbers in parentheses are from Ref. [13].

Table 8.4 Environmental and health properties of some candidate working fluids [14].

Fluid	Formula	Toxicity	Flammability	ODP	GWP
R-12	CCl_2F_2	non-toxic	non-flam.	1.0	4,500
R-114	$C_2Cl_2F_4$	non-toxic	non-flam.	0.7	5,850
Propane	C_3H_8	low	very high	0	3
i-Butane	i-C_4H_{10}	low	very high	0	3
n-Butane	C_4H_{10}	low	very high	0	3
i-Pentane	i-C_5H_{12}	low	very high	0	3
n-Pentane	C_5H_{12}	low	very high	0	3
Ammonia	NH_3	toxic	lower	0	0
Water	H_2O	non-toxic	non-flam.	0	-

differences between the two approaches. However, the main result is the ranking of the working fluids by relative sizes of the turbines, and in this respect, the two approaches are in agreement. By way of comparison, steam would have a relative exit area of about 120.

8.3.3 Health, safety, and environmental considerations

Lastly, the environmental, safety and health properties of potential working fluids must be considered. These include flammability, toxicity, ozone depletion potential (ODP), and global warming potential (GWP). Table 8.4 summarizes these properties for the fluids in Table 8.1, plus two chlorofluorocarbons that used to be considered candidate working fluids.

The ODP is normalized at 1.0 for refrigerants R-11 and R-12, which are the worst in this regard. The GWP is normalized at 1.0 for carbon dioxide. Owing to their very high ODP and GWP, both R-12 and R-114 have been banned from use by the Copenhagen Amendment (effective as of 1994) to the Montreal Protocol (signed in 1987, effective as of 1989). The original binary plant at Paratunka in Russia that was installed in 1967 used R-12 as its working fluid.

The contribution of the hydrocarbons to global warming comes about mainly through the carbon dioxide that is a byproduct of their decomposition. All of the hydrocarbon candidate fluids obviously are flammable and necessitate appropriate fire protection equipment on site, over and above the usual requirements for any power plant.

8.4 Advanced binary cycles

In this section we will discuss innovative and complex binary cycles, but before we tackle this subject, let us consider the theoretical optimum binary cycle for use with a geothermal hot water resource. This will give us a basis of comparison for all other types of binary plant.

8.4.1 *Ideal binary cycle*

Basic binary plants have low thermal efficiencies mainly due to the small temperature difference between the heat source and the heat sink. The thermodynamic ideal cycle operating between a heat source at a temperature T_H and a heat sink at a temperature T_L is a Carnot cycle consisting of an isothermal heat addition process at T_H, followed by an isentropic expansion process, an isothermal heat rejection process at T_L, and an isentropic compression process to return the working fluid to its initial state [15].

The Carnot efficiency is the highest possible efficiency for any cycle operating between these two temperatures and is given by

$$\eta_{CC} = \eta_{\max} = 1 - \frac{T_L}{T_H} \tag{8.26}$$

where the temperatures must be in kelvins or degrees Rankine. For a geothermal binary plant using a brine at 150°C (423.15 K) and a heat sink at 40°C (313.15 K), the maximum ideal Carnot efficiency 26%.

However, the brine is not an isothermal heat source, but in fact cools as it transfers heat to the working fluid. Thus, a more realistic ideal cycle for a geothermal binary plant is a triangular cycle consisting of an isobaric (constant pressure) heat addition process up to the brine inlet temperature T_H, followed by an isentropic expansion, and an isothermal heat rejection process at T_L to complete the cycle. It is easy to show [16] that the efficiency for the triangular cycle is given by

$$\eta_{TRI} = \frac{T_H - T_L}{T_H + T_L} \tag{8.27}$$

For the same temperatures used in the above example, the triangle cycle yields an efficiency of 15%.

These two ideal cycles are shown in temperature-entropy coordinates in Fig. 8.9.

In recognition of the inherently low thermal efficiency of basic binary plants, there are several variations on the basic cycle aimed at achieving higher efficiencies. When a cycle has an efficiency of say 10%, an improvement of only one percentage point represents a 10% improvement, and this may make the difference between an economically viable project and one that is not. In the next sections we will discuss some of these innovative systems.

8.4.2 *Dual-pressure binary cycle*

A dual-pressure cycle is designed to reduce the thermodynamic losses incurred in the brine heat exchangers of the basic cycle. These losses arise through the process of transferring heat across a large temperature difference between the hotter brine and the cooler working fluid; see Fig. 8.7, for example. By maintaining a closer match

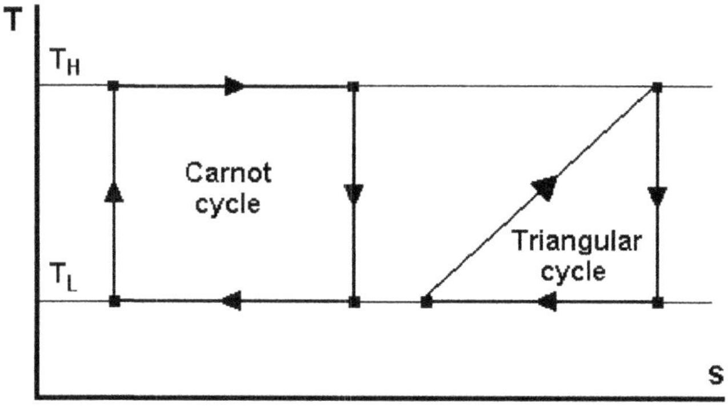

Fig. 8.9 Two ideal thermodynamic cycles.

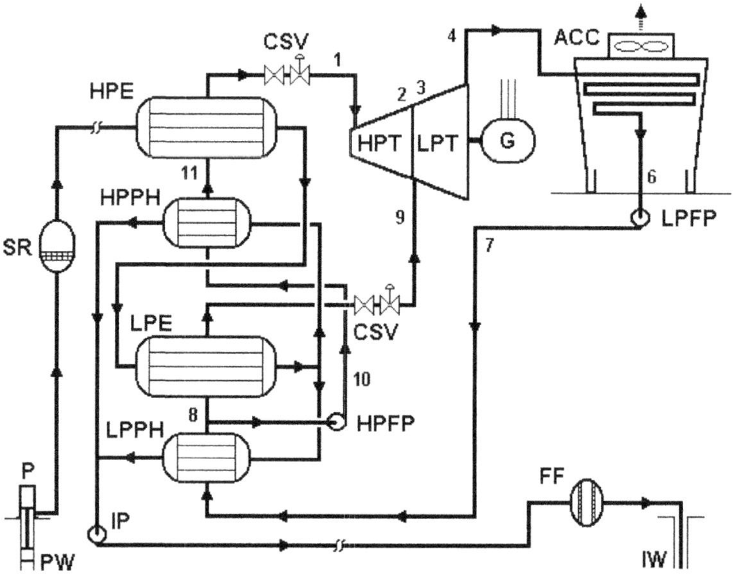

Fig. 8.10 Dual-pressure binary plant: simplified schematic flow diagram.

between the brine cooling curve and the working fluid heating/boiling curve these losses can be reduced.

The dual-pressure cycle has a two-stage heating/boiling process that allows the two fluids to achieve a smaller average temperature difference than the one-stage process used in a basic cycle. A schematic of a dual-pressure plant is given in Fig. 8.10 and the corresponding process diagram is shown in Fig. 8.11.

A dual-admission turbine is shown in Fig. 8.10 in which the low-pressure saturated vapor (state 9) is admitted to the turbine to mix with the partially-expanded high-pressure vapor (state 2) to form a slightly superheated vapor (state 3). Given the small size of turbines using organic working fluids, practical considerations may lead to an alternative design using two separate turbines; see Fig. 8.12.

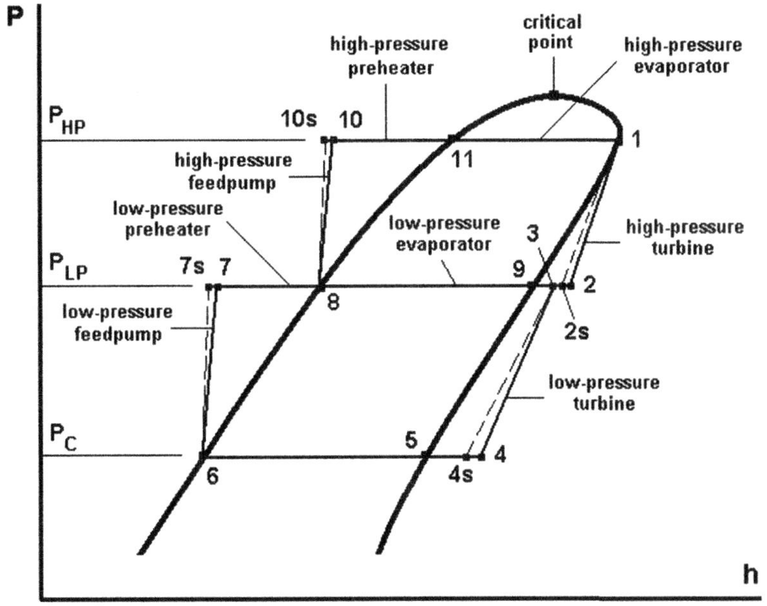

'ig. 8.11 *Dual-pressure binary plant: pressure-enthalpy process diagram.*

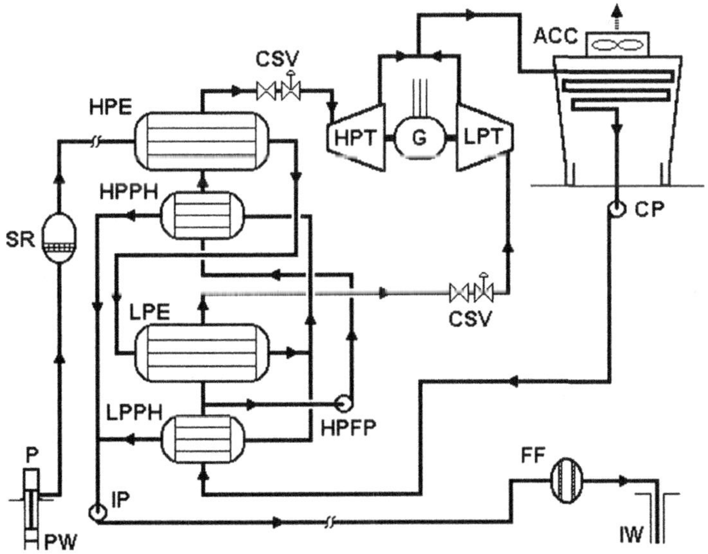

'ig. 8.12 *Dual-pressure binary plant: separate high- and low-pressure turbines.*

Unlike dry-steam and flash-steam plants, binary plants do not have steam conden-sate to serve as makeup for a water cooling tower. Thus binary plants need a separate cooling medium, either fresh water or air. An air-cooled condenser ACC is depicted here since makeup water for cooling towers is scarce at many geothermal sites.

Table 8.5 Comparison of efficiencies of single- and dual-pressure binary cycles [17].

Working fluid	Brine temperature	Thermal efficiency, %		Utilization efficiency, %	
		Basic	Dual-press	Basic	Dual-press
i-C_4H_{10}	93°C (200°F)	5.5	4.6	31.9	39.7
i-C_5H_{12}	93°C (200°F)	5.2	4.2	30.5	37.0
i-C_4H_{10}	149°C (300°F)	10.3	9.8	48.8	56.9
i-C_5H_{12}	149°C (300°F)	9.8	8.8	44.6	51.5
i-C_5H_{12}	204°C (400°F)	13.7	13.1	57.7	61.2

Note: The condensing and dead-state temperatures were both taken as 38°C (100°F).

The analysis of a dual-pressure cycle follows the same methodology as for a basic cycle but is merely longer. A detailed comparison of basic cycles (single-pressure) and the dual-pressure cycles has been conducted by Khalifa and Rhodes [17] for two different working fluids; their results are summarized in Table 8.5. The results are quite interesting. In all cases, the thermal efficiency for a dual-pressure cycle is actually lower than for a basic cycle, but the utilization efficiency for a dual-pressure cycle is significantly higher than for a basic cycle, ranging from a 6% advantage at the highest brine temperature to 24% at the lowest.

The thermal efficiency depends on the amount of heat added to the cycle but makes no distinction between high-exergy heat and low-exergy heat, and ignores the temperature difference between the fluids. Thermodynamics requires that the higher the average temperature of the heat added to a cycle, the higher will be the thermal efficiency, for the same heat sink temperature. Since a significant amount of heat is needed to evaporate the fluid from 8–9 (see Fig. 8.11) at a relatively low temperature, this adversely affects the thermal efficiency of the dual-pressure cycle. However, the utilization efficiency depends on how effectively the exergy of the brine is used. By more closely matching the brine cooling curve with the heating and boiling curves, the average temperature difference between the two fluids is made smaller and the irreversibilities are reduced. This allows more exergy from the brine to enter the cycle and leads to a higher overall utilization efficiency.

The 5 MW Raft River dual-boiling plant in Idaho, U.S., was the first to make use of the dual-pressure concept [18]. It was operated as a demonstration plant from 1981–82 by the Idaho National Engineering Laboratory for the U.S. Dept. of Energy; see Chap. 20.

8.4.3 Dual-fluid binary cycle

It may sound odd but the first commercial binary plant in the United States was a remarkably advanced design, the Magmamax plant at East Mesa in California's Imperial Valley. It was a 12.5 MW plant that began operations in 1979 using a dual-fluid cycle in which two different hydrocarbons were used in interlocking Rankine cycles, one a subcritical cycle and one a supercritical cycle [19, 20]. We will present a detailed case study of this plant in Chap. 18; here we will describe the thermodynamic principles that underlie the design of such a system.

The dual-fluid binary cycle shown in Fig. 8.13 features a heat recuperator E2 that nks the upper cycle having fluid 1 and the lower cycle having fluid 2.

The process diagram is given in Fig. 8.14 in temperature-entropy coordinates. Note ıat there are two entropy axes, one for each working fluid; the saturation curves are rawn in a convenient location to illustrate the relationship between the two cycles. If ıe fluids are selected judiciously according to their thermodynamic properties, they ʒill complement one another to create synergy in the overall dual cycle.

As with the dual-pressure cycle, the motivation here is to create a good match etween the brine and the working fluid heating-boiling curves. The temperature-heat ·ansfer diagram, Fig. 8.15, shows this relationship. The discontinuity between state oints 5 and 11 arises from the internal heat transfer between the working fluids and oes not involve the brine. From the diagram it can be seen that the pinch point ccurs between state b on the brine cooling curve and state 6, the bubble point for uid 1. The near parallelism between the brine and the working fluids in the prehea- ers means that the thermodynamic irreversibilities will be low, as will the loss of xergy during the heat transfer process in those components. Since the average tem- erature difference in the fluid 1 evaporator is relatively large, there will be a higher ɔss of exergy there.

Note that all cycles considered so far in this chapter have been subcritical cycles, i.e., ıe pressure of the working fluids in the brine heat exchangers is less than the critical ressure. If fluid 1 is raised to a supercritical pressure before entering its preheater, the ɜmperature-heat transfer diagram would change dramatically; see Fig. 8.16. The sharp orner at state 6 denoting the bubble point for fluid 1 has vanished. Fluid 1 now has smooth heating curve that takes the fluid from a cool compressed liquid to a hot upercritical vapor. There will still be a point of closest approach between the two urves, but it will be far less pronounced. This now allows a very good match between

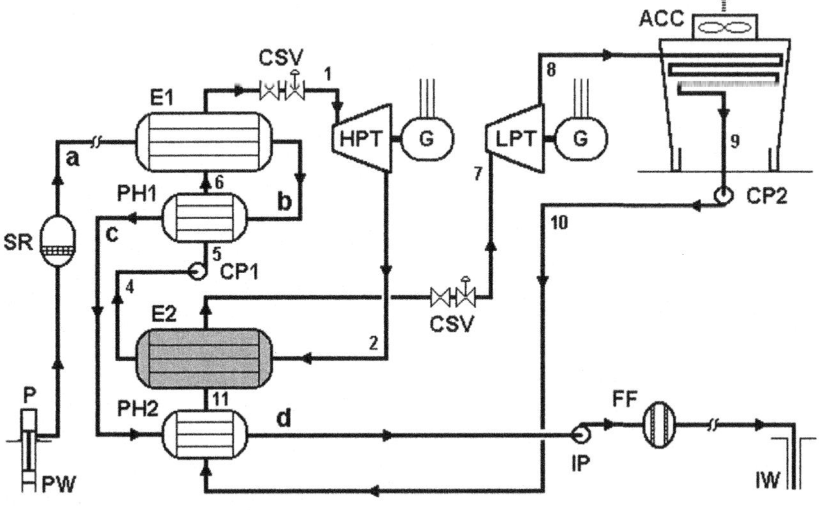

ig. 8.13 Dual-fluid cascaded binary cycle featuring a heat recuperator.

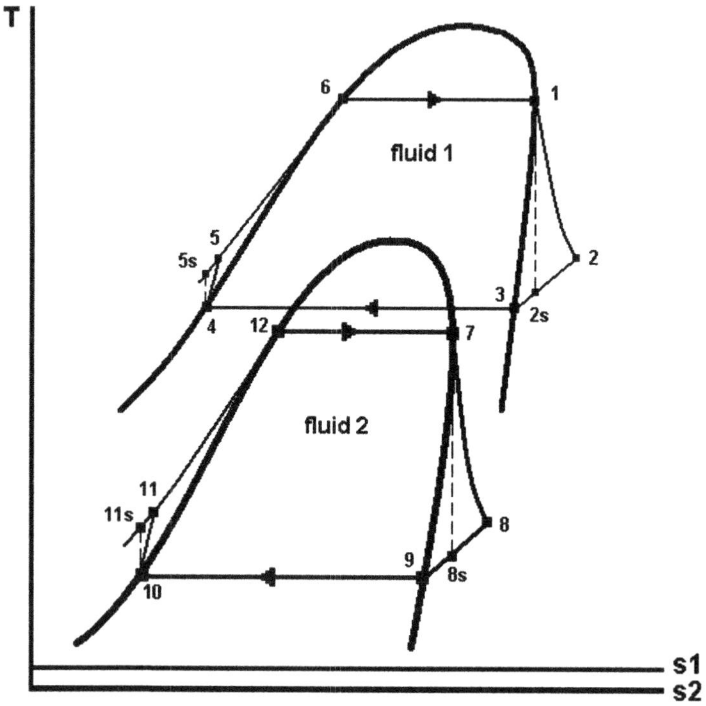

Fig. 8.14 Dual-fluid process diagram in temperature-entropy coordinates.

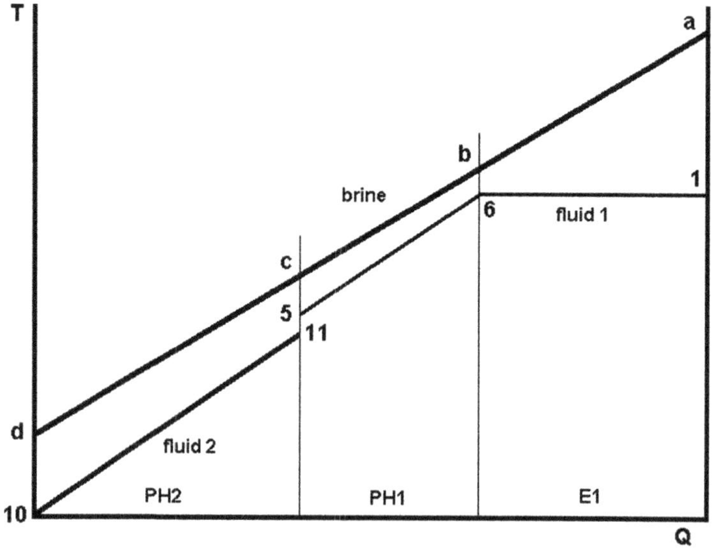

Fig. 8.15 Dual-fluid binary plant: temperature-heat transfer diagram for brine heat exchangers with subcritical working fluid pressures.

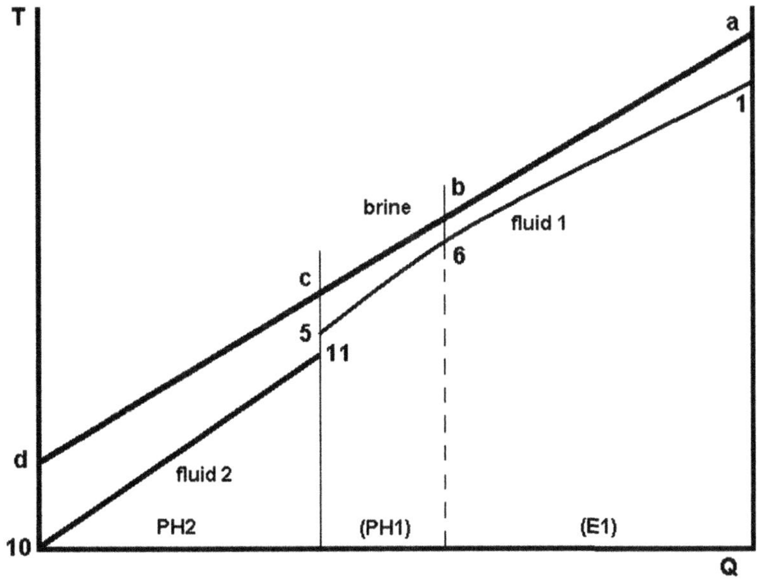

'ig. 8.16 Dual-fluid binary plant: temperature-heat transfer diagram for brine heat exchangers with supercritical ressure for working fluid 1.

he brine and the working fluids that will result in lower exergy losses and a higher utilization efficiency for the cycle.

The cycle thermal efficiency (net power/heat input) for a supercritical vapor generator will exceed that for a subcritical cycle. Using isobutane as the working fluid with a urbine inlet temperature of 420 K and a condensing temperature of 320 K, a turbine sentropic efficiency of 85%, and a feed pump isentropic efficiency of 80%, the following results are obtained for two different vapor generator pressures:

4.0 MPa (supercritical): $\eta_{th} = 12.3\%$, $w_P = 16\%$ of $w_{net,cycle}$
3.0 MPa (subcritical): $\eta_{th} = 11.0\%$, $w_P = 11\%$ of $w_{net,cycle}$

'he supercritical cycle has a 12% higher thermal efficiency. Notice, however, that in he supercritical cycle, the pump work is a greater fraction of the net cycle work and s 45% higher than for the subcritical cycle.

There are practical difficulties, however, with a supercritical cycle. The higher pressures may require thicker, more costly tubing in the heat exchangers unless the fluid pressures on each side are made nearly equal. Thicker tubes offer more resistance to heat transfer and require longer, more expensive heat exchangers. There is no distinction between the preheater and the evaporator since there is no conventional phase transition. Rather, a continuous process of increasing temperature occurs as the working fluid receives heat from the brine. In a traditional fossil-fueled power plant using supercritical water/steam, this transition occurs in a long series of tubes within the furnace, in what is called a "once-through" steam generator. In geothermal binary plants, shell-and-tube heat exchangers having the brine in the tubes and the working fluid within the shell are appropriate for subcritical operation, but are ill-suited for supercritical operation. This was a main reason that the original Magmamax plant [19] placed the supercritical isobutane inside the tubes and the brine on the shell side of the heat exchangers.

8.4.4 Kalina binary cycles

Water-ammonia mixtures have long been used in absorption refrigeration cycles, but it was not until A. Kalina patented his Kalina cycle that this working fluid was used for power generation cycles. A typical Kalina cycle, KCS-12, is shown schematically in Fig. 8.17. The features that distinguish the Kalina cycles (there are several versions) from other binary cycles are these:

- The working fluid is a binary mixture of H_2O and NH_3
- Evaporation and condensation occur at variable temperature
- Cycle incorporates heat recuperation from turbine exhaust
- Composition of the mixture may be varied during cycle in some versions.

As a consequence, Kalina cycles show improved thermodynamic performance of heat exchangers by reducing the irreversibilities associated with heat transfer across a finite temperature difference. The heaters are so arranged that a better match is maintained between the brine and the mixture at the cold end of the heat transfer process where improvements in exergy preservation are most valuable.

A reheater RH is needed because the water-ammonia mixture has a normal saturated vapor line, i.e., $dT/ds < 0$, leading to wet mixtures in the turbine. The plant relies on good heat exchangers because more heat is transferred than in a supercritical binary plant of the same power output. Bliem and Mines [21] showed that the Kalina

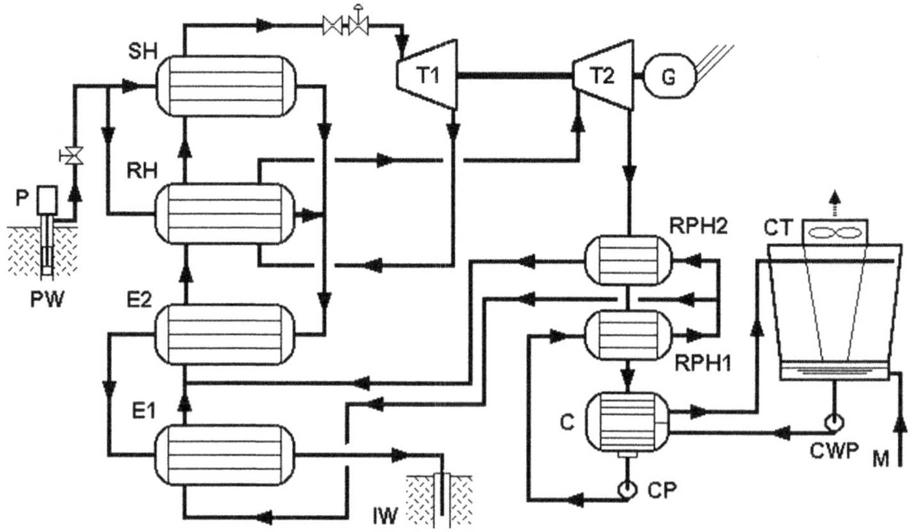

Fig. 8.17 Typical Kalina cycle employing a reheater and two recuperative preheaters.

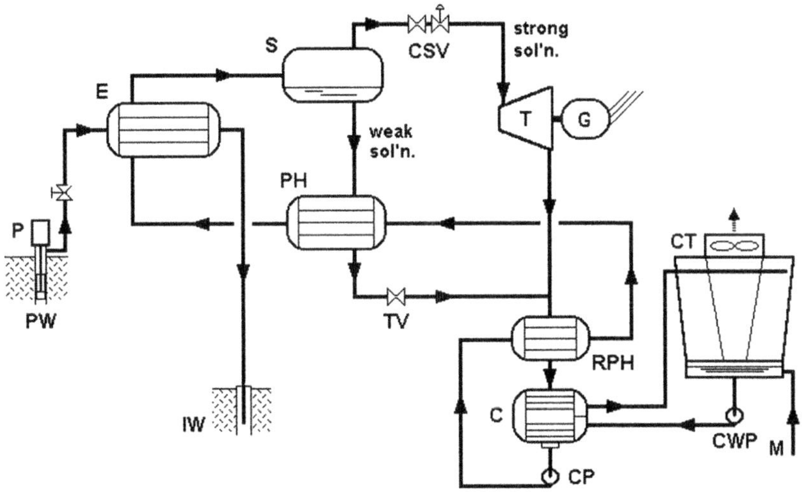

'ig. 8.18 Kalina cycle with variable composition of the water-ammonia working fluid.

ycle of Fig. 8.17 requires about 25% more heat transfer. A possible advantage to ısing the recuperative preheaters is that they reduce the heat load on the condenser nd cooling tower. The lower capital cost for a smaller condenser and cooling tower nust be compared to the extra cost for the recuperators; over the long haul, the resulting ıigher efficiency should mean lower operating costs.

The physical plant is more complex than a basic binary plant, particularly when a listillation column is used to vary the mixture composition. The simplest configuration ıf Kalina cycle with variable working fluid composition is shown in Fig. 8.18. The eparator S allows a saturated vapor that is rich in ammonia to flow to the turbine, hus permitting a smaller and less costly turbine than for a hydrocarbon working fluid. 'he weak solution, a liquid rich in water, is used in the preheater and then throttled lown to the turbine exhaust pressure before mixing with the strong solution to restore he primary composition. The mixture is then used in a recuperative preheater RPH ›rior to being fully condensed.

A possible difficulty for the Kalina cycle, one that is common to all cycles that strive or high efficiency, is maintaining very tight pinch-point temperature differences in the ıeat exchangers. Also, the advantage of variable-temperature condensation is lessened ›ecause the condensing isobars of the ammonia-rich NH_3-H_2O mixtures used in power ycles are concave upward, leading to a pinch point. Thus, there are relatively large emperature differences at the start of and at the end of the condensing process.

3.5 Example of binary cycle analysis

Ve will illustrate the analysis of a binary cycle using the simple case shown in 'ig. 8.19. We will assume the working fluid is isopentane, i-C_5H_{12}, and that the cycle ıas a subcritical boiler pressure. The net cycle power is 1,200 kW, a typical value for

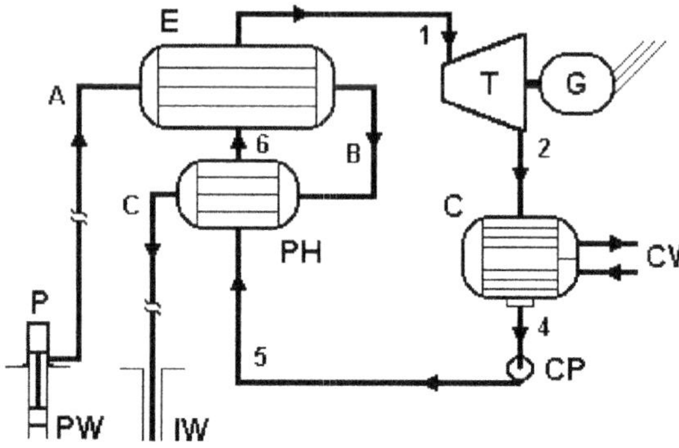

Fig. 8.19 Basic binary plant with state-point notations.

this type of plant. Pressure losses in all heat exchangers and piping will be assumed negligible.

The cycle specifications are as follows:

- Brine inlet temperature, $T_A = 440$ K
- Brine specific heat, $c_b = 4.19$ kJ/kg·K = constant
- Brine density = 56 lbm/ft^3
- Pinch-point temperature difference = 5 K
- Preheater-evaporator pressure, $P_5 = P_6 = P_1 = 2.0$ MPa
- Condensing temperature, $T_4 = 320$ K
- Turbine isentropic efficiency = 85%
- Feed pump isentropic efficiency = 75%.

Our objectives in this example will be to determine the following quantities:

1. Specific work of the turbine, w_T, in kJ/kg i-C_5H_{12}.
2. Heat rejected to the cooling water, q_C, in kJ/kg i-C_5H_{12}.
3. Specific work of the feed pump, w_P, in kJ/kg i-C_5H_{12}.
4. Heat transferred to the working fluid, q_{IN}, in kJ/kg i-C_5H_{12}.
5. Cycle thermal efficiency, η_{th} in %.
6. Mass flow rate of i-C_5H_{12}, $\dot{m}_{i-C5}$, in kg/s.
7. Mass flow rate of brine, $\dot{m}_b$, in kg/s.
8. Brine outlet temperature, T_C, in K.
9. Number of wells needed if a typical well can produce 800–900 GPM.
10. Utilization efficiency for the dead-state temperature of 25°C.

While this may appear a tall order, it will be seen that a systematic approach will yield results without much difficulty.

We begin by translating the description of the plant into two thermodynamic process diagrams: a pressure-enthalpy diagram and a temperature-entropy diagram; see Fig. 8.20. The former will be very useful because there exist scale-drawn property charts for isopentane in *P-h* coordinates, and both are helpful in visualizing the cycle.

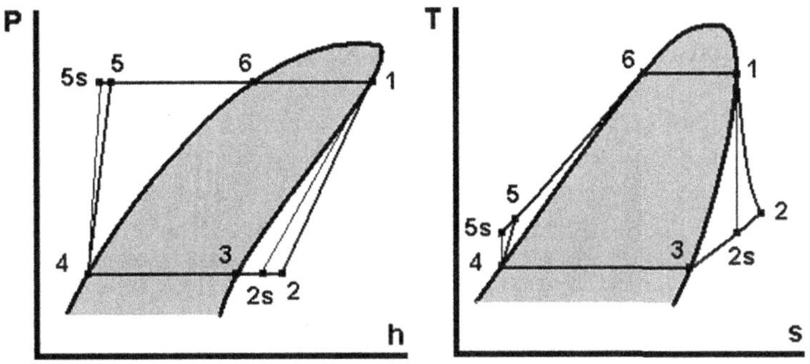

'ig. 8.20 *Pressure-enthalpy and temperature-entropy diagrams.*

From the analysis presented in Sect. 8.2, it is clear that we must determine the nthalpy values for the isopentane at the six state points in the cycle. This is facilitated y using a table to keep track of the calculations. The numbers in **bold face** are given r obvious. The other numbers have been found using the procedure described below.

tate	*P* MPa	*T* K	*v* m^3/kg	s kJ/kg · K	*h* kJ/kg	Comments for *h*
[illegible]	**2.0**	**427.1**		2.2022	741.08	Direct from table
[illegible]	**0.1866**				663.38	From η_t
s	**0.1866**			2.2022	649.66	Interpolation
[illegible]	**0.1866**	**320**		1.9887	578.16	Direct from table
[illegible]	**0.1866**	**320**	0.001686		249.50	Direct from table
[illegible]	**2.0**				253.58	From η_p
s	**2.0**				252.56	From $v \times \Delta P$
[illegible]	**2.0**				545.73	Direct from table

'he properties of isopentane were obtained from Reynolds' data book [9], which ncludes a scale-drawn *P-h* diagram.

State 1: Saturated vapor at 2 MPa: $s_1 = 2.2022$ kJ/kg · K, $h_1 = 741.08$ kJ/kg.

State 2: Actual turbine outlet state; must find isentropic outlet state first.

State 2s: Isentropic turbine outlet state; $s_{2s} = s_1$, $P_2 = P_{sat}$ for $T_3 = 320$ K. By interpolation, $h_{2s} = 649.66$ kJ/kg.

State 2: Using the definition of the turbine efficiency, eq. (5.9), we can find h_2: $h_2 = h_1 - \eta_t(h_1 - h_{2s}) = 663.38$ kJ/kg.

State 3: Saturated vapor at $T_3 = 320$ K: $s_3 = 1.9887$ kJ/kg · K, $h_3 = 578.16$ kJ/kg.

State 4: Saturated liquid at $T_3 = 320$ K: $v_4 = 0.001686$ m^3/kg, $h_4 = 249.50$ kJ/kg. We record the specific volume v_4 because we can use a very good approximation to find the enthalpy at state 5s.

State 5: Actual pump outlet state; must find isentropic outlet state first.

State 5s: Isentropic pump outlet state: $P_{5s} = 2$ MPa. Because the liquid is very nearly incompressible (i.e., constant density), to a high degree of accuracy, the value of h_{5s} can be found from: $h_{5s} \approx h_4 + v_4(P_{5s} - P_4) = 252.56$ kJ/kg.

State 5: The definition of the pump efficiency is similar to that for the turbine but the numerator is the isentropic work and the denominator is the actual work; thus, we can find h_5: $h_5 = h_4 + (h_{5s} - h_4)/\eta_p = 253.38$ kJ/kg.

State 6: Saturated liquid at 2 MPa: $h_6 = 545.73$ kJ/kg.

With all the enthalpy values in hand, we can easily find the first six of our objectives:

1. Specific work of the turbine: $w_t = h_1 - h_2 = 77.70$ kJ/kg i-C_5H_{12}.
2. Heat rejected to the cooling water: $q_c = h_2 - h_4 = 413.88$ kJ/kg i-C_5H_{12}.
3. Specific work of the feed pump: $w_p = h_5 - h_4 = 3.06$ kJ/kg i-C_5H_{12}.
4. Heat transferred to the working fluid: $q_{IN} = h_1 - h_5 = 487.50$ kJ/kg i-C_5H_{12}.
5. Cycle thermal efficiency: $\eta_{th} = (w_t - w_p)/q_{IN} = 1 - (q_C/q_{IN}) = 15.1\%$.
6. Mass flow rate of i-C_5H_{12}: $\dot{m}_{i-C_5} = \dot{W}_{net}/(w_t - w_p) = 1200/73.62 = 16.3$ kg/s.

To continue with the analysis, it is necessary to turn to the temperature-heat transfer diagram, Fig. 8.21. Although we do not know the location of the pinch point, we tentatively show it at the isopentane bubble point; we will verify that assumption later.

7. Mass flow rate of brine: The First Law for the evaporator is: $\dot{m}_b c_b(T_A - T_B) = \dot{m}_{i-C_5}(h_1 - h_6)$, where $T_B = T_6 + 5$ K. Thus we can find the brine flow rate:

$$\dot{m}_b = \dot{m}_{i-C_5} \frac{h_1 - h_6}{c_b(T_A - T_B)} = 96.2 \text{kg/s}$$

8. Brine outlet temperature, T_C: From the enthalpies for isopentane, we notice that 60% of the total heat transfer from the brine occurs in the preheater. Using similar triangles, we can find the temperature of the brine at the cold end:

$$\frac{T_A - T_C}{T_A - T_B} = \frac{h_1 - h_5}{h_1 - h_6} \quad \text{or} \quad T_C = T_A - (T_A - T_B)\left[\frac{h_1 - h_5}{h_1 - h_6}\right] = 420.3 \text{ K}$$

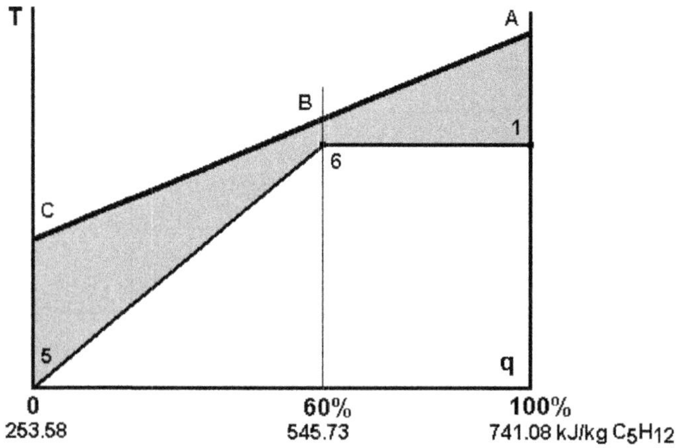

Fig. 8.21 Temperature-heat transfer diagram for preheater and evaporator.

At this point we must check our assumption about the location of the pinch point. Since the feed pump outlet temperature T_5 is not much higher than the pump inlet temperature T_4 (320 K), it is clear that the pinch point in fact does occur at the bubble point. Indeed we notice that the brine temperature does not drop very much as it passes through the plant, losing only about 20°C.

9. Number of wells needed if a typical well can produce 800–900 GPM: The number of wells, N_W, is easily found from the previously obtained results: $N_w = \dot{m}_b/\dot{m}_{per\ well}$. The calculation involves merely attending to units conversions; let us convert the brine flow rate from kg/s to GPM:

$$96.2\frac{\text{kg}}{\text{s}} \times \frac{2.204\ \text{lbm}}{1\ \text{kg}} \times \frac{1\ \text{ft}^3}{56\ \text{lbm}} \times \frac{1\ \text{gal}}{0.1337\ \text{ft}^3} \times \frac{60\ \text{s}}{1\ \text{min}} = 1,700\ \text{GPM}$$

Therefore, $N_W = 1700/850 = 2$ wells.

10. Utilization efficiency for a dead-state temperature of 25°C: Finally the utilization efficiency can be found: $\eta_u = \dfrac{\dot{W}_{net}}{\dot{m}_b[h_A - h_O - T_O(s_A - s_O)]}$. Assuming the brine behaves as pure water we can find the brine enthalpy and entropy values from Reynolds' data book [9]:

$$h_A = 705.0\ \text{kJ/kg},\ s_A = 2.0096\ \text{kJ/kg}\cdot\text{K},\ h_o = 103.93\ \text{kJ/kg},\ s_o = 0.36384\ \text{kJ/kg}\cdot\text{K}$$

Thus the result is

$$\eta_u = \frac{1,200}{96.2 \times 110.39} = \frac{1,200}{10,619.2} = 0.113 \text{ or } 11.3\%$$

Discussion: The reader will observe that the thermal efficiency is fairly high for a simple binary plant (15% vs. typ. 10–13%), but that the utilization efficiency is very poor, even lower than the thermal efficiency. Why? With the pinch point located at the bubble point and with so much heat required to preheat the isopentane, the irreversibilities in the preheater are very large due to the very large average temperature difference. If the boiler pressure were lowered, the evaporating line 6–1 in Fig. 8.18 would drop and point 6 would move down and to the left, providing a more balanced heat load between the evaporator and the preheater. This would also lower the brine outlet temperature, and result in a lower brine flow rate. Thus the brine exergy rate would be less for the same net power output and the utilization efficiency would rise. Can you imagine the effect this would have on the cycle thermal efficiency? Problem 8.9 at the end of the chapter will allow you to explore this problem further.

8.6 Environmental impact of binary cycles

Geothermal binary plants are among the most benign of all power plants regarding environmental impact. A review of the binary plant flow diagrams in this chapter will reveal that the only impact on the environment takes place at the heat rejection side of the plant. Since the geofluid is pumped from the reservoir and returns entirely to the reservoir after passing through heat exchangers, the potentially harmful geofluid never sees the light of day. Furthermore, the cycle working fluid is contained completely within pipes, heat exchangers, and the turbine, so that it too never comes in chemical or physical contact with the environment.

The only possible form of pollution from a binary plant might be called thermal pollution, i.e., the amount of heat that must be rejected from the cycle in accordance with the laws of thermodynamics. As was mentioned in Sect. 5.4.6, geothermal plants of all types discharge more waste heat per unit of power output than other thermal power plants. In the case of a basic binary plant, the amount of thermal power that needs to be absorbed by the surroundings is about nine times the useful power delivered by the plant [22]. Even this effect can be minimized if there is a beneficial use for the waste heat such as soil or greenhouse heating. It is interesting to note that the first binary plant at Paratunka on the cold Russian Kamchatka peninsula used the waste heat to assist farmers in extending their growing season [23].

8.7 Equipment list for basic binary plants

Binary plants differ considerably in their equipment requirements from flash- or dry-steam plants. Where they do not differ, the item is simply shown as a bullet item; otherwise it carries an annotation.

8.7.1 *Downwell pumps and motors*

- Multistage centrifugal pumps, lineshaft-driven from surface-mounted electric motors or submersible electric pumps

8.7.2 *Brine supply system*

- Sand removal system
  Solids knock-out drum

8.7.3 *Brine/working fluid heat exchangers*

- Preheater
  Horizontal cylinder, liquid-liquid, shell-and-tube type with brine on tube side and working fluid on shell side, or vertical, corregated plate type
- Evaporator/superheater
  Horizontal cylinder or kettle-type boiler
  Superheater section (optional)
  Brine on tube side, working fluid on shell side

8.7.4 *Turbine-generator and controls*

- Working fluid turbine (axial or radial flow), generator, and accessories

8.7.5 *Working fluid condenser, accumulator and storage system*

- Condenser
- Dump tank and accumulator
  - Holding tank large enough to store full capacity of working fluid charge
- Evacuation pumps to remove working fluid to storage during maintenance

8.7.6 *Working fluid feed pump system*

- Condensate pumps
- Booster pumps (as needed)

8.7.7 *Heat rejection system*

- Wet cooling system
  - Water cooling tower with external source of make-up water
  - Cooling water pumps and motors
  - Cooling water treatment system (as needed)
- Dry cooling system (if a source of make-up water is not available)
  - Air-cooled condensers with manifolds and accumulator
  - Induced draft fans and motors

8.7.8 *Back-up systems*

- Standby power supply

8.7.9 *Brine disposal system*

- Brine return pumps and piping
  - Horizontal, variable-speed, motor-driven units
  - High-head, high-volume flow design

8.7.10 *Fire protection system (if working fluid is flammable)*

- High-pressure sprinkler system
- Flare stack

"A colossal column of cloud towered to a great height in the air immediately above the crater, and the outer swell of every one of its vast folds was dyed with a rich crimson luster, which was subdued to a pale rose tint in the depressions between. It glowed like a muffled torch..."

Mark Twain, describing an eruption of Kilauea – 1872

References

[1] DiPippo, R., *Geothermal Energy as a Source of Electricity: A Worldwide Survey of the Design and Operation of Geothermal Power Plants*, U.S. Dept. of Energy, DOE/RA/28320-1, U.S. Gov. Printing Office Washington, DC, 1980.

[1a] Kraml, M., K. Kessels, U. Kalberkamp, N. Ochmann and C. Stadtler, "The GEOTHERM Programme of BGR, Hannover, Germany: Focus on Support of the East African Region," *The 1st African Geothermal Conference*, Addis Ababa, Ethiopia, 2006, http://www.bgr.de/geotherm/ArGeoC1/pdf/50%20%20Kraml,%20M.%20GEOTHERM%20programme.pdf.

[1b] Hochstein, M.P., "Geothermal Resources of the Western Branch of the East African Rift System," *IGA NEWS — Newsletter of the International Geothermal Association*, Qtr. No. 42, Oct.-Dec. 2000 pp. 10–12.

[1c] Mudiay T., Gisele, "Geothermal Heat Sources and Thermal Minerals in the Democratic Republic of the Congo," Interregional Expert Meeting, UNIDO/ICS Africa-Mexico Cooperation Programme on the Geothermal Area for Productive Uses, April 6–15, 2011.

[1d] Omenda, P. and M. Teklemariam, "Overview of Geothermal Resource Utilization in the East African Rift System," *Short Course V on Exploration for Geothermal Resources*, Lake Bogoria and Lake Naivasha Kenya, Oct. 29–Nov. 19, 2010.

[2] ENEL, *Larderello Field Trip: Electrical Power Generation*, World Geothermal Congress, Florence, Italy May 18–31, 1995, p. 1.

[3] Anon., "East Mesa 18.5 MW × 2 Double Flash Cycle Geothermal Power Plant," Mitsubishi Heavy Industries, Ltd., Tokyo, Japan (undated).

[4] DiPippo, R., "Geothermal Power Systems," Sect. 8.2 in *Standard Handbook of Powerplant Engineering, 2nd ed.*, T.C. Elliott, K. Chen and R.C. Swanekamp, Eds., McGraw-Hill, Inc., New York, 1998 pp. 8.27–8.60.

[5] Pitts, D.R. and L.E. Sissom, *1000 Solved Problems in Heat Transfer*, Schaum's Outline Series, McGraw-Hill, New York, 1991.

[6] Hicks, T.G., Ed., *Standard Handbook of Engineering Calculations*, McGraw-Hill, New York, 1972.

[7] Anon., *Engineering Data Book*, Gas Processors Suppliers Association, Tulsa, OK, 1972.

[8] Incropera, F.P. and D.P. DeWitt, *Fundamentals of Heat and Mass Transfer, 4th Ed.*, John Wiley & Sons, New York, 1996.

[9] Reynolds, W.C., *Thermodynamic Properties in SI: Graphs, Tables and Computational Equations for 40 Substances*, Dept. of Mechanical Engineering, Stanford University, Stanford, CA, 1979.

[10] Gallagher, J.S., D. Linsky, G. Morrison and J.M.H. Levelt Sengers, *Thermodynamic Properties of a Geothermal Working Fluid; 90% Isobutane-10% Isopentane*, NBS Tech. Note 1234, National Bureau of Standards, U.S. Gov. Printing Office, Washington, DC, 1987.

[11] Demuth, O.J., C.J. Bliem, G.L. Mines and W.D. Swank, "Supercritical Binary Geothermal Cycle Experiments With Mixed-Hydrocarbon Working Fluids and a Vertical, In-tube, Counterflow Condenser," EGG-EP-7076, Idaho National Engineering Laboratory, 1975.

[12] IAPWS, *Guideline on the Tabular Taylor Series Expansion (TTSE) Method for Calculation of Thermodynamic Properties of Water and Steam Applied to IAPWS-95 as an Example*, International Association for the Properties of Water and Steam, Vejle, Denmark, August 2003.

[13] Milora, S.L. and J.W. Tester, *Geothermal Energy as a Source of Electric Power: Thermodynamic and Economic Criteria*, The MIT Press, Cambridge, MA, 1976.

[14] Anon., *1997 ASHRAE Handbook Fundamentals*, Chap. 18, American Society of Heating, Refrigeration and Air-Conditioning Engineers, Inc., Atlanta, GA, 1997.

[15] Moran, M.J. and H.N. Shapiro, *Fundamentals of Engineering Thermodynamics, 5th Ed.*, John Wiley & Sons, New York, 2004.

[16] DiPippo, R., "High-Efficiency Geothermal Plant Designs," *Geothermal Resources Council TRANSACTIONS*, V. 21, 1997, pp. 393–398.

[17] Khalifa, H.E. and B.W. Rhodes, "Analysis of Power Cycles for Geothermal Wellhead Conversion Systems," EPRI AP-4070, Electric Power Research Institute, Palo Alto, CA, 1985.

[18] Bliem, C.J. and L.F. Walrath, "Raft River Binary-Cycle Geothermal Pilot Power Plant Final Report," EGG-2208, Idaho National Engineering Laboratory, Idaho Falls, ID, 1983.

[19] Hinrichs, T.C. and B.W. Dambly, "East Mesa Magmamax Power Process Geothermal Generating Plant: A Preliminary Analysis," EPRI TC-80-907, *Proceedings Fourth Annual Geothermal Conf. and Workshop*, Electric Power Research Institute, Palo Alto, CA, 1980, pp. 5-1–5-14.

[20] Hinrichs, T.C., "Magmamax Power Plant – Success at East Mesa," EPRI AP-3686, *Proceedings Eighth Annual Geothermal Conf. and Workshop*, Electric Power Research Institute, Palo Alto, CA, 1984, pp. 6-21–6-30.

[21] Bliem, C.J. and G.L. Mines, "Advanced Binary Performance Power Plants: Limits of Performance," EGG-EP-9207, Idaho National Engineering Laboratory, Idaho Falls, ID, 1991.

[22] DiPippo, R., "Geothermal Energy: Electricity Generation and Environmental Impact," *Renewable Energy: Prospects for Implementation*, T. Jackson, Ed., Stockholm Environment Inst., Sweden, 1993, pp. 113–122.

[23] Moskvicheva, V.N. and A.E. Popov, "Geothermal Power Plant on the Paratunka River," *Geothermics – Special Issue 2, U.N. Symposium on the Development and Utilization of Geothermal Reseources*, Pisa, V. 2, Pt. 2, 1970, pp. 1567–1571.

Nomenclature for figures in Chapter 8

ACC	Air-cooled condenser
C	Condenser
CP	Condensate pump
CSV	Control & stop valves
CT	Cooling tower
CWP	Cooling water pump
E	Evaporator
FF	Final filter
	Saturated liquid
g	Saturated vapor
HPE	High-pressure evaporator
HPFP	High-pressure feed pump
HPPH	High-pressure preheater
HPT	High-pressure turbine
IP	Injection pump
IW	Injection well
LPE	Low-pressure evaporator
LPFP	Low-pressure feed pump
LPPH	Low-pressure preheater
LPT	Low-pressure turbine
M	Make-up water
P	Pump
PH	Preheater
PW	Production well
RH	Reheater
RHP	Recuperative preheater
S	Separator
SH	Superheater
SR	Sand remover
T/G	Turbine/generator
TV	Throttle valve

Problems

8.1 A counterflow, double-pipe heat exchanger is used in a geothermal binary plant to heat isobutane, $i\text{-}C_4H_{10}$, from 20 to 70°C. The $i\text{-}C_4H_{10}$ flows at 100 kg/s and has a $c_p = 0.55$ Btu/lbm°F; the brine enters at 100°C and leaves at 45°C. The overall heat transfer coefficient is $U = 100$ Btu/h·ft²·°F. Calculate:
   (a) Required brine mass flow rate in kg/s.
   (b) Required heat transfer surface area in ft².
   (c) Second Law efficiency in %, if $T_o = 15$°C.

8.2 A hydrocarbon turbine in a binary plant receives isobutane, $i\text{-}C_4H_{10}$ at 600 lbf/in² and 290°F. The turbine exhausts at 40 lbf/in². The isentropic efficiency is 88%. Calculate:
   (a) Work output of the turbine.
   (b) Temperature of the $i\text{-}C_4H_{10}$ at turbine exhaust.
   (c) Ratio of exhaust-to-inlet volume.
   (d) Heat that must be removed from the $i\text{-}C_4H_{10}$ after expansion (at P = constant) before condensation begins, i.e., the desuperheat in the turbine exhaust.

8.3 Consider a simple Rankine cycle using R-12 as the working fluid. Hot geothermal liquid is used to heat the R-12. The turbine inlet conditions are saturated vapor at 180°F; the turbine has an isentropic efficiency of 82%, and the condensing temperature is 100°F. The heat exchanger pinch-point temperature difference is 10°F.
   (a) Calculate the Rankine cycle thermal efficiency, η_{th}.
   (b) If the geofluid is available at 210°F, calculate the required mass flow rate of geofluid to generate 500 kW of <u>net</u> power from the cycle, assuming the pump is isentropic.
   (c) Determine the location of the pinch point in the heat exchanger.
   (d) Calculate the geothermal resource utilization efficiency η_u, and compare it with η_{th}.

8.4 Consider the brine/R-12 heat exchanger in Problem 8.3.
   (a) Calculate the log mean temperature difference, *LMTD*, separately (i) for the liquid R-12 heater and (ii) for the evaporator portions.
   (b) Estimate the required total heat exchanger surface area.
   (c) Calculate the efficiency of the heat exchanger based on the transfer of exergy.

8.5 Consider a simple geothermal binary plant consisting of a heat exchanger to vaporize the working fluid (R-12), a turbine ($\eta_t = 85\%$), a condenser, and a simple feed pump ($\eta_p = 75\%$). The turbine inlet conditions are $P_1 = 400$ lbf/in² and $T_1 = 300$°F; the turbine exhaust pressure is $P_2 = 100$ lbf/in². The condenser outlet condition is $P_3 = 100$ lbf/in², saturated liquid. The pump outlet and heat exchanger inlet pressure is $P_4 = 450$ lbf/in².
   Calculate the following with the aid of R-12 property tables:
   (a) Specific work of the turbine, w_t, Btu/lbm.
   (b) Specific work of the pump, w_p, Btu/lbm.
   (c) Specific heat added, q_a, Btu/lbm.
   (d) Specific heat rejected, q_c, Btu/lbm.
   (e) Cycle thermal efficiency (First Law), η_t, %.

8.6 The plant described in Problem 8.5 is supplied with brine at $T_A = 330°F$. The heat exchanger pinch-point temperature difference is 20°F. The brine may be taken as having a constant specific heat, $c = 0.98$ Btu/lbm°F.
Calculate the following with the aid of a *P-h* diagram for R-12:
(a) The ratio of the brine mass flow rate to the R-12 mass flow rate.
(b) The brine exit temperature, T_B, °F.
(c) The Second Law efficiency of the heat exchanger based on the exergy given up the brine and the exergy received by the R-12. Use a dead-state temperature $T_o = 60°F$.
If the plant must produce a net cycle power output of 10,000 kW, calculate:
(d) The mass flow rate of brine, in lbm/h.
(e) The mass flow rate of R-12, in lbm/h.
(f) The plant Second Law efficiency for the above-stated net cycle power.
The plant parasitics include (i) well pumping power and (ii) cooling system pumps and fans. If the brine pumps require 4.2 Btu/lbm brine, and the total cooling system takes 500 kW, calculate:
(g) The bottom-line, net power, and the overall plant Second Law efficiency.

8.7 The original Larderello geothermal power station employed binary-type units in which geothermal steam (saturated vapor) was used to heat and evaporate pure water which circulated in a close, simple Rankine cycle. See the simplified plant schematic below.

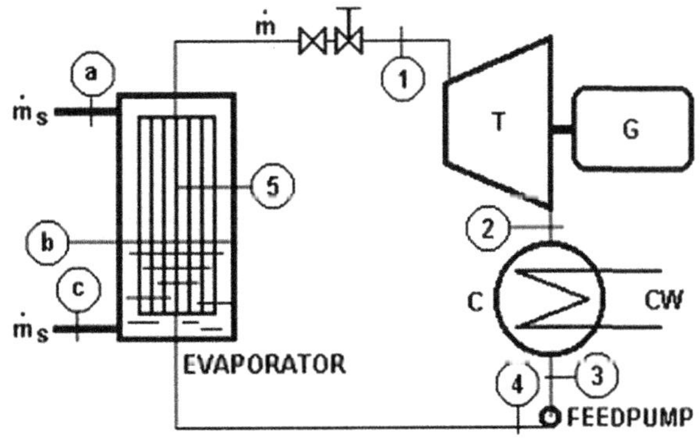

The following data are given:

$P_a = P_b = P_c = 1.0$ MPa	$T_c = 130°C$ (subcooled liquid)
turbine wet efficiency = 0.77	$T_1 = 160°C$ (saturated vapor)
feed pump efficiency = 1.00	$T_3 = 50°C$ (saturated liquid)

(a) Sketch the plant processes in a *T-s* diagram. Label all states using the notation given in the schematic.
(b) Calculate (in kJ/kg for pure water):
1. Turbine work, w_t.
2. Feed pump work, w_p.
3. Net cycle work, w_{net}.

(c) Calculate the net cycle thermal efficiency, η_{th} (in %).
(d) If the plant is to produce 10,000 kW of net power, calculate:
   1. Mass flow rate of pure water, in kg/s.
   2. Mass flow rate of geothermal steam, in kg/s.
(e) For a dead-state temperature of 25°C, find the Second Law resource utilization efficiency, η_u (in %).

8.8 A modular binary power plant uses propane, C_3H_8, as the cycle working fluid The geothermal brine is pumped from the reservoir and reaches the plant at a temperature $T_A = 420$ K. The cycle employs a supercritical pressure in the main heat exchanger, where the pinch-point temperature difference is 5°C. The inlet conditions at the propane turbine are $T_1 = 400$ K and $P_1 = 5.0$ MPa; the turbine exhausts at $P_2 = 1.0$ MPa. The turbine has an isentropic efficiency of 85% and the feed pump has an isentropic efficiency of 80%. The gross turbine output is 1,500 kW. You may ignore pressure losses in heat exchangers and piping; the specific heat of the brine may be taken as constant and equal to 4.1 kJ/kg · K.
(a) Prepare a system schematic and the corresponding temperature-entropy process diagram.
(b) Sketch the temperature-heat transfer diagram for the supercritical heat exchanger showing the brine cooling line and the propane heating curve.
(c) Determine the brine outlet temperature from the heat exchanger.
(d) Calculate the mass flow rates in kg/s for (i) propane and (ii) brine.
(e) Calculate the net cycle thermal efficiency in %.
(f) Calculate the net plant utilization efficiency for a dead-state at 20°C in %.

Hints: Using a *P-h* diagram for propane, lay out the propane heating curve to scale in a *T-q* diagram. Knowing the brine inlet temperature, adjust the slope of the brine cooling curve to obtain the correct pinch-point temperature difference. The brine outlet temperature may then be read from the diagram.

8.9 Repeat the example worked in Sect. 8.5 with the following change to the given data: Instead of a boiler pressure of 2 MPa, use 0.8 MPa, otherwise use all the other data as given. Compare your results to those found in the example and discuss your findings.

Chapter 9

Advanced Geothermal Energy Conversion Systems

Geothermal Power Plants: Principles, Applications, Case Studies and Environmental Impact, Third Edition

© 2012 Elsevier Ltd. All rights reserved.

"Why should we expect nature to be interested either positively or negatively in the purposes of human beings, particularly purposes of such unblushingly economic tinge?"

Percy W. Bridgman, on the thermodynamic prohibition against perpetual motion machines – 1941

9.1 Introduction

Geothermal resources are like human fingerprints: no two are exactly alike. Therefore energy conversion systems must be chosen and often adapted to suit the particular resource. We have seen how single- and double-flash systems are used for moderate- and high-temperature liquid-dominated resources, how dry-steam plants can be used at the unique dry-steam resources, and how binary plants utilize the lower-temperature liquid-dominated resources.

By way of review, the four basic types of geothermal energy conversion systems that were covered in Chapters 5–8 are summarized in Figs. 9.1 and 9.2 and in Table 9.1.

There are some geothermal resources that demand more sophisticated energy conversion systems than the basic ones we have considered so far. Furthermore, energy conversion systems have evolved to fit the needs of specific developing fields by integrating different types of power plant into a unified, complex enterprise.

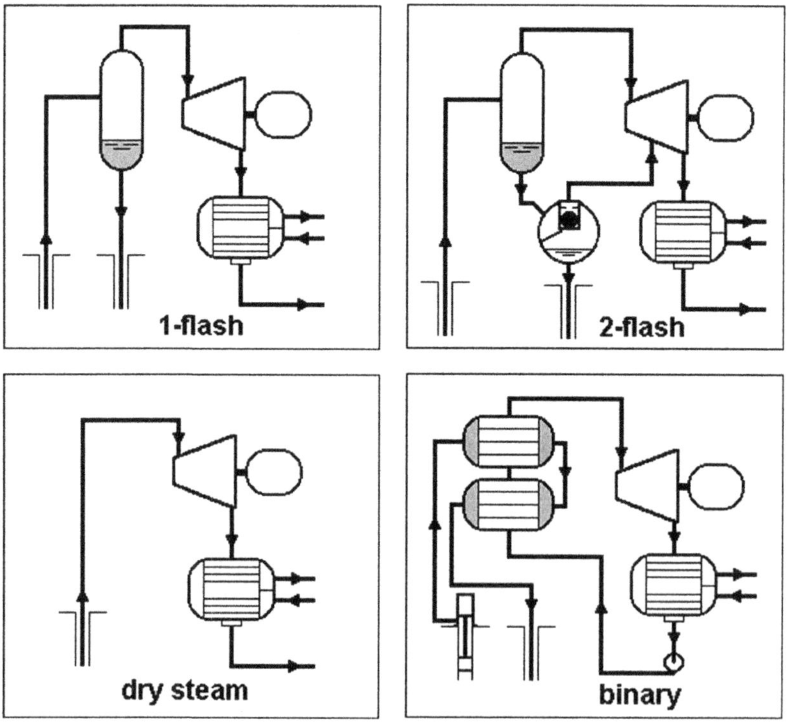

Fig. 9.1 Basic geothermal energy conversion systems: simplified plant layouts.

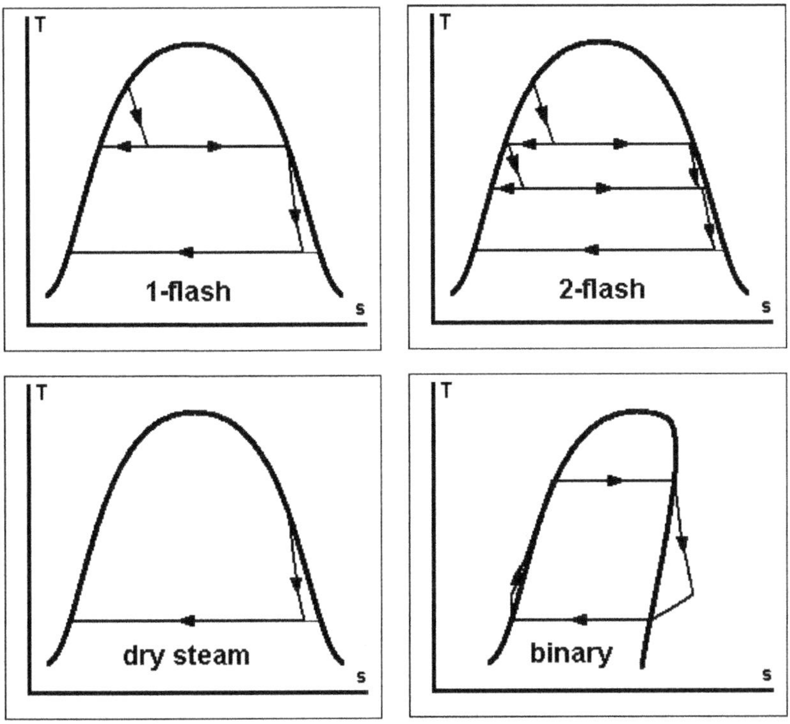

ig. 9.2 Basic geothermal energy conversion systems: simplified process diagrams.

able 9.1 Comparison of basic geothermal energy conversion systems.

ype of plant	Reservoir temperatures, °C	Utilization efficiency, %	Plant cost and complexity	Current usage
ingle-flash	200–260	30–35	moderate	widespread
ouble-flash	240–320	35–45	moderate → high	widespread
ry-steam	180–300+	50–65	low → moderate	special sites
asic binary	125–165	25–45	moderate → high	widespread

In this chapter we will look at some of the advanced systems for geothermal power eneration. Many of these have already achieved commercial status by improving the fficiency of resource utilization. Others are in various stages of development but hold romise to meet future geothermal power needs.

.2 Hybrid single-flash and double-flash systems

et us begin by considering how two of the systems we have already studied can e combined to form a hybrid type of power plant. Given the relative simplicity and relibility of single-flash plants, they are often the first type of plant installed at a newly develped field. However, their utilization efficiency is lower than that of a double-flash plant, nd there usually comes a time in the life of a field when expansion of the generating capacity becomes possible. When this happens, say because step-out wells have been successful or the electricity demand rises, it is logical to add another power unit.

Since single-flash plants have a significant amount of waste liquid from their separators that is still fairly hot, typically 150–170°C, the question arises as to whether this could be used to generate more power instead of being directly reinjected. At several fields around the world, the answer has been "Yes," and combined single- and double-flash plants have been built. Here we will deal with the thermodynamics of such arrangements, leaving discussion of actual plants to the third part of the book.

9.2.1 Integrated single- and double-flash plants

When the geofluid reservoir temperature is about 220–240°C and single-flash units have been built and have been operating for some time, the addition of one more flash using the separated brine allows for a lower pressure unit. The arrangement shown in Fig. 9.3 consists of two single-flash units, Units 1 and 2, and a third unit, Unit 3. Taken by itself, Unit 3 appears to be simply another single-flash unit, but the power plant as a whole is an integrated single- and double-flash facility since the original geofluid experiences two stages of flashing.

The advantage to this arrangement is that no new wells need to be drilled to supply the third unit. Unit 3 serves as a bottoming unit to recover some of the wasted potential from the still-hot brine. The thermodynamic process diagram is given in Fig. 9.4 in temperature-entropy coordinates.

One possible thermodynamic drawback to this arrangement lies in the selection of the pressure (or equivalently, the temperature) for the second flash process 3–6 in Fig. 9.4. Recall that we found the optimum flash conditions for a true double-flash plant were

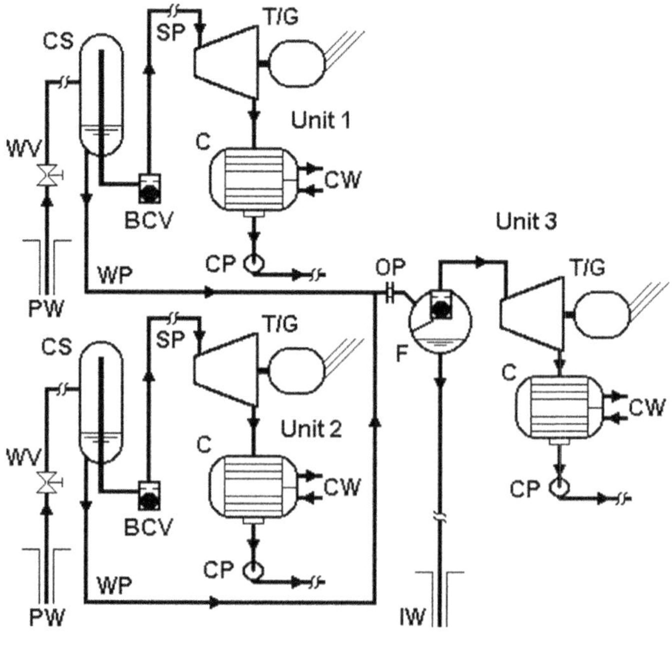

Fig. 9.3 Integrated single- and double-flash facility [1].

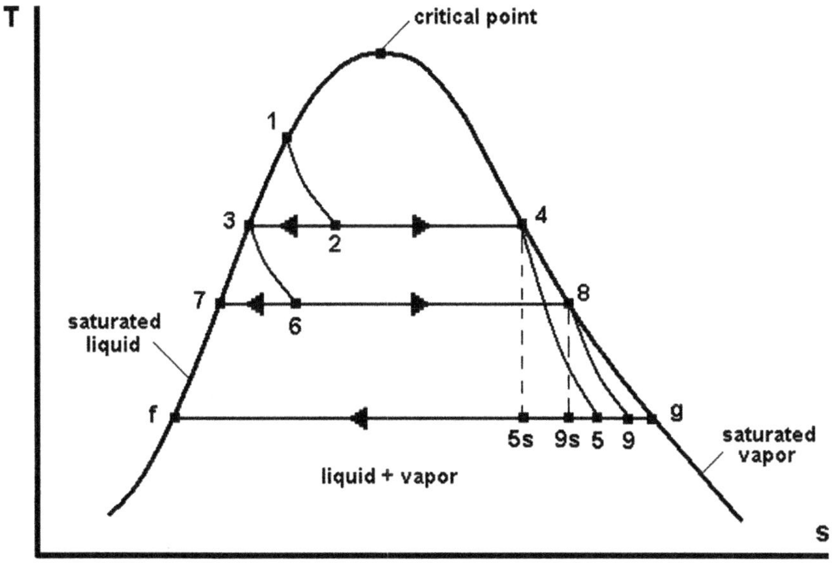

ig. 9.4 *Thermodynamic temperature-entropy process diagram for integrated single- and double-flash plant shown* ı *Fig. 9.3.*

uch that the temperature differences between the reservoir, the first and second flashes, nd the condenser were all roughly equal. If the flash temperatures for the first two units assumed identical) had been optimized according to this rule, then they will be too low vhen the third unit is added. Furthermore it may not be feasible to redesign Units 1 and ' owing to the changes that would have to be made to the existing equipment. Nevertheless, the flash temperature for Unit 3 can still be selected to optimize the new ombination, given the existing brine condition at state 3.

.2.2 Combined single- and double-flash plants

Vhen the resource temperature is equal to or greater than say 240°C, it may be possi- le to augment the single-flash units with a true double-flash bottoming cycle, as seen n the schematic flow diagram, Fig. 9.5, and in the process diagram, Fig. 9.6. For this ase, the waste brine from the first units is subjected to two more flashes, resulting in wo additional low-pressure steam flows. It could be argued that this constitutes a riple-flash if one simply counts the number of flashes experienced by the geofluid, but iven the sequential timing of the construction of the units, the use of the term "com- ined single- and double-flash" seems appropriate.

Although the thermodynamics of this arrangement are favorable, i.e., a higher esource utilization efficiency than for the original single-flash plant, there may be pro- lems with chemical scaling due to silica precipitation at the low temperatures associ- ted with the last flash. This was discussed in Sect. 6.6. This arrangement would not e a good choice unless there is no possibility of silica precipitation or the plant owner s willing to invest in chemical treatment of the low temperature brine to prevent or ontrol scale formation.

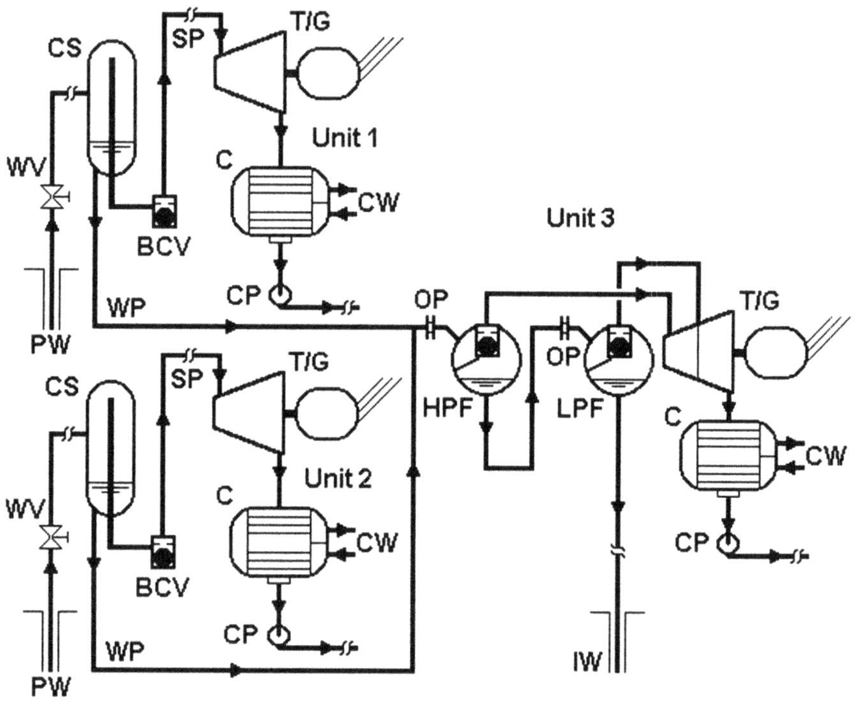

Fig. 9.5 Combined single- and double-flash plant [1].

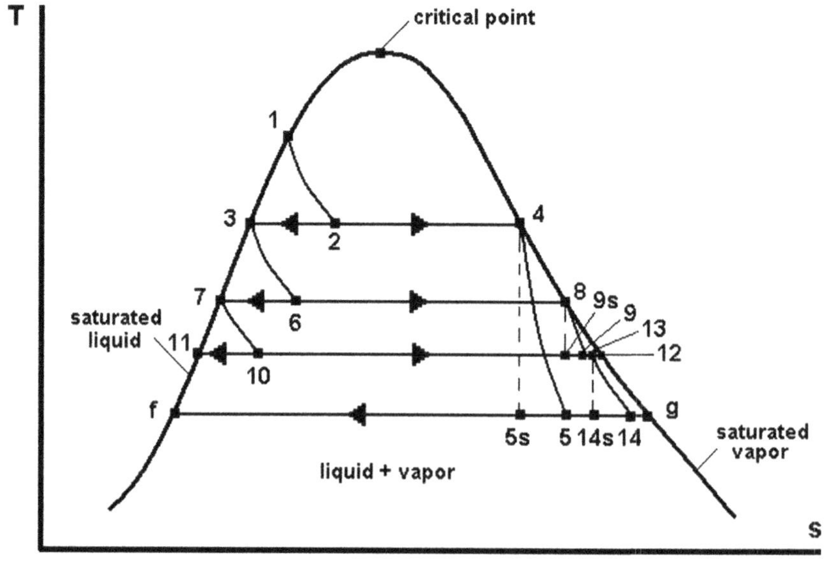

Fig. 9.6 Process diagram for combined single- and double-flash plant.

.3 Hybrid flash-binary systems

n attractive alternative to the use of bottoming flash plants at existing single-flash lants is to add a binary cycle. Combined flash-binary plants are in operation at everal plant sites around the world (see Appendix A).

A different approach is to design a plant, from scratch, as an integrated flash-binary lant, thereby taking advantage of the best features of both units. We will examine nese two strategic alternatives in the sections that follow.

.3.1 Combined flash-binary plants

or this case, we assume that a single-flash plant has been running for some time, sually a few years, and the reservoir has shown itself capable of sustaining operations or many more years. The power output can be raised by adding a binary unit between he separators and the reinjection wells. A simplified schematic of such an arrangement given in Fig. 9.7.

Initially the single-flash plant operated by itself and the waste liquid from the yclone separators CS was sent directly to the injection wells IW. The binary cycle is nserted as shown to tap into the reinjection pipeline where it extracts some heat and hereby lowers the temperature of the waste brine prior to injection. The additional ower generated by the binary cycle is gained without any new production wells.

The thermodynamic process diagram is presented in Fig. 9.8 in temperature-ntropy coordinates to facilitate comparison with the cycles in the previous sections. he power units are coupled thermodynamically through the preheater PH and the vaporator E. The First Law gives the relationship between the brine flow rate $\dot{m}_b$ from he wells (state 1) and that of the binary cycle working fluid $\dot{m}_{wf}$:

$$\dot{m}_b(1 - x_2)c_b(T_3 - T_7) = \dot{m}_{wf}(h_a - h_e) \tag{9.1}$$

his equation expresses the fact that the heat extracted from the waste brine is equal to he heat absorbed by the binary cycle working fluid, assuming perfect insulation on the eat exchangers. Solving for the working fluid flow rate, we find

$$\dot{m}_{wf} = \dot{m}_b(1 - x_2)\left[\frac{c(T_3 - T_7)}{h_a - h_e}\right] \tag{9.2}$$

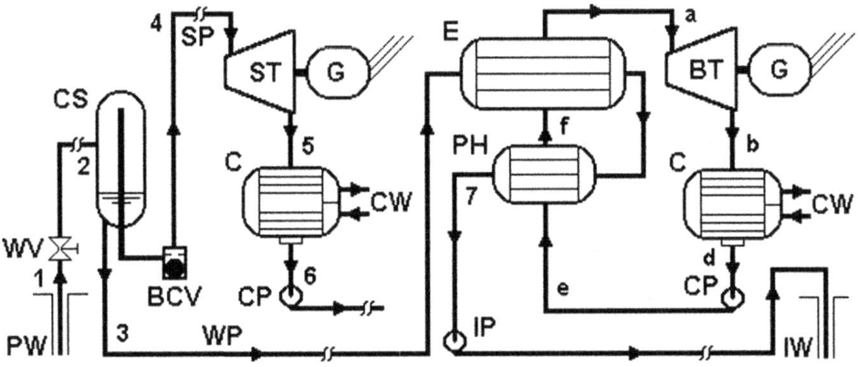

ig. 9.7 Combined single-flash and basic binary plant; after [1].

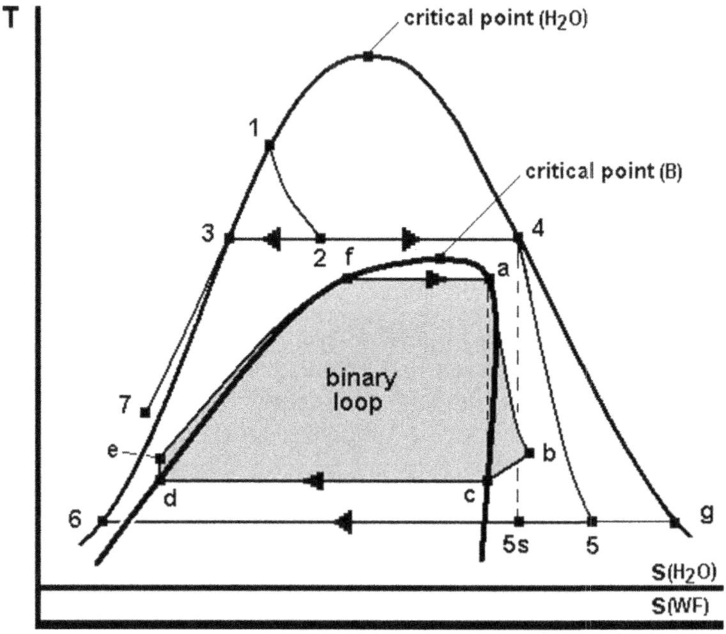

Fig. 9.8 Combined single-flash/binary plant: temperature-entropy process diagram.

Since the state points 1, 2, and 3 for the flash unit are fixed and the new state point 7 is subject to the constraint imposed by silica precipitation, only the binary cycle parameters are open for assignment. These, in turn, are subject to the constraint imposed by the pinch-point temperature difference in the heat exchangers. The analysis presented in Sect. 8.2 can be applied to this situation.

The fact that the waste brine is simply cooled rather than flashed to supply the bottoming cycle tends to avoid the silica scaling problem. With reference to Fig. 6.15, let us assume that the brine temperature is 240°C in the reservoir. The corresponding silica concentration is about 425 ppm. If the separator operates at 150°C, the silica concentration in the waste brine rises to 526 ppm owing to the removal of 19% (mass) of the geofluid in the form of steam. As can be seen from Fig. 6.15, the brine is safely undersaturated with respect to amorphous silica.

When the brine is cooled, its silica concentration remains the same and it becomes saturated when it cools to roughly 133°C. This temperature then is the safe lower limit at state 7; any further cooling places the brine in the supersaturated region and invites silica precipitation in the cool end of the heat exchanger, the reinjection piping, or wells. Therefore, the available temperature drop is about 17°C. If 1000 kg/s of waste brine are available, then the bottoming binary cycle should be able to generate about 7.3 MW assuming a conservative 10% cycle thermal efficiency.

.3.2 Integrated flash-binary plants

Vhen a binary cycle is integrated with a flash plant, the result is a plant with practically ero emissions. Where environmental concerns are significant, such plants have great ppeal. An integrated single-flash/binary plant is shown schematically in Fig. 9.9; the rocess diagram is given in two parts: Fig. 9.10 for the main, upper portion and Fig. 9.11 or the bottoming binary portion.

The geothermal steam first drives the back-pressure steam turbine and then is ondensed in the upper binary cycle's evaporator E. The two turbines in the upper art of the plant may be connected to a common generator, as shown. The separated rine (state 3) is used to preheat and evaporate the working fluid in the lower binary ycle. The noncondensable gases flow with the steam through the steam turbine T and into the evaporator where they are isolated, removed, and compressed for ecombination with the waste brine just before being reinjected. The brine holding ank BHT collects all the steam condensate, waste brine, and compressed gases that o back into solution.

In principle, this plant has no emissions to the surroundings. The only environmental mpact is the heat rejected to the atmosphere from the binary cycle condensers. The chematic shows water-cooled condensers but air-cooling is an option.

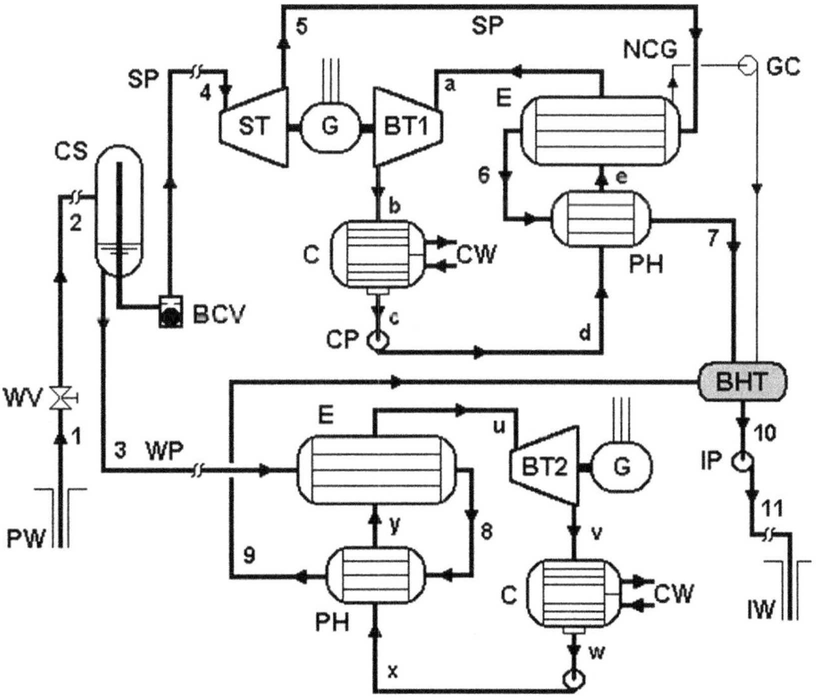

ig. 9.9 *Integrated single-flash/binary plant.*

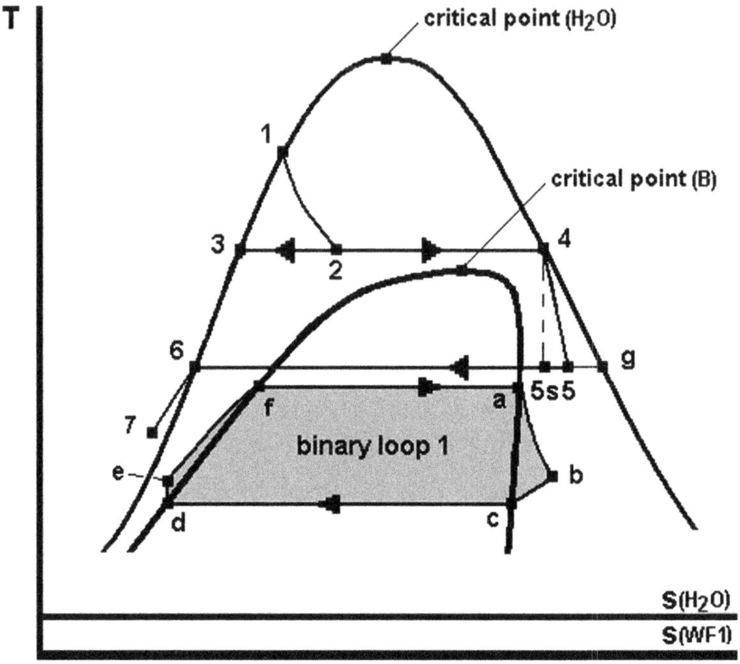

Fig. 9.10 Process diagram for an integrated single-flash/binary plant: upper plant.

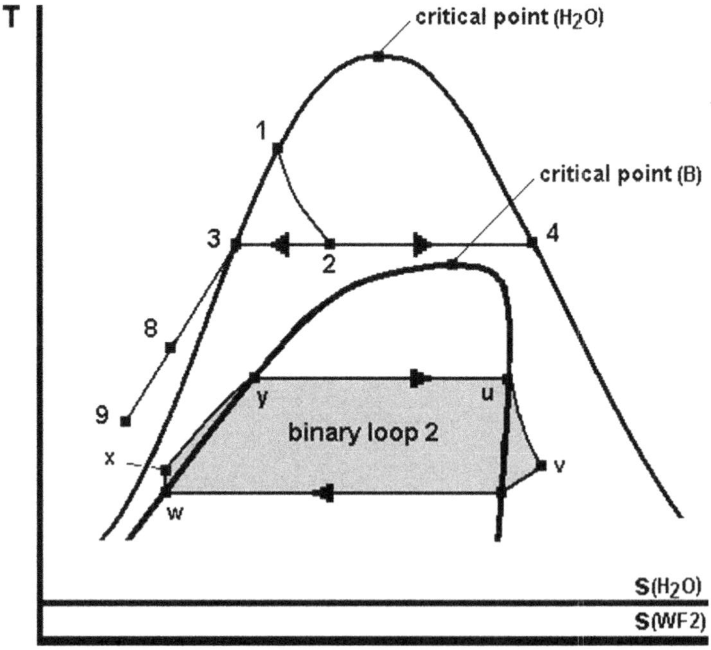

Fig. 9.11 Process diagram for the bottoming binary portion of an integrated single-flash/binary plant: main portion omitted for clarity; see Fig. 9.10.

There is a trade-off involved in this arrangement in that the steam expansion process is cut short relative to the combined flash-binary plant. The turbine is called a back-pressure turbine because its exhaust pressure is much higher than would be the case if it exhausted to a water-cooled condenser. The work that would be obtained from the lower-pressure stages of the turbine is forfeited in favor of using the heat of condensation (from process 5–6) to vaporize the working fluid in the binary loop. The bottoming binary cycle is a means of capturing some of the potential of the hot, separated brine, and raises the plant total output without the need for additional wells.

In the following section we will present a numerical example to illustrate the analysis of this relatively complex system.

9.4 Example: Integrated flash-binary hybrid system

This example is drawn on the case just studied and will use Figs. 9.9–9.11. Although we will select arbitrary values for the geofluid conditions and the cycle parameters, the results will be fairly realistic. Here are the assumed conditions:

- Reservoir temperature, $T_1 = 250°C$
- Brine main flow rate, $\dot{m}_b = 500$ kg/s
- Dead-state temperature, $T_0 = 25°C = 298.15$ K
- Flash separator temperature, $T_2 = 185°C$
- Steam turbine exhaust pressure, $P_5 = 1.5$ bar
- Steam condensate outlet temperature, $T_7 = 65°C$
- Brine outlet temperature, $T_9 = 165°C$
- Binary cycle working fluid (both loops): n-pentane, C_5H_{12}
- Loop 1 turbine inlet temperature, $T_a = 380$ K (saturated)
- Loop 2 turbine inlet temperature, $T_u = 420$ K (saturated)
- Loop 1 & 2 condensing temperature, $T_d = T_w = 310$ K
- Steam turbine isentropic efficiency, $\eta_{st} = 0.80$
- C_5 turbine isentropic efficiency, $\eta_{bT1} = \eta_{bT2} = 0.85$
- C_5 feed pump isentropic efficiency, $\eta_{CP1} = \eta_{CP2} = 0.75$

Our objectives are to find:

1. Power output of the steam turbine, $\dot{W}_{ST}$
2. Net power output of the upper binary cycle (loop 1), $\dot{W}_{B1,net}$
3. Net power output of the lower binary cycle (loop 2), $\dot{W}_{B2,net}$
4. Cycle thermal efficiency for both loops, $\eta_{b1,th}$ and $\eta_{b2,th}$
5. Overall plant utilization efficiency, η_u.

The thermodynamic properties for the brine and steam are found from the *Steam Tables* of Keenan *et al* [2] and those of n-pentane were taken from Reynolds' data book [3]. The table on page 195 summarizes the important properties at all the state points in the plant.

The methodology used in the example of Sect. 8.5 has been applied here. The n-pentane mass flow rate in loop 1 can be found from the coupling equation on the preheater-evaporator combination:

$$\dot{m}_{C_5} = \dot{m}_b\, x_2 \left[\frac{h_5 - h_7}{h_a - h_e}\right] \tag{9.3}$$

where we will use the actual enthalpy values for the brine instead of the constant-specific-heat approximation shown in eq. (9.2). The mass flow rate of brine from the separator is 75.11 kg/s and from this we find $\dot{m}_{C_5} = \mathbf{360.35\ kg/s}$. It is now possible to calculate our first objective:

1. Power output of the steam turbine:

$$\dot{W}_{ST} = \dot{m}_4(h_4 - h_5) = \mathbf{20{,}974\ kW} \tag{9.4}$$

Next we can assess the upper binary cycle, loop 1. The turbine power is found from

$$\dot{W}_{BT1} = \dot{m}_{C_5}(h_a - h_b) = 27{,}862\ \text{kW} \tag{9.5}$$

and the feed pump power is found from

$$\dot{W}_{CP1} = \dot{m}_{C_5}(h_e - h_d) = 461\ \text{kW} \tag{9.6}$$

Thus the second objective may be determined:

2. Net power output of the upper binary cycle (loop 1):

$$\dot{W}_{B1,net} = \dot{W}_{BT1} - \dot{W}_{CP1} = \mathbf{27{,}401\ kW} \tag{9.7}$$

Next we assess the lower binary cycle, loop 2, in the same fashion:

$$\dot{m}_{C_5} = \dot{m}_b(1 - x_2)\left[\frac{h_3 - h_9}{h_u - h_x}\right] \tag{9.8}$$

Since the separated brine flow rate is 424.89 kg/s (i.e., 500 − 75.11), we find the n-pentane flow for the lower cycle is $\dot{m}_{C_5} = \mathbf{71.18\ kg/s}$. Then the turbine and pump power follow directly:

$$\dot{W}_{BT2} = \dot{m}_{C_5}(h_u - h_v) = 6{,}232\ \text{kW} \tag{9.9}$$

and

$$\dot{W}_{CP2} = \dot{m}_{C_5}(h_x - h_w) = 91\ \text{kW} \tag{9.10}$$

Thus the third objective can be found:

3. Net power output of the lower binary cycle (loop 2):

$$\dot{W}_{B2,net} = \dot{W}_{BT2} - \dot{W}_{CP2} = \mathbf{6{,}141\ kW} \tag{9.11}$$

tate	*P* MPa	*T* °C or K	*x*	*s* kJ/kg·K	*h* kJ/kg	Comments
team processes						
	3.973	**250°C**	0		1085.36	Saturated liquid
	1.123	**185°C**	0.1502		1085.36	$h_2 = h_1$
	1.123	**185°C**	0		785.37	Saturated liquid
	1.123	**185°C**	1	6.5465	2782.4	Saturated vapor
s	**0.15**		0.8831	6.5465	2433.34	$s_{5s} = s_4$
	0.15		0.9145		2503.15	From η_{ST}
	0.15		0		467.11	Saturated liquid
	0.15	**65°C**	n.a.		272.06	Approx. value
	0.15	459.3°C	n.a.			From *T-q* diag.
	0.15	**165°C**	n.a.		697.34	Approx. value
inary loop 1 – n-pentane processes						
	0.6869	**380 K**	1		609.07	Saturated vapor
s	**0.1047**		n.a.		518.1	Approx. value
	0.1047		n.a.		531.75	From η_{BT1}
	0.1047	**310 K**	1		497.27	Saturated vapor
	0.1047	**310 K**	0		142.75	Saturated liquid
s	**0.6869**		n.a.		143.71	From $v_d \, \Delta P$
	0.6869		n.a.		144.03	From η_{CP1}
	0.6869	**380 K**	0		319.76	Saturated liquid
inary loop 2 – n-pentane processes						
	1.505	**420 K**	1		669.51	Saturated vapor
	0.1047		n.a.		581.95	From η_{BT2}
v	**0.1047**	**310 K**	0		142.75	Saturated liquid
	1.505		n.a.		145.83	From η_{CP2}
	1.505	**420 K**	1		435.22	Saturated liquid

Vote: Numbers in boldface are given in the specifications or are obvious from them.

The cycle thermal efficiencies are the ratio of the net power output to the thermal power input for each loop. The two heat transfer terms are found from:

$$\dot{Q}_{IN1} = \dot{m}_{C_5}(h_a - h_e) = 167,577 \text{ kW} \tag{9.12}$$

and

$$\dot{Q}_{IN2} = \dot{m}_{C_5}(h_u - h_x) = 37,403 \text{ kW} \tag{9.13}$$

We can now find our fourth objective:

4. Cycle thermal efficiency for both loops:

$$\eta_{B1,th} = \frac{\dot{W}_{B1}}{\dot{Q}_{IN1}} = \mathbf{0.1635} \text{ or } \mathbf{16.4}\% \tag{9.14}$$

and

$$\eta_{B2,th} = \frac{\dot{W}_{B2}}{\dot{Q}_{IN2}} = \mathbf{0.1642} \text{ or } \mathbf{16.4}\% \tag{9.15}$$

The last objective requires us to find the exergy of the original reservoir fluid, taken to be a saturated liquid at the reservoir temperature. The specific exergy is

$$e_1 = h_1 - h_0 - T_0[(s_1 - s_0)] = 257.37 \text{ kJ/kg} \tag{9.16}$$

and the rate at which exergy is produced from the reservoir is

$$\dot{E}_1 = \dot{m}_1 e_1 = 128,683 \text{ kW} \tag{9.17}$$

The net power of the whole plant (ignoring parasitic loads such as cooling tower fans and pumps) is

$$\dot{W}_{plant} = \dot{W}_{ST} + \dot{W}_{B1,net} + \dot{W}_{B2,net} = 54,516 \text{ kW} \tag{9.18}$$

Thus, we find our last objective:

5. Overall plant utilization efficiency:

$$\eta_u = \frac{\dot{W}_{plant}}{\dot{E}_1} = \mathbf{0.4236} \text{ or } \mathbf{42.4\%} \tag{9.19}$$

Discussion: Given the magnitude of the power output for the upper binary cycle, it is likely that several smaller units would be deployed rather than one large turbine-generator. Probably nine units of roughly 3 MW capacity would work out well. Similarly, the lower loop would probably have two units. Smaller units can be modular, largely factory assembled and tested, and ready for rapid installation.

The cycle thermal efficiencies are somewhat on the high side but remember that we did not account for the power needed to run the fans for air-cooled condensers or for the cooling towers, depending on which are used. For a plant of this size it is likely that several hundred kilowatts would be needed in either case. The fact that both cycles turned out to have nearly the same efficiency is merely a coincidence.

The utilization efficiency of 42.4% is very respectable and compares closely with what one might obtain from a double-flash plant.

A plant similar to this design is operating on Leyte in The Philippines [4]. The photo, Fig. 9.12, shows that air-cooled condensers occupy much more area than would water cooling towers, but they have the advantage of not requiring an external water supply to satisfy the makeup water needs of wet cooling towers.

9.5 Total-flow systems

The notion of a total-flow system arises from a desire to avoid the irreversibilities associated with the flashing processes needed for either a single- or double-flash plant. Also the separation process, while not inherently lossy, requires large pressure vessels and separate outlet pipelines for steam and water, all of which adds to the capital cost of the plant. If a way could be devised to use the geofluid directly as it emerges from the well in the prime mover, be it a turbine or some other specially designed device, significant savings would be achieved.

There have been three major development efforts aimed at producing such a system. One involved a single-stage pure impulse turbine [5], another used a positive-displacement

Fig. 9.12 Upper Mahiao integrated flash-binary 125 MW power plant; photo courtesy of Ormat [4] [WWW].

expander (a helical screw expander) [6,7], and the third was a novel concept of a rotary separator turbine [8]. The first one was a project of the U.S. Dept. of Energy in the late 1970s and the other two were private projects that each received funding support from government and industrial organizations.

9.5.1 Axial-flow impulse turbine

The fundamental simplicity of a total-flow machine can best be visualized using a temperature-entropy process diagram; see Fig. 9.13.

If machinery could be designed and built to follow the processes in Fig. 9.13, the power plant would consist of only a turbine (2–3) and its generator, a condenser (3–4), and a cooling tower. An artist's conception of such a plant is given in Fig. 9.14 [9].

The analysis of this plant is simple; the turbine power output is

$$\dot{W}_{TF} = \dot{m}_2(h_2 - h_3) = \dot{m}_2\eta_{TF}(h_2 - h_{3s}) \tag{9.20}$$

The crucial factor is the isentropic total-flow turbine efficiency:

$$\eta_{TF} = \frac{h_2 - h_3}{h_2 - h_{3s}} \tag{9.21}$$

A typical steam turbine has an efficiency of at least 80%. For the total-flow system to compete with a conventional flash plant the efficiency of the new machine must be roughly 40–50%. This can be seen from the following argument; see Fig. 9.15. It is

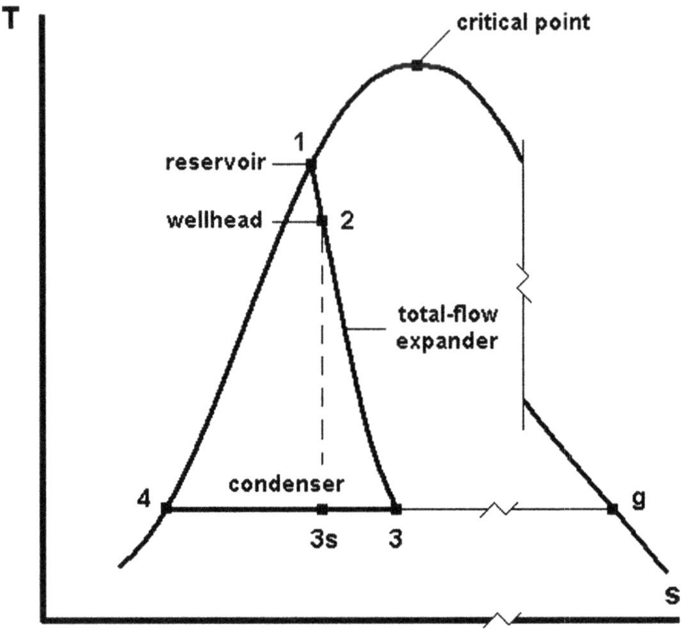

Fig. 9.13 Total-flow expander concept in temperature-entropy coordinates.

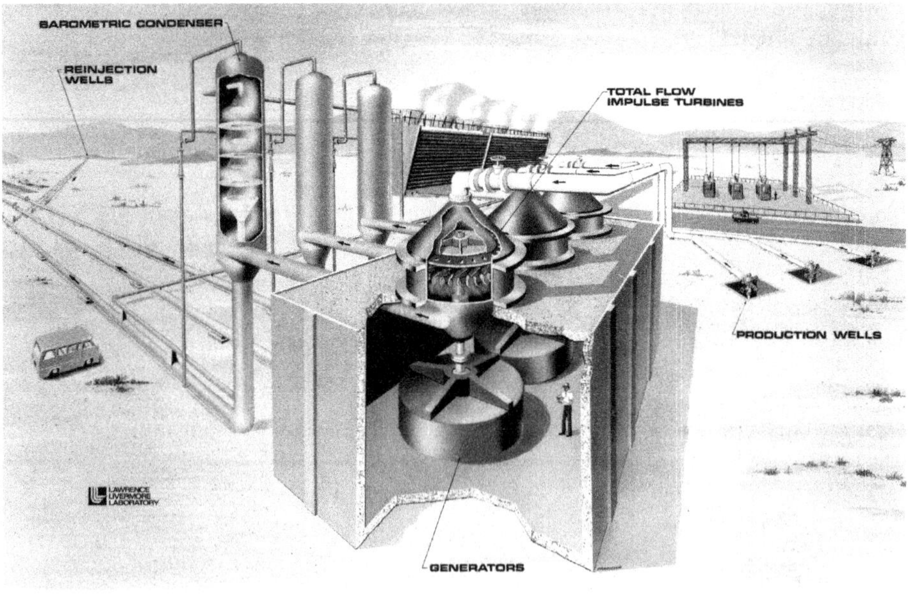

Fig. 9.14 Artist's concept of a total-flow geothermal power plant [9].

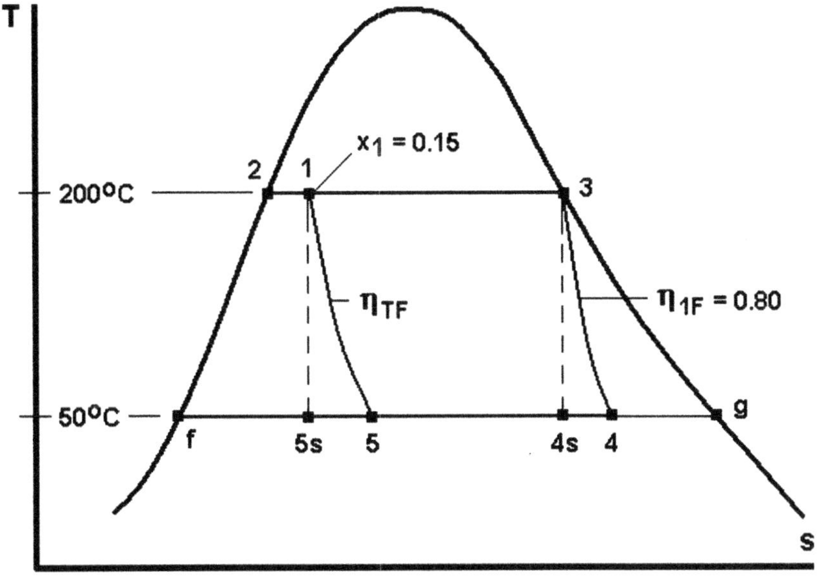

'ig. 9.15 *Comparison between total-flow and single-flash turbines.*

10t possible to make an absolute comparison owing to the wide variation in geother-nal fluid conditions and condensing temperatures, so a set of typical conditions were :hosen for this comparison and are shown in the figure. For the two turbines to be equivalent, they must produce the same power output:

$$\dot{W}_{TF} = \dot{m}_1(h_1 - h_5) = \dot{m}_1\eta_{TF}(h_1 - h_{5s}) = \dot{W}_{1F} \\ = \dot{m}_1 x_1(h_3 - h_4) = \dot{m}_1 x_1 \eta_{1F}(h_3 - h_{4s}) \tag{9.22}$$

Solving for the total-flow efficiency, we obtain

$$\eta_{TF} = \eta_{1F} x_1 \left[\frac{h_3 - h_{4s}}{h_1 - h_{5s}}\right] \tag{9.23}$$

vhere the two isentropic turbine outlet enthalpies are found from

$$h_{5s} = h_f + \left[\frac{s_1 - s_f}{s_{fg}}\right] h_{fg} \tag{9.24}$$

and

$$h_{4s} = h_f + \left[\frac{s_3 - s_f}{s_{fg}}\right] h_{fg} \tag{9.25}$$

For the selected conditions, the total-flow machine needs an efficiency of 42% to equal the output of the 80% efficient single-flash steam turbine. If the wellhead quality is).10, then the break-even efficiency becomes 49.2%. These two cases would corre-spond roughly to reservoir temperatures of 262°C and 242°C, respectively. The best value obtained on a prototype, single-nozzle, total-flow impulse turbine (see Fig. 9.16)

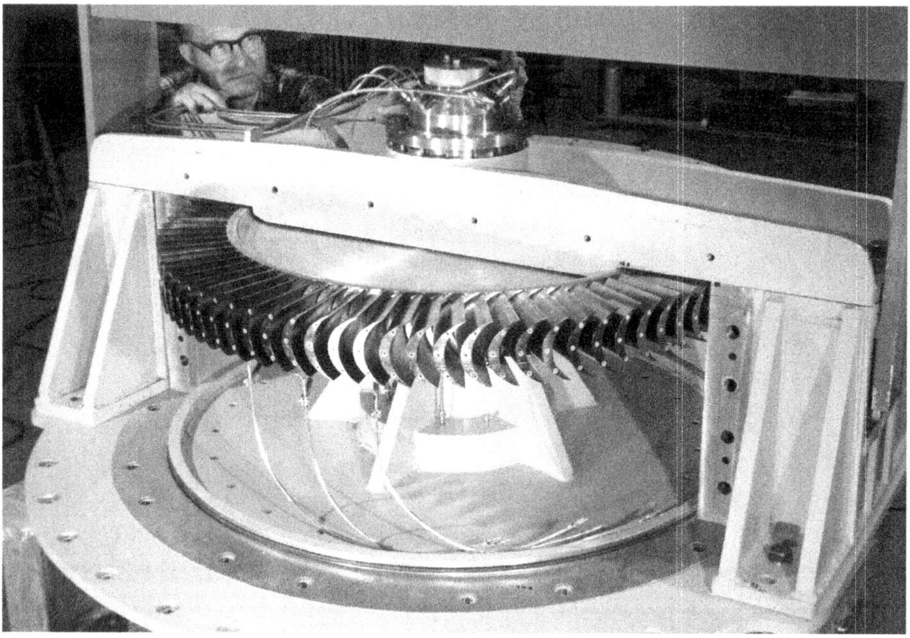

Fig. 9.16 Lawrence Livermore Laboratory prototype impulse total-flow turbine [9] [WWW].

was 23% in laboratory tests, although it was estimated that a full-admission turbine would operate at about 45% efficiency [5]. The prototype was never installed at a geothermal site nor developed further.

9.5.2 Rotary separator turbine

At about the same time that the total-flow turbine was being developed at the Lawrence Livermore Laboratory, a novel prime mover was also being designed and tested, a rotary separator turbine (RST) or the Biphase Turbine, after the company that invented it [8]. The features of the RST that distinguished it from other geothermal turbines included the ability to generate power using a two-phase, liquid-vapor stream, while simultaneously separating and pressurizing the liquid in preparation for reinjection. It did all this by means of five principal components:

- Two-phase nozzle
- Rotary separator
- Liquid turbine
- Liquid-transfer rotor
- Stationary diffuser.

A cross-section of the RST is shown in Fig. 9.17. In a sense, the RST generates power from the kinetic energy of the moving two-phase flow by capturing the liquid energy while allowing the phases to separate by centrifugal force. One could imagine

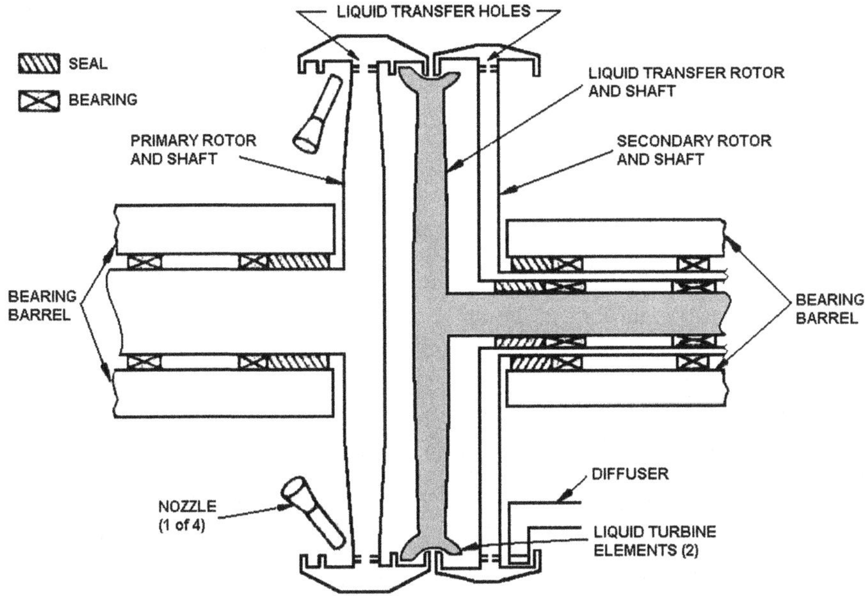

Fig. 9.17 Cross-section of the rotary separator turbine from Biphase Turbines [10].

a vertical cyclone separator mounted on a turntable connected to a generator; as the two-phase fluid swirled around the inner surface, the liquid drag would cause the vessel and the generator to turn. Of course that idea is impractical, whereas the RST is compact and has performed reasonably well in several field tests.

In its later field tests the unit was operated in conjunction with an upstream cyclone separator. The RST was packaged on a skid with a conventional steam turbine and a common generator. The liquid from the cyclone separator was sent to the RST while the steam went directly to the steam turbine. The quality of the fluid entering the two-phase nozzle is a critical parameter and this was controlled by feeding a small amount of separated steam into the nozzles. To assure equal flow to each of the four nozzles, the quality-adjusted geofluid flow was passed through a mixer-splitter unit before being admitted to the nozzles. The nozzles accelerate the flow to a high velocity and direct it onto the inner rim of the primary rotor. The liquid moves by centrifugal action and passes through the rim (liquid transfer holes) and impinges on the liquid transfer rotor where it gives up most of its kinetic energy. The shaft of this rotor is connected to the steam turbine shaft through a speed-increasing gearbox and so contributes to the power output of the packaged power plant. The spent liquid is transferred to the secondary rotor (again by passing through transfer holes in the web of the turbine rotor) where it is picked up by a stationary diffuser. This imparts a pressure boost to the liquid which may be sufficient to carry it to the injection wells without the need for booster pumps. The steam that is separated within the RST is removed through the RST casing and delivered to the steam turbine where it is admitted at a stage appropriate for its lowered pressure.

Extensive tests were conducted in 1981–82 at Roosevelt Hot Springs, Utah, U.S. the results are reported in Ref. [8]. The bottom-line machine efficiency was quoted as 27% at part load versus a 35% expected design value at full load.

Total-flow machines may also find application in vapor-compression refrigerators in place of the throttling control device where the power generated both offsets to some degree the compressor power requirements and increases the cooling effect in the evaporator [11].

9.5.3 Helical screw expander

The helical screw expander was invented by A. Lysholm [12] of Sweden who intended it for use as a compressor. It is a positive displacement machine in which a trapped volume of fluid is conveyed between two meshing, helical, counter-rotating elements. It is easier to visualize as a compressor in which the rotors are driven by an external motor and in which the volume decreases in the axial direction, thereby compressing the fluid to a higher pressure. However, if high-pressure fluid is admitted to the device at the small-volume side, it will expand through the rotors causing them to rotate and will produce work. A cross-section of the machine is shown in Fig. 9.18 [13].

In 1973, Sprankle won a patent for the use of a Lysholm expander with geothermal liquid-dominated resources [14]. Sprankle's company, Hydrothermal Power Company, built and tested helical screw expanders at several geothermal fields. These machines are rugged and handle difficult fluids, i.e., those that tend to deposit scale, quite well. The scrubbing action of the rotors against the inner surface of the casing tends to remove excess scale buildup; the scale that remains actually serves a useful purpose, namely, isolating the chambers to prevent fluid leakage in the axial direction.

The main obstacle that positive displacement expanders (PDEs), of any design, must overcome is that of a limited volume ratio, i.e., the ratio of the exit volume to the inlet volume. The engine tested in Ref. [13] could handle a maximum volume ratio of 5.3. When saturated liquid water is admitted to a PDE, it enters with a very low specific volume (high density), but as soon as it begins to expand it flashes into steam forming a two-phase mixture with a much greater specific volume. A study of this limitation [15]

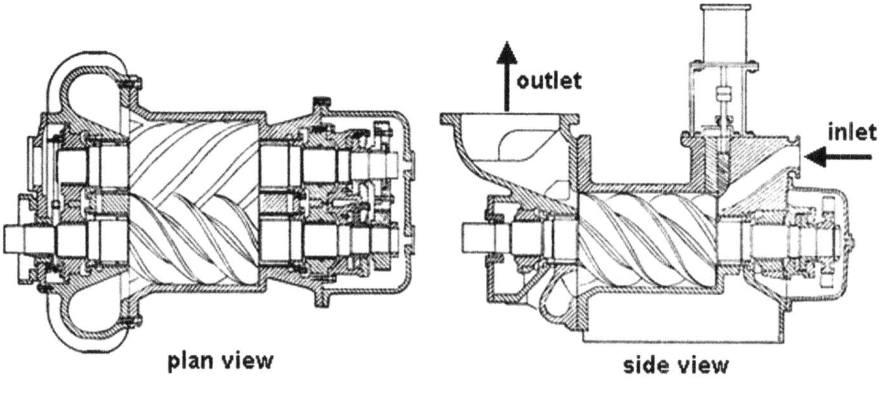

Fig. 9.18 Helical screw expander for geothermal applications [13].

oncluded that PDEs could not be competitive thermodynamically with double-flash lants unless PDEs could accommodate volume ratios of 10 or more. For example, at a esource with wellhead conditions of 200°C and 10% quality, a PDE with a volume atio of 5 would need to demonstrate a machine efficiency of 81.5% to be equivalent to double-flash steam plant under the same conditions. However, if the PDE can handle volume ratio of 10, then it only needs to show an efficiency of 43.2% to match the ouble-flash plant.

A way to mitigate this limitation is to introduce a small amount of steam to the nlet of the PDE. This will raise the inlet specific volume and for the same volume atio permit the expansion to go "deeper" into the two-phase region, producing a reater enthalpy drop and more work [15,16]. Two ways to accomplish this are (1) to re-flash the liquid through a small pressure drop or (2) mix in some steam from a vellhead separator. In both cases, the PDE would be used to supplement the output of standard flash-steam plant, most probably a double-flash plant. This is the same pproach described earlier for the RST.

Tests conducted in 1986 in New Zealand on the Model 76-1 of the Hydrothermal ower Company showed a maximum efficiency of 43% at an effective volume ratio of bout 18. The machine suffered from leakage because the geofluid was non-scaling 17]; it appears that helical screw expanders thrive on scaling fluids but have more rouble maintaining tight clearances with pure fluids. Sprankle indicated that machine fficiencies of 75% should be achievable under specific power ratings and inlet geofluid uality; the higher the inlet quality, the more power that can be generated, and the asier it is to achieve an efficiency of 75% [17].

.5.4 Conclusions

o total-flow device has achieved commercial status as a stand-alone power unit. Those hat have been tested have not reached levels of machine efficiency that are sufficient to varrant their use even as supplements to standard geothermal plants. Each one examined n this section has advantages – simplicity, elegance, ingenuity, or scale-resistance – and vith further development may eventually become commercially successful.

.6 Hybrid fossil-geothermal systems

Currently, the world is striving to develop as much renewable energy as is feasible to ope with growing demand and to meet environmental standards. One way to gain upport for renewables is to combine them with conventional energy sources so as oth to extend the life of the depletable resources and to create new applications for enewables in the existing marketplace.

Hybrid power plants combine two different sources of energy in a single plant so as o achieve higher overall utilization efficiencies than separate plants. One way to do his is to combine fossil and geothermal energy inputs in such a way as to yield a plant hat outperforms two individual state-of-the-art plants, one using the fossil fuel and ne a pure geothermal plant. In this section, we will examine several hybrid designs of his general type.

9.6.1 Fossil-superheat systems

The idea of using fossil fuels to enhance geothermal resources is not a new idea. It was reported in 1924 that a Frenchman P. Caufourier had proposed a hybrid power system in which hot water from a geothermal spring would be flashed successively to generate steam and a fossil-fired superheater would raise the steam temperature prior to its admission to a multi-pressure turbine [18]. His system today would be called a 4-stage flash-steam plant with fossil superheating; a schematic of the system is shown in Fig. 9.19. Note that the waste brine was put to use in therapeutic baths, making this plant not only hybrid but multi-use as well.

Using reasonable assumptions, a thermodynamic analysis was carried out [19] that showed the hybrid plant was capable of a very high resource utilization efficiency 65.3%, but the bulk of the output came from the geothermal contribution. A fossil-fuel effective efficiency can be defined as

$$\eta_{ff} = \frac{\dot{W}_{hyb} - \dot{W}_{geo}}{\dot{Q}_{ff}} \tag{9.26}$$

where the numerator is the total plant output minus the output that could be obtained by using the geofluid by itself, in this case, in a 4-stage flash-steam plant. The denominator is the thermal input from burning the fossil fuel. For the proposed plant $\eta_{ff} = 0.20$, a slightly low figure for its time, and certainly very low by today's fossil plant standards. For a hybrid plant to make thermodynamic sense, it must be synergistic, i.e., capable of producing more than two separate, state-of-the-art plants using the same input from the two energy sources. Thus, this proposal from 1924 failed that test with regard to the fossil fuel usage. The analysis ignored the benefits of the hot water used in the spa.

A new proposal was put forth in the early 1980s for a hybrid fossil-geothermal plant at The Geysers in Northern California. A detailed analysis of the thermodynamics and economics of two schemes was carried out by the California Energy Commission [20].

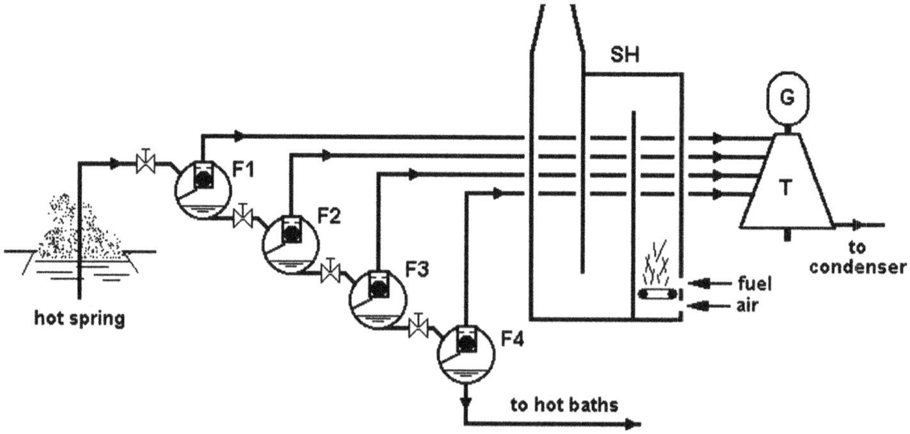

Fig. 9.19 Caufourier's 4-stage flash plant with fossil superheating; after [19].

One used a simple superheater fired by natural gas to add superheat to the saturated steam produced by wells at The Geysers. The other used a natural gas-fired gas turbine with an exhaust heat recovery superheater to accomplish the same objective. Clearly the latter alternative produced much more power than the first, and in fact did so in a synergistic fashion. The fossil fuel used at The Geysers in this plan would have the highest utilization of any fossil fuel then being burned for power generation in California [20]. The simple superheating by direct firing of natural gas was found to be uneconomic even though it produced 57% more power than the basic dry-steam plant.

General studies of fossil superheating in geothermal plants were conducted at Brown University and the interested reader may consult Refs. [21–22] for more information.

9.6.2 Geothermal-preheat system

In the previous section we considered how fossil fuels might be used to enhance the performance of geothermal plants. It is also possible to use geothermal fluids to enhance the performance of fossil-fueled power plants. The geothermal-preheat systems do just that.

One of the earliest suggestions for this type of system is found in a 1961 paper by Hansen [23]. The idea is to use hot geothermal liquid as the heating medium in the feedwater heaters of a conventional fossil-fired power plant, thereby supplanting some of the extraction steam in a regenerative Rankine cycle. This would allow more power to be generated in the plant since the previously extracted steam would now be available to flow through the turbines. If the geofluid by itself was of low potential for use in a flash-steam plant, this would allow it to be used effectively in the hybrid system.

It is evident that the geothermal resource must be located close to the site of the fossil plant for this hybrid scheme to be practical. The City of Burbank, California, conducted a survey of possible sites for such a plant and concluded that Roosevelt Hot Springs, Utah, would be an economically viable site. There are nearby coal fields that could support a mine-mouth plant at which geothermal hot water could relieve the feedwater heating load. An engineering design verification study confirmed the plant's synergistic advantages [24]. The plant design was based on theoretical studies of geothermal-preheat systems at Brown University and reported in Ref. [25].

9.6.3 Geopressure-geothermal hybrid systems

Along the northern shore of the Gulf of Mexico, there is a potent energy resource called "geopressure." Oil and gas drilling in the coastal areas of Louisiana and Texas has encountered fluids with pressures greater than hydrostatic and approaching lithostatic. Hydrostatic pressure increases with depth in proportion to the weight of water, i.e., at about 0.465 lbf/in^2 per ft. However in formations where the fluid plays a supportive role in maintaining the structure of the reservoir, the weight of the solid overburden increases the gradient to the lithostatic value of 1.0 lbf/in^2 per ft.

In the geopressured reservoirs of the Gulf, sufficiently high pressure had been experienced to cause cessation of drilling for oil and gas. With improved drilling techniques, these zones can now be safely drilled.

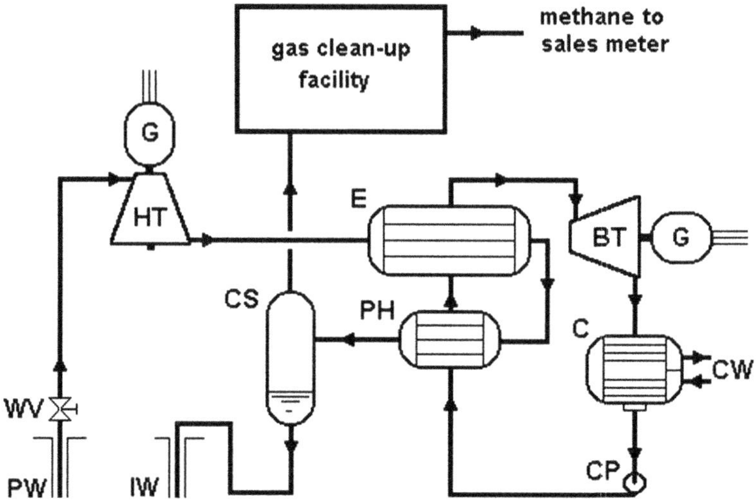

Fig. 9.20 Conceptual hybrid plant utilizing a geopressured geothermal resource.

Besides having very high pressure, these fluids are also hot and contain dissolved methane and other gases. A cleverly designed geothermal power plant can use each of these energy sources to advantage. A drastically simplified schematic flow diagram of such a plant is shown in Fig. 9.20.

First the high pressure fluid drives a back-pressure, hydraulic turbine HT, generating electricity. Next it flows through the heat exchangers of a binary cycle, generating more electricity. Finally it enters a separator where the dissolved gases are liberated. The gases continue to a clean-up facility that would include scrubbers and contactors to purify the methane gas for sale. The waste brine from the separator is reinjected to help prevent subsidence which would be potentially much more serious for this type of system than a conventional hydrothermal plant.

The U.S. Dept. of Energy built and operated a small demonstration plant near Pleasant Bayou, Texas. The plant was rated at 1 MW and ran for one year in 1989–90. Instead of selling the methane gas, the plant burned it on-site to drive a small gas turbine. The plant incorporated a binary cycle power generator, but no hydraulic turbine could be obtained to fit the brine conditions. A suitable hydraulic turbine might have increased the output by about 7–8%.

The feasibility of geopressured geothermal plants turns strongly on the resource characteristics: the pressure must be high enough, the fluid hot enough, and most importantly from an economic viewpoint, there must be sufficient dissolved methane. The solubility of methane in water solutions depends on pressure, temperature, and salinity. The solubility obviously increases with pressure, but at the pressures expected in geopressured reservoirs, the solubility dependency on temperature exhibits a minimum between 160–200°F. Furthermore, if the fluid is heavily mineralized, as is expected for these brines, the solubility decreases with increasing salinity. For example, at 250°F, the solubility of methane decreases 3–4% for each 10,000 ppm of total dissolved solids [26]. Lastly, the presence of other noncondensable gases reduces the solubility of methane.

Thus, the conditions need to be especially favorable for a geopressured resource to be commercially viable, and so far these conditions have not been found. Nevertheless, there are geopressured resources in several countries and researchers continue to explore them [27,28]. Finally, a recent study by Griggs at Louisiana State University [29] concluded that geopressured geothermal resources could be economically viable with the right set of natural gas and electricity prices, but that in the near-term it is unlikely that these conditions would occur. Meanwhile, there needs to be more work done in characterizing the resources, in optimizing the power facility including finding more efficient binary cycles, in conducting detailed economic analyses, and in clarifying legal and political issues including ownership of subsurface mineral rights [29].

9.7 Combined heat and power plants

In many places it is common to combine both power generation and direct heat usage in a single geothermal plant. By capturing some of the waste heat in the leftover brine before it is reinjected, the overall utilization efficiency of the resource is enhanced. Furthermore, when that heat is provided to the community adjacent to the plant, it demonstrates to the community that the plant is indeed a "good neighbor." There are two countries where this is widely practiced: Iceland and Japan. We will cover specific case histories in Part 3, but here we will present the basic principles of combined geothermal power and heating applications, using a single-flash steam plant as the basis for the analysis.

Figure 9.21 illustrates a single-flash power plant in which a side-stream of the separated brine is sent to a water-heating facility to supply the needs of various end users.

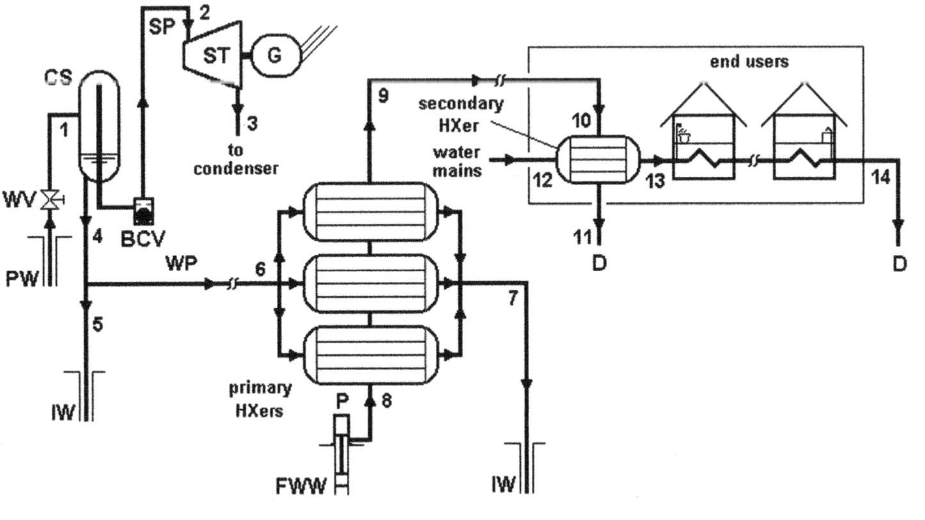

Fig. 9.21 Combined geothermal heat and power plant; after [30].

The bank of primary heat exchangers would be located close to the power plant to avoid excessive heat loss from the brine and to allow it to be reinjected into the reservoir. A supply of fresh water is shown being delivered to the primary heat exchangers from well pumps. Secondary heat exchangers would be located close to the end users to permit the water from the mains to be heated for domestic consumption or heating purposes.

The overall utilization efficiency of the plant including power generation and heating is the ratio of the sum of all beneficial output and effects to the exergy of the geothermal fluid under reservoir conditions. In terms of the states identified in Fig. 9.21, this leads to the following equation:

$$\eta_u = \frac{\dot{W}_{ST} + \Delta\dot{E}_{13,14}}{\dot{E}_R} \tag{9.27}$$

where the useful heating effect, in exergy terms, is given by the difference in exergy across the end users,

$$\Delta\dot{E}_{13,14} = \dot{E}_{13} - \dot{E}_{14} = \dot{m}_{13}[h_{13} - h_{14} - T_0(s_{13} - s_{14})] \tag{9.28}$$

Since the fluid passing through the end users' facilities is a liquid, we can simplify eq. (9.28):

$$\Delta\dot{E}_{13,14} = \dot{m}_{13}c\left[T_{13} - T_{14} - T_0 \ln\frac{T_{13}}{T_{14}}\right] \tag{9.29}$$

The reservoir fluid exergy is given by

$$\dot{E}_R = \dot{m}_1(h_1 - T_0 s_1 - e_0) \tag{9.30}$$

where

$$e_0 = h_0 - T_0 s_0 \tag{9.31}$$

is the exergy of the fluid at the dead state.

For a 240°C reservoir, a 160°C separator temperature, a 50°C condensing temperature, a 125°C inlet to the secondary heat exchangers, and other reasonable assumptions, it was shown in Ref. [30] that the plant had a 33.4% utilization efficiency. Without the direct heat benefit, the single-flash plant would have a 31.2% utilization efficiency. Thus the combined plant has a 7% higher utilization plus the intangible benefits from good relations with the plant's neighbors.

To optimize such a plant, one would try to maximize the sum of the power output from the steam turbine $\dot{W}_{ST}$ and the exergy of the fluid sent to the primary heat exchangers $\dot{E}_6$. Both terms depend on the choice of separator temperature and the latter depends on the fraction of brine sent to the primary heat exchangers. Usually there is a requirement for a certain amount of reinjection at state 5 to help maintain reservoir pressure and temperature, so that degree of freedom can be eliminated in favor of fixed amounts to reinjection and to the primary heat exchangers. Although the problem now is a simple one-dimensional optimization on thermodynamic grounds, there is one other parameter of importance in the economic sense: the monetary value of a unit of heat relative to a unit of work output. Several cases were examined in Ref. [30]

and it was seen that the optimum flash temperature depends strongly on the ratio of the heat-to-work monetary values. As this ratio increases, so does the optimum flash temperature because this lowers the power output in favor of sending more exergy to the heat exchangers. Clearly, the local prices for alternative fuels and electricity will play a critical role in deciding whether or not to build a combined geothermal heat and power plant, and in the selection of the system operating parameters.

9.8 Power plants for hypersaline brines

Since geothermal fluids are products of nature, it is sometimes difficult to use them as working fluids in a power plant. In fact some fluids are so challenging that it is nearly impossible to produce them from their reservoir. Geothermal brines have been found that are more acidic than battery acid, that can clog a production well casing in a matter of a few days, or so contaminate surface vessels as to render them useless.

One of the most notorious geothermal resources is located in the Imperial Valley of Southern California, near the southeastern shore of the Salton Sea. The resource was recognized in the 1850s when explorers moving west came upon hot pools and mud volcanoes in an otherwise barren desert [31]. The latter persist to this day, driven by carbon dioxide gas emissions; see Fig. 9.22.

In 1905 the Colorado River broke though its banks and flooded the Salton Sink, creating what we know today as the Salton Sea and partially inundating the thermal manifestations. In 1925 interest was renewed in this area as a possible geothermal resource, and plans were even drawn up for a 25,000 kW power plant [32]. However, the area remained undeveloped as a source of electric power for nearly another fifty years, but was for a period of time a source of carbon dioxide for a thriving dry ice industry.

Drilling for power production began in the 1960s but the early wells were all plugged and abandoned. Some wells drilled in the 1970s are still in operation today, but the fluids that were produced resisted exploitation for power generation because of severe scaling and corrosion problems. The temperatures were high, up to 680°F (360°C), and

Fig. 9.22 The author at mud volcano site near the Salton Sea in California's Imperial Valley [WWW].

Table 9.2 Geothermal fluid composition for Magmamax No. 1 well, Salton Sea [33].

Element	Values
Brine:	*mg/l*
Chlorides	128,500
Sodium	40,600
Calcium	21,400
Potassium	11,000
Manganese	681
Strontium	440
Ammonia	360
Iron	315
Silicon	246
Zinc	244
Lithium	180
Barium	142
Magnesium	105
Lead	52
Copper	3
Total dissolved solids	219,000
pH	5.3
Oxidation reduction potential	+25
Gases:	*% by volume*
Carbon dioxide	98.14
Methane	0.68
Hydrogen sulfide	0.18
Nitrogen	0.02

the total dissolved solids reached as much as 300,000 ppm, placing these fluids in the hypersaline category. The chemical analysis of the fluid produced from the Magmamax No. 1 well that was drilled in 1972 about 1.5 mi from the southeast shore of the Salton Sea is given in Table 9.2 [33].

As the result of an extensive research effort carried out starting in the 1970s that was funded by the U.S. Dept. of Energy, the Electric Power Research Institute, and several private companies, techniques were devised that now permit these fluids to be used for the generation of electricity in a reliable and cost-effective manner [34]. Two approaches for dealing with these aggressive brines have been used with reasonable success: (1) flash-crystallizer/reactor-clarifier (FCRC) and (2) pH modification (pH-Mod) systems. We will sketch the principles underlying these two methods.

9.8.1 Flash-crystallizer/reactor-clarifier (FCRC) systems

In the FCRC approach, clean steam is generated in a train of separators and flash vessels, similar to what we have seen for standard flash-steam plants, but in which the separated brine is seeded with material that induces precipitation. A simplified schematic of a FCRC power plant is shown in Fig. 9.23 [34]. The seed material is obtained from the waste stream of highly concentrated brine. In this way, the unstable, supersaturated solids precipitate on the seed particles, rather than on the surfaces

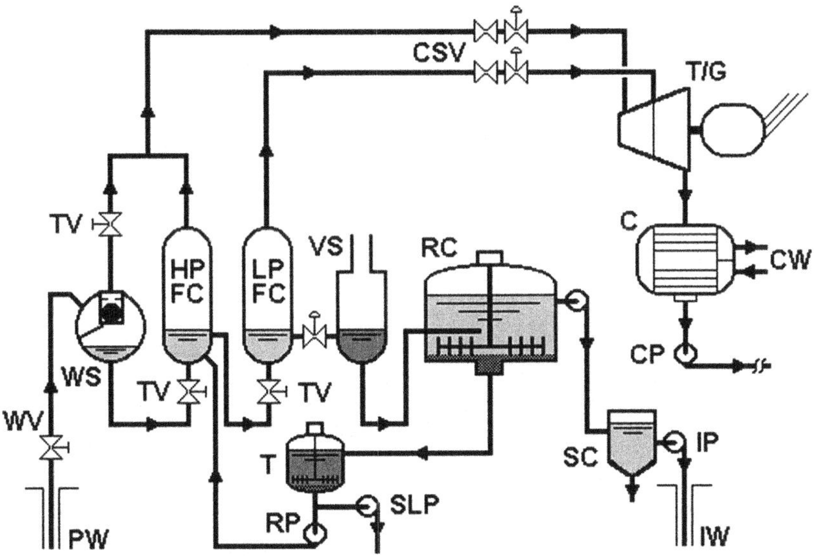

ʼig. 9.23 Simplified schematic flow diagram for a FCRC power plant; after [1,34].

ɔf the vessels and piping. The particulate matter is eventually allowed to settle in ɩ reactor-clarifier vessel. The slurry from the reactor-clarifier is thickened and a por-ion of it is recirculated as seed material. The clarified liquid is pumped to a secondary :larifier and then sent to reinjection wells.

The most recent power plant to use this approach, Salton Sea Unit 5, has a riple-pressure turbine that receives the high-pressure steam separated at the ʋellhead separators and expands it through the first four stages of the turbine [35]. 'his eliminates the throttling loss from the pressure-letdown throttle valve TV hown in Fig. 9.23.

).8.2 pH modification (pH-Mod) systems

n Sect. 6.6 we discussed the problem of silica scaling and how it might be controlled.)ne approach is to modify the pH of the brine to alter the kinetics of the precipitation ɔrocess. The technique of acidifying the brine has been used in some Salton Sea plants ɩs an alternative to the FCRC approach. By reducing the pH of the geofluid the solubil-ty of silica is increased, the kinetics of the reaction are slowed, and it is possible to ɩvert precipitation, at least until the separated liquid has been processed to generate he flash steam needed for the turbine. A highly simplified flow diagram for a pH-Mod ɔlant is shown in Fig. 9.24.

The addition of hydrochloric acid to the brine requires appropriate corrosion-ʻesistant materials. However, pH-Mod plants are much simpler than FCRC plants in erms of the number of vessels needed and the operating procedures to be followed.)mitted from Fig. 9.24 is the processing of the discharge waste brine D. If the injection ɔf the treated brine could lead to problems in the pipelines or the injection wells, then

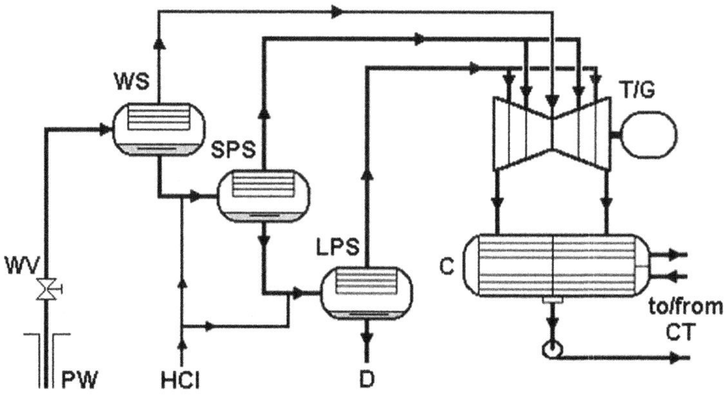

Fig. 9.24 Flow diagram for pH-Mod power plant.

the brine pH could be raised. A reactor-clarifier might then be used to remove the silica to assure that the waste brine can be safely reinjected.

With the use of the FCRC and pH-Mod methods for coping with the Salton Sea brines, it has been possible to construct ten power plants with a total capacity of 312 MW. A new 185 MW triple-flash plant is scheduled to be built soon. This plant will represent the largest single geothermal power unit ever constructed, surpassing the 133 MW Unit 13 at The Geysers (now renamed Big Geysers and de-rated to 78 MW). [Note added in proof: This project has been modified; the Hudson Ranch facility, a staged development of three 50 MW (nom.) plants is underway. The first plant should be online in 2012]. With a significant portion of the geothermal field lying offshore beneath the Salton Sea itself, it is believed that the ultimate potential of this resource is far from being fully exploited. The technologies described in this section should allow this valuable resource to eventually reach its full potential.

9.9 Solar-geothermal hybrid plants

9.9.1 Basic concept

As with any hybrid power plant, the basic idea is to achieve a synergistic outcome wherein the hybrid plant can outperform two individual state-of-the-art plants using the solar and geothermal energy sources separately; see Fig. 9.25. That is, the hybrid plant (H) must be so designed that the separate energy sources are used to augment each other, i.e., the hybrid plant H-power must exceed the sum of the S-power and the G-power, for the same energy inputs. In Sect. 9.6 it was shown how this could be achieved with geothermal and fossil energy sources. Since both of those sources are available on a continuous basis, it is relatively simple to devise means to achieve synergy.

For the hybrid systems being considered in this section, one of the two energy sources is not available all the time; the geothermal may be assumed on hand continuously, but the solar input disappears at night and during cloudy conditions. This places serious practical constraints on possible hybrid designs.

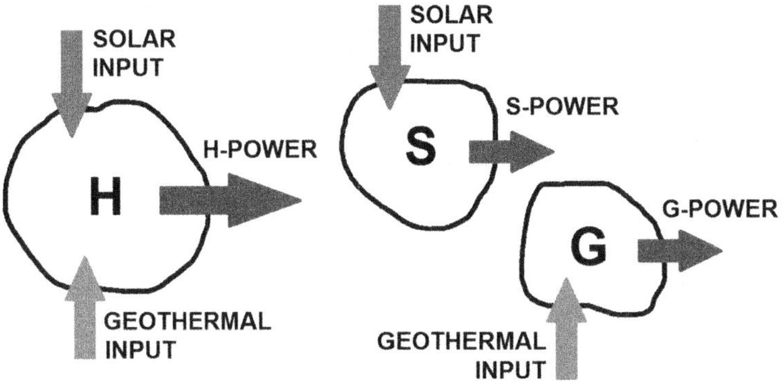

Fig. 9.25 *Comparison between a hybrid solar-geothermal plant and two separate solar and geothermal plants.*

On the economic side, solar and geothermal plants have dramatically different unit costs, with solar electricity being considerably more expensive than electricity from conventional geothermal plants. Thus, when taking into account both thermodynamic and economic performance, it is hard to avoid reaching an optimum configuration that favors all geothermal and no solar.

Nevertheless, since economics depend on local circumstances (i.e., labor costs, cost of money, material costs, financial incentives, government subsidies, etc.) and these can vary over time, it is useful to consider some possible solar-geothermal hybrid configurations that may become both thermodynamically and economically viable in the future, even if not at the present time.

9.9.2 Geothermal-augmented solar thermal plants

As mentioned in Sect. 9.6.2, the two energy sources must be collocated or at least within easy access. In the case of a solar-assisted geothermal plant, this constraint is not so important since the solar input is available anywhere and can be used at any geothermal site, albeit with varying levels of solar insolation. However, if it is desired to augment a solar plant with geothermal energy, then collocation can be a problem. In this section we consider just such a design.

Solar thermal power plants typically use concentrating parabolic trough collectors to focus the incoming solar rays onto tubes carrying a heat transfer fluid, which in turn is used to heat and boil a secondary working fluid, usually water, to drive a Rankine cycle, usually one designed for high efficiency. Such plants resemble conventional fossil-fueled plants and have at least one feedwater heater.

If low-to-moderate temperature geothermal fluid can be found near the solar thermal plant, a synergistic design is possible wherein the geofluid is pumped through one of the feedwater heaters, eliminating a steam extraction point from the turbine, thus increasing steam flow through the last stages of the turbine and raising the power output. The system described in Sect. 9.6.2 is an illustration of this approach for a fossil-geothermal hybrid plant. The utilization efficiency of the geothermal energy is very high in this application. In the Southwest sector of the U.S. there is excellent insolation and many hot springs that would allow this approach to be considered.

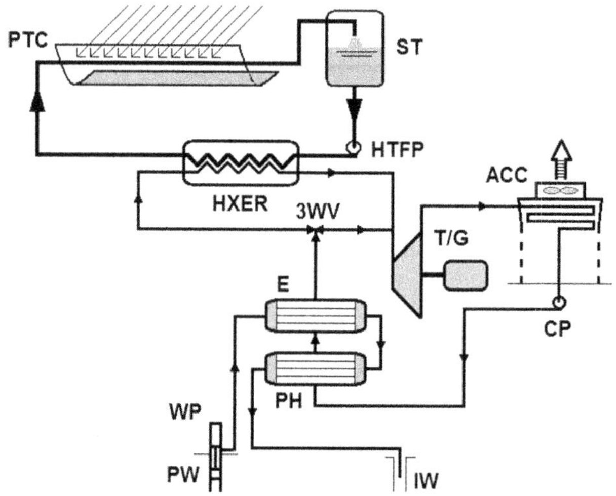

Fig. 9.26 Solar-geothermal binary plant with superheating of the binary working fluid.

9.9.3 Solar-augmented binary plants

If solar energy is used to augment a basic geothermal plant of the binary type, the intermittent nature of the solar input must be factored into the design. The obvious way to alleviate this limitation is to incorporate thermal storage for the solar heat transfer fluid (HTF), as shown schematically in the flow diagram in Fig. 9.26. The effect of the solar energy in this design is to superheat the binary cycle working fluid while the sun is shining, thereby improving the exergy of the working fluid as it enters the turbine [36]. If the binary working fluid is one with normal condensing properties (see Fig. 8.8), adding superheat can assure a fully dry expansion process and a higher turbine efficiency. After the storage system has been depleted, say 5–7 hours after sunset, bypass valves are activated and the system operates as a simple geothermal binary plant, albeit at a lower efficiency.

Whether or not the added expense of the solar system is justified by the increased output during the day will depend on local conditions, as mentioned earlier.

9.9.4 Solar-augmented flash plants

The notion of using solar energy to assist geothermal flash plants is not new. Hiriart and Gutiérrez [37] describe a combination flash-steam system in which the steam line is passed through a parabolic solar collector to impart about 100°C of superheat before entering the turbine. This is enough to produce 4 MW of additional power above the base case of 20 MW for a pure geothermal flash plant. Besides offering thermodynamic advantages, the proposal appeared to have economic merit as well.

There are several ways to add solar energy to a geothermal flash plant. Fig. 9.27 illustrates the possibilities for a base double-flash plant, but the ideas can easily be applied to single-flash plants. The parabolic trough collector (PTC) field and storage system has a circulating heat transfer fluid (HTF) that can be routed to any of several heat exchangers (HXER) to augment the flash plant. The one shown is used to

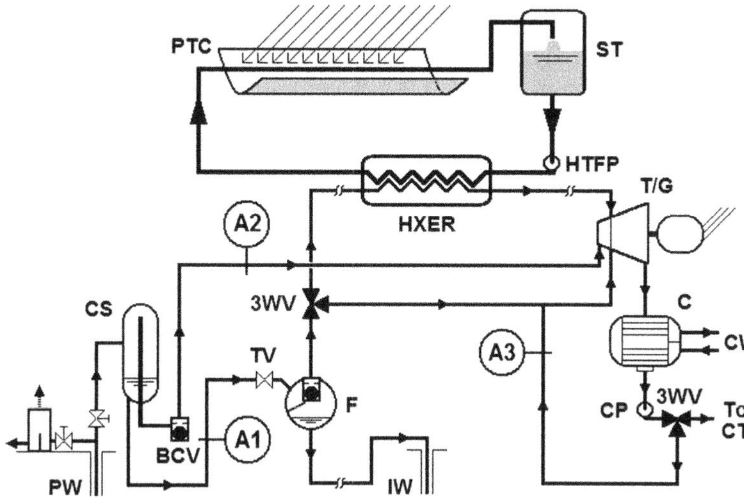

Fig. 9.27 *Solar-geothermal double-flash plant.*

superheat the low-pressure steam from the flash vessel. The HXER could also be located at sites A1, A2, and/or A3. At A1, the separated brine from the cyclone separator is heated prior to being admitted to the flash vessel, which will generate a higher fraction of low-pressure steam. At A2, the high-pressure steam is superheated prior to turbine entry. And at A3, a portion of the steam condensate is recirculated through a solar HXER for reheating and boiling, much like a conventional Rankine cycle, and returned to the low-pressure steam line [38].

Various solar-geothermal hybrid schemes can be imagined. For example, a combined cycle (flash-binary) is amenable to hybridization [36]. Field testing of several schemes has been carried out at the Ahuachapán geothermal flash plant in El Salvador [39,40]. The main difficulty in implementing any of these arrangements lies in the intermittent nature of the solar energy. On the geothermal side, any arrangement that requires periodic shutting down of wells or electromechanical equipment would be unacceptable. Thus, without significant thermal storage, it may not be feasible to deploy such ingenious schemes.

References

[1] DiPippo, R., "Geothermal Power Systems," Sect. 8.2. in *Standard Handbook of Powerplant Engineering, 2nd ed.*, T.C. Elliott, K. Chen and R.C. Swanekamp, Eds., pp. 8.27–8.60, McGraw-Hill, Inc., New York, 1998.

[2] Keenan, J.H., F.G Keyes, P.G. Hill and J.G. Moore, *Steam Tables: Thermodynamic Properties of Water Including Vapor, Liquid, and Solid Phases (International Edition – Metric Units)*, John Wiley & Sons, Inc., New York, 1969.

[3] Reynolds, W.C., *Thermodynamic Properties in SI: Graphs, Tables and Computational Equations for 40 Substances*, Dept. of Mechanical Engineering, Stanford University, Stanford, CA, 1979.

[4] Forte, N. "The 125 MW Upper Mahiao Geothermal Power Plant," *Geothermal Resources Council BULLETIN*, V. 25, No. 8, 1996, pp. 299–304.
[5] Austin, A.L. and A.W. Lundberg, "The LLL Geothermal Energy Program: A Status Report on the Development of the Total-Flow Concept," UCRL-500-77, Lawrence Livermore Laboratory, Livermore CA, 1978.
[6] McKay, R., "Helical Screw Expander Evaluation Project: Final Report," DOE/ET-28329–1, JPL Pub 82–5, Jet Propulsion Laboratory, Pasadena, CA, 1982.
[7] Carey, B., "Total Flow Power Generation from Geothermal Resources Using a Helical Screw Expander," *Proc. 5th New Zealand Geothermal Workshop*, 1983, pp. 127–132.
[8] Cerini, D.J. and J. Record, "Rotary Separator Turbine Performance and Endurance Test Results," *Proc Seventh Annual Geothermal Conf. and Workshop*, EPRI AP-3271, Electric Power Research Institute, Palo Alto, CA, 1983, pp. 5-75–5-86.
[9] Austin, A.L., "Status of the Development of the Total Flow System for Electric Power Production from Geothermal Energy," Sect. 4.4 in *Sourcebook on the Production of Electricity from Geothermal Energy* Kestin, J., Ed. in Chief, R. DiPippo, H.E. Khalifa and D.J. Ryley, Eds., U.S. Dept. of Energy, DOE/RA 4051–1, U.S. Gov. Printing Office, Washington, DC, 1980.
[10] Hughes, E.E., "Summary Report: Rotary Separator Turbine," Final Report, EPRI AP-4718, Electric Power Research Institute, Palo Alto, CA, 1986.
[11] Stoecker, W.F. and J.W. Jones, *Refrigeration and Air Conditioning, 2nd Ed.*, McGraw-Hill, New York, 1982.
[12] Lysholm, A., *Rotary Compressor*, U.S. Patent No. 2,243,874, June 3, 1941.
[13] Weiss, H., R. Steidel and A. Lundberg, "Performance Test of a Lysholm Engine," Rep. No. UCRL-51861, Lawrence Livermore Laboratory, Livermore, CA, July 1975.
[14] Sprankle, R.S., *Electric Power Generating System*, U.S. Patent No. 3,751,673, August 7, 1973.
[15] DiPippo, R., "The Effect of Expansion-Ratio Limitations on Positive-Displacement, Total-Flow Geothermal Power Systems," *Geothermal Resources Council TRANS.*, V. 6, 1982, pp. 343–346.
[16] DiPippo, R., "The Effect of Expansion-Ratio Limitations on Positive-Displacement, Total-Flow Geothermal Power Systems," LA-UR-82-596, Rep. No. GEOFLO/15, Div. of Engineering, Brown University Providence, RI, 1982.
[17] Sprankle, R., "Helical Screw Expander Power Plant Model 76–1 Test Result Analysis," Hydrothermal Power Co., 1986.
[18] Anon., "Recent Developments in the Utilization of the Earth's Heat," *Mechanical Engineering*, V. 46, No. 8, 1924, pp. 448–449.
[19] DiPippo, R., "An Analysis of an Early Hybrid Fossil-Geothermal Power Plant Proposal," *Geothermal Energy Magazine*, V. 6, No. 3, 1978, pp. 31–36.
[20] Janes, J., *Evaluation of a Superheater Enhanced Geothermal Steam Power Plant in the Geysers Area*, Rep. P700-84-003, California Energy Commission, Siting and Environmental Div., 1984.
[21] Kestin, J., R. DiPippo and H.E. Khalifa, "Hybrid Geothermal-Fossil Power Plants," *Mechanical Engineering*, V. 100, 1978, pp. 28–35.
[22] DiPippo, R., H.E. Khalifa, R.J. Correia and J. Kestin, "Fossil Superheating in Geothermal Steam Power Plants," *Geothermal Energy Magazine*, V. 7, No. 1, 1979, pp. 17–23.
[23] Hansen, A., "Thermal Cycles for Geothermal Sites and Turbine Installation at The Geysers Power Plant, California," Paper G/41, *Proc. Conf. on New Sources of Energy*, Rome, United Nations, V. 3, 1961, pp. 365–379.
[24] Anon., *System Design Verification of a Hybrid Geothermal/Coal Fired Power Plant*, The Ralph M. Parsons Company, Project 5905, 1978.
[25] Khalifa, H.E., R. DiPippo and J. Kestin, "Geothermal Preheating in Fossil-Fired Steam Power Plants," *Proc. 13th Intersociety Energy Conversion Engineering Conference*, V. 2, 1978, pp. 1068–1073.
[26] Swanson, R.K., "Geopressured Energy Availability," Final Report, EPRI AP-1457, Electric Power Research Institute, Palo Alto, CA, 1980.
[27] Árpási, M., Á. Lorberer and S. Pap, "High Pressure and Temperature (Geopressured) Geothermal Reservoirs in Hungary," *Proc. World Geothermal Congress 2000*, International Geothermal Association, pp. 2511–2514.
[28] He, L. and L. Xiong, "Extensional Model for the Formation of Geopressured Geothermal Resources in the Yinggehai Basin, South China Sea," *Proc. World Geothermal Congress 2000*, International Geothermal Association, pp. 1211–1216.

[29] Griggs, J., *A Re-Evaluation of Geopressured-Geothermal Aquifers as an Energy Resource,* Masters Thesis, Louisiana State University, Craft and Hawkins Dept. of Petroleum Engineering, August 2004.

[30] DiPippo, R., "Exergy Analysis of Combined Electricity and Direct-Heat Geothermal Flash-Steam Plants," *Geothermal Resources Council TRANS.*, V. 11, 1987, pp. 411–416.

[31] LeConte, J.L., "Account of Some Volcanic Springs in the Desert of the Colorado, in Southern California," *The American J. of Science and Arts, Second Series,* V. XIX, May 1855, pp. 1–6.

[32] Siegfried, H.N., "Further Possibilities for Development in Hot Areas Adjacent to the Transmission Lines of the Southern Sierras Power Company," *Geothermal Exploration in the First Quarter Century,* D.N. Anderson and B.A. Hall, Eds., Spec. Report No. 3, Geothermal Resources Council, Davis, CA, 1973, pp. 145–171.

[33] Lombard, G.L., "Operational Experience at the San Diego Gas & Electric ERDA Niland Geothermal Loop Experimental Facility," *Proc. EPRI Annual Geothermal Program Project Review and Workshops,* EPRI ER-660-SR, Electric Power Research Institute, Palo Alto, CA, 1978, pp. 3-11–3-16.

[34] Featherstone, J., S. Butler and E. Bonham, "Comparison of Crystallizer Reactor Clarifier and pH Mod Technologies at the Salton Sea Geothermal Field," *Proc. World Geothermal Congress 1995,* V. 4, International Geothermal Association, pp. 2391–2396.

[35] Anon., "CalEnergy Company, Inc., U.S.A. Salton Sea Unit 5 Geothermal Power Plant 1 × 58.32 MW," Brochure GEC 82–14, Fuji Electric Co., Ltd., Tokyo, Japan, 2001.

[36] Greenhut, A.D., J.W. Tester, R. DiPippo, R. Field, C. Love, K. Nichols, C. Augustine, F. Batini, B. Price, G. Gigliucci and I. Fastelli, "Solar-Geothermal Hybrid Cycle Analysis for Low Enthalpy Solar and Geothermal Resources," *Proc. World Geothermal Congress 2010,* Paper 2615, Bali, Indonesia, 25–29, 2010.

[37] Hiriart L.B., G. and L.C.A. Gutiérrez N., "Calor del Subsuelo Para Generar Electricidad: Combinacion Solar-Geotermia," *Ingenieria Civil,* V. 313, May 1995, pp. 13–22 (in Spanish).

[38] Lentz, A. and R. Almanza, "Parabolic Troughs to Increase the Geothermal Wells Flow Enthalpy," *Solar Energy,* V. 80, 2006, pp. 1290–1295.

[39] Handal, S., Y. Alvarenga and M. Recinos, "Geothermal Steam Production by Solar Energy," *Geothermal Resources Council TRANS.*, V. 31, 2007, pp. 503–510.

[40] Alvarenga, Y., S. Handal and M. Recinos, "Solar Steam Booster in the Ahuachapán Geothermal Field," *Geothermal Resources Council TRANS.*, V. 32, 2008, pp. 395–399.

Nomenclature for figures in Chapter 9

ACC	Air-cooled condenser
BCV	Ball check valve
BHT	Brine holding tank
BT	Binary turbine
C	Condenser
CP	Condensate pump
CS	Cyclone separator
CT	Cooling tower
CW	Cooling water
D	Drain
E	Evaporator
F	Flash vessel
FWW	Fresh water well
f	Saturated liquid
G	Generator
GC	Gas compressor
g	Saturated vapor
HPF	High-pressure flash vessel

HPFC	High-pressure flash crystallizer
HT	Hydraulic turbine
HTFP	Heat transfer fluid pump
HXER	Heat exchanger
IP	Injection pump
IW	Injection well
LPF	Low-pressure flash vessel
LPFC	Low-pressure flash crystallizer
LPS	Low-pressure separator
NCG	Noncondensable gases
OP	Orifice plate
P	Pump
PH	Preheater
PTC	Parabolic trough collector
PW	Production well
RC	Reactor-clarifier
RP	Recirculating pump
SH	Superheater
S	Separator
SC	Secondary clarifier
SLP	Sludge pump
SP	Steam piping
SPS	Standard-pressure separator
ST	Steam turbine
T	Thickener
T/G	Turbine/generator
TV	Throttle valve
VS	Vent silencer
WP	Water piping; Well pump
WS	Wellhead separator
WV	Wellhead valve

Problems

9.1 Consider a hybrid fossil-geothermal power plant of the geothermal-preheat type shown in the schematic on the following page. The following are given:

$$P_1 = 2400\ \text{lbf/in}^2,\quad T_1 = 1000°\text{F},\ \eta_t = 0.80,\ \eta_{pumps} = 1.0,$$
$$P_2 = P_3 = 1.0\ \text{lbf/in}^2,\ T_a = 300°\text{F}.$$

State 5 is saturated liquid, there are no pressure losses in heat exchangers or pipelines, and the temperature difference in the geothermal heat exchanger is a constant 10°F. With reference to a base fossil-fueled power plant consisting of the same elements as the hybrid plant minus the geothermal heat exchanger and the second pump (i.e., the base plant has its condensate pumped to P_6 in one pump and more sensible heating is required in the fossil-fueled boiler), determine the following:

(a) Net work of hybrid plant, w, in Btu/lbm steam.
(b) Net work of base fossil plant, w_f, in Btu/lbm steam.
(c) Maximum practical geothermal work output, w_g, in Btu/lbm steam.
(d) Overall hybrid figure of merit, F.
(e) Fossil-fuel figure of merit, F_f.
(f) Geothermal-energy figure of merit, F_g.
(g) For a net hybrid plant output of 250 MW, estimate the number of geothermal wells needed to supply the plant if each well can produce 500,000 lbm/h.
(h) Discuss the feasibility of the plant.

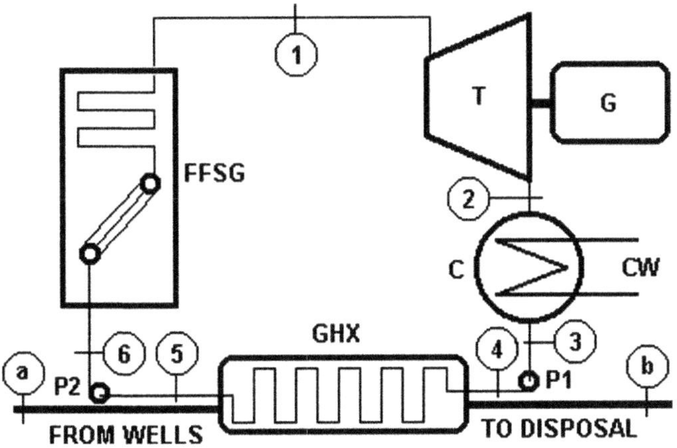

9.2 A "total flow" turbine is proposed that will use a geothermal liquid at 200°C (saturated liquid). What isentropic turbine efficiency must the total flow machine have in order for it to match the thermodynamic performance of: (a) a 1-stage flash plant, and (b) a 2-stage flash plant? Assume the "equal-temperature-split" rule for flash points and a condenser temperature of 50°C. Use a steam turbine efficiency of 75% for the flash plants and use separate cylinders (no mixing) for the 2-stage flash plant.

9.3 A single-stage total-flow machine is capable of handling a 2-phase mixture of geothermal fluid as it comes from the well. The geofluid temperature is 380°F and the wellhead quality is 25%. The condenser operates at 120°F. Determine the required isentropic efficiency of the total-flow machine for it to be thermodynamically equivalent to a single-flash plant having a steam turbine with an efficiency of 77% and operating with the same wellhead and condenser conditions.

9.4 With reference to Fig. 9.7, the following conditions are given:

Reservoir fluid = saturated liquid at 200°C

Geofluid mass flow rate at wellhead = 100 kg/s

$T_2 = 140°C$, $T_6 = T_d = 40°C$, $P_a = P_e = P_f = 3.0$ MPa
Binary cycle working fluid = isobutane, i-C_4H_{10}
Pinch-point ΔT in brine-isobutane heat exchanger, E-PH = 5°C
State 4 and state a are saturated vapors
Dry turbine efficiency = 85%; wet turbine efficiency from Baumann rule
Neglect temperature rise in isobutane feed pump

Calculate the following quantities:
(a) Mass flow rate of isobutane, in kg/s.
(b) Power output of (i) steam turbine and (ii) isobutane turbine, in kW.
(c) Power to run the isobutane feed pump, in kW.
(d) Overall plant utilization efficiency based on the exergy of the reservoir geofluid; dead-state temperature is 25°C.

9.5 A very simple geothermal-preheat type of fossil-geothermal hybrid plant is shown in the schematic below.

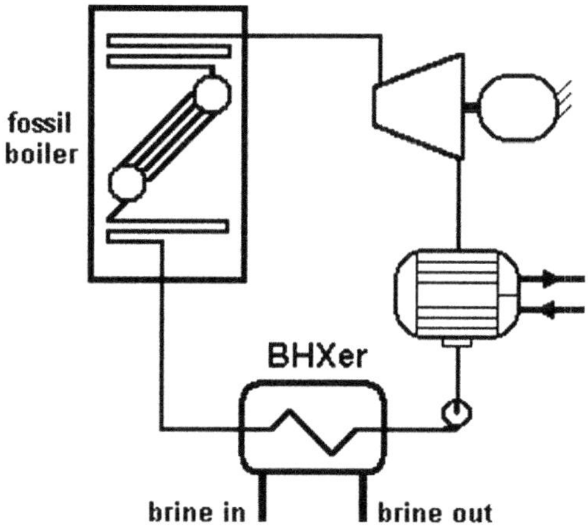

The following data are given:
Boiler pressure = 2400 lbf/in^2,a
Turbine inlet temperature = 1000°F
Condenser pressure = 2.0 lbf/in^2,a
Steam flow rate = 500, 000 lbm/h
Brine inlet temperature = 380°F
Terminal temperature difference at BHXer = 20°F

Calculate the following quantities:
(a) Brine flow rate, in lbm/h.
(b) Thermal efficiency of the hybrid plant.

Assuming that the state-of-the-art thermal efficiency of a fossil steam plant is 35% and that for a geothermal binary plant is 15%, determine the following quantities:
(c) Overall figure of merit for the hybrid plant.

(d) Fossil-fuel figure of merit.
(e) Geothermal figure of merit.

The respective figures of merit are defined as: Part (c) – the ratio of the actual hybrid plant output to the best that could be obtained using two separate state-of-the-art plants, Part (d) – the ratio of the portion of the hybrid output attributable to the fossil fuel to the output from a state-of-the-art fossil plant, and Part (e) – the ratio of the portion of the hybrid output attributable to the geothermal heat to the output from a state-of-the-art geothermal binary plant.

9.6 A simple fossil-superheat type of fossil-geothermal hybrid plant is being proposed. Geothermal steam from a separator is available as a saturated vapor at 380°F. A gas-fired superheater would raise the steam temperature to 1000°F. The condenser operates at an absolute pressure of 2 lbf/in^2. You may assume that the turbines are isentropic. Calculate the following quantities:
(a) Work output of the hybrid plant, in Btu/lbm geosteam.
(b) Work output of a geothermal single-flash plant, in Btu/lbm geosteam.
(c) Fossil-fuel figure of merit if the fossil heat could be used in a state-of-the-art fossil plant with a thermal efficiency of 40%.

9.7 Use the geothermometers described in Sect. 2.3.5 to analyze the data given in Table 9.2 to estimate the geofluid reservoir temperature for the Magmamax No. 1 well at the Salton Sea. You may assume that the data has been corrected for reservoir conditions. Discuss the results and any differences among the values found from the various formulations.

Chapter 10

Exergy Analysis Applied to Geothermal Power Systems

"The Second Law of thermodynamics holds the supreme position among the laws of Nature. If someone points out to you that your pet theory of the universe is in disagreement with Maxwell's equations – then so much the worse for Maxwell's equations. If it is . . . contradicted by observation – well, these experimentalists do bungle things sometimes. But if your theory is found to be against the Second Law of thermodynamics, I can give you no hope; there is nothing for it but to collapse in deepest humiliation."

Sir Arthur S. Eddington, Gifford Lectures – 1927

10.1 Introduction

This chapter offers an introduction to the fundamental principles of exergy analysis, which is based on the Second Law of thermodynamics. The presentation is restricted

Geothermal Power Plants: Principles, Applications, Case Studies and Environmental Impact, Third Edition

© 2012 Elsevier Ltd. All rights reserved.

to open systems operating in steady state. These are the applications encountered in all types of power plants, including geothermal power plants, once they reach their design operating conditions. These principles also may be used to assess refrigeration plants, air-conditioning systems, and many other common engineering process plants.

After the basic working equations are derived, we will examine some of the components usually found in geothermal plants to illustrate the use of these principles. We will find that exergy analysis is the best tool for finding those elements within a plant that are most in need of redesign from an efficiency point of view.

10.2 First Law for open, steady systems

Although exergy analysis relies strongly on the Second Law of thermodynamics, it also is grounded in the First Law, namely, that energy is conserved during any process while it is transformed from one form to another. We begin with the basic First Law equation for open systems operating in steady state. Briefly, an open system is one in which matter crosses the boundary of the system during the process, and in steady operation, the values of all thermodynamic properties at any point in the system remain constant with time. We adopt the macroscopic viewpoint in which properties are measured by instruments which by design average out microscopic fluctuations.

The general First Law working equation is

$$\dot{Q} - \dot{W}_s = -\sum_{i=1}^{n} \dot{m}_i (h_i + 0.5 v_i^2 + g z_i) \tag{10.1}$$

where each term is defined as follows:

$\dot{Q}$ ... rate of heat transfer (thermal power) between the system and its surroundings (+ when heat enters the system)
$\dot{W}_s$ ... rate of work transfer (mechanical power) between the system and the surroundings (+ when work is delivered to the surroundings by the system)
i ... an index that accounts for all inlets and outlets of the system
n ... total number of inlets and outlets
$\dot{m}_i$ ... mass flow rate crossing each inlet or outlet
h_i ... specific enthalpy of the fluid at each inlet or outlet
v_i ... velocity of the fluid at each inlet or outlet
z_i ... elevation of each inlet or outlet relative to an arbitrary datum
g ... local gravitational acceleration.

The principle of conservation of mass in steady state requires that

$$\sum_{i=1}^{n} \dot{m}_i = 0 \tag{10.2}$$

Note that when using eqs. (10.1) and (10.2) mass flows are taken as positive when entering the system and negative when leaving. This sign convention allows us to write the equations in summation form and must be scrupulously applied.

10.3 Second Law for open, steady systems

It is instructive to write down the general Second Law formulation for open systems before specializing it to steady systems.

$$\dot{\theta}_p = \frac{dS}{d\tau} - \sum_{i=1}^{n} \dot{m}_i s_i - \int_{\tau_1}^{\tau_2} \frac{1}{T}\frac{dQ}{d\tau} \tag{10.3}$$

where each new term is defined as follows:

$\dot{\theta}_p$ ... rate of entropy production for the system caused by irreversibilities
S ... entropy of the system
τ ... time
s_i ... specific entropy of the fluid at each inlet or outlet
T ... absolute temperature (in K or °R) associated with the heat transfer Q.

The physical meaning of the integral in the last term in eq. (10.3) is that it represents a summation taken over the whole surface area of the open system, for the duration of the process, of all the incremental heat transfer rates divided by their corresponding absolute temperatures. In general, there may be several heat reservoirs interacting with the open system, but, as we will see shortly, this term can be drastically simplified for the present study.

Since we will be dealing only with steady systems, time derivatives of thermodynamic properties will all vanish and the working equation reduces to:

$$\dot{\theta}_p = -\sum_{i=1}^{n} \dot{m}_i s_i - \int_{\tau_1}^{\tau_2} \frac{1}{T}\frac{dQ}{d\tau} \tag{10.4}$$

because the entropy S is a system property and the heat Q is not.

10.4 Exergy

10.4.1 General concept

The basic concept of exergy is that it is the maximum work (or power) output that could theoretically be obtained from a substance at specified thermodynamic conditions relative to its surroundings. While this definition is comprehensive, here we are only interested in open, steady systems. A system may receive (or discharge) fluids from (or to) the surroundings, and exchange heat and work with the surroundings.

We will seek the maximum power output from the operation of the system. To achieve this ideal outcome, there are two thermodynamic conditions that must be met:

1. All processes taking place within the system must be perfectly reversible.
2. The state of all fluids being discharged from the system must be in thermodynamic equilibrium with the surroundings.

The first condition means that no losses occur because of friction, turbulence, or any other source of irreversibility. The second condition means that the leaving fluids have

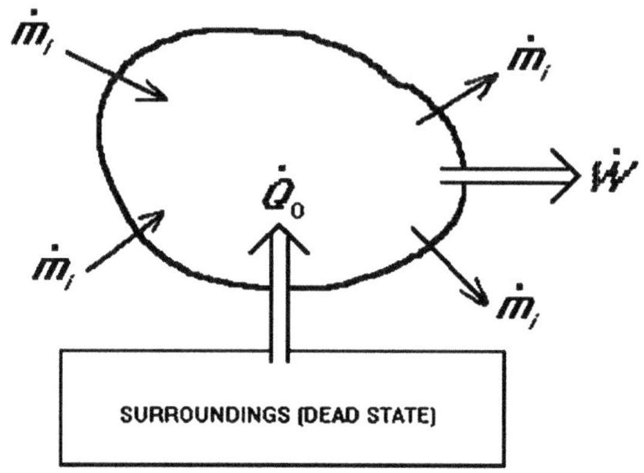

Fig. 10.1 Generalized system-surroundings interactions.

no more potential to do work relative to the surroundings. For this reason, we refer to the state of the surroundings as the "dead state" because when a fluid is in equilibrium with the surroundings it may be considered "dead."

A consequence of condition 1 is that $\dot{\theta}_p$ in eq. (10.4) vanishes, i.e.,

$$-\sum_{i=1}^{n} \dot{m}_i s_i - \int_{\tau_1}^{\tau_2} \frac{1}{T}\frac{dQ}{d\tau} = 0 \tag{10.5}$$

In general for any system of the type we are considering, all heat transfer can be viewed ultimately as an interaction solely with the surroundings. A generalized schematic of the system and its interactions with the surroundings is given in Fig. 10.1.

10.4.2 Exergy of fluid streams

Next we use the two conditions given above to specialize the First and Second Law equations and arrive at an exergy working equation. As a simplification, let us first consider the simplest system that can undergo a steady, open process, namely, one with only two channels, an inlet and an outlet. Also, let us for the moment ignore the effects of kinetic and potential energy; thus eq. (10.1) becomes:

$$\dot{Q} - \dot{W} = \dot{m}(h_2 - h_1) \tag{10.6}$$

where the subscript s on the power has been dropped for simplicity. Since the only heat transfer is between the system and the dead state, eqs. (10.6) and (10.5) can be rewritten, respectively, as:

$$\dot{Q}_0 - \dot{W} = \dot{m}(h_2 - h_1) \tag{10.7}$$

$$-\dot{m}(s_1 - s_2) - \frac{\dot{Q}_0}{T_0} = 0 \tag{10.8}$$

Solving eq. (10.8) for $\dot{Q}_o$ and substituting into eq. (10.7), one obtains:

$$\dot{W} = \dot{m}[h_1 - h_2 - T_0(s_1 - s_2)] \tag{10.9}$$

inally, we use the condition 2 to stipulate that the exit state 2 be identical to the ead state, and so obtain the maximum power output:

$$\dot{W}_{max} = \dot{m}[h_1 - h_0 - T_0(s_1 - s_0)] \quad (10.10)$$

his maximum output is given a special name, exergy, $\dot{E}$. The expression in brackets is he specific exergy, e:

$$e_1 \equiv h_1 - h_0 - T_0(s_1 - s_0) \quad (10.11)$$

his equation may be used to find the specific exergy of any fluid stream at a temperaure T_1 and pressure P_1, relative to a given set of ambient conditions T_0 and P_0. f kinetic or potential energy effects are important, the enthalpy h_1 should be augnented by $0.5{V_1}^2$ or gz_1, as appropriate. Since state 1 is really arbitrary, we can drop he subscript and obtain a general expression for the specific exergy:

$$e = h - h_0 - T_0(s - s_0) \quad (10.12)$$

vhere the properties at the dead state are evaluated at T_0 and P_0. When the fluid vould be a liquid at the dead state conditions, it is sufficiently accurate to take the nthalpy and entropy values for a saturated liquid at T_0. When calculating e, T_0 must e in absolute units, either kelvins or degrees Rankine.

It is sometimes convenient to introduce a new quantity, b, called the availability unction, defined as:

$$b \equiv h - T_0 s \quad (10.13)$$

he usefulness of the availability function is evident when considering a process etween two given states, neither of which is the dead state. Then the maximum possile specific work output for such a process is simply:

$$e_{1,2} = h_1 - h_2 - T_0(s_1 - s_2) = b_1 - b_2 \quad (10.14)$$

estin [1] called this the available work, w°, since it involves the best one can do for wo specified states.

A few more words of explanation are needed regarding exergy. First, exergy is not a hermodynamic property. For a quantity to be a thermodynamic property it must be a state-function, i.e., its value must be a function strictly of the state of the system. t can only depend on other system properties [2]. It is clear from the definition of xergy, eq. (10.12), that the value of the exergy of a system depends on the dead state. A different dead state for the same system condition will yield a different value of the xergy. For this reason there are no "exergy tables"; property tables such as *Steam Tables* or *Gas Tables* [3] only include actual properties. Once the dead state is defined, nowever, it is an easy matter to calculate exergy values for any set of system conditions.

The second point is also important; the dead state is not arbitrary [4]. It represents he local surroundings in which the system finds itself. The designation of the dead state s fundamentally different from setting the reference level for energy, enthalpy, or any other thermodynamic potential. In those cases, when we take differences between, say he energy at one state and that at another state, the reference value cancels out. This loes not happen with exergy, as can be seen from eq. (10.14) since the dead-state temerature remains in the equation. The practical way to treat the dead state is to take nto account the nature of the processes under study and the environment in which the processes take place, and to choose the dead state accordingly. For example, if the system is a geothermal power plant equipped with a water cooling tower, an appropriate choice would be the design wet-bulb temperature for the cooling tower. This temperature is the ideal lower limit on the water leaving the tower. If the plant uses an air-cooled condenser, then the design dry-bulb temperature would be appropriate.

Lastly, the dead state as we have defined it here is sometimes called a restricted dead state [5]. Thermodynamic equilibrium requires that the system be in mechanical equilibrium (i.e., equal pressure), thermal equilibrium (i.e., equal temperature), and chemical equilibrium (i.e., equal reaction potential) with its surroundings. Since chemical equilibrium does not play a role in a practical sense in most geothermal plants, we will disregard it and only require mechanical and thermal equilibrium at the dead state in calculating the exergy.

10.4.3 Exergy for heat transfer

Whenever heat is transferred from one system to another, a certain amount of exergy is also transferred. The important point in such a case is this:

The absolute amount of exergy given up by the hotter system is, in reality, always greater than the exergy received by the cooler system; only in the ideal case of reversible heat transfer are the two exergy amounts equal.

The magnitude of the exergy involved with the heat transfer process can be found from the basic concept of exergy, namely, what is the maximum work that could be produced on a continuous basis from the given amount of heat. The answer is:

The work that could be obtained using a reversible Carnot cycle operating between the temperature from which the heat is derived and the lowest practical available temperature, i.e., the ambient or dead-state temperature.

If a quantity of heat Q is transferred from a system at a temperature T to another system at some lower temperature within an environment having a dead-state temperature T_0, then the exergy E_Q associated with the heat Q is given by:

$$E_Q = \left[1 - \frac{T_0}{T}\right] Q \qquad (10.15)$$

where the factor in brackets is the familiar Carnot efficiency for an ideal cycle operating between T and T_0. When applying eq. (10.15), it is customary to use the magnitude of the exergy, i.e., to use the magnitude of the heat transfer term Q irrespective of whether the heat is entering or leaving the system since the direction of the exergy flow will be obvious from the context.

10.4.4 Exergy for work transfer

Whenever a system experiences a work transfer process with its surroundings, exergy is also transferred. The relationship between the work and the exergy associated with it is simple:

The maximum work that can be delivered in the absence of any dissipative phenomena is exactly the amount of work itself.

'hus the exergy E_W associated with the transfer of an amount of work W is given by:

$$E_W = W \tag{10.16}$$

: should be noted that both eqs. (10.15) and (10.16) may be written as rate quations involving the thermal power or mechanical power, respectively. Then the xergy calculated from a rate equation will be the rate of exergy transfer or exergetic ower.

0.5 Exergy accounting for open, steady systems

n this section we will come up with equations that express how closely a system omes to meeting thermodynamic ideality. The technique involves isolating the system as a "free-body" in the language of mechanics) and performing an exergy accounting nalysis. The system is assumed to operate in steady state with several mass flows into nd out of the system; it receives a flow of thermal power from a heat reservoir at a emperature T and delivers power to an end user residing in the surroundings, characerized by a dead-state temperature T_0.

The exergy accounting for the open system will involve the exergy associated with ach flow stream and each exergy transfer term. Unlike an energy accounting, which y the First Law must always balance, an exergy accounting will in reality always how that less exergy leaves the system than enters it. In other words, some exergy vill always be destroyed. If all dissipative phenomena could be eliminated and all proesses carried out reversibly, then no exergy would be lost, but this could only occur n an ideal world. In our accounting we will denote the lost exergy as Δe (per unit nass).

The exergetic power input to the system is given by:

$$\dot{E}_{IN} = \dot{E}_Q + \sum_{i=1}^{n} \dot{m}_i e_i \tag{10.17}$$

vhere the summation is taken over all *incoming* streams, $i = 1, 2, \ldots, n$.

The exergetic power output from the system is given by:

$$\dot{E}_{OUT} = \dot{E}_W + \sum_{j=1}^{k} \dot{m}_j e_j \tag{10.18}$$

vhere the summation is taken over all outgoing streams, $j = 1, 2, \ldots, k$.

Thus the exergy accounting equation is simply:

$$\Delta\dot{E} = \dot{E}_{IN} - \dot{E}_{OUT} = \dot{E}_Q + \sum_{i=1}^{n} \dot{m}_i e_i - \dot{E}_W - \sum_{j=1}^{k} \dot{m}_j e_j \tag{10.19}$$

The exergy loss term $\Delta\dot{E}$ must always be numerically positive. Note that the terms ssociated with heat and work can be either positive or negative depending on their lirection in accordance with our sign conventions.

10.6 Exergy efficiencies and applications to geothermal plants

10.6.1 Definitions of exergy efficiencies

It is useful to define efficiencies based on exergy, sometimes called Second Law efficiencies or utilization efficiencies. There is no standard set of definitions in the literature but we will describe two different approaches: "brute-force" and "functional."

- A "brute-force" exergy efficiency for any particular system is defined as the ratio of the sum of all output exergy terms to the sum of all input exergy terms.
- A "functional" exergy efficiency for any particular system is defined as the ratio of the exergy associated with the desired energy output to the exergy associated with the energy expended to achieve the desired output.

The brute-force definition can be applied in a straightforward manner, irrespective of the nature of the system, once all exergy flows have been determined. The functional definition, however, requires judgment and an understanding of the purpose of the system before the working form of the efficiency equation can be formulated. In the remaining sections of this chapter we will examine the systems usually found in geothermal plants and apply both exergy definitions to each of them. We will also include numerical examples to illustrate their use.

10.6.2 Exergy efficiencies for turbines

Figure 10.2 shows a simple turbine. Incoming fluid drives a set of blades to rotate a shaft, thereby producing work output. There is exergy associated with the work and with both streams. It is reasonable to neglect any heat transfer between the turbine and the surroundings. This analysis is general and applies to any kind of turbine.

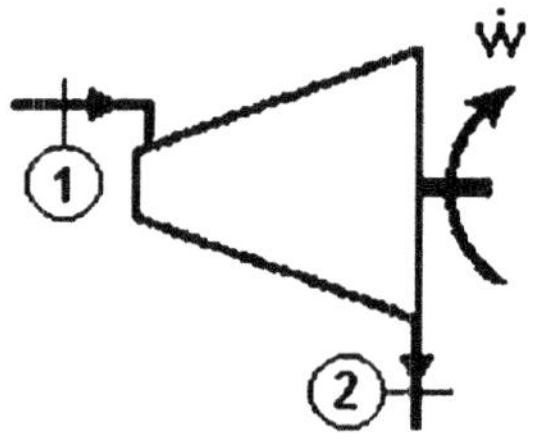

Fig. 10.2 Simple turbine.

The exergy accounting equation for the turbine is:

$$\Delta \dot{E}_t = \dot{m}_1 e_1 - \dot{m}_2 e_2 - \dot{W} \quad (10.20)$$

where the two mass flow rates must be equal by conservation of mass and the steady flow assumption. Thus, we obtain:

$$\Delta \dot{E}_t = \dot{m}(e_1 - e_2) - \dot{W} \quad (10.21)$$

or

$$\Delta e_t = e_1 - e_2 - w \quad (10.22)$$

written per unit mass flow through the turbine.

The Second Law or exergy efficiencies may be expressed as:

$$\eta^{II}_{t,BF} = \frac{e_w + e_2}{e_1} = \frac{w + e_2}{e_1} \quad \text{(brute-force)} \quad (10.23)$$

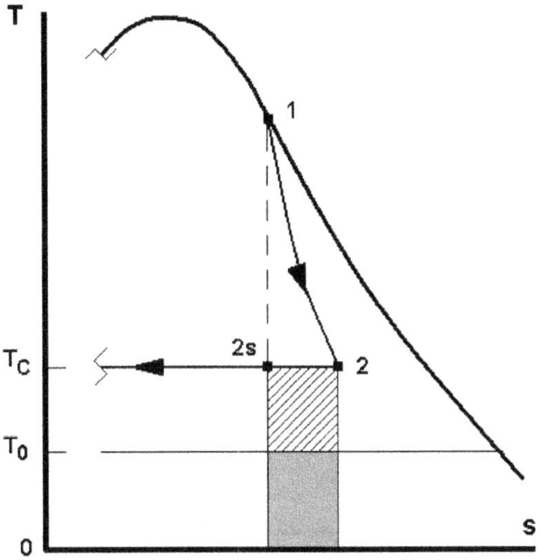

Fig. 10.3 *Irreversibility in temperature-entropy coordinates.*

$$\eta_{t,FUN}^{II} = \frac{e_w}{e_1 - e_2} = \frac{w}{e_1 - e_2} \quad \text{(functional)} \tag{10.24}$$

The isentropic turbine efficiency is the ratio of the actual work to the work obtainable from an isentropic (reversible and adiabatic) expansion process:

$$\eta_t = \frac{w}{w_{s=const}} = \frac{h_1 - h_2}{h_1 - h_{2s}} = \frac{h_1 - h_2}{h_1 - h_2 + (h_2 - h_{2s})} \tag{10.25}$$

If we expand eq. (10.24) we can compare the functional exergy efficiency with the isentropic efficiency, eq. (10.25):

$$\eta_{t,FUN}^{II} = \frac{h_1 - h_2}{h_1 - h_2 + T_0(s_2 - s_1)} \tag{10.26}$$

From the temperature-entropy diagram, Fig. 10.3, it is clear that

$$h_2 - h_{2s} > T_0(s_2 - s_1) \tag{10.27}$$

since the right-hand side is the shaded area and the left-hand side is the shaded area plus the cross-hatched area. Thus, we conclude that

$$\eta_{t,FUN}^{II} > \eta_t \tag{10.28}$$

Since the shaded area represents the difference between the ideal work output $(e_1 - e_2)$ and the actual work output $(h_1 - h_2)$, it is called the irreversibility, I:

$$I \equiv T_0(s_2 - s_1) \tag{10.29}$$

Numerical example: Suppose a turbine receives geothermal steam at $P_1 = 0.80$ MPa as a saturated vapor; the steam leaves at $T_2 = 50°C$. The isentropic efficiency is 80%.

Determine the exergy loss and both exergy efficiencies. The dead-state temperature is $T_0 = 20°C$.

From *Steam Tables* [6], $h_1 = 2769.12$ kJ/kg and $s_1 = 6.6628$ kJ/kg·K. The ideal isentropic outlet state 2s has $s_2 = s_1$ and thus $h_{2s} = 2135.20$ kJ/kg. Using the definition of the turbine isentropic efficiency, eq. (10.25), we find that $h_2 = 2261.99$ kJ/kg. The entropy at the turbine outlet is $s_2 = 7.0551$ kJ/kg·K. The specific work $w = 507.13$ kJ/kg.

We need the specific exergy of the steam entering and leaving; to find these we also need h_0 and s_0. These are taken from the *Steam Tables* under saturated liquid at the temperature T_0, namely: $h_0 = 83.96$ kJ/kg and $s_0 = 0.2966$ kJ/kg·K. Now we can find e_1 and e_2, using eq. (10.11): $e_1 = 818.91$ kJ/kg and $e_2 = 196.77$ kJ/kg. Next we can determine the exergy loss and the exergetic efficiencies:

$$\Delta e = e_1 - e_2 - e_w = e_1 - e_2 - w \tag{10.30}$$

which yields $\Delta e = 115.00$ kJ/kg. Lastly, using eqs. (10.23) and (10.24), the two efficiencies are: $\eta^{II}_{t,BF} = 0.860$ or 86.0%, and $\eta^{II}_{t,FUN} = 0.815$ or 81.5%.

Notice that the loss of exergy in the turbine amounts to about 14% of the incoming exergy in the geothermal steam (i.e., $115.00/818.89 = 0.140$). Also note that the functional efficiency turned out to be somewhat larger than the isentropic efficiency, as expected.

10.6.3 Exergy efficiencies for heat exchangers

Figure 10.4 shows a shell-and-tube heat exchanger in which a hotter fluid (stream a–b) transfers heat to a cooler fluid (stream 1–2). The fluids may be gases, liquids, or mixtures of these two phases. We will assume that the shell of the heat exchanger is perfectly insulated, i.e., adiabatic. This type of system occurs in binary geothermal plants as preheaters, evaporators, superheaters and condensers, and in flash- and dry-steam plants as condensers.

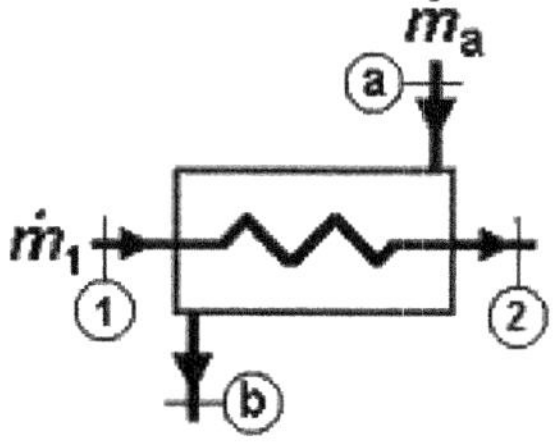

Fig. 10.4 Heat exchanger

The exergy accounting for the entire heat exchanger is:

$$\begin{aligned}\Delta \dot{E}_{HXer} &= \sum_{IN} \dot{E}_{HXer} - \sum_{OUT} \dot{E}_{HXer} = \dot{m}_a e_a + \dot{m}_1 e_1 - (\dot{m}_a e_b + \dot{m}_1 e_2) \\ &= \dot{m}_a (e_a - e_b) - \dot{m}_1 (e_2 - e_1)\end{aligned} \tag{10.31}$$

where we have incorporated the obvious fact that the mass flow rates of each fluid are the same entering and leaving the heat exchanger. Notice that in general $\dot{m}_a$ and $\dot{m}_1$ will not be equal. Also, eq. (10.31) can be read as saying the exergy destroyed in the heat exchanger is the difference between the exergy given up by the hotter fluid and the exergy acquired by the cooler one during the heat transfer process.

The ratio of the two flow rates is found from the First Law as follows:

$$\dot{m}_1 (h_2 - h_1) = \dot{m}_a (h_a - h_b) \tag{10.32}$$

where we have noted that there is no work transfer and that all of the heat transfer takes place within the heat exchanger and does not cross the system boundary.

The brute-force exergy efficiency is given by:

$$\eta^{II}_{HXer,BF} = \frac{\dot{m}_1 e_2 + \dot{m}_a e_b}{\dot{m}_1 e_1 + \dot{m}_a e_a} \tag{10.33}$$

The functional exergy efficiency is not as obvious. Functional exergy efficiencies for heat exchangers will vary depending on the nature and purpose of the unit. Suppose, for example, it is used to preheat a stream of isobutane prior to entering an evaporator and that the heat comes from a stream of hot geothermal brine. Then we might take the ratio of the exergy gained by the isobutane to the exergy given up by the brine stream as a measure of the functional efficiency of the exchanger:

$$\eta^{II}_{HXer,FUN1} = \frac{\dot{m}_1(e_2 - e_1)}{\dot{m}_a(e_a - e_b)} \tag{10.34}$$

Alternatively, we might observe that the brine at state b is simply disposed of by means of reinjection back into the formation with no further use made of it, in which case we might use the following definition:

$$\eta^{II}_{HXer,FUN2} = \frac{\dot{m}_1(e_2 - e_1)}{\dot{m}_a e_a} \tag{10.35}$$

Numerical example – Preheater-evaporator: Suppose a stream of geothermal brine is used to heat a stream of isobutane from a liquid to a saturated vapor. With reference to Fig. 10.4, let $\dot{m}_a$ be the brine flow rate and $\dot{m}_1$ be that for the isobutane. The heat exchanger consists of two sections: the preheater PH which raises the liquid to its boiling point and the evaporator E which changes the saturated liquid to a saturated vapor. We will assume that both the brine and the isobutane pass through the heat exchanger at constant pressure. The following data are given:

$$T_2 = 370\ \text{K},\ T_a = 400\ \text{K},\ \Delta T_{pp} = 5\ \text{K},\ T_0 = 293.15\ \text{K},\ \dot{m}_a = 85\ \text{kg/s}$$

The pinch point where the brine and the isobutane will come closest in temperature, $\Delta T_{pp} = 5$ K, will most likely occur at the isobutane bubble point. We will assume this to be true and verify it as we proceed. Let us use the state point z for the brine at the pinch point, and y for the isobutane at the pinch point. Furthermore, we will assume that the following isobutane properties are known:

$$T_1 = 320\ \text{K},\ h_1 = 275.58\ \text{kJ/kg},\ s_1 = 1.0600\ \text{kJ/kg}\cdot\text{K}$$

Property values for both brine (pure water) and isobutane will be taken from Reynolds' property book [3].

Using eq. (10.32) applied to the evaporator section only, we find the isobutane flow rate:

$$\dot{m}_1 = \dot{m}_a \frac{h_a - h_z}{h_2 - h_y} = 85 \times \frac{532.7 - 427.3}{640.09 - 418.58} = 40.45\ \text{kg/s} \tag{10.36}$$

The brine outlet temperature may now be found assuming linearity of the brine cooling curve: $T_b = 358.9$ K. So the cold-end temperature difference is 38.9 K and the pinch point does indeed occur at the bubble point.

We now have everything needed to find the rate of exergy loss using eq. (10.31):

$$\Delta\dot{E} = 11{,}997.5 - 11{,}355.0 = 642.5\ \text{kW} \tag{10.37}$$

Note that this amounts to only 5.4% of the total exergy given up by the brine.

The exergy efficiencies may now be found for the heat exchanger as a whole and for the evaporator and preheater sections separately. The results are shown in the table below.

Efficiency	Formula	Value
$\eta^{II}_{HXer,BF}$	$(\dot{m}_1 e_2 + \dot{m}_a e_b)/(\dot{m}_1 e_1 + \dot{m}_a e_a)$	0.9221
$\eta^{II}_{HXer,FUN1}$	$\dot{m}_1(e_2 - e_1)/\dot{m}_a(e_a - e_b)$	0.8216
$\eta^{II}_{HXer,FUN2}$	$\dot{m}_1(e_2 - e_1)/\dot{m}_a e_a$	0.4793
$\eta^{II}_{PH,BF}$	$(\dot{m}_1 e_y + \dot{m}_a e_b)/(\dot{m}_1 e_1 + \dot{m}_a e_z)$	0.9509
$\eta^{II}_{PH,FUN1}$	$\dot{m}_1(e_y - e_1)/\dot{m}_a(e_z - e_b)$	0.7674
$\eta^{II}_{PH,FUN2}$	$\dot{m}_1(e_y - e_1)/\dot{m}_a e_z$	0.1693
$\eta^{II}_{E,BF}$	$(\dot{m}_1 e_2 + \dot{m}_a e_z)/(\dot{m}_1 e_y + \dot{m}_a e_a)$	0.9616
$\eta^{II}_{E,FUN1}$	$\dot{m}_1(e_2 - e_y)/\dot{m}_a(e_a - e_z)$	0.8500

The brute-force and the first functional efficiency indicate a reasonable performance; the second functional efficiency is very poor because we do not attribute any exergetic value to the brine as it leaves the unit. This is probably not a good basis for judgment since the brine does return to the reservoir and may eventually return to the plant after reheating in the formation. In the case of the evaporator, the second functional efficiency does not apply since the brine immediately passes to the preheater.

We can calculate the log mean temperature differences for the heat exchanger as a whole and for each section using eq. (8.12). The results are as follows:

$$LMTD_{HXer} = 34.3°C, \quad LMTD_{PH} = 16.5°C, \quad LMTD_E = 14.0°C$$

It will be noted that the brute-force exergy efficiencies vary inversely with the *LMTD* values, i.e., the smaller the *LMTD*, the closer the match between the heating and cooling curves, the lower the thermodynamic irreversibility, and the higher the efficiency of exergy transfer.

Numerical example – Mixed working fluid condenser: One of the advantages of using a mixture as the working fluid in a binary plant is the fact that the mixture changes phase at variable temperature instead of isothermally as does a pure fluid. Consider the case of a 10% isopentane-90% isobutane mixture condensing against cooling water. The properties of the mixture are available courtesy of the National Institute of Standards (NIST), the successor to the National Bureau of Standards (NBS) in Ref. [7].

In this example we will assume the working fluid mixture enters the condenser at its dew point (i.e., the first drop of liquid is about to form) and leaves at its bubble point (i.e., the last bubble of vapor has just condensed). We will take the mixture pressure to be 0.50 MPa and assume the mass flow rate is 1 kg/s (i.e., we will solve the problem per unit mass flow rate of working fluid). The cooling water enters the condenser at 25°C and leaves at 40°C. Water properties will be taken from Ref. [6]. The dead state is assumed to be 20°C = 293.15 K.

From Ref. [7] we find the following property values for the mixture:

- Dew point: $T_1 = 317.84$ K, $h_1 = -18.1$ kJ/kg, $s_1 = 0.1270$ kJ/kg · K
- Bubble point: $T_2 = 313.91$ K, $h_2 = -337.8$ kJ/kg, $s_2 = -0.8865$ kJ/kg · K.

From Ref. [6] we find the following property values for the cooling water:

- Inlet: $h_a = 104.89$ kJ/kg, $s_a = 0.3674$ kJ/kg · K
- Outlet: $h_b = 167.57$ kJ/kg, $s_b = 0.5725$ kJ/kg · K.

The mass flow rate of cooling water comes from eq. (10.32):

$$\dot{m}_{CW} = \dot{m}_{WF} \frac{h_1 - h_2}{h_b - h_a} = 1 \times \frac{-18.1 + 337.8}{167.57 - 104.89} = 5.10 \text{ kg/s}$$

The exergy destroyed during the heat transfer process is the difference between the exergy given up by the mixture and that acquired by the cooling water:

$$\Delta \dot{E}_{COND} = \dot{m}_{WF}[h_1 - h_2 - T_0(s_1 - s_2)] - \dot{m}_{CW}[h_b - h_a - T_0(s_b - s_a)] = 9.562 \text{ kW}$$

It is interesting to compare this loss of exergy to that for the case of a pure fluid condensing under similar conditions. Let us say that 1 kg/s of pure isobutane enters the condenser at the same temperature as the mixture in the example. We will keep the cooling water temperatures the same. Of course, the isobutane would leave the condenser at the same temperature as it entered. The new cooling water flow rate is 4.93 kg/s, and the loss of exergy is 11.258 kW. This is 17.7% higher (i.e., worse) than for the mixture, again because of the better match between the mixed working fluid and cooling water. The *LMTD* in this case is 10.4°C vs. 9.1°C; it is remarkable that a mere 1.3°C increase in the *LMTD* leads to 17.7% greater loss of exergy, underscoring the need to design heat exchangers with care to achieve the greatest thermodynamic performance.

10.6.4 Exergy efficiencies for flash vessels

All geothermal flash-steam plants generate vapor from a stream of compressed or saturated liquid. The figures below show a typical flash tank and the temperature-entropy diagram for the processes.

The throttle valve (TV) creates a flow restriction which maintains a pressure drop from state 1 to state 2. State 2 is an intermediate state consisting of a two-phase mixture of saturated liquid and saturated vapor. The vessel is constructed so that the two phases are separated by centrifugal and gravity effects, producing a stream of saturated liquid at state 3 and one of saturated vapor at state 4. We assume the vessel is perfectly insulated and that the throttle operates isenthalpically.

Using the so-called "lever rule" of thermodynamics for two-phase mixtures, it is easy to find the quality at state 2:

$$x_2 = \frac{h_2 - h_3}{h_4 - h_3} = \frac{h_1 - h_3}{h_4 - h_3} \tag{10.38}$$

The specific entropy at state 2 is then found from:

$$s_2 = s_3 + x_2(s_4 - s_3) \tag{10.39}$$

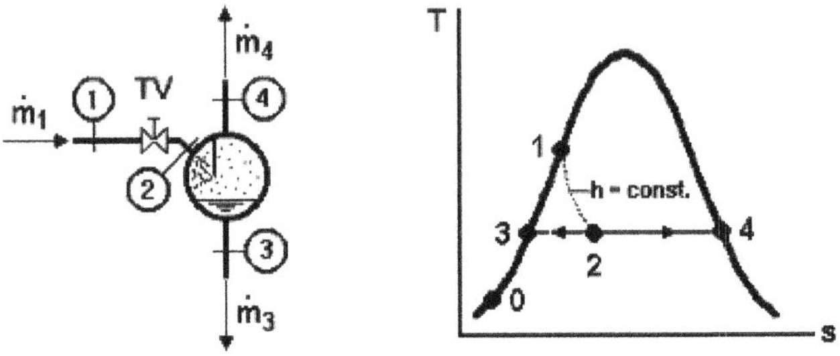

Fig. 10.5 Flash vessel and process diagram.

We may define Second Law efficiencies for this system using the brute-force and functional forms:

$$\eta^{II}_{FV,BF} = \frac{\dot{m}_3 e_3 + \dot{m}_4 e_4}{\dot{m}_1 e_1} \tag{10.40}$$

and

$$\eta^{II}_{FV,FUN} = \frac{\dot{m}_4 e_4}{\dot{m}_1 e_1} \tag{10.41}$$

Whereas the form of eq. (10.40) follows directly from the brute-force definition, that of eq. (10.41) expresses the ratio of the flow rate of exergy of the steam generated to that of the incoming liquid. The reason for the latter is that the purpose of the flasher is to produce a steam flow for use in some process, in this case, a geothermal steam turbine, and therefore the exergy of the steam produced is the desired output.

The exergy loss is found from the exergy accounting equation:

$$\Delta \dot{E}_{FV} = \dot{m}_1 e_1 - \dot{m}_3 e_3 - \dot{m}_4 e_4 \tag{10.42}$$

There are no exergy terms for work or heat transfer since only flows are involved. The mass flows can be related to the incoming mass flow using the quality as follows:

$$\dot{m}_3 = (1 - x_2)\dot{m}_1 \tag{10.43}$$

and

$$\dot{m}_4 = x_2 \dot{m}_1 \tag{10.44}$$

Numerical example: Suppose an incoming liquid is a saturated liquid at 2.5 MPa and flows steadily at 100 kg/s. The throttle valve creates a pressure drop leading to $P_2 = 0.70$ MPa. Let us assume that the pressure loss within the flash vessel is negligible, i.e., $P_2 = P_3 = P_4$. The dead state is at 20°C, 1 atm. With the aid of *Steam Tables* [6], the following state-point data may be obtained; all entries except the bold ones are found directly:

State	P (MPa)	T (°C)	x	s (kJ/kg · K)	h (kJ/kg)	e (kJ/kg)
1	2.5	223.99	0	2.5547	962.11	**216.19**
2	0.7	164.97	**0.1282**	**2.5967**	962.11	**203.88**
3	0.7	164.97	0	1.9922	697.22	**116.19**
4	0.7	164.97	1	6.7080	2763.3	**799.84**
0	0.1	20	0 (assumed)	0.2966	83.96	0

Using eqs. (10.38) and (10.39), the quality and entropy at state 2 can be alculated: $x_2 = 0.1282$ and $s_2 = 2.5967$ kJ/kg · K. Then the specific exergy values can e found.

The mass flow rates are:

$$\dot{m}_1 = 100 \text{ kg/s}, \quad \dot{m}_3 = 87.18 \text{ kg/s}, \quad \text{and} \quad \dot{m}_4 = 12.82 \text{ kg/s}$$

hus, the Second Law efficiencies and the exergy loss may be calculated:

$$\eta^{II}_{FV,BF} = 0.943, \text{ or } 94.3\%$$

$$\eta^{II}_{FV,FUN} = 0.474, \text{ or } 47.4\%$$

$$\Delta \dot{E}_{FV} = 1235.6 \text{ kW}$$

Since the brute-force efficiency is 94.3%, the exergy loss is 5.7% of the exergy carried by the incoming stream. Notice also that only 47.4% of the incoming exergy is reserved in the steam flow (i.e., 12.82 × 799.84/21,619), and that 52.6% of the xergy leaves with the separated liquid. If no further use is made of this liquid, then he exergetic performance of the flash-steam plant will suffer.

Lastly, the reader may wish to verify that the separation process that occurs after he flash preserves all the exergy in the post-flash two-phase fluid (within numerical ound-off error).

0.6.5 Exergy efficiencies for compressors

As the next illustration, consider a gas or vapor compressor. Geothermal plants routinely have air compressors to operate pneumatic devices and some condensing plants have compressors to remove noncondensable gases. The purpose of a compressor is to increase the pressure of the fluid, be it a gas uch as air (i.e., a perfect gas) or a vapor (not a perfect gas). n both cases, the Second Law theoretical analysis is the ame; the implementation of the analysis differs.

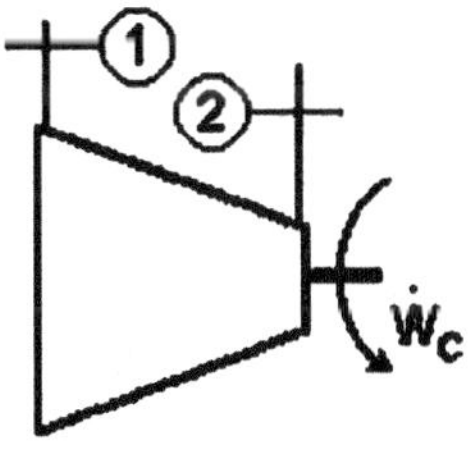

Fig. 10.6 Gas compressor.

Figure 10.6 is a schematic of a compressor. It is assumed o operate adiabatically, but irreversibly, being characterized y an isentropic efficiency, η_c. This efficiency is defined in a nanner similar to that for a turbine, but so as to give a value less than one, i.e., it is he ratio of the ideal, isentropic work to the actual work.

In terms of the state points shown in the figure, the compressor isentropic efficiency s given by:

$$\eta_c = \frac{h_{2s} - h_1}{h_2 - h_1} \tag{10.45}$$

vhere the state 2s refers to the ideal outlet state that would follow an isentropic proess from state 1.

The specific loss of exergy during the process is given by:

$$\Delta e_c = e_1 + |w_c| - e_2 \tag{10.46}$$

where we show the absolute magnitude of the work to emphasize that it is an exergy *input*. The reader will recall that the exergy associated with work is the work itself.

The Second Law efficiencies are defined as follows:

$$\eta_{c,BF}^{II} = \frac{e_2}{e_1 + |e_{w_c}|} = \frac{e_2}{e_1 + |w_c|} \tag{10.47}$$

and

$$\eta_{c,FUN}^{II} = \frac{e_2 - e_1}{|e_{w_c}|} = \frac{e_2 - e_1}{|w_c|} \tag{10.48}$$

The functional definition is based on the view (from the standpoint of the Second Law) that the role of a compressor is to increase the exergy of the fluid and that it does so at the expense of the exergy associated with the work of compression.

Note that we have been able to ignore the mass flow rate of the fluid since it is the same at the outlet as at the inlet. Had this not been the case (say, if there was an extraction point between the inlet and the outlet), we would have to multiply each specific exergy by its appropriate flow rate. Also we would then need to use the power to run the compressor, instead of the specific work.

Numerical example: We will consider an air compressor with the following given conditions: air is compressed from state 1 where the pressure is 0.110 MPa and the temperature is 275 K to state 2 where the pressure is 1.40 MPa, the compressor having $\eta_c = 0.725$ or 72.5%. The dead state is at 20°C, 1 atm.

In solving this example, we will use the air data tables in Reynolds' data book [3], implying the air obeys the perfect gas equation of state, and that the air has specific heats, c_p and c_v, that depend on temperature. The process diagram is given in Fig. 10.7.

Note that the dead state 0 is at 20°C (293.15 K), which is higher than the inlet temperature of 275 K. State 2s is the state that would have been reached if the compressor were ideally isentropic. The properties at state 2s can be found using the relative pressure, p_r, defined as follows for isentropic processes only:

$$\frac{p_{r2s}}{p_{r1}} = \frac{P_{2s}}{P_1} \tag{10.49}$$

The values for the relative pressure are tabulated in the air tables as a function of temperature.

The air tables also include an entropy function, ϕ, that is related to the entropy change by the equation

$$s_j - s_i = \phi_j - \phi_i - R \ \ln \left[\frac{P_j}{P_i} \right] \tag{10.50}$$

If the pressure is constant during a process, then the change in the entropy s is the same as the change in the entropy function ϕ. However, for a compressor this is

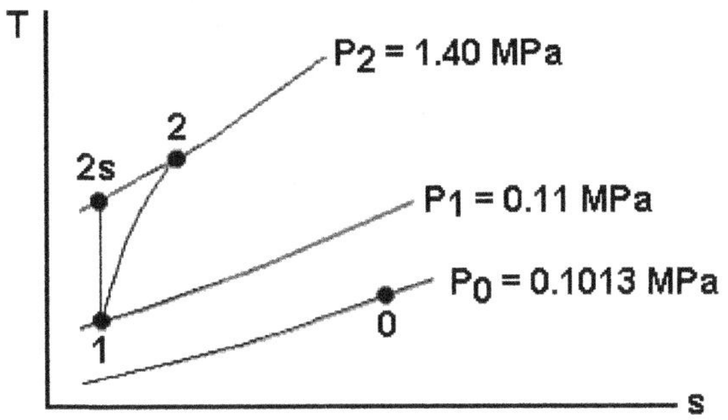

ig. 10.7 Gas compressor process diagram.

learly not the case and the pressure ratio term in eq. (10.50) must be retained. From he given data together with the above equations, the results in the table below can be btained:

tate	P (MPa)	T (K)	p_r	ϕ (kJ/kg · K)	h (kJ/kg)	e (kJ/kg)
	0.110	275	54.14	7.7559	435.0	7.504
s	1.40		689.05	8.48606	729.567	302.04
	1.40			8.66734	841.30	360.63
	0.1013	293.15		7.81997	453.21	0

his is the sequence of calculations: (i) p_{r2s} from Eq. (10.49), (ii) ϕ_{2s} and h_{2s} from nterpolation in air tables, (iii) h_2 from eq. (10.45), and (iv) e_1 and e_2 from qs. (10.12) and (10.50). Thus, the working equations for the exergy terms are

$$e_i = h_i - h_0 - T_0\left[\phi_i - \phi_0 - R\ln\left(\frac{P_i}{P_0}\right)\right], \quad i = 1, 2 \tag{10.51}$$

vhere the gas constant for air is $R = 0.287$ kJ/kg · K

The work to run the compressor, $h_2 - h_1$, comes to 406.3 kJ/kg. Then, the brute-orce Second Law efficiency is 0.872 or 87.2%, the functional Second Law efficiency is).869 or 86.9%, and the exergy loss is 53.17 kJ/kg.

The same type of analysis can be carried out for a mixture of noncondensable gases rovided the composition is known. It is acceptable to treat the mixture as a mixture of ideal gases as long as the pressure is not very high, i.e., Dalton's law of additive par-ial pressures and Amagat's law of additive partial volumes may be applied to a good pproximation [8]. Where more accurate results are demanded, one may turn to Kay's ule [9] using pseudo-critical properties or more elaborate equations of state such as he Benedict-Webb-Rubin equation or others like it [2].

10.6.6 Exergy efficiencies for pumps

umps play a vital role in geothermal plants. They move cooling water from the basins of cooling towers to condensers, they lift geofluid from reservoirs to the surface, they orce waste brine back into the reservoir, they repressurize working fluids in binary cycles, and they boost steam or working fluid condensate from the condenser hot wel to the top of cooling towers. The power required to run pumps is a major parasitic load for geothermal plants and their efficient operation can help improve overall plant performance. Fig. 10.8 is a schematic of a pump receiving low-pressure liquid at state 1 and discharging it at a higher pressure at state 2. From a thermodynamic analysis perspective the internal details of the pump are unimportant, and only the properties at the inlet and outlet matter.

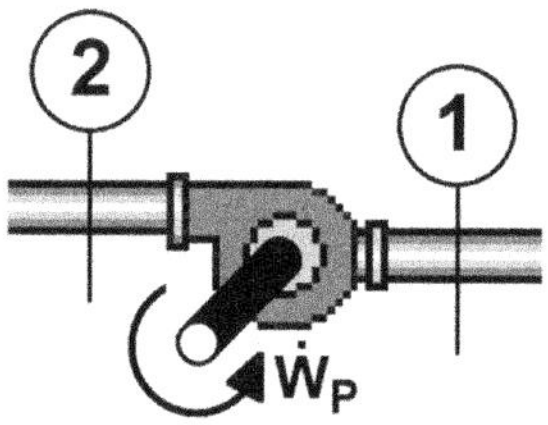

Fig. 10.8 Liquid pump.

As with a compressor, the role of the pump is to increase the pressure of the fluid passing through it, in this case a liquid. From the point of view of the Second Law, its role is to raise the liquid's exergy to a desired level by expending as little work (exergy) as possible. In terms of the state points shown in the figure, the pump isentropic efficiency is given by:

$$\eta_p = \frac{h_{2s} - h_1}{h_2 - h_1} \qquad (10.52)$$

where the state 2s refers to the ideal outlet state that would follow an isentropic process from state 1.

The specific loss of exergy during the process is given by:

$$\Delta e_p = e_1 + |w_p| - e_2 \qquad (10.53)$$

where the absolute magnitude of the work is used to emphasize that it is an exergy input. The reader will recall that the exergy associated with work is the work itself.

The Second Law efficiencies are defined as follows:

$$\eta_{p,BF}^{II} = \frac{e_2}{e_1 + |e_{w_p}|} = \frac{e_2}{e_1 + |w_p|} \qquad (10.54)$$

and

$$\eta_{p,FUN}^{II} = \frac{e_2 - e_1}{|e_{w_p}|} = \frac{e_2 - e_1}{|w_p|} \qquad (10.55)$$

The functional definition is based on the job of a pump to increase the exergy of the fluid and that it does so at the expense of the exergy associated with the work of pressurization.

Numerical example: We will consider a condensate pump in a supercritical isopentane binary cycle with the following given conditions: state 1 is a saturated liquid at a

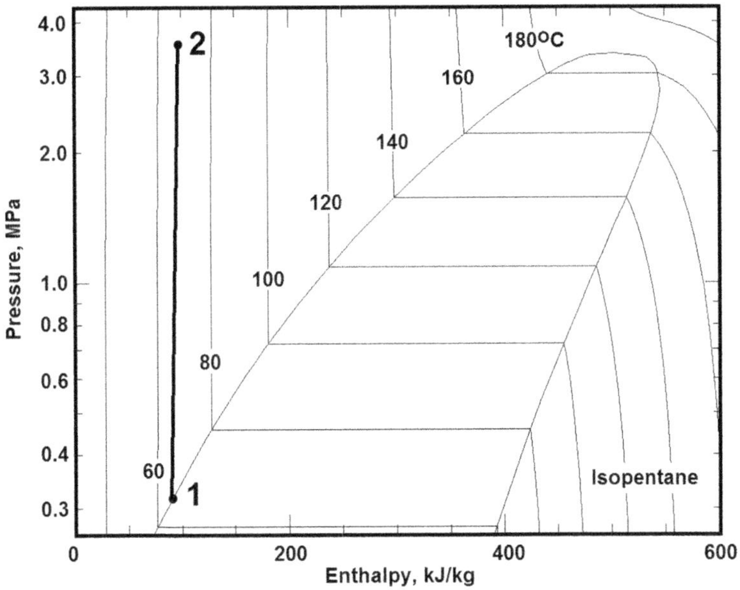

ig. 10.9 Supercritical pump process diagram.

emperature of 65°C, state 2 has a pressure of 3.5 MPa, and the pump isentropic fficiency is 81.5%. The dead state is at 20°C, 1 atm.

The properties of isopentane will be taken from REFPROP. The process diagram is iven in Fig. 10.9, to scale for isopentane.

From the given data, the results in the table below can be obtained:

State	P (MPa)	T (°C)	s (kJ/kg · K)	h (kJ/kg)	e (kJ/kg)
1	0.31267	65	0.27768	89.070	7.904
2s	3.5	66.492	0.27768	94.618	
2	3.5	66.999	0.28138	95.877	13.626
0	0.1	20.0	−0.059756	−17.753	0

The work to run the pump, $h_2 - h_1$, comes to 6.807 kJ/kg. Then, the brute-force Second Law efficiency is 0.926 or 92.6%, the functional Second Law efficiency is).841 or 84.1%, and the exergy loss is 1.085 kJ/kg.

10.6.7 Exergy analysis for production wells

The last illustration of exergy principles will be a production well. Two cases will be examined: (A) artesian flow and (B) pumped flow.

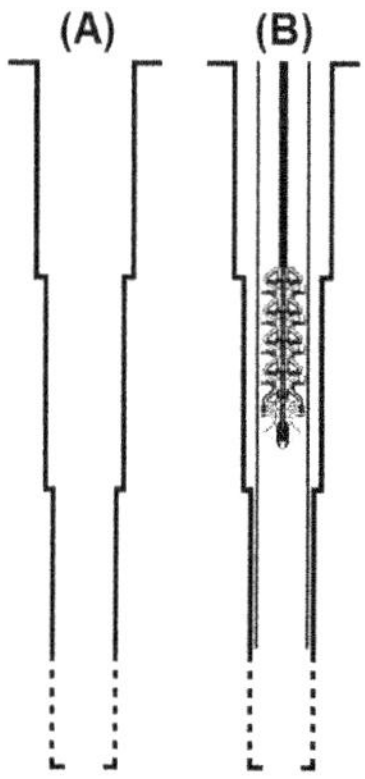

Fig. 10.10 Production wells: (A) artesian flow, (B) pumped flow.

The geofluid enters the well at the bottom (state 1) and exits at the surface (state 2). When a pump is used, the fluid is kept in a liquid state under pressure all the way to the surface but under artesian flow the fluid is assumed to flash into a 2-phase mixture at some point along the well. Thermodynamically, the well should deliver the highest exergy geofluid with the least loss of exergy during the process. A pump will enhance the well mass flow rate but at the expense of the power to run the pump.

The specific loss of exergy during the well flow process is given by:

$$\Delta e_w = e_1 - e_2 \tag{10.56}$$

if the well flow is artesian, and

$$\Delta e_w = e_1 + |w_p| - e_2 \tag{10.57}$$

if the well is pumped. Since the flow rates will generally not be the same for these two cases, the total exergy flow rate will be obtained by multiplying the specific exergy terms by the appropriate mass flow rate.

The Second Law efficiencies are defined as follows:

Case (A):

$$\eta^{II}_{w,BF} = \eta^{II}_{w,FUN} = \frac{e_2}{e_1} \tag{10.58}$$

Case (B):

$$\eta^{II}_{w,BF} = \frac{e_2}{e_1 + |e_{w_p}|} = \frac{e_2}{e_1 + |w_p|} \tag{10.59}$$

and

$$\eta^{II}_{w,FUN} = \frac{e_2 - e_1}{|e_{w_p}|} = \frac{e_2 - e_1}{|w_p|} \tag{10.60}$$

For case (A), both definitions are the same and yield 100% efficiency if the produced exergy is equal to the exergy that entered the well from the reservoir. For case (B), the focus is on how much additional exergy is being produced with respect to the amount of exergy expended in running the pump.

Numerical example – Case (A): Consider a reservoir at 180°C and 1.10 MPa, and a well capable of flowing at 50 kg/s under artesian conditions. The wellhead pressure is maintained at 0.250 MPa. The dead state is at 20°C, 1 atm.

Using REFPROP and the given conditions, the results in the table below are obtained.

State	*P* (MPa)	*T* (°C)	*s* (kJ/kg·K)	*h* (kJ/kg)	*x*	*e* (kJ/kg)
1	1.1	180	2.1391	763.10	—	138.87
2	0.25	127.41	2.1757	763.05	0.1044	128.09
0	0.1	20	0.29646	84.06	—	0

Thus, the brute-force and functional Second Law efficiencies are 0.922 or 92.2%, and the exergy loss is 10.78 kJ/kg. The total rate of exergy production is 6.40 MW.

Numerical example – Case (B): Consider the same reservoir at 180°C and 1.1 MPa, but now fitted with a downhole pump capable of producing 75 kg/s at a wellhead pressure of 1.25 MPa and temperature of 182°C. The dead state is at 20°C, 1 atm.

Using REFPROP and the given conditions, the results in the table below are obtained.

State	*P* (MPa)	*T* (°C)	*s* (kJ/kg·K)	*h* (kJ/kg)	*x*	*e* (kJ/kg)
1	1.1	180	2.1391	763.10	1.00	138.87
2	1.25	182	2.1583	771.99	—	142.13
0	0.1	20	0.29646	84.06	—	0

Thus, the specific work to run the pump, $h_2 - h_1$, comes to 8.89 kJ/kg and the power required is 666.8 kW. The brute-force efficiency is 0.962 or 96.2% and the functional efficiency is 0.367 or 36.7%. The total rate of exergy production is 10.66 MW (gross) and, after subtracting the pump power, 9.99 MW (net).

Discussion of numerical examples: The pumped well yields 3.59 MW more exergy production than allowing the well to self-flow. Assuming a conversion efficiency of 40% from exergy to electricity, case (B) should yield about 1.4 MW more electrical power output than case (A), for the conditions stipulated.

As a final note, most of the basic principles presented in this chapter were published by the author in Ref. [10] where they were applied to specific geothermal binary plants operating at low-temperature resources.

References

[1] Kestin, J., "Available Work in Geothermal Energy," Chap. 3 in *Sourcebook on the Production of Electricity from Geothermal Energy*, Kestin, J., Ed. in Chief, R. DiPippo, H.E. Khalifa and D.J. Ryley, Eds. U.S. Dept. of Energy, DOE/RA/4051-1, U.S. Gov. Printing Office, Washington, DC, 1980.

[2] Bejan, A., *Advanced Engineering Thermodynamics, 2nd Ed.*, Wiley-Interscience, John Wiley & Sons, New York, 1997.

[3] Reynolds, W.C., *Thermodynamic Properties in SI: Graphs, Tables and Computational Equations for 40 Substances*, Dept. of Mechanical Engineering, Stanford University, Stanford, CA, 1979.

[4] Wepfer, W.J. and R.A. Gaggioli, "Reference Datums for Available Energy," Chap. 5 in *Thermodynamics: Second Law Analysis*, R.A. Gaggioli, Ed., ACS Symposium Series 122, American Chemical Society, Washington, DC, 1980.

[5] Moran, M.J., *Availability Analysis: A Guide to Efficient Energy Use, Corrected Edition*, ASME Press American Society of Mechanical Engineers, New York, 1989.

[6] Keenan, J.H., F.G Keyes, P.G. Hill and J.G. Moore, *Steam Tables: Thermodynamic Properties of Water Including Vapor, Liquid, and Solid Phases (International Edition – Metric Units)*, John Wiley & Sons, Inc. New York, 1969.

[7] Gallagher, J.S., D. Linsky, G. Morrison, and J.M.H. Levelt Sengers, *Thermodynamic Properties of a Geothermal Working Fluid; 90% Isobutane-10% Isopentane*, NBS Technical Note 1234, U.S. Dept. of Commerce, 1987.

[8] Çengel, Y.A. and M.A. Boles, *Thermodynamics: An Engineering Approach, 4th Ed.*, McGraw-Hill New York, 2002.

[9] Kay, W.C., Sc.D. Thesis, Massachusetts Institute of Technology, 1937.

[10] DiPippo, R., "Second Law Assessment of Binary Plants for Power Generation from Low-Temperature Geothermal Fluids," *Geothermics*, V. 33, 2004, pp. 565–586.

Problems

10.1 Calculate and plot lines of constant exergy, e = constant, for pure water in the two-phase, liquid-vapor region, starting from the saturated liquid line and covering the range of temperature from 50°C to 350°C in 50° intervals. Use 20°C as the dead-state temperature. Plot the results on a temperature-entropy diagram (to scale) showing the saturation curve, and lines of constant enthalpy on the same figure. The h = constant lines should begin at the saturated liquid line from the same points as the e = constant lines.

10.2 A single-flash steam plant operates at a reservoir having a temperature of 250°C. The condenser runs at 50°C; T_o = 20°C. The separator temperature is found from the "equal-temperature-split" rule. The isentropic efficiency of the turbine is 82%.

(a) Find the specific exergy (in kJ/kg) of the geofluid at all points in the plant.

(b) Calculate the loss of exergy (in kJ/kg) for the geofluid in traveling from the reservoir to the separator (assuming saturated liquid in the reservoir).

(c) Calculate the loss of exergy (in kJ/kg) for the geofluid as it passes through the turbine.

(d) If the cooling water enters the shell-and-tube condenser at 30°C and leaves at 40°C, calculate (i) the brute-force exergy efficiency of the condenser and (ii) the loss of exergy (in kJ/kg of geofluid) as it passes through the condenser.

0.3 A regenerator is used in an isopentane binary plant. Isopentane from the turbine exhaust at 350 K enters the regenerator and leaves at 325 K; the pressure is 0.18 MPa. The isopentane from the feed pump enters the other side of the regenerator at approximately 2 MPa, 320 K, and an enthalpy of 251.5 kJ/kg. The dead state temperature is 20°C and the regenerator is perfectly insulated.

(a) Determine the outlet state of the isopentane on the high-pressure side.
(b) Calculate the exergy lost in the heat exchanger, in kJ/kg.
(c) Calculate the log mean temperature difference, in °C.

Part 3

Geothermal Power Plant Case Studies

- Larderello, Tuscany, Italy
- The Geysers, California, USA
- Cerro Prieto, Baja California, Mexico
- Hatchobaru, Kyushu, Japan
- Mutnovsky, Kamchatka, Russia
- Miravalles, Guanacaste, Costa Rica
- Heber Binary Plants, California, USA
- Magmamax, California, USA
- Nesjavellir and Hellisheidi Plants, Iceland
- Raft River Plants, Idaho, USA
- Geothermal Power Plants in Turkey
- Enhanced Geothermal Systems – Projects and Plants
- Environmental Impact of Geothermal Power Plants

"Geothermal electricity, unlike fossil or nuclear, cannot be ordered: it must be developed, for there is nothing more hazardous than a premature order for conversion equipment."

Joseph Kestin – July 1979

The third part of the book offers case studies of the planning, design, and performance of selected geothermal power plants, chosen to illustrate all of the traditional energy conversion systems covered in the second part. Included are plants that have played important historical roles. Some of the cases show integrated usage of geothermal energy, for heat and power, and for waste heat conversion to generate added electricity. Many more plants could have been cited but space limitations have forced us to narrow our selection to fourteen sites. An entire chapter in this edition is devoted to Turkey where geothermal energy is rapidly being developed. The new Enhanced Geothermal Systems technology is highlighted in a new chapter with several sites covered.

Nesjavellir 60 MWe and 200 MWt geothermal plant, Iceland.
Photo: Ballzus, C. et al, "The Geothermal Power Plant at Nesjavellir, Iceland," Proc. World Geothermal Congress 2000, *Japan, International Geothermal Association, pp. 3109–3114 [WWW].*

Chapter 11

Larderello Dry-Steam Power Plants, Tuscany, Italy

"We can ... give expression at this time to the expectation of conquering ... the vast Plutonian forces which formerly manifested themselves in such spectacles as Vesuvius, Hekla, Krakatoa, and of utilizing at least a small portion of this nearly unlimited store of energy for the benefit of mankind."

George A. Orrok – November 12, 1925

11.1 History of development

The case studies section of this book begins at the birthplace of geothermal energy as a source of electricity – Larderello, Italy. It was here in this Tuscan village in 1904 that Prince Piero Ginori Conti fashioned the first electro-mechanical device that converted the energy of the indigenous steam, issuing from the earth for centuries, into electricity – enough to illuminate five light bulbs in his boric acid factory.

Students of ancient history are well acquainted with Julius Caesar and the Romans. The Romans knew places like Larderello very well, having built magnificent baths at many of them throughout their empire from England to Turkey, as we know them today. Certainly Larderello is a part of the ancient history of Tuscany, situated in gently rolling hills, roughly equidistant from the famous cities of Siena and Volterra.

Perhaps the best way to tell the story of ancient Larderello is with an annotated timeline, Fig. 11.1 [1, 2]. The information contained in what follows was found in the booklet produced by ENEL, Ente Nazionale per L'Energia Elettrica (the electric authority of Italy) [1], and in Burgassi's thorough contribution to *Stories from a Heated Earth* [2].

Geothermal Power Plants: Principles, Applications, Case Studies and Environmental Impact, Third Edition

© 2012 Elsevier Ltd. All rights reserved.

Fig. 11.0 Valle Secolo geothermal power plant, Larderello, Italy; photo courtesy of Enel GreenPower.

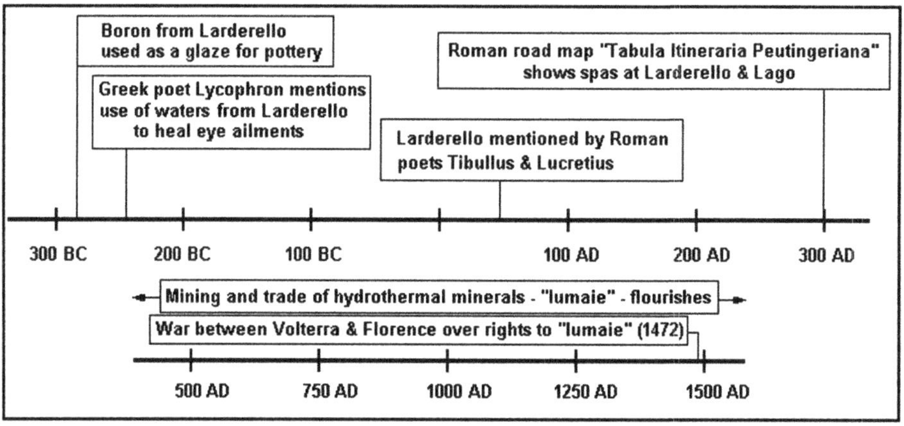

Fig. 11.1 Timeline of ancient history of Larderello and the Boraciferous Region [1, 2].

The main events beginning in the late eighteenth century are listed in Table 11.1 [1, 2]. The destruction of practically all of the geothermal power stations at Larderello in 1944 during World War II marked the low point in production from the field [3]. The power plants were a strategic target during the Italian Campaign. Following the landing of the Allied Forces at Salerno and Anzio in late 1943 and early 1944, respectively, the battle line moved relentlessly northward toward Rome, and after its capture in June of 1944, farther north toward Florence. Starting in February 1944, the geothermal power plants at Castelnuovo and Larderello came under attack by Allied fighter-bombers. At first the plants sustained only light damage and continued to

able 11.1 Chronology of events in the history of Larderello [1, 2].

ear	Event
777	Discovery of boric acid in "lagoni" (thermal pools) by F.U. Höfer.
810	First try at using steam from lagoni for industrial purposes.
812	Timber was cut and burned to supply heat to evaporate water during boron recovery.
818	Francesco de Larderel establishes his company to recover boric acid on a commercial scale at Montecerboli; eight more factories are built throughout the region over the next 17 years.
827	Invention of the "lagone coperto" (covered pool), replaced wood with geothermal steam to provide heat for the mineral recovery process, greatly improving the efficiency of the operation.
828	F. Larderel laid the basis for drilling for geothermal steam.
842	Using efficient counterflow heat exchangers (Adriane boilers) devised by A. Larderello, boric acid production reached 1000 tons; in 1829 it had been 125 tons.
865	Steam-driven pumps replaced animal- or water-wheel devices to pump the boraciferous waters.
894	Invention of the tubular boiler using geothermal fluids by F. Raynant of the Larderel factory.
904	On July 4, P.G. Conti ran his ¾-horsepower reciprocating steam engine at his factory in Larderello, marking the first production of electricity from geothermal energy.
905	A 20 kW dynamo was commissioned, driven by a 40-horsepower reciprocating steam engine to serve the Larderel factory.
912	Construction began for the first commercial geothermal power station.
914	On March 10, a 250 kW turbo-alternator was connected to the power distribution system of Volterra and Pomarance, marking the first commercial generation of electricity from geothermal energy.
944	Prior to WWII, the installed capacity at Larderello had reached 136.8 MW and yearly produced more than 900 GW · h. Practically all power plants were destroyed as a result of the war; only one small generating unit survived.
948	Reconstruction of the Larderello plants that had begun immediately after the war was completed, and the original station was retired to make way for more modern plants.
959	Larderello power capacity reached 300 MW; annual generation exceeded 2,000 GW · h.
971	Successful drilling of Travale 22, the most powerful geothermal well in Italy, sparked new interest in the Travale area.
983	Reinjection was deemed necessary to combat the decline in reservoir performance. Standardized 60 MW turbines were designed.
1988	"Program 2000"was launched to triple the electrical generation from geothermal energy from 3,000 to 9,000 GW · h by the year 2000.
1993	Two new 60 MW units commissioned at Valle Secolo; see Fig. 11.0.
2000	Start of major renovation of power plants; over the next 5 years, 10 new plants, 254.5 MW, are installed, replacing less efficient, obsolescent plants.
2004	Deep drilling program initiated in area of Larderello-Radicondoli-Montieri. A plant at Latera, 90 km SE of Larderello, is opposed on environmental grounds causing the cancellation of the plant that had been under construction.
2008	Larderello deep drilling program completed.
2010	22 "AMIS" H_2S and Hg removal systems are in operation.

generate electricity and to produce boric acid. From April until June, the plants were repeatedly bombed and strafed, despite being defended by six anti-aircraft batteries. It was during this period that the plants were put out of commission. Remarkably, the personnel at both power plants and at the boric acid factory managed until June 28 to continue working an 8-hour shift, albeit at night. Once the plants were no longer in the line of fire, they were quickly rebuilt. Since then the geothermal facilities have continued to grow in capacity and output as older units are replaced with modern, more efficient ones.

[Note: The events during WWII related here are based on the contents of a diary kept by Mr. Geo Desi, an Italian eyewitness to the attacks at Castelnuovo and Larderello. The contents of Desi's diary, in Italian, was provided to the author by Mr. Andrea Rossetto (Torreglia, Italy), through the good offices of Mr. Steve Cole (Memphis, Tennessee). Mr. Cole maintains a comprehensive web site "The Italian Campaign of World War II September 1943 – May 1945," http://members.aol.com/ItalyWW2/History.htm.]

11.2 Geology and reservoir characteristics

The anomalous geothermal activity in the southern Tuscany region arises from the collision of edges of two of the earth's tectonic plates, the European and the African. Over some 30 million years this region has been subjected to alternating periods of compressive and tensional effects. The result has been the creation of fault systems, tectonic highs, and the emplacement of magma bodies relatively close to the surface, as shallow as 7 km [4].

The geothermal areas in Italy include: Larderello and its several adjacent areas such as Lago and Travale-Radicondoli; Monte Amiata and its two foci of development at Bagnore and Piancastagnaio; and Monti Volsini and its fields at Latera and Torre Alfina near Lake Bolsena; see Fig. 11.2. In this section, we will concentrate on the main area of geothermal power development, namely, the greater Larderello area.

Throughout the 400 km^2 Larderello-Travale-Radicondoli-Lago geothermal area, the geologic stratigraphy consists of a sequence of rock formations that can be described as follows, starting with the youngest at the surface and moving downwards [4]:

- Neoautochthonous complex, recent volcanics that cover the surface in most places.
- Flysch facies formation, consisting mainly of shales but also of limestones and sandstones; this layer forms the cap rock above the reservoir.
- Nonmetamorphic Tuscan unit or Tuscan Nappe, having evaporitic carbonate rocks at the bottom and a terrigenous sequence at the top; a permeable reservoir is hosted primarily in this group and the underlying Verrucano Group.
- Verrucano Group, formed as tectonic wedges consisting of metamorphic and sedimentary rocks.
- Metamorphic Tuscan units, consisting of the Amphibolitic Gneisses Group and Micaschists-Phyllites Group; this is the basement of the formation.

Underlying and embedded in the basement are the magmatic intrusions that supply the heat for the system.

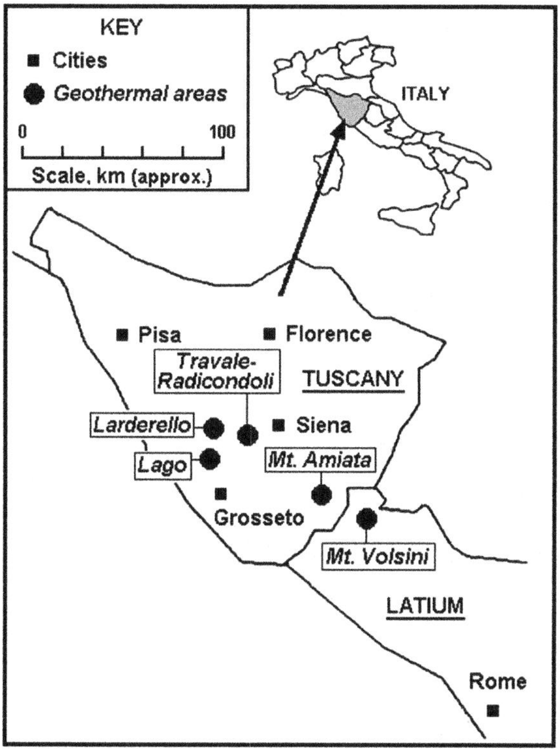

ig. 11.2 General location map of geothermal areas in Tuscany and northern Latium; adapted and expanded om [5].

The sites of the once numerous thermal manifestations correspond to structural ighs of the permeable reservoir rocks, often lying no more than a few hundred meters eep. These areas are now marked by large patches of altered ground and a few ctive fumaroles, and include sites at Monterotondo, Sasso Pisano, Lago, Castelnouvo, errazzano, Lustignano, and, of course, Larderello (originally Montecerboli, later enamed for Francesco de Larderel). It is reasonably certain that the greater Larderello egion was characterized by massive thermal discharge dating back at least 2,300 ears, and probably much longer. Recalling from Chapter 7 one of the important equirements for the creation of a dry-steam reservoir, namely, extensive long-term elease of geothermal fluids that lowers the water table, we see that this condition was ertainly met at Larderello.

The areas having the thermal manifestations were obviously attractive sites for the rst wells. These were drilled only to depths sufficient to win steam for either boric acid roduction or later to drive electric generators. Only after decades of exploitation that gnificantly depleted the shallow reservoir was the deep reservoir sought and explored.

The permeable reservoir rocks, the Tuscan Nappe, outcrop importantly in many laces throughout the southern half of the greater Larderello region, providing an asy entry for cold recharge to the reservoir. Fortunately, the low temperatures

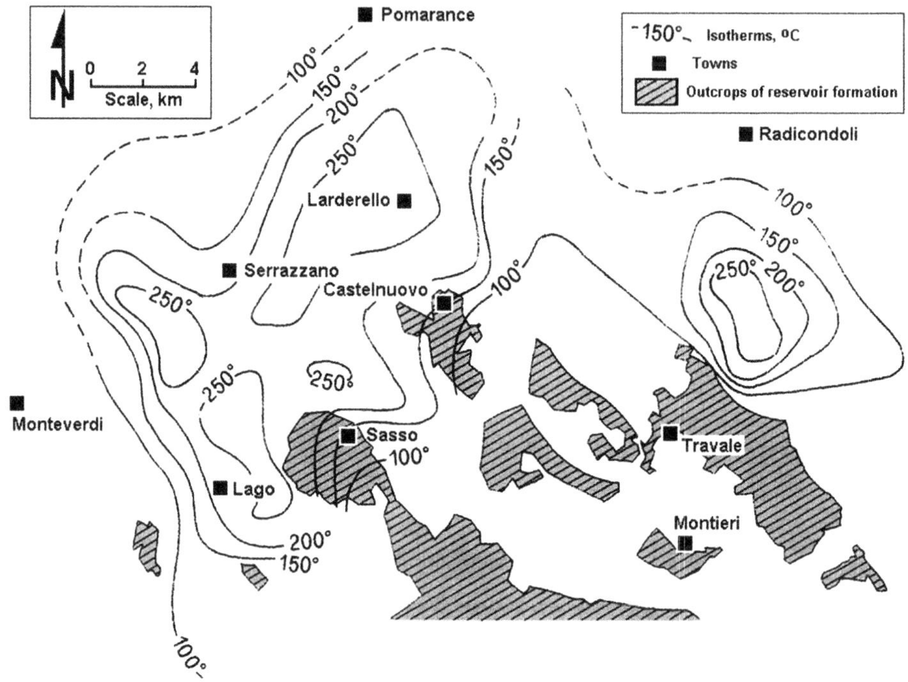

Fig. 11.3 Isotherms at the top of the reservoir in the greater Larderello area; after [4].

associated with this infiltration are detected only at shallow levels owing to the vertically variable permeability within the productive reservoir that effectively isolates the deeper layers from the more shallow ones.

This can be seen very clearly in the three isotherm distribution plots in Figs. 11.3–11.5, first at the top of the reservoir (which generally is less than 1,000 m deep), next at 2,000 mbsl, and the last one at 3,000 mbsl [4, 6]. The influence of the cold recharge is evident in Fig. 11.3 in the areas surrounding the outcrops of reservoir rocks. However, from Fig. 11.4, it is clear that the high temperature isotherms at depth form a more or less continuous pattern indicating a broad zone of reservoir having temperatures in excess 250°C.

In Fig. 11.5 we see the results of the most recent and deepest drilling. The hottest part of the reservoir lies toward the southwest in the vicinity of Lago. At 3,000 mbsl, the 400°C isotherm approximates a 6.5 km circle centered a little north of Lago. A small, nearly circular lake lies close to Lago (from which it took its name) and is believed to be the remains of a phreatic explosion. Temperatures of 450°C are present below about 3,200 mbsl in the middle of the hottest area [7]. Furthermore, since the isotherms are fairly closely stacked here, heat is transferred mainly by conduction. This could indicate the proximity of a magmatic intrusion. Granitic intrusions indeed have been encountered in some of the deepest wells drilled in the region.

The most promising and so far successful exploration method for the deep reservoir is 3D reflection seismic surveys [8]. Small explosive charges, about 3 kg of TNT, are detonated in shallow holes, about 12 m deep, and the reflected waves are picked up by

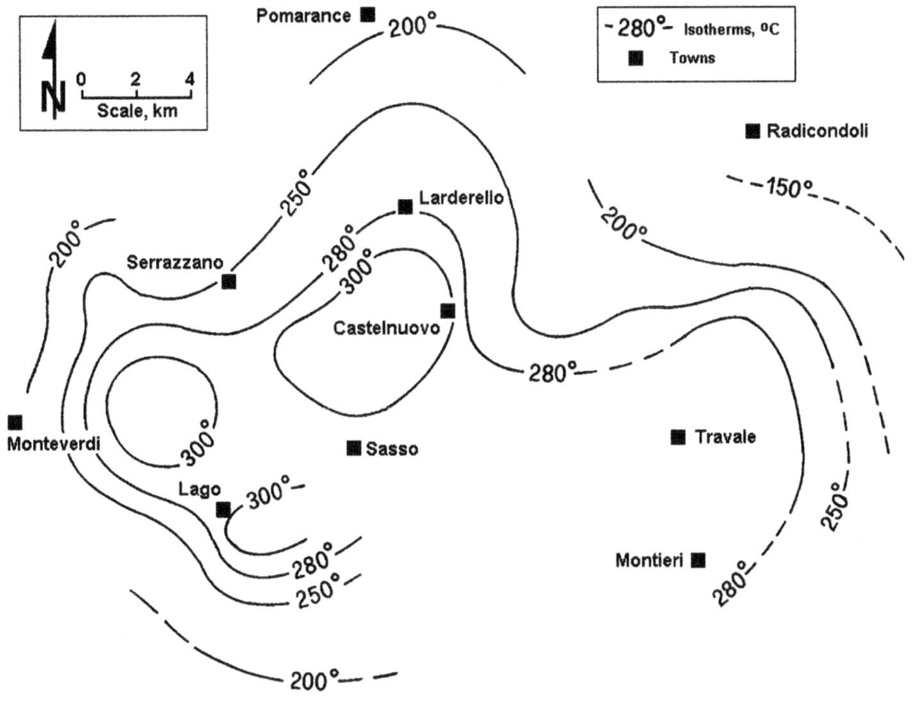

Fig. 11.4 Isotherms at 2,000 mbsl in the greater Larderello area; after [4].

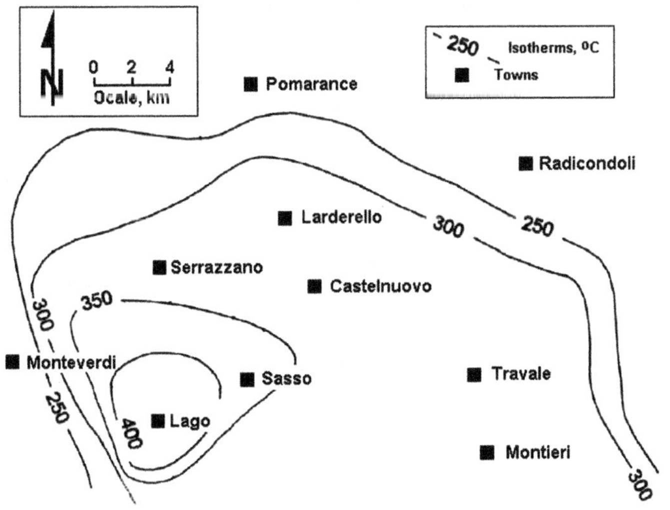

Fig. 11.5 Isotherms at 3,000 mbsl in the greater Larderello area; highly simplified from [6].

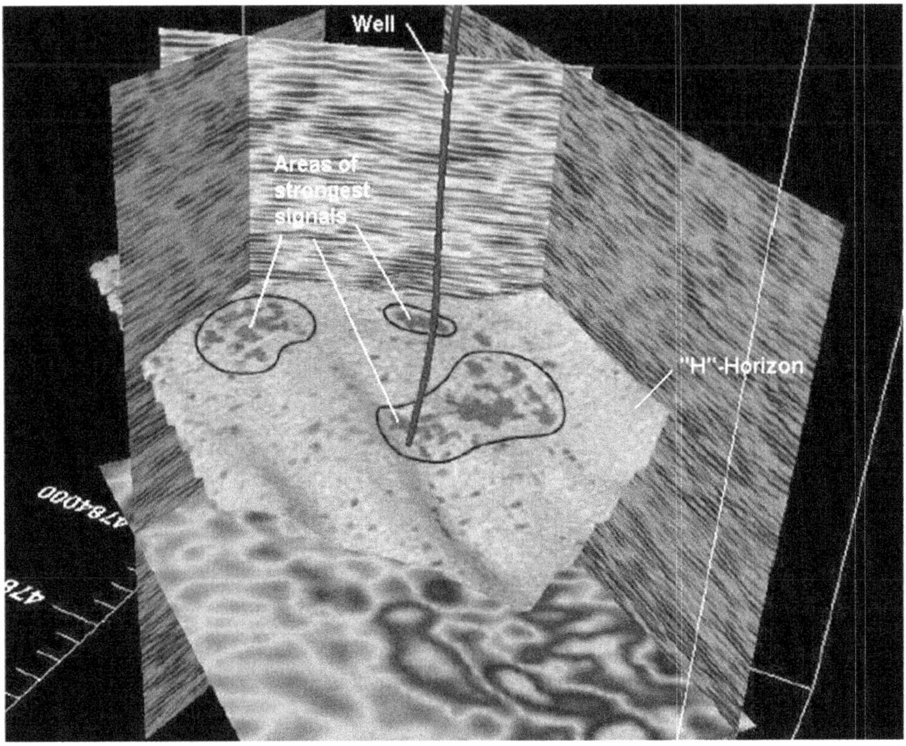

Fig. 11.6 3D plot of the "H"-Horizon as determined by reflective seismic surveys. The darkest colored areas are those with the highest RMS amplitudes and could be interpreted as promising drilling targets. Excerpted from [8] [WWW].

a large array of geophones. Rock layers that have high permeability are characterized by lower density and lower wave propagation velocity. The returns from several lines of receivers are processed into 3D plots that reveal the most promising targets for the deep wells. A portion of such a plot is shown in Fig. 11.6 [8].

The most important finding of the new deep exploration program is that the greater Larderello geothermal field is unified at depth. A reservoir of enormous extent exists below 3,000 mbsl that feeds the entire region. There is continuity of the hottest isotherms that encompass an area of about 400 km^2. This will lead to more drilling in the outlying areas where no thermal manifestations are seen and below areas characterized by poor thermal gradients caused by the surface infiltration of cold meteoric water. This discovery should lead to an expansion of the geothermal power capacity of the Larderello area.

11.3 Power plants

We will now discuss the power plants of Larderello, beginning with the early designs (1904–1968), the modern designs (1968–1995), and the recent designs (1995–present).

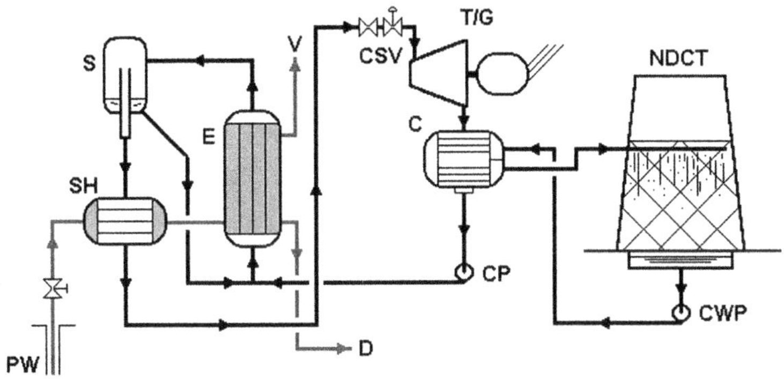

Fig. 11.7 Indirect-cycle for power generation from "soffioni" at Larderello.

11.3.1 Early power plants

The design of the earliest power plants at Larderello was driven by three main considerations: (1) the steam was of low pressure; (2) it contained acidic gases and other chemical impurities that precluded its use in the prime movers of the time, either reciprocating steam engines or rotodynamic steam turbines; and (3) the geofluid had to be processed to recover the valuable minerals it contained. The latter use was the original commercial application of the fluids from the "Boraciferous Region" and would remain a lucrative venture for many decades.

Although Prince Piero Ginori Conti initially fed the geothermal steam directly from the "soffioni" into a hydraulic turbine adapted for geothermal use, and later tried to use the natural steam in his historic reciprocating engine (see Fig. 7.1), it was feared that the machines could not withstand the corrosive attack from the natural fluid. Conti even developed a method of securing somewhat higher pressure steam by boring 0.2–0.4 m diameter holes to depths of 100–180 m. This produced steam flows of 5–20 t/h at pressures from 2–3 bar and temperatures from 150–180°C, but the steam was burdened with significant amounts of noncondensable gases (NCG). Thus, a new method would be needed for this steam to power electrical generators [9].

Thus was born the so-called "Indirect Cycle". The geothermal steam was sent to heat exchangers where it was used to boil pure water obtained from a supply of fresh water. The drained condensed geothermal steam was then sent to the chemical recovery plant, while the clean steam powered the electrical generator; see Fig. 11.7.

The evaporators were a complex set of shell-and-tube heat exchangers that were known as Kestner evaporators. An elevation view is shown in Fig. 11.8 [10]. The geothermal steam from the "soffioni" entered the unit at the pure steam superheater where it flowed through aluminum tubes while the pure steam passed over the tubes. It continued into the steam manifold and was distributed to the shell side of the vertical evaporators, arranged in pairs. The spent steam drained from the bottom of the evaporators and the condensate was finally used to preheat the incoming pure water in another shell-and-tube heat exchanger. The condensate of the natural steam was then transported to the mineral recovery plant. The evaporator tubes were 7 m long, 30 mm in diameter, and also made of aluminum, which was the only metal that could

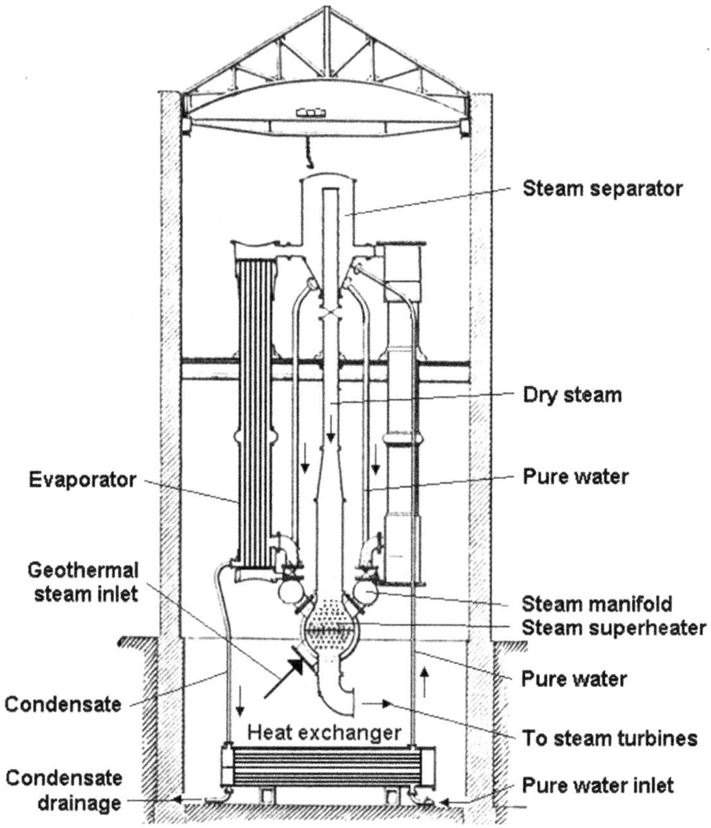

Fig. 11.8 Kestner evaporators for raising pure steam from "soffioni"; after [10].

withstand the corrosive attack of the sulfuric-acid-laden geothermal steam. The mechanical strength of the aluminum was less than desired and vibration problems were encountered. Also the tubes suffered somewhat from corrosion, probably at the tube sheet connections, allowing geofluid and NCG to mix with the pure water and steam. This overburdened the condenser air removal system, raising the condenser pressure, and lowering the turbine power output.

The "pure" steam was used in three 2.5 MW turbo-generator units. Photographs of the turbine hall and of one of the rotors are shown in Figs. 11.9 and 11.10 [11]. The turbine was a double-flow unit with seven groups of blades; the first five groups had brass blades and the last two low-pressure groups were made of 5% nickel steel.

The pressure of the pure steam leaving the superheater was about 2 bar, a, but the turbine was designed to produce its rated power at an inlet steam pressure of only 1.2 bar, a. The steam flow rate was about 9.7 kg/s (77,000 lbm/h) per turbine [11].

Integrated with the generation of electrical power was the recovery from the natural steam condensate of boric acid, H_3BO_3, borax, $Na_2B_4O_7 \cdot 10H_2O$, and ammonium bicarbonate, NH_4HCO_3 (the main ingredient of "smelling salts"). The recovery method

'ig. 11.9 Three 2.5 MW steam turbine-generator units at Larderello, c. 1916 [11].

'ig. 11.10 Rotor from a 2.5 MW Larderello unit, c. 1916 [11].

nvolved the use of the so-called Adriane boiler, a series of inclined trays through vhich the condensate flowed while being heated from below by the natural steam ›btained from the covered pools. The natural concentration of the boric acid in the ;eofluid was about 2,500 ppm or 0.25%; this was increased to about 8% during the

Table 11.2 Technical specifications for three original indirect-cycle units at Larderello.

Start-up year	1916	*Turbine:*	
Type	indirect cycle	type	reaction, axial flow
No. of units	3	cylinders/unit	1
Rating, MW	9 (3 × 3)	flows/turbine	2
Net power, MW	7.5 (3 × 2.5)	stages/flow	14
Geosteam flow rate, kg/s	29 (total, est.)	inlet press., bar, a:	1.25
Working fluid	pure water	mass flow/unit, kg/s	9.7
		speed	3000
Preheaters:			
No. per unit	5 (5 × 1)	*Generator:*	
type	horizontal shell & tube	type	3-phase, 50 Hz
cond. outlet temp., °C	63	power factor	0.7 @ 4,500 V
Evaporators:		*Condenser:*	
No. per unit	10 (5 × 2)	No. per unit	1
type	vertical shell & tube	type	horizontal shell & tube
No. of tubes	300	No. of tubes	3,000
tube length, m	7	tube length, m	4.5
tube diameter, mm	30	tube diameter, mm	22
tube area/unit, m^2	150	tube area, m^2	1,050
Superheaters:		*NCG removal system:*	
No. per unit	5 (5 × 1)	type	water-jet ejector
type	horizontal shell & tube	*Plant performance:*	
geosteam inlet temp., °C	150–180	net SSC, (kg/s)/MW	3.9
geosteam inlet press., bar, a	2	net utilization efficiency, %:	40.6
pure steam flow rate, kg/s	15.2 (total, max)	net thermal efficiency, %:	10.3
Cooling tower:			
No. per unit	1		
type	natural draft		
range, °C	10		
CW flow rate, kg/s	333		

process. When the solution was cooled and processed, nearly pure crystals of boric acid could be obtained.

Operating data, culled mainly from technical journals of the 1920s, are given in Table 11.2 [9–15]. Some inconsistencies were found in the literature, so the author exercised his judgment as to which data were more likely to be correct. Although there were five sets of evaporators, each set with two evaporators, available to serve each 3 MW turbine, in fact only four sets were needed to meet the average demand of 2.5 MW. The claimed efficiency for the steam generator unit is 88% [10]. Only two of the three units actually were put into operation at Larderello, the original unit having been moved to Lago where improvements on the cycle were tested.

The remarkable thing about this data is that the utilization and thermal efficiencies are very similar to those found in many of today's geothermal power plants. The 10% thermal efficiency is about normal for a simple binary cycle today. Nevertheless, Prince Conti and his technical staff conducted an ongoing program of research and development aimed at improving the operation. Over a period of time, the problematic Kestner boilers were replaced, eliminating the contamination of the turbine working

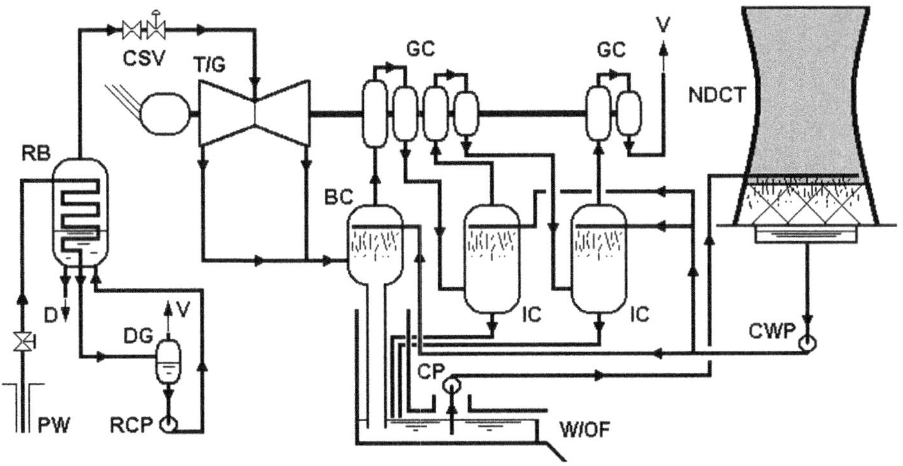

Fig. 11.11 Cycle 2 plant schematic showing reboiler RB, degasser DG, recirculation pump RCP, NCG vent V, and drain for boron-rich geofluid D. An elaborate NCG compressor train was still needed because some NCG remained in the turbine steam.

fluid. They were replaced by what Conti called "depurators" [16] or what we would call reboilers. The system that evolved is referred to as the Cycle 2 power plant design and is shown schematically in Fig. 11.11.

Finally, it is worth remembering that this early development was taking place during World War I when there were extreme shortages of conventional fuels in Italy. The price for a ton of coal sold on the wharves of Genoa was reported to be as much as $97.33 in 1918 dollars [9]. Obviously there was enormous incentive for the Italians to develop an indigenous source of energy since the country had no supplies of coal or petroleum, relying on imports of fossil fuels and their own hydropower plants for the generation of electricity. Thus, the development of the geothermal resources in Larderello beyond the mineral recovery applications to electric power production was driven by the necessity that wartime often brings.

11.3.2 Power plants of the modern era

The last of the indirect-cycle plants was taken out of service in 1968 when the plant at Castelnuovo was converted to a direct-cycle process [15]. The indirect-cycle plants had become overly expensive to operate and maintain, and the recovery of boric acid and borax was no longer economical. In fact, the mining of borax from Southern California, an area that would become another important geothermal territory, essentially put the Boraciferous Region out of the boric acid and borax business. Furthermore, the materials used in the turbo-machinery had been improved through metallurgical advances and were now capable of functioning in the presence of the natural steam, eliminating the need for the intermediate heat exchangers.

Two types of plant became the standards for geothermal plants in Italy: (a) direct-intake, exhausting-to-atmosphere units, and (b) direct-intake, condensing units. These plants are referred to as "Cycle 1" and "Cycle 3," respectively, in the literature [15].

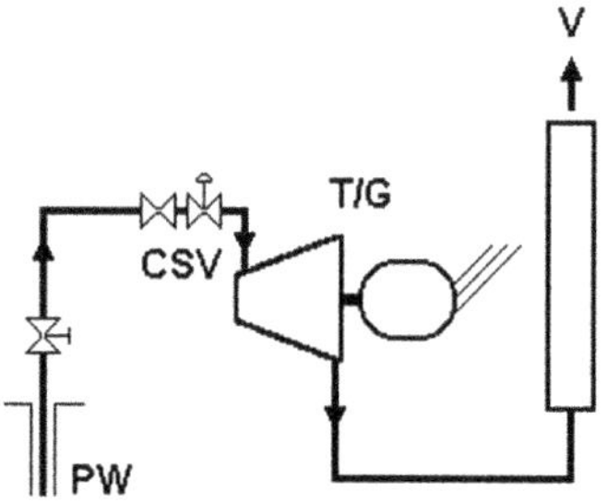

Fig. 11.12 Larderello "Cycle 1" or direct-intake, exhaust-to-atmosphere plant.

11.3.2.1 Direct-intake, exhausting-to-atmosphere units

Figure 11.12 shows the ultra simple Cycle 1 flow diagram. Cycle 1 plants were used mainly where the geothermal steam contained excessive amounts of noncondensable gases such that it was uneconomical to extract them from the condenser owing to the high cost of installing and operating the gas compressors. The largest plant to operate as a Cycle 1 plant was the 15 MW unit at Piancastagnaio near Mt. Amiata, some 72 km (45 mi) southeast of Larderello. Several plants of this type were installed within the greater Larderello area, the largest there being the 7 MW Sasso 1 plant [17].

When boric acid recovery was still conducted, the exhaust steam was directed to the Adriane boilers and assisted in evaporating the liquid geofluid. While these plants may have been optimal from an economic standpoint, they were thermodynamically wasteful of the potential of the geothermal steam. Since the subatmospheric portion of the turbine expansion is absent in these units, the power output is roughly half of what a condensing turbine could produce. For example, the Piancastagnaio plant had only a 24% utilization efficiency and a specific steam consumption of 4.92 (kg/s)/MW [17].

11.3.2.2 Direct-intake, condensing units

All plants operating on the shallow reservoir in the greater Larderello area used steam that was accompanied by significant amounts of noncondensable gases (NCG). When the concentration exceeded about 5% by weight of steam, Cycle 1 plants were used. Otherwise, it was feasible to install condensing units of the general arrangement shown in the schematic diagram in Fig. 11.13.

As can be seen, it was necessary to use several stages of gas compression to remove the NCG from the condenser to maintain the vacuum. These compressors were sometimes driven by electric motors or directly from the turbine (as shown in Fig. 11.13). The compressors were fitted with intercoolers to minimize the amount of work consumed. The Cycle 3 plant that was designed to handle the highest concentration of NCG was the 26 MW Castelnuovo unit shown in Fig. 11.14.

This unit was one of four in the Castelnuovo power station; it received the steam with the highest temperature (188°C), pressure (4.2 bar, a) and NCG (12–14%). The natural steam flow rate was 47.2 kg/s. The centrifugal turbocompressors handled 85.9 m^3/s and required 2.27 MW. The net specific geosteam consumption was 2.06 (kg/s)/MW and the net utilization efficiency was 62.8%, assuming the geosteam to be pure steam. If one allows for the roughly 15% degradation of the exergy owing

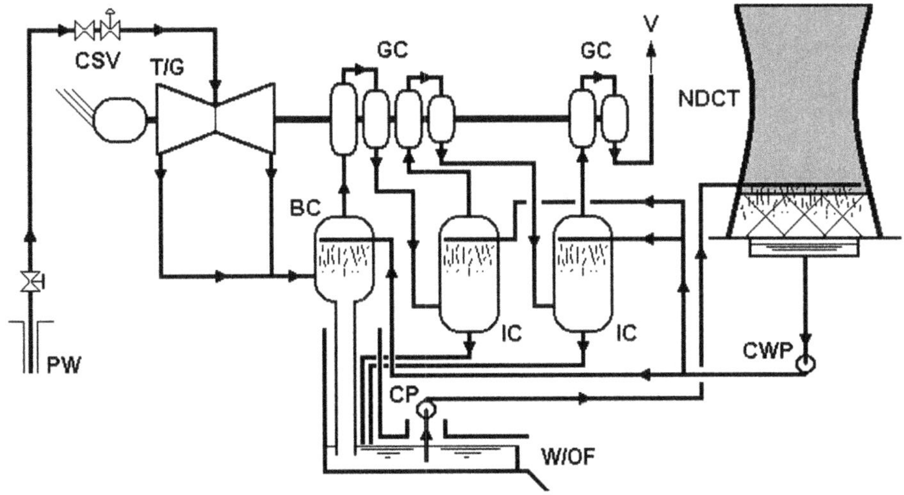

Fig. 11.13 Larderello "Cycle 3" or direct-intake, condensing plant.

Fig. 11.14 Castelnuovo turbine hall showing 26 MW turbine-generator, gas compressors, intercoolers, and aftercondenser [17, 18].

to the presence of the NCG in the steam [19], assuming it is all CO_2, the utilization efficiency was even more impressive at 73.9%.

The technical details and many photographs of all the Cycle 3 plants may be found in the author's earlier book [17].

11.3.3 Recent power plant designs

Following the discovery of the deep geothermal reservoir about 20 years ago, there has been a revolution in the way geothermal power stations are developed in Italy. Many of the older Cycle 3 plants had outlived their economic lifetimes, and Cycle 1 plants were no longer acceptable from an environmental standpoint. Instead of designing each plant to suit the conditions of the geosteam found at each site throughout the region, it was decided to adopt standardized modular plants of a fixed nominal capacity, but with internal flexibility to allow the plants to be adjusted over their lifetimes should the steam conditions change. The modularity and flexibility also allow the plants to be moved from one site to another as conditions warrant.

The standard power ratings for the new plants are 10, 20, and 40 MW, and the steam mass flow rates are nominally 80, 130, and 250 t/h, respectively [20]. Many of the new plants are located in the old power houses, replacing the obsolete units, and as such had to conform to the dimensions of the existing structures. This has made for some interesting compromises, such as the arrangement of the condenser relative to the turbine. To save vertical space, the new turbines exhaust axially which permits direct entry into the condenser on the same level. However, owing to space restrictions, the condenser axis is perpendicular to the turbine axis and the condenser is placed outside the power house [20]. Figure 11.15 shows the arrangement.

The geothermal steam remains as aggressive as ever, containing typically 5% NCG consisting of CO_2, H_2S, and various silicates, sulfates, boric acid, and chlorides. It becomes necessary to scrub the steam prior to turbine entry when the chloride content

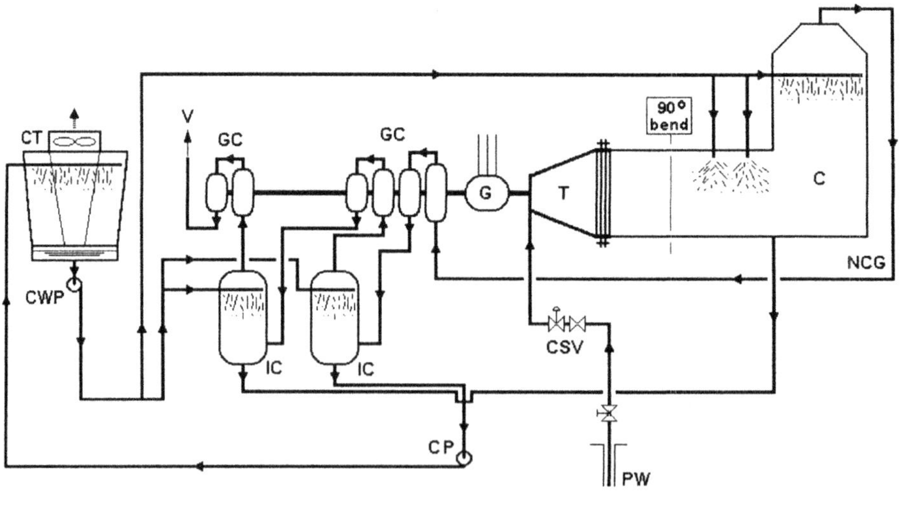

Fig. 11.15 Modular plant layout for new Larderello units.

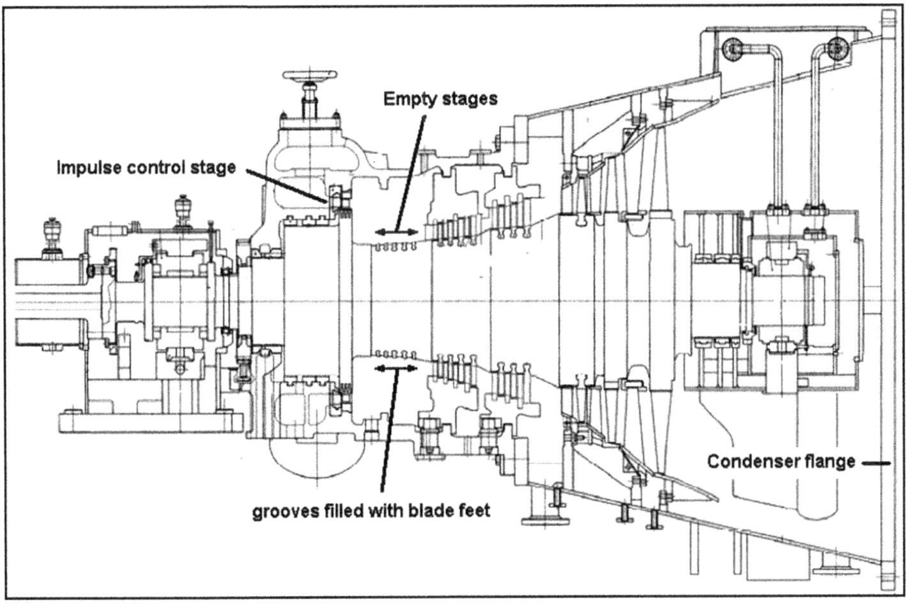

Fig. 11.16 Modular turbine design; courtesy of General Electric Oil & Gas, Florence, Italy [20].

Table 11.3 Utilization efficiency for selected recent plants at Larderello.

Item	Plant			
	Lago	Molinetto	Gabbro	Travale
Steam flow, kg/s	22.22	36.11	40.28	69.44
Inlet steam press., bar, a	2.5	6.5	6.5	14.0
Inlet steam temp., °C	127.4	190	162.0	195.1
NCG, % (wt)	1.7	4	12	5
Gross power, kW	8,855	19,210	19,005	43,230
Net power, kW	8,305	17,945	16,515	40,750
$\eta_{u,g}$, %	62.1	68.3	68.7	73.1
$\eta_{u,n}$, %	58.3	63.8	59.7	68.9

exceeds 200–300 ppm to reduce the concentration to no more than 35 ppm. The train of gas compressors and intercoolers is essentially the same as in the Cycle 3 plants. Mechanical draft cooling towers have replaced the imposing natural draft towers that gave the Larderello landscape its distinctive appearance for many decades.

The turbines are fabricated with standard-sized rotors designed to handle the maximum expected steam flow, but can be specialized for each site by removing certain rows of nozzles and blades. To maintain a smooth steam path, the empty rows are filled with blade feet; see Fig. 11.16 [20].

Table 11.3 shows the utilization efficiencies for several of the new plants; the raw data from Santini [20] was averaged for the minimum and maximum values cited in the reference. The exergy of the inlet steam has been corrected for the gas content. The

Travale plant has the benefit of the highest inlet steam pressure and temperature of the plants shown, and it is the most efficient in converting the incoming exergy into useful power output. The Lago plant is at the other end of the spectrum. The Gabbro plant with the highest level of NCG, suffers the most in going from the gross to the net output, dropping some 13% in efficiency. Nevertheless, all the plants exhibit excellent utilization efficiencies, verifying the effectiveness of the modular design strategy.

11.4 Mitigation of environmental impact

The extensive program to upgrade the older plants has been accompanied by more attention being paid to mitigating the environmental impact of the plants. As has been the case for many years, all excess geothermal condensate is reinjected. Furthermore, since 1994 there has been in operation a 39 km pipeline that delivers about 7,600 m^3/day of groundwater and geothermal fluids to the Larderello area from other geothermal sites.

A novel system for the abatement of both mercury and hydrogen sulfide from the geofluid has been designed, tested, and implemented by ENEL on several plants. The system is called AMIS (Abbattiamento Mercurio e Idrogeno Solforato). Special sorbents are used to remove the Hg, and the H_2S is catalytically oxidized to sulfur dioxide which in turn is scrubbed by a side stream of cooling water. The system will be deployed at other plants as needed [21].

New plants are painted so as to blend in with the natural colors near the plants. An example of this is shown in Fig. 11.17 where the two 20 MW units at Carboli can be seen.

Fig. 11.17 Carboli 2 × 20 MW power units; photo courtesy of Enel GreenPower [WWW].

Finally, although it may not first appear to reduce the environmental impact, the use of modern, highly efficient power plants reduces the amount of geofluid that must be extracted to generate a fixed amount of electrical power. During 2002–2004, ten new plants were started up in the greater Larderello area, and of these, seven replaced obsolescent plants. From 1999–2003 during the renovation period, the total installed power capacity remained about constant but the electricity generated increased by 37 GW·h or by more than 20% [21]. The old plants would have had to extract and process 20% more geofluid to generate the same amount of electricity.

References

[1] ENEL, *The History of Larderello*, Public Relations and Communications Dept., Ente Nazionale per L'Energia Elettrica, May 1993.

[2] Burgassi, P.D., "Historical Outline of Geothermal Technology in the Larderello Region to the Middle of the 20th Century," Chap. 13 in *Stories from a Heated Earth: Our Geothermal Heritage*, R. Cataldi, S.F. Hodgson and J.W. Lund, Eds., Geothermal Resources Council and International Geothermal Association, Sacramento, CA, 1999, pp. 194–219.

[3] Lund, J.W., "100 Years of Geothermal Power Production," *Geo-Heat Center Quarterly Bulletin*, V. 25, N. 3, Sept. 2004, pp. 11–19.

[4] ENEL, *Geothermal Energy in Tuscany and Northern Latium*, Generation and Transmission Dept., Relations and Communications Dept., Ente Nazionale per L'Energia Elettrica, May 1995.

[5] ENEL, *Larderello Field Trip: Electrical Power Generation*, Generation and Transmission Dept., Relations and Communications Dept., Ente Nazionale per L'Energia Elettrica, May 1995.

[6] Fiordelisi, A., J. Moffatt, F. Ogliani, M. Casini, S. Ciuffi and A. Romi, "Revised Processing and Interpretation of Reflection Seismic Data in the Travale Geothermal Area (Italy)," *Proc. World Geothermal Congress 2005*, Paper No. 0758, Int'l. Geothermal Ass'n., Antalya, Turkey, 2005.

[7] ENEL, *Larderello Field Trip: Geology and Geothermal Drilling*, Generation and Transmission Dept., Relations and Communications Dept., Ente Nazionale per L'Energia Elettrica, May 1995.

[8] Cappetti, G., A. Fiordelisi, M. Casini, S. Ciuffi and A. Mazzotti, "A New Deep Exploration Program and Preliminary Results of a 3D Seismic Survey in the Larderello-Travale Geothermal Field (Italy)," *Proc. World Geothermal Congress 2005*, Paper No. 0759, Int'l. Geothermal Ass'n., Antalya, Turkey, 2005.

[9] Anon., "Electrical Energy from the Volterra 'Soffioni'," *Power*, V. 47, N. 15, 1918, p. 531.

[10] Anon., "Steam from the Earth," *Power*, V. 51, June 1, 1920, p. 889.

[11] Anon., "A 'Natural Steam' Turbine Plant," *Power*, V. 48, Dec. 24, 1918, p. 935.

[12] Hahn, E., "Some Unusual Steam Plants in Tuscany," *Power*, V. 57, N. 23, June 5, 1923, pp. 882–885.

[13] Anon., "The Utilization of Volcanic Steam in Italy," *Scientific American*, August 1924, p. 97.

[14] Anon., "Using Volcanic Steam for the Production of Electrical Energy," *Scientific American*, V. 112, January 30, 1915, pp. 97–98.

[15] ENEL, *Larderello and Monte Amiata: Electric Power by Endogenous Steam*, Ente Nazionale per L'Energia Elettrica, Compartimento de Firenze, Direzione Studi e Ricerche, Roma, 1970.

[16] Conti, P.G., "The Larderello Natural Steam Power Plant," *First World Power Conf.*, London, 1924; reprinted in *Geothermal Exploration in the First Quarter Century*, Spec. Rep. No. 3, D.N. Anderson and B.A. Hall, Eds., Geothermal Resources Council, 1973.

[17] DiPippo, R., *Geothermal Energy as a Source of Electricity: A Worldwide Survey of the Design and Operation of Geothermal Power Plants*, U.S. Dept. of Energy, DOE/RA/28320-1, U.S. Gov. Printing Office, Washington, DC, 1980.

[18] Villa, F.P., "Geothermal Plants in Italy: Their Evolution and Problems," *Proc. Second U.N. Symposium on the Development and Use of Geothermal Resources*, San Francisco, V. 3, 1975, pp. 2055–2064.

[19] DiPippo, R., "Geothermal Power Cycle Selection Guidelines," Part 2 of *Geothermal Information Series*, DCN 90–213–142–02–02, Electric Power Research Institute, Palo Alto, CA, 1990.

[20] Santini, P., "Modular Geothermal Plants in Italy: Technical Characteristics and Operation Results," *Proc. World Geothermal Congress 2005*, Paper No. 1327, Int'l. Geothermal Ass'n., Antalya, Turkey, 2005.

[21] Cappetti, G. and L. Cappatelli, "Geothermal Power Generation in Italy 2000–2004 Update Report," *Proc. World Geothermal Congress 2005*, Paper No. 0159, Int'l. Geothermal Ass'n., Antalya, Turkey, 2005.

Nomenclature for figures in Chapter 11

BC	Barometric condenser
C	Condenser
CP	Condensate pump
CSV	Control & stop valves
CW	Cooling water
CWP	Cooling water pump
D	Discharge to mineral recovery
DG	Degasser
E	Evaporator
G	Generator
GC	Gas compressor
IC	Intercooler
NCG	Noncondensable gases
NDCT	Natural draft water cooling tower
PW	Production well
RB	Reboiler
RCP	Recirculating pump
S	Separator
SH	Superheater
T	Turbine
T/G	Turbine/generator
V	Vent
W/OF	Weir/overflow

Chapter 12

The Geysers Dry-Steam Power Plants, Sonoma and Lake Counties, California, USA

"The problem of developing a steam plant at The Geysers has been considered ... The permanence of the steam supply may be questioned, but the origin seems to indicate a long life to the project. The marketing of the energy is not difficult and at the present time it appears as if a power plant will be installed in the near future."

J.D. Galloway – November 12, 1925

12.1 History and early power plants

The Geysers, the world's largest geothermal power complex, has been the subject of numerous books, articles, technical papers, theses, conferences, and workshops. As with Larderello, there is a visitor's center complete with displays and artifacts depicting the history and development of the field, and the general public can tour a part of the field and see a power plant. It is impossible, and unnecessary, to recount here all of the details of The Geysers' history; we will summarize the highlights and refer the interested reader to more complete documents.

Geothermal chronicler Susan Hodgson has divided the history of The Geysers into five distinct eras beginning with the time when it was in its natural state [1]. Native

Geothermal Power Plants: Principles, Applications, Case Studies and Environmental Impact, Third Edition
© 2012 Elsevier Ltd. All rights reserved.

Fig. 12.0 The McCabe plant, formerly Units 5 and 6, as seen from Vista Point looking across Big Sulphur Creek. This is one of 23 power plants that have been built in the rugged Mayacamas Mountains in Northern California. Photo by author [WWW].

American Indians were well acquainted with the place and used its hot springs and fumaroles for many domestic and health-related activities. The modern history at The Geysers is reputed to have begun when a recent settler to California, William Bell Elliott, came across the impressive thermal manifestations while hunting bear in 1847. He was the person who misnamed the place "The Geysers" because there were no actual geysers at The Geysers.

Although the field covers about 30 square miles, thermal manifestations as observed in historic times were limited to essentially two areas: Geyser Canyon and the Little Geysers. Allen and Day of the Carnegie Institution studied these areas in the late 1920s and produced a classic reference work that describes the field in considerable technical detail and includes photos prior to major exploitation [2]. Eight shallow wells however, had already been sunk by the time they surveyed the site. The most significant manifestations consisted of boiling pools, hot and muddy springs, fumaroles, and steaming and altered ground that were confined to the very narrow Geyser Canyon that was about one-quarter mile long, and a few smaller areas not farther than 600 feet distant.

Figure 12.1 is a map of the central thermal area as it appeared in 1927 in Allen and Day's book [2]; it has been modified from its original form for simplicity and clarity. One of the current power plants, the McCabe plant (formerly Units 5 and 6), is situated on a knoll overlooking what was once the center of thermal activity (see chapter opening photo and upper left corner of Fig. 12.1). At the present time, none of the manifestations shown in Fig. 12.1 remain active, the resort buildings are gone, and the canyon itself has largely been filled in. Today, small thermal pools are found only at the Little Geysers area, about five miles or so upstream (to the southeast) along the Big Sulphur Creek.

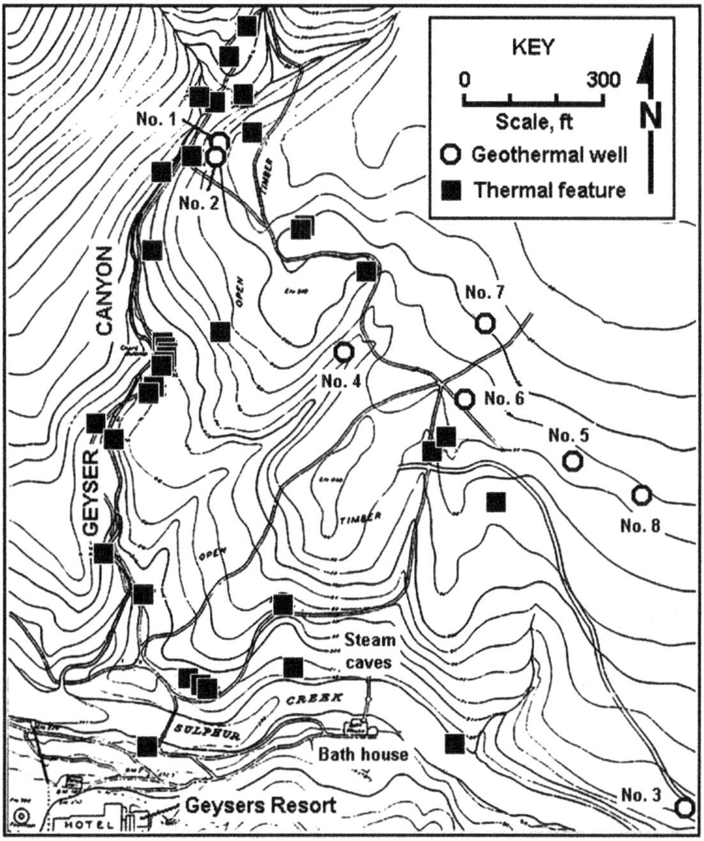

Fig. 12.1 Geyser Canyon, Geysers Resort, thermal features, and original steam wells at The Geysers. Modified from the original map from Allen and Day [2].

A view of the thermal area in Geyser Canyon and the resort is shown in Fig. 12.2 [3]. The abundance of steam issuing from various vents is obvious and apparently fascinating to the formally dressed visitors. The spa business thrived during the last half of the nineteenth century, achieving an international reputation. But like most thermal spas in the United States, its popularity rapidly declined in the early twentieth century. Remnants of the once flourishing operation remained until 1980 when the buildings were dismantled by UNOCAL, then The Geysers operator, at the request of the owner [1].

The first steam wells were completed at The Geysers in 1921 by J.D. Grant. He sited his well very close to one of the hottest thermal springs, the so-called Witches' Cauldron, at the upper end of Geyser Canyon; the well blew out at a shallow depth. Undeterred, he moved his homemade cable-tool rig across the canyon and successfully completed wells No. 1 and 2; see Fig. 12.1 [1]. Both wells were able to produce abundant steam, as can be seen from Fig. 12.3 in which No.2 is blowing across the canyon [2].

Fig. 12.2 A view looking south down the Geyser Canyon toward the Geysers Resort. From the private collection of B.C. McCabe, Magma Power Company, pub. in [3].

Fig. 12.3 J.D. Grant's first two successful wells at The Geysers. No. 2 is blowing across Geyser Canyon; No. 1 is 40 feet farther up the canyon rim. From Allen and Day [2].

Fig. 12.4 The first power house at The Geysers. It was built about 80 feet east of the No. 1 and 2 wells, and used their steam to produce 35 kW. Photo courtesy of Geothermal Resources Council, pub. in [1].

Using reciprocating steam engines built by General Electric, Grant assembled the first electrical generating plant close to his first two wells; see Fig. 12.4. Of course, such a small power plant was viewed more as a curiosity than a serious source of electricity. Later, electricity was generated for the resort hotel in another tiny power house located next to the resort; see Fig. 12.5.

Given the low cost of hydropower and the difficult access to the remote and mountainous site, further development did not take place at The Geysers until the 1950s when B.C. McCabe began to drill wells using modern equipment and techniques. McCabe's company was Magma Power Company. McCabe's partner was Dan McMillan, Jr. whose company was Thermal Power Company. As a joint venture, Magma-Thermal played the pioneering role in developing the steam field that today (August 2011) supports 26 electrical generators with a total installed capacity of 1477 MW.

12.2 Geographic and geologic setting

The Geysers steam field lies some 80 miles north of San Francisco in the Mayacamas Mountains. Productive wells have been drilled covering a surface area roughly 4 miles by 7 miles, oriented NW-SE, and aligned with the regional fault system.

The reservoir is hosted in fractured greywacke, characterized by a matrix porosity associated with veins, that lies above a deep felsite intrusion [4]. The porosity is low, on the order of 1–5%, but excellent productivity is observed when a well finds an intensely fractured region of the formation. The top of the productive zone has been well documented by the more than 500 wells drilled, and Fig. 12.6 depicts the results [5].

Fig. 12.5 The Geysers power plant that provided electricity to the Geysers Hotel; photographer unknown [WWW].

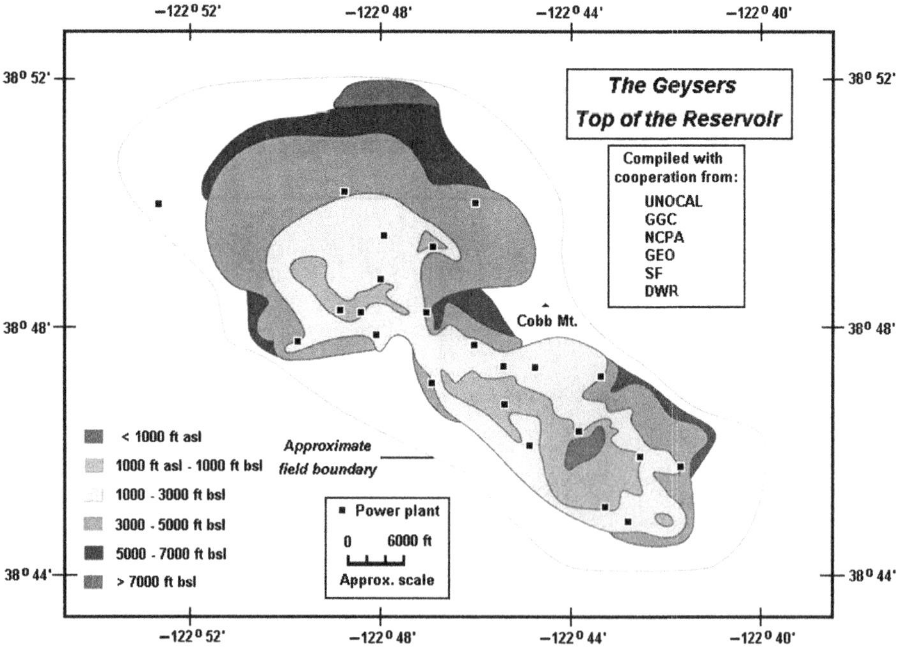

Fig. 12.6 The top of The Geysers reservoir by depth interval, measured above sea level (asl) and below sea level (bsl). The heat source is believed to lie beneath Cobb Mountain. Modified from [5].

The shallowest parts of the reservoir lie in the southeast area where the reservoir can be reached at levels less than 1000 ft above sea level within a small area. Although the elevation of the ground surface varies considerably, it is generally between 2000–3000 ft above sea level within the productive field.

Recalling the conditions needed to create a dry steam reservoir (see Sect. 7.2), the requirement to lower the water table to extreme depths is one of the most important. It appears that there was much prehistoric thermal activity across the field, manifested today by extensive areas of altered ground. The discharge of fluids must have been prolific for long periods of time because no well so far has encountered a liquid-dominated zone beneath the steam reservoir, including wells deeper than 3000 m. Furthermore, the entire thickness of the productive formation has yet to be penetrated.

The areal extent of the reservoir is also evident from Fig. 12.6. It appears to be bounded along the southwest and the northeast by the Mercuryville and Collayomi fault systems, but the actual limits of production lie somewhat within those boundaries [6]. Along the southwest, there is a thin vertical plane separating productive from nonproductive wells, indicating a tight lateral boundary to the steam reservoir. This is yet another of the requirements mentioned in Sect. 7.2 to preserve a steam reservoir that is underpressured from being inundated by cold water from the sides.

Drilling in the north and northwest areas has identified a very high-temperature, deep reservoir; see Fig. 12.6. Temperatures of 575°F (300°C) have been found but the steam is highly superheated and acidic, containing hydrogen chloride, HCl, causing it to be extremely corrosive. So far this steam has not been used in any power plant.

12.3 Well drilling

Several problems confront the well driller at The Geysers. The terrain is rugged, making it difficult to site drilling pads. The ground can be unstable, either from landslides or thermal activity, creating a challenge for the civil engineer who must build roads, set foundations for power plants and cooling towers, and create level areas for drill sites that must support large drill rigs.

Since the formation derives its productivity from swarms of tiny fractures in otherwise hard rock, drilling in the reservoir must be conducted with air as the drilling fluid to avoid clogging the fractures. This increases the erosion on the tubular elements of the drill string, the well casing, the blowout preventer, and other surface equipment as the rock cuttings are transported to the surface from the bottom of the hole in a high-speed air jet. Mud can be used as the drilling fluid for the shallow sections of the well above the reservoir, but then another potentially serious problem can occur.

If the well passes through a permeable region at relatively shallow depth, total loss of circulation (LOC) will be encountered, necessitating the use of LOC materials or the setting of a cement plug in order to continue drilling ahead. The use of LOC materials is often a stop-gap solution that is ineffective in the long term because the formation is below hydrostatic pressure. Therefore, when the casing is cemented, the LOC material may give way under the hydrostatic pressure of the cement leaving a void behind the casing and thus endangering the life of the well.

Typical well profiles are shown in Fig. 12.7 [7,8]. The well at left is a deviated well; many wells at The Geysers are drilled from a single pad to conserve space in the

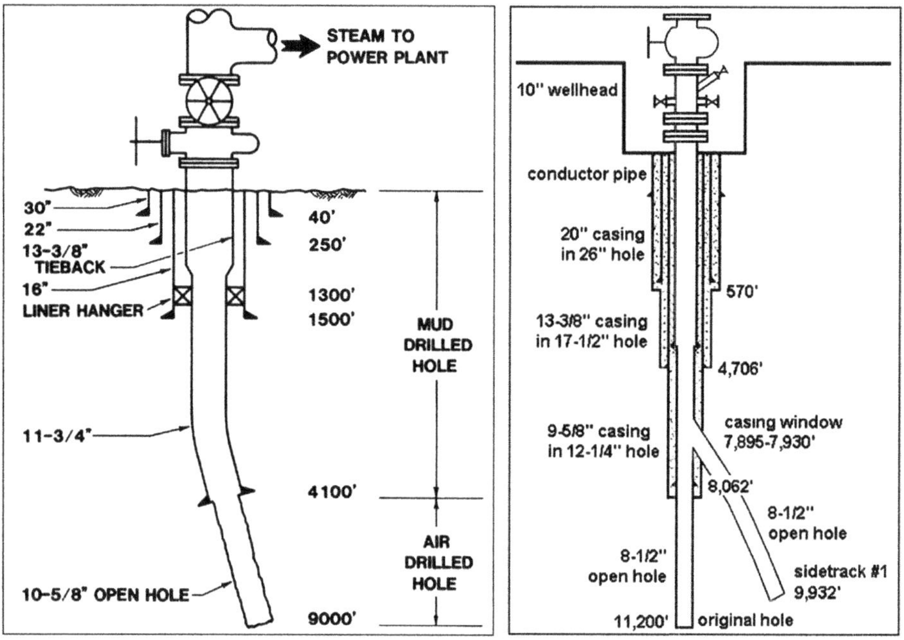

Fig. 12.7 A typical deviated well (left) and forked well (right) at The Geysers [7,8].

rugged landscape. The well at right is a forked well that allows steam to be gathered from a wider zone of the reservoir and brought to a single wellhead for ease of transmission to the power plant. Notice that the wells are finished as open holes since there is little need for a liner given the competent nature of the rock in the formation.

Recent practice includes the drilling of wells with three or more forks from a single trunk. For example, the Aidlin No. 9 well has four sidetracks besides the original well-bore; the original bore and one of the sidetracks were abandoned, but the three active sidetracks produce 116,000 lbm/h of steam compared to 75,000 lbm/h for the original well, a 55% increase in productivity. Such wells are extremely difficult to complete; in the case of Aidlin No. 9, it took 97 days to finish the sidetrack redrilling operation [8]. In 2004, new injection wells associated with the Santa Rosa Geysers Recharge (SRGR) project (see Sect. 12.6) were deviated nearly horizontally to distribute the injectate more uniformly in the reservoir.

12.4 Steam pipeline system

In Sect. 7.3, we discussed in general the steam gathering system for a dry steam plant, and the reader may wish to review that section before proceeding. Here we will show diagrams of the piping used at The Geysers between the wells and the power house. Reference [9] may be consulted for more details.

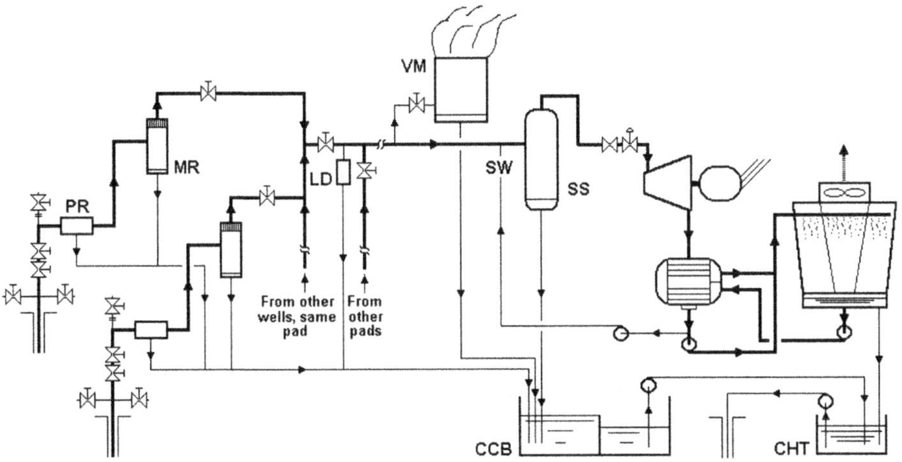

Fig. 12.8 Steam and condensate piping system for a typical Geysers power plant [9].

In Fig. 12.8, steam from several wells at one pad is collected after passing through a wellhead particulate remover (PR) and, if the steam is wet, through a wellhead moisture remover (MR). The steam from each well is joined in a single pipeline and is connected to the main steam line, where it merges with the steam from other well pads. Along the steam lines, liquid condensate that forms from heat loss is drained (LD) to a central condensate collection basin (CCB). A vent muffler (VM) is used during emergency situations such as turbine trips to prevent damage to the wells. Before the steam enters the final steam separator (SS), it receives a spray of water (SW) that acts as a scrubber to help remove any silica that may be entrained in the steam; this prevents buildup of scale on the turbine nozzles and blades. Excess water from the cooling tower is collected in a condensate holding tank (CHT); the collected condensate drained from all sources is pumped over from the CCB to the CHT, and then is reinjected.

12.5 Power plants

There have been 23 separate power plants built at The Geysers since 1960. Figure 12.9 shows the location of each one, including those that have been retired and dismantled.

The first unit, PG&E Geysers Unit 1, was a small unit, 11 MW, situated just across Big Sulphur Creek about 1500 ft east of the site of the old Geysers Hotel. In October 1985, it was designated a National Historic Mechanical Engineering Landmark by the American Society of Mechanical Engineers [10]. Figure 12.10 shows the original power house that held both Units 1 and 2. Since the plant was so close to Big Sulphur Creek, it was able to take make-up and fire protection water from it.

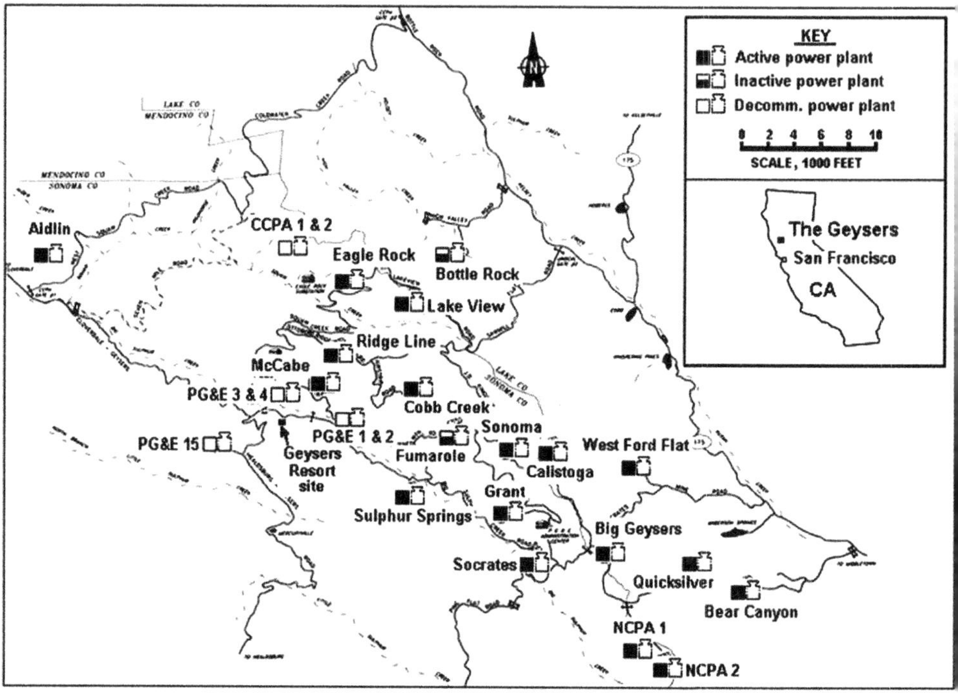

Fig. 12.9 The Geysers location map showing power plants, active, inactive, and decommissioned since 1960; modified from the Geysers Geothermal Association [1].

In 1992, after a successful run of 32 years, the old plant was shut down in favor of the more efficient modern units, and dismantled. A view of the site in 1997 is given in Fig. 12.11. In January 2004, a new well, GDC-31, was completed on the old power house site. It turned out to be the most powerful Geysers well drilled to date, capable of about 10 MW. So potent was the well that it required a separate pipeline to convey its steam to the McCabe unit [11,12].

Table 12.1 gives a summary of information on all Geysers plants that have ever been in operation.

12.5.1 Plant design under PG&E

Beginning in the 1960s, the Pacific Gas and Electric Company (PG&E) owned and operated the power plants at The Geysers, and the steam was developed and delivered to the plants by a number of different companies. PG&E paid the steam suppliers according to a negotiated agreement that did not encourage the power plant owner to make the most efficient use of the steam. The formula to determine the price was based

Fig. 12.10 Historic photo of PG&E Geysers Units 1 and 2 looking generally east; Unit 1 is at the rear. The access road ran between the cooling tower and the power house. Big Sulphur Creek lies to the right of the plant, parallel to the main axis of the cooling tower. Note the steep terrain, approaching 45° behind the plant. Photo from [10].

Fig. 12.11 Site of original Units 1 and 2 after demolition and prior to drilling of well GDC-31. The power house was on the left and the cooling tower was on the right. Photo by author [WWW].

Table 12.1 Summary of plant data for The Geysers.

Plant name	Original name	Start-up	Rated MW	No. units	Total MW	Owner	Notes and comments
Unit 1	(same)	1960	11	1	11	PG&E	Dismantled
Unit 2	(same)	1963	13	1	13	PG&E	Dismantled
Unit 3	(same)	1967	27	1	27	PG&E	Dismantled
Unit 4	(same)	1968	27	1	27	PG&E	Dismantled
McCabe	Units 5&6	1971	2 × 53	2	106	Calpine	
Ridge Line	Units 7&8	1972	2 × 53	2	106	Calpine	
Fumarole	Units 9&10	1973	2 × 53	2	106	Calpine	Dismantled
Eagle Rock	Unit 11	1975	106	1	106	Calpine	
Cobb Creek	Unit 12	1979	106	1	106	Calpine	
Big Geysers	Unit 13	1980	78	1	78	Calpine	
Sulphur Springs	Unit 14	1980	65	1	65	Calpine	
Unit 15	(same)	1979	59	1	59	PG&E	Dismantled
WGP Unit 1	*See Note 1*	2013	26	1	26	WGP	Under const.
Quicksilver	Unit 16	1985	113	1	113	Calpine	
Lake View	Unit 17	1982	113	1	113	Calpine	
Socrates	Unit 18	1983	113	1	113	Calpine	
Calistoga	Santa Fe	1984	2 × 40	2	80	Calpine	
Grant	Unit 20	1985	113	1	113	Calpine	
Bottle Rock	(same)	1985	15	1	15	USRG[2]	Reactivated 3/07
Sonoma	SMUDGEO#1	1983	72	1	72	Calpine	
NCPA 1	(same)	1983	2 × 55	2	110	NCPA	
NCPA 2	(same)	1985,86	2 × 55	2	110	NCPA	
Coldwater Creek	CCPA 1&2	1988	2 × 65	2	130	CCPA	Dismantled
Bear Canyon	(same)	1988	2 × 11	2	22	Calpine	
West Ford Flat	(same)	1988	2 × 14.5	2	29	Calpine	
J.W. Aidlin 1&2	(same)	1989	2 × 10	2	20	Calpine	
TOTAL:				**35**	**1890**		
Active:				**26**	**1477**		

Note 1: Being built on the same site as Unit 15 by Western GeoPower.
Note 2: US Renewables Group.

on PG&E's costs to generate electricity from its fossil-fuel and nuclear power plants. The formula used in the 1970s is given in eq. (12.1):

$$C_S = \frac{2.11\, E_F(C_F/C_F^{\text{o}})(MHR/MHR^{\text{o}}) + E_N C_N}{E_F + E_N} \tag{12.1}$$

where C_S is the price in year n of the geothermal steam in mills/kW·h, i.e., 10^{-3} \$/kW·h, and the other terms are as follows:

E_F = electricity generated from fossil fuel in year $n-1$
E_N = electricity generated from nuclear fuel in year $n-1$
C_F = average cost of fossil fuel in year $n-1$
C_F^{o} = average cost of fossil fuel in year 1968
C_N = average cost of nuclear fuel in year $n-1$
MHR = minimum heat rate for fossil-fuel plants in year $n-1$
MHR^{o} = minimum heat rate for fossil-fuel plants in year 1968.

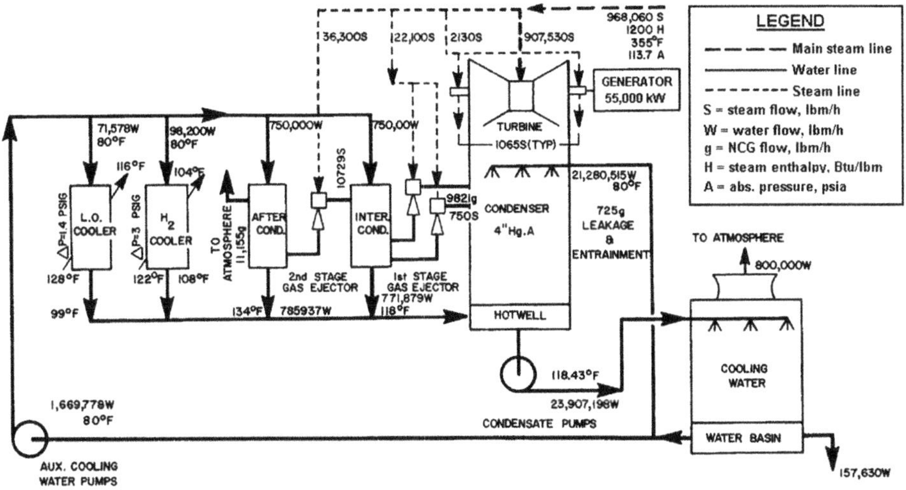

Fig. 12.12 Heat balance diagram for PG&E Geysers Unit 5 [13].

Since the amount of money paid out by PG&E was proportional to the amount of electricity generated from the geothermal steam, there was no incentive for PG&E to design their plants for high efficiency. It would cost more for them to build such plants and the extra investment could not be justified. Steam brought in "over the fence" that was not used in the plant could be "wasted" without penalty to the plant owner. It was only later when the steam purchase agreements were revised to treat the steam as a valuable commodity that more efficient plants were built.

The heat balance diagram for Unit 5, the oldest unit still in operation, is shown in Fig. 12.12 [13]. It is striking in its simplicity. At the time there was no need to abate the noncondensable gases since the amount of hydrogen sulfide being discharged to the atmosphere was within the allowable limits.

The inlet steam is about 18°F (10°C) superheated. The gross output is 55,000 kW and the gross utilization efficiency based on the design wet-bulb temperature of 65°F (18.3°C) is 54.2%. There are two main parasitic power loads: water circulating pumps, 930 kW, and cooling tower fan motors, 605 kW. All other loads amount to 445 kW, leaving a net power of 53,020 kW; this gives a net utilization efficiency of 52.2%. These values indicate that the plant operated relatively efficiently, compared to flash-steam plants, owing to the availability of high-exergy steam as the primary resource.

12.5.2 SMUDGEO #1 plant design

To illustrate the impact of the steam purchase agreement on the design of a geothermal plant, we will consider the case of the 72 MW SMUDGEO #1 plant, now called Sonoma. In 1978, the Sacramento Municipal Utility District (SMUD) signed a steam purchase agreement with Aminoil USA, then a steam developer at The Geysers, to purchase 1,100,000 lbm/h of steam at a pressure of 115 lbf/in^2 with not more than 200 ppm of H_2S. The price of the steam was $1.00/1000 lbm, or $1,100/h at full flow rate [14].

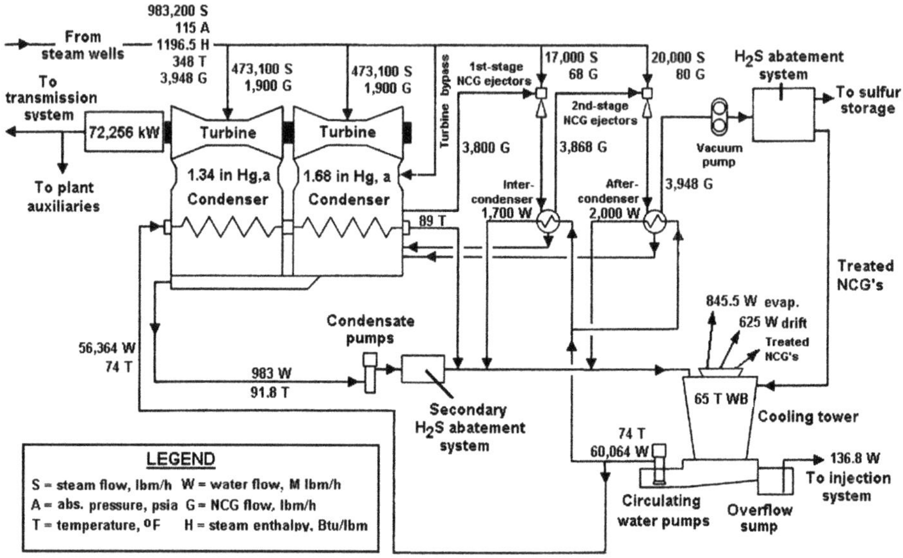

Fig. 12.13 Heat balance diagram for the SMUDGEO #1 power plant at The Geysers [15].

It is clear that under this arrangement, SMUD should try to generate as much electricity as is feasible from each pound of steam. This drove the design to include many features that could not be economically justified under the PG&E type of contract. The design that emerged is shown in the heat balance diagram of Fig. 12.13 [15].

The turbine was larger than previous machines, having 25 in (635 mm) last stage blades, four exhaust ends, with moisture removal sections after each stage, and Stellite© strips to protect the leading edges of the last two rows of blades [16]. The condenser was a two-zone design with dual vacuum conditions for the two turbines. The vacuum was the lowest seen at The Geysers at the time, 1.34 and 1.68 in Hg. A highly efficient means of removing the NCGs was employed: a 2-stage steam jet ejector followed by a liquid-ring vacuum pump. In selecting the suppliers for the other auxiliary systems, including pumps, transformers, H_2S abatement system, cooling tower, etc., a penalty was applied to any bidder for inefficient use of steam and electricity [17]. The end result was the most efficient geothermal power plant ever built: SMUDGEO #1 had a gross utilization efficiency of 70.1%, and a specific steam consumption of 13.6 (lbm/h)/KW or 1.71 (kg/s)/MW.

12.5.3 Power plant operations under Calpine ownership

With the exception of the two NCPA plants (Northern California Power Agency), all the power plants at The Geysers and their steam resources have been owned and operated by Calpine Corporation since 1999. This unified ownership has created even more incentive for efficient use of the steam resource. The electricity generated is now sold into the power exchange where it must compete with every other source of electricity on the open market. An extensive and ongoing program of modernization and performance nprovement is underway to identify areas of inefficiency and to correct them. Integrated eld management has resulted in interconnections of steam lines serving various units, llowing for balancing the supply of steam to the units and diverting steam to other units n the event of an outage. Areas of the field, such as the northwest sector which is known o be productive but with high NCG content, may be exploited now that advanced echnology is available to permit such fluids to be utilized efficiently. Finally, new wells re being drilled, such as well GDC-31 mentioned earlier, that indicate the field may be ven more extensive than most believed, and may lead to the building of additional power lants after a hiatus of some 15 years.

NCPA has reaped similar benefits since acquiring the rights to the steam field that upplies its two power plants in the Southeast Geysers area.

12.6 Recharging the reservoir

One of the problems in the management of a dry steam field is the replenishment of the luid that is withdrawn through production. Natural recharge usually occurs at a rate ar less than the rate of removal, leading to pressure reduction and a drying out of the eservoir. The contrast between a dry steam (DS) resource and a liquid-dominated (LD) one is dramatic. Although vastly more mass is withdrawn from a LD reservoir compared o a DS one to generate the same power, 70–80% of that mass can be reinjected for the LD case, whereas only the excess steam condensate from the cooling tower, roughly 10–15%, is available in the DS case.

The trends in production and reinjection from the start of operations in 1960 hrough 2010 are shown in the series of graphs in Figs. 12.14–12.17 [18]. After a period of rapid growth in the 1980s, spurred by the two oil shocks of 1973 and 1979,

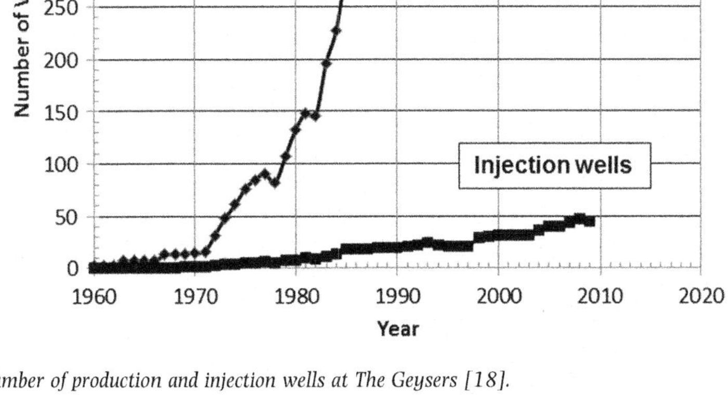

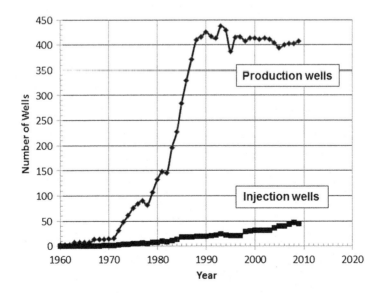

Fig. 12.14 Number of production and injection wells at The Geysers [18].

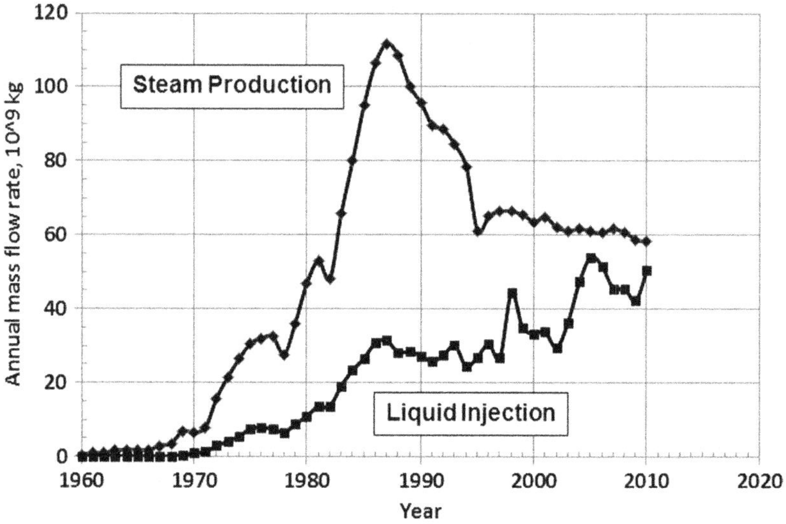

Fig. 12.15 Annual steam production and liquid injection at The Geysers [18].

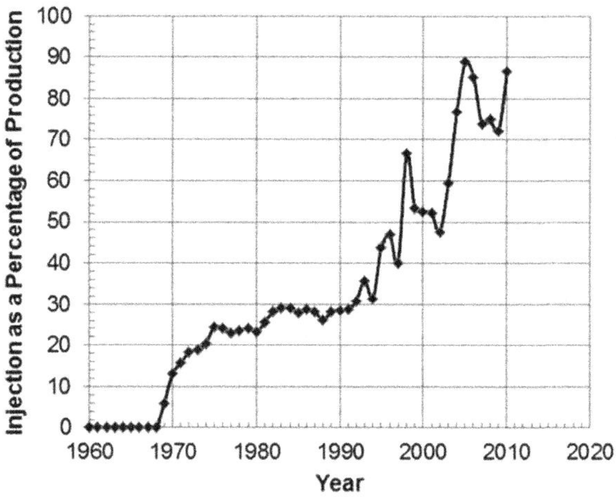

Fig. 12.16 Mass of injectate as a percentage of mass removed at The Geysers [18].

the number of production wells drilled declined somewhat and has recently stabilized at about 405 (Fig. 12.14). For the first nine years of operation, no reinjection was carried out at all. Thereafter, the number of injection wells has gradually grown to reach 45–50 nowadays. Likewise, the percentage of fluid injected relative to the mass of steam removed has grown rapidly in the last decade and approaches 90% (Fig. 12.16). Besides cooling tower overflow, water has been taken from Big Sulphur Creek during the wet season, and rainwater is captured in holding ponds and then injected [19]. The field has yielded over 2.5×10^{12} kg of steam in 50 years of operation, while slightly over 1.0×10^{12} kg of liquid – 40% of production – has been returned to the reservoir.

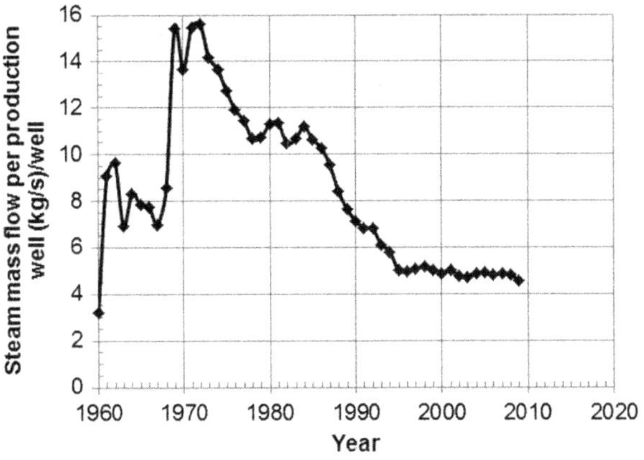

ig. 12.17 Average steam production rate per production well at The Geysers [18].

The average annual steam mass flow rate per production well was about 8 kg/s er well in the 1960s (Fig. 12.17). This jumped to nearly 16 kg/s per well in the arly 1970s as newer parts of the field were brought into production by a variety of evelopers. However, this rate was not sustainable and a general decline ensued until 995 when this value reached about 5 kg/s per well. Since then, it appears the rate as stabilized at 4.5–5 kg/s, albeit with a slight annual decline. Roughly it takes about kg/s of steam to produce one MW of power. The stabilization is related to two innovaive wastewater-to-electricity projects: the Lake County Effluent Pipeline (LCEP), also nown as the Southeast Geysers Pipeline project, and the Santa Rosa Geysers Recharge SRGR) project [19–23]. These are the first projects of their kind in the world.

The LCEP was put into operation on September 25, 1997 [19]. It carries secondaryreated municipal wastewater from several communities along the north side of Clear ake, south to Middletown where a second water treatment facility is located. From here, about 2.8 billion gallons per year (10.5×10^9 kg) is pumped through 53 miles 85 km) of 16- to 20-inch pipe into the Southeast Geysers field, gaining about 2000 ft 610 m) in elevation. Besides helping to recharge the reservoir at The Geysers, the njected water has resulted in lower NCGs in the steam, reducing the parasitic power eeds. Furthermore, the project has several environmental benefits, including the creaion of a wetlands habitat for wildlife that also serves to polish the effluent prior to eing pumped to The Geysers, and the elimination of the discharge of treated and ntreated waste water into Clear Lake, which is used as the source of drinking water or some communities. It is planned to extend the pipeline along the southern shore of lear Lake to complete Project "Full Circle" [23].

The SRGR project is patterned on the successful LCEP and started up on December 5, 003, two years later than scheduled and at a cost double what was expected, 200 million vs. $102 million [21]. It is a 41-mile (66 km) pipeline from the City of anta Rosa and surrounding communities to the central and northwest sectors of the eld. It will carry 4 billion gallons per year (15×10^9 kg) of tertiary-treated wastewater, ssentially drinking water, to help recharge the other end of the field from the LCEP injection area [24]. The elevation gain is about 3000 ft (915 m). Since the water is so pure, some of it may be used as cooling water at those plants with high levels of H_2S in the steam condensate, eliminating the need to treat the condensate.

Between the two recharge projects, the end result has been a boost in power output due to increased steam flow amounting to over 160 MW, compared to the anticipated power without the projects. Assuming an electricity cost of \$0.05/kW · h and a conservative average system capacity factor of 90%, the extra power would bring in additional revenues of \$67 million per year. Future benefits may derive from the ability to use the deep, 575°F (300°C) reservoir beneath the northwest sector of the field that produces highly superheated but extremely corrosive steam. If the pure water can dilute the native steam, then a whole new area of the field will be opened to development.

In November 2004, Calpine Corporation was honored by the U.S. DOE and the U.S. EPA by being selected for a "Green Power Leadership Award" for the SRGR project. The award recognizes companies for "Innovative Use of Renewable Energy Technology" [25].

12.7 Toward sustainability

The Geysers' history of power generation demonstrates that geothermal energy may be used in a manner that allows it to approach sustainability, or it can be overexploited resulting in a premature demise of the resource. Lovekin [26] has described the life cycle of a geothermal field as passing through four distinct phases or periods: (1) development, (2) sustaining, (3) declining, and (4) renewable; see Fig. 12.18. Only during the last phase does a geothermal resource approach the ideal of a sustainable and renewable resource, and to attain it requires prudent management of the resource.

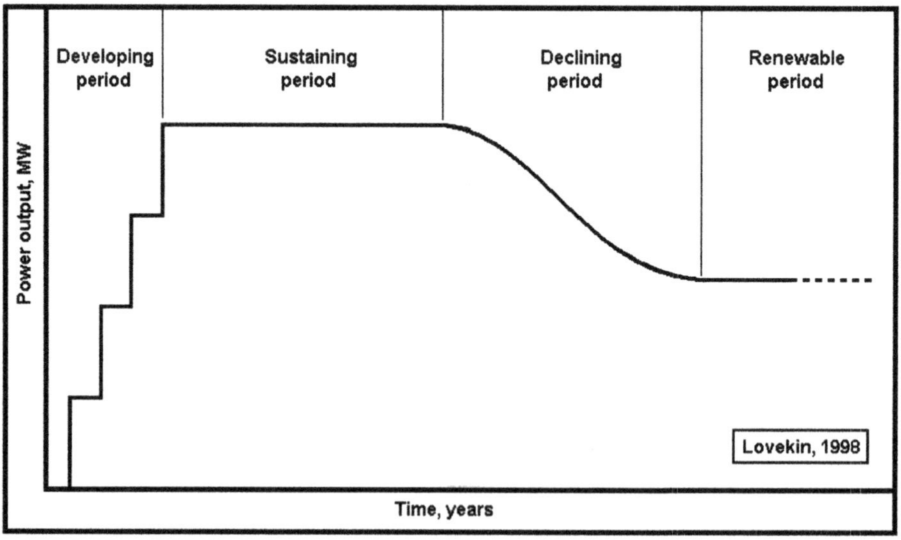

Fig. 12.18 Life cycle of a geothermal field, according to Lovekin [26].

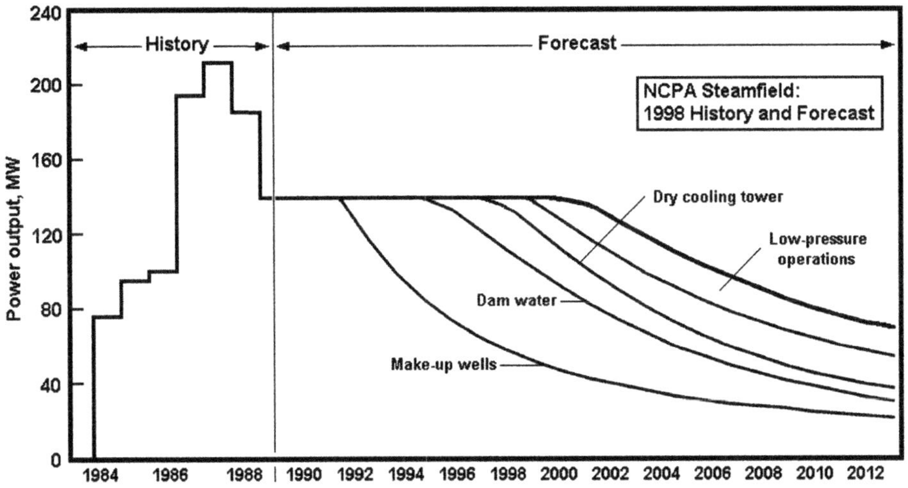

Fig. 12.19 *History and forecasted behavior of the NCPA steamfield [19].*

Once a field reaches the point where the fluid extraction matches the combined natural and augmented recharge, then the plant can continue to operate essentially indefinitely, given proper maintenance and a market for its electricity at a profitable price.

In phase (1), the field is developed incrementally, building up its power capacity in a number of plants. In phase (2), a reasonable steady state is achieved since the wells at first are operated at some fraction of their full capability. In phase (3), make-up wells are drilled to compensate for the decline in well outputs but the thermodynamic properties of the resource begin to decline under continued exploitation. By scaling back the output of the power system, a sustainable level can be achieved in phase (4), but at the sacrifice of some capital equipment that will no longer be useful. The remaining power plants can still be useful and profitable for a very long time, but at a considerably lower power capacity than what was originally installed.

The history of the two NCPA power plants illustrates this behavior; see Fig. 12.19. The units are identical 110 MW units that reached their rated output in 1987 but began their decline soon thereafter. By various means, the sustaining period could be stretched out for several years. In 2003, the LCEP had restored about 22 MW to the output relative to the projected decline without it, i.e., 142 MW vs. 120 MW [19]. Although the forecast does not extend into what could be the renewable period, it might be possible to achieve it given the correct economic environment and the commitment of management to maintain the facilities even though the plant will be running at a fraction of its rated capacity.

References

[1] Hodgson, S.F., *A Geysers Album: Five Eras of Geothermal History*, Pub. No. TR49, CA Dept. of Conservation, Div. of Oil, Gas, and Geothermal Resources, Sacramento, 1997.

[2] Allen, E.T. and A.L. Day, *Steam Wells and Other Thermal Activity at "The Geysers" California*, Carnegie Institution of Washington, Washington, DC, 1927.

[3] Anderson, D.N. and B.A. Hall, eds., *Geothermal Exploration in the First Quarter Century*, Geothermal Resources Council, Spec. Rep. No. 3, 1973.
[4] Gunderson, R.P., "Porosity of Reservoir Graywacke at The Geysers," *Monograph on The Geysers Geothermal Field*, Spec. Rep. No. 17, Geothermal Resources Council, Davis, CA, 1992, pp. 89–93.
[5] *Monograph on The Geysers Geothermal Field*, Spec. Rep. No. 17, Geothermal Resources Council, Davis, CA, 1992, appended map.
[6] Truesdell, A.H., W.T. Box, Jr., J.R. Haizlip and F. D'Amore, "A Geochemical Overview of The Geysers Geothermal Reservoir," *Monograph on The Geysers Geothermal Field*, Spec. Rep. No. 17, Geothermal Resources Council, Davis, CA, 1992, pp. 121–132.
[7] Pye, D.S. and G.M. Hamblin, "Drilling Geothermal Wells at The Geysers Field," *Monograph on The Geysers Geothermal Field*, Spec. Rep. No. 17, Geothermal Resources Council, Davis, CA, 1992, pp. 229–235.
[8] Henneberger, R.C., M.C. Gardner and D. Chase, "Advances in Multiple-Legged Well Completion Methodology at The Geysers Geothermal Field, California," *Proc. World Geothermal Congress*, V. 2, Florence, Italy, 1995, pp. 1403–1408.
[9] Veizades, H. and W.J. Cain, "Design of Steam Gathering Systems at The Geysers: A State-of-the-Art Review," *Monograph on The Geysers Geothermal Field*, Spec. Rep. No. 17, Geothermal Resources Council, Davis, CA, 1992, pp. 245–250.
[10] Anon., *The Geysers Unit 1: Pacific Gas and Electric Company*, Pub. No. 61-8232, American Society of Mechanical Engineers, October 1985.
[11] Box, T. and M. Stark, "Old Steam Field ... BIG New Well," *Geothermal Resources Council BULLETIN*, V. 33, No. 3, 2004, pp. 113–114.
[12] Talkington, K., Calpine Corporation, personal communication, December 9, 2004.
[13] DiPippo, R., *Geothermal Energy as a Source of Electricity: A Worldwide Survey of the Design and Operation of Geothermal Power Plants*, U.S. Dept. of Energy, DOE/RA/28320-1, U.S. Gov. Printing Office, Washington, DC, 1980.
[14] Anon., *Sacramento Municipal Utility District SMUDGEO #1, The Geysers, Sonoma County, California*, undated report by Stone & Webster Engineering Corporation, Boston, MA.
[15] Anon., *SMUDGEO #1 Application for Certification*, Sacramento Municipal Utility District, Sacramento, CA, June 1980.
[16] Tucker, R.E., P.V. Kleinhans Jr. and L.R. Keilman, "SMUDGEO #1 Economic Impacts on Geothermal Power Plant Design," *Geothermal Resources Council TRANSACTIONS*, V. 4, 1980, pp. 533–536.
[17] Kleinhans, P.V., Jr. and D.L. Prideaux, "Design, Start-up and Operation of SMUDGEO #1," *Proc. Ninth Annual Geothermal Conf. and Second IIE-EPRI Workshop Geothermal Conf. and Workshop*, EPRI AP-4259-SR, Electric Power Research Institute, Palo Alto, CA, 1987, pp. 13-1–13-7.
[18] Anon., *2010 Annual Report of the State Oil & Gas Supervisor*, CA Dept. of Conservation, Div. of Oil, Gas, and Geothermal Resources, Sacramento, 2011: http://www.conservation.ca.gov/dog/geothermal/manual/Pages/production.aspx.
[19] Grande, M., S. Enedy, B. Smith, J. Counsil and S. Jones, "NCPA at The Geysers," *Geothermal Resources Council BULLETIN*, V. 33, No. 4, 2004, pp. 155–161.
[20] Fraser, G., "NCPA Gives Good Grade to Geysers Pipeline Project," *Geothermal Resources Council BULLETIN*, V. 28, No. 6, 1999, pp. 189–191.
[21] Clutter, T.J., "Picking Up Steam," *Geothermal Resources Council BULLETIN*, V. 30, No. 3, 2001, pp. 106–110.
[22] Clutter, T.J., "Recharging The Geysers," *Geothermal Resources Council BULLETIN*, V. 32, No. 6, 2003, p. 242.
[23] Dellinger, M. and E. Allen, "Lake County Success," *Geothermal Resources Council BULLETIN*, V. 33, No. 3, 2004, pp. 115–119.
[24] Brauner, E., Jr. and D.C Carlson, "Santa Rosa Geysers Recharge Project: GEO-98-001," Final Report, Rep. No. 500-02-078V1, California Energy Commission, October 2002.
[25] Anon., "Calpine Receives 2004 Green Power Leadership Award From DOE and EPA," *Power Engineering Web Exclusive*, November 9, 2004: http://pe.pennwellnet.com/Articles/Article_Display.cfm?Section=Archi&Subsection=Display&ARTICLE_ID=215520&KEYWORD=The%20Geysers.
[26] Lovekin, J., "Sustainable Power and the Life Cycle of a Geothermal Field," *Geothermal Resources Council TRANSACTIONS*, V. 22, 1998, pp. 515–519; also, *Geothermal Resources Council BULLETIN*, V. 28, No. 3, 1999, pp. 95–99.

Chapter 13

Cerro Prieto Power Station, Baja California Norte, Mexico

"Our planet Earth has been generous with humankind, providing an infinite number of resources, and in the case of geothermal energy, it has not only made us participants in its deep warmth, but also of related benefits... Human beings are responsible for searching and taking rational advantage of this treasure, using both the energy and the salts ... in the geothermal fluids."

Luis F. De Anda Flores, on the occasion of his receipt of the GRC Pioneer Award – October 13, 2003

13.1 Overview of Mexican geothermal development

Mexico was among the earliest countries to operate a commercial geothermal power plant at a liquid-dominated resource, commissioning its first single-flash plant in 1973. Currently, Mexico is the world's fourth largest generator of electricity from geothermal energy with a total installed capacity of 983 MW; see Appendix A for further details. The plants are clustered at the five locations shown in Fig. 13.1; all of them are owned and operated by the Mexican electric authority, the Comisión Federal de Electricidad (CFE).

Los Azufres and Los Humeros lie within the east-west, Neovolvanic Axis that includes the capital, Mexico City, and Guadalajara, the country's second largest city. Los Azufres is a large field supporting 188 MW from a variety of plant types comprising seven exhausting-to-atmosphere 5 MW units, four 25 MW and one 50 MW single-flash condensing units, and two binary units of 1.5 MW each. Los Humeros has been

Geothermal Power Plants: Principles, Applications, Case Studies and Environmental Impact, Third Edition
© 2012 Elsevier Ltd. All rights reserved.

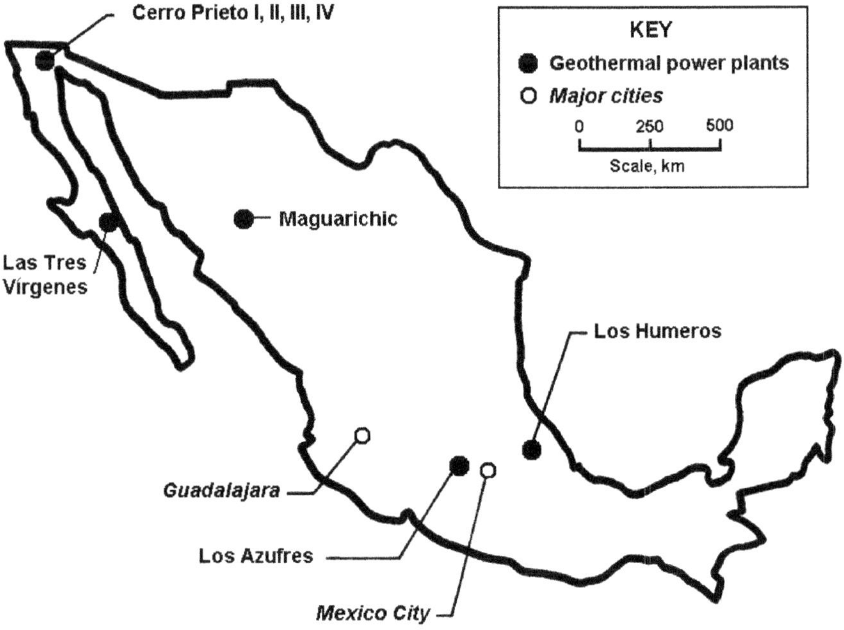

Fig. 13.1 Geothermal power plant locations in Mexico.

developed with eight exhausting-to-atmosphere 5 MW units, for a total capacity of 40 MW, and a single-flash condensing 25 MW unit. A 300 kW binary unit serves the rural community of Maguarichic. Las Tres Virgenesis is about halfway down the Baja California peninsula in a volcanic complex, and generates 10 MW from two 5 MW units. Cerro Prieto, the focus of this chapter, lies south of California's Imperial Valley, just over the border from the United States, and has an installed capacity of 720 MW. As a system, the geothermal power plants generated 7,047 GWh of electricity in 2008, up from 6,281.66 GWh in 2003, produced about 3% of all electricity generated in Mexico, and had an overall capacity factor of 84% for the year [1].

13.2 Cerro Prieto geographical and geological setting

Cerro Prieto is situated 32 km (20 mi) south of the Mexico-United States border (see Fig. 13.2), not far from the geothermal fields at Heber and East Mesa in Southern California (see Chapters 17 and 18).

Unlike its geothermal neighbors to the north, Cerro Prieto was an area with numerous surface thermal manifestations, such as boiling springs, mud volcanoes, and altered ground. Not surprisingly, the first exploratory drilling was in the area of the thermal activity. The field lies just east of the Cerro Prieto volcano, an eroded edifice that stands only 225 m (740 ft) asl, within the otherwise extremely flat Mexicali Valley; see Fig. 13.3 [3,4]. An elevation cross-sectional view of the western part of the field is shown in highly schematic form in Fig. 13.4 [2].

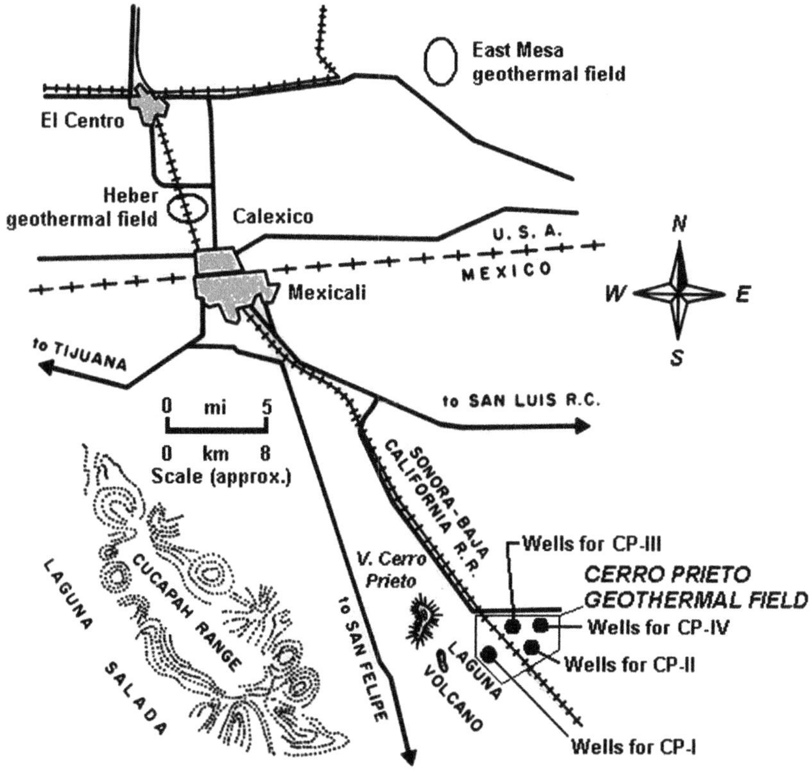

ig. 13.2 Location map for Cerro Prieto. Note proximity to the East Mesa and Heber geothermal fields in the Jnited States; modified from [2].

The Cerro Prieto volcano lies directly on the Cerro Prieto fault that is the apparent western boundary of the reservoir. The field lies between that fault and the Imperial fault to the east within a deep down-thrown graben; see Fig. 13.5. The Michoacan fault runs parallel to the Cerro Prieto fault, and its intersection with one or more of he transverse faults gives rise to high permeability and access to 350°C (660°F) fluids at depths of about 2000 m (6560 ft).

The reservoir formation consists of alternating layers of deltaic sandstones and sediments of continental origin, and shales that act as barriers between the permeable layers of sandstones. The region is part of a spreading center that has given rise to transverse faults that are more or less orthogonal to the Cerro Prieto, Michoacan, and Imperial faults. These transverse faults are created by tension and thus are open to fluid flow. Permeability and high temperatures are found at relatively shallow depths at the west end, and at progressively deeper depths as one moves to the east.

The hot geothermal fluids rise along the open tension faults, Hidalgo, Patzcuaro, and Delta (see Fig. 13.5), in the center of the field, and mix with colder fluids entering from the northeast from the direction of Jalapa, and from the west as recharge from he Sierra de los Cucapahs. The southwestern boundary of the field is clearly seen in

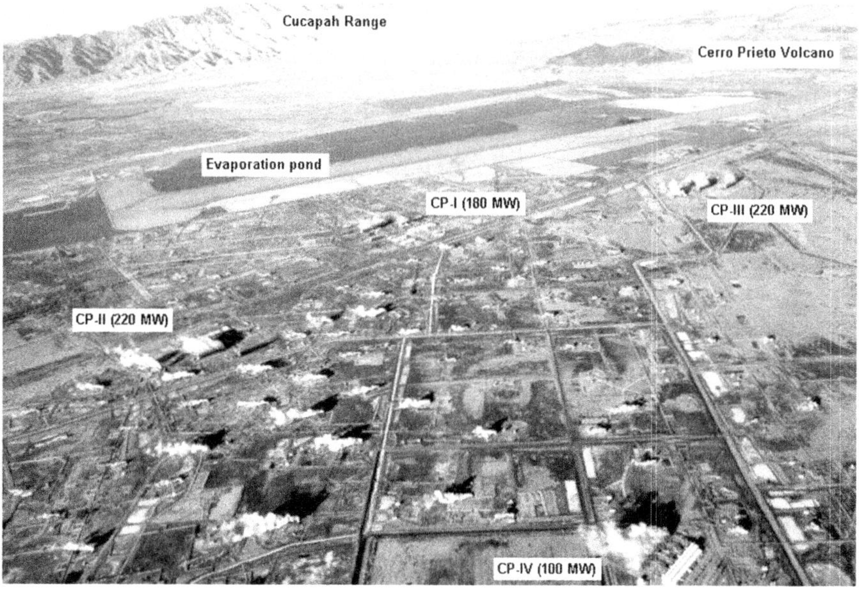

Fig. 13.3 An overview of part of the well field supporting four Cerro Prieto power stations. Photo courtesy of Luis Gutierrez Negrin, CFE [WWW].

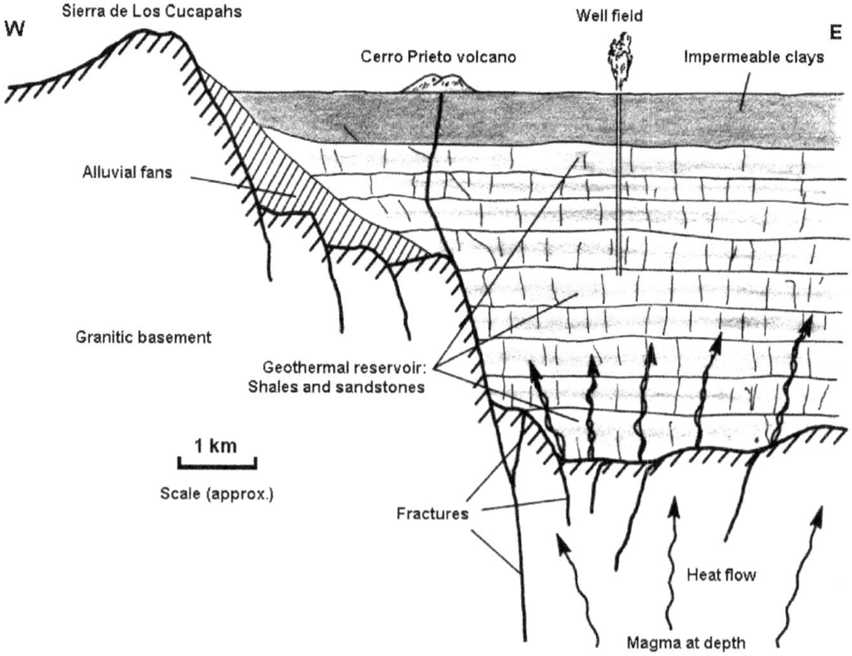

Fig. 13.4 Conceptual geologic cross-section looking generally north across the western portion of the Cerro Prieto reservoir [2].

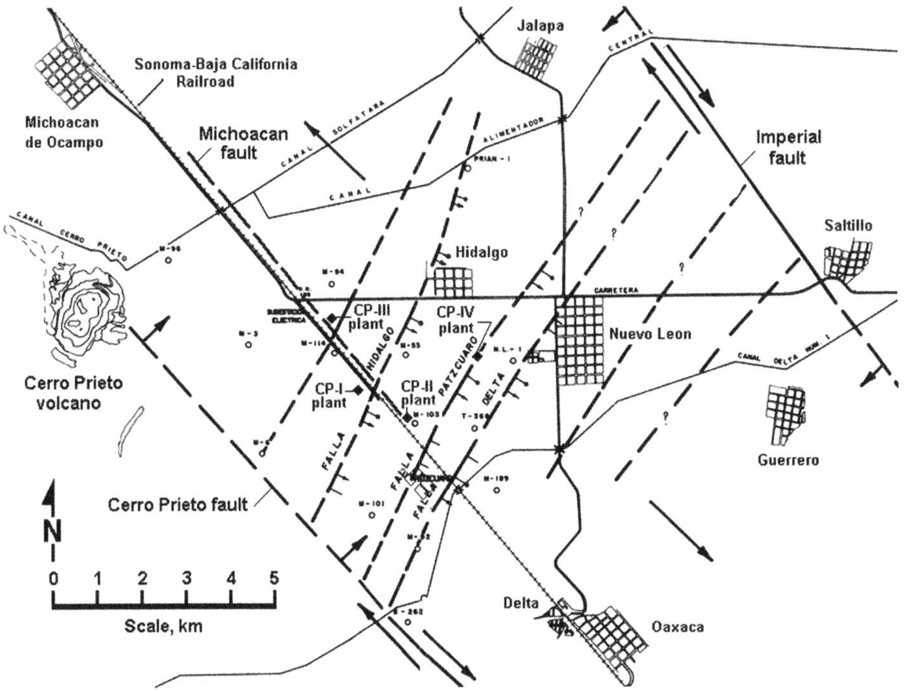

Fig. 13.5 Plan view of the Cerro Prieto geothermal field showing major faults. The area between the Cerro Prieto fault and the railroad tracks contains the wells for CP-I, Units 1–5. The area between the railroad tracks and the communities of Hidalgo and Nuevo Leon supplies CP-II, Units 1 and 2, and CP-III, Units 1 and 2. CP-IV is in the eastern-most section of the field. Map modified from [5].

the steep drop-off in temperature in wells drilled near the Cerro Prieto fault. The fall-off in temperature is more gradual to the northeast. It appears that temperatures up to 350°C (660°F) are accessible at depths of less than 3000 m (9850 ft) over a surface area of about 5 km^2 (2.0 mi^2) lying between the site of the first power plant, CP-I, and the communities of Hidalgo and Nuevo Leon [6].

Cerro Prieto is one of the most studied reservoirs in the world. Every imaginable geoscientific tool has been applied to obtain a good understanding of the field; for example, see Refs. [7] and [8] for a collection of papers on the field. However, the early exploration was done "by the drill bit." As mentioned, it was fairly obvious from the thermal manifestations where the hot fluid would be found, and the first wells discovered fluids of 180°C (360°F) at relatively shallow depths, 600–900 m (2000–3000 ft). Moving to the northeast and drilling somewhat deeper, 800–1400 m (2600–4600 ft), yielded fluids with temperatures of between 250–280°C (480–535°F). It was only with the drilling of step-out wells farther east that the hottest fluids were found, albeit at greater depths.

The mass of scientific studies have served to define the field and allow the drilling results to fit a consistent model of the reservoir. The scientific work includes: geohydrologic surveys, hydrothermal geochemistry, gravity and precision gravity, gas geothermometry, resistivity, magnetics, active and passive seismic, self-potential, magnetotellurics, noble gas tracing, radon and ammonia detection, subsidence monitoring, and microearthquake detection.

Reservoir models have been constructed based on this wealth of information One numerical model was developed to assess the possibility of adding an 80 MW plant to the then-existing three plants using wells to the east of those that were supplying CP-III (see Figs. 13.2 and 13.5) [9]. The simulation indicated that the proposed plant could operate for 30 years with a properly designed reinjection strategy. On this basis, in the year 2000, CFE constructed the CP-IV plant, rated at 100 MW.

13.3 Cerro Prieto power plants

The power units at Cerro Prieto were developed sequentially over a period of 27 years starting with CP-I Unit 1 in 1973 and culminating in 2000 with the installation of the four-unit, 100 MW plant at CP-IV.

13.3.1 Cerro Prieto I – Units 1–5

Cerro Prieto I (CP-I) has four single-flash units each of 37.5 MW and a double-flash unit of 30 MW. The five units were commissioned from 1973 to 1981. The history of these units exemplifies one of the characteristics of geothermal reservoirs, namely, evolution and change. The CP-I plant is an example of a combined single- and double-flash system described in Sect. 9.2.2.

Figure 13.6 shows a highly simplified flow diagram for each of the first four units at CP-I. It is a basic single-flash design that includes the discharge of the waste brine at the surface into a large evaporation pond; the photo in Fig. 13.3 shows the proximity of the pond to CP-I units. Reinjection was not initially adopted for several reasons. First, there was not enough information available on the mechanics of reservoir behavior in the early days to be sure that interference could be avoided. Second, owing to the high temperatures of the geofluid, the silica

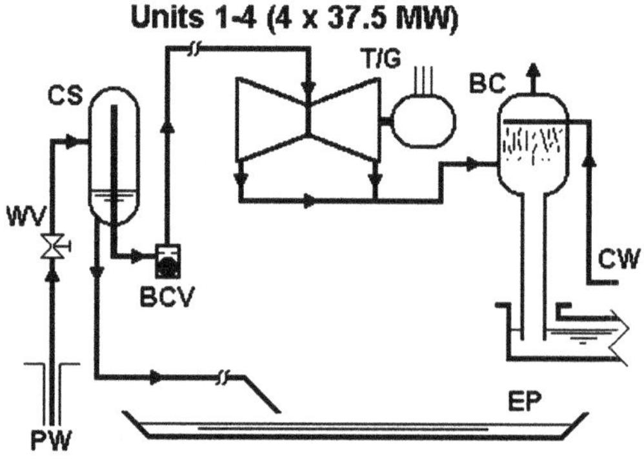

Fig. 13.6 Simplified flow diagram for CP-I, Units 1–4.

oncentration was high enough to cause concern about clogging reinjection wells, ven though the brine exit temperature from the separators was high enough to void supersaturation. Third, the area to the west of the plant was geothermally nproductive, dotted with mud pots and mud volcanoes, unusable for either grazng or agriculture, and therefore offered an easy solution to waste brine disposal. 'oday, the evaporation pond covers an area of roughly 14 km^2. As knowledge of he reservoir has increased from the many scientific studies and simulations, reinection is now a part of the field management along with disposal to the pond; in 2003, nine injection wells were in use [1].

The gathering system consists of cyclone separators at each well with steam lines o the steam header at the power house, and liquid lines to the evaporation pond. A typical wellhead is shown in Fig. 13.7; Fig. 13.8 shows the steam header and moisture eparators outside the CP-I power house.

The turbine-generators for Units 1–4 were supplied by the Toshiba Corporation; he technical specifications are given in Table 13.1. Shortly after the startup of Unit 4, t was decided to augment the output of the station by flashing the waste brine in a 2-stage flash process to yield two additional steam flows. This would not require dditional wells and would increase the utilization efficiency of the overall plant. The low diagram for the new combined system is shown in Fig. 13.9.

Unit 5 is a true double-flash design since it is fed liquid at essentially saturated onditions and produces medium- and low-pressure steam to drive a dual-admission urbine. The turbine was supplied by Mitsubishi Heavy Industries, Ltd.; the design pecifications for the unit are shown in Table 13.2. Although Unit 5 by itself has

Fig. 13.7 Well M-114 serving CP-I. Photo by author [WWW].

Fig. 13.8 CP-I separated steam arrives through the pipelines from the right, enters the steam header, and passes through the moisture remover prior to being admitted to the turbine. Note the barometric condensers at the rear left and the two tall stacks that are discharging the noncondensable gases along with a small amount of steam. Photo by author [WWW].

a utilization efficiency of only 31.7%, as compared to 34.5% for each of the single-flash units, by adding Unit 5, the utilization efficiency for the whole plant increased to 41.4% – an example of synergy in design.

Since Unit 5 depended on a stable supply of waste brine from the separators of Units 1–4, if that quantity were to change during the years of operation, it would affect the balance of steam that could be supplied to Unit 5. The reservoir pressure did in fact drop and the steam fraction at the wellhead increased, leaving less brine for the double-flash unit. In 2003, while Units 1–4 were able to operate at a capacity factor of 87.5%, Unit 5 could manage only 13.1% [1]. Plans are now underway to move the flash station from next to the evaporation pond, and to have two of them situated closer to the CP-II and CP-III well fields in order to use the waste brine as well as excess low-pressure steam from those plants to bring Unit 5 back to its rated output [11].

13.3.2 Cerro Prieto II – Units 1–2 and Cerro Prieto III – Units 1–2

The next four units to come online at Cerro Prieto consisted of identical 110 MW units, two each located in separate power houses designated as Cerro Prieto II and III (CP-II and CP-III); see Fig. 13.5 for locations. The turbines are the largest double-flash machines at any geothermal plant and were built by the Toshiba Corporation; see Fig. 13.10. The turbine has two cylinders, arranged in tandem fashion, each

'able 13.1 Technical specifications for Cerro Prieto I, Units 1–4; data are per unit [2].

lant	Units 1–4
tart-up year	1973 (1 + 2), 1979 (3 + 4)
'ype	single flash
.ating, MW	37.5
urbine:	
cylinders	1
flows/turbine	2
stages/flow	6
inlet pressure, bar,a	6.2
inlet temperature, °C	160 (sat.)
steam mass flow, kg/s	79.25
exhaust pressure, kPa	11.85
last-stage blade height, mm	508
speed, rpm	3600
;enerator:	
output, kVA	44,200
voltage, kV	13.8
frequency, Hz	60
power factor	0.85
cooling medium	hydrogen
:ondenser:	
type	elevated, barometric, direct-contact
CW flow, kg/s	2974
CW temperatures, °C:	
inlet	32.0
outlet	45.3
VCG system:	
steam ejector:	yes
stages	2
steam flow, kg/s	6.68
:ooling system:	
type	cross-flow, mechanical-draft cooling tower
No. cells	3 (1 + 2), 5 (3 + 4)
water flow rate, kg/s	3213
pumping power, kW	835
'lant performance	
SSC, (kg/s)/MW	2.11
utilization efficiency, %:	34.5 [(1)]

[1)]Based on geofluid temperature of 250°C, separation at 7 bar, a, and dead state at 25°C.

vith a double-flow steam path. The rotors are interchangeable, with three HP and 'our LP stages per flow. The turbine is equipped with moisture removers between :he stages to minimize erosion from wet steam. Table 13.3 gives some data on Units :P-II and CP-III.

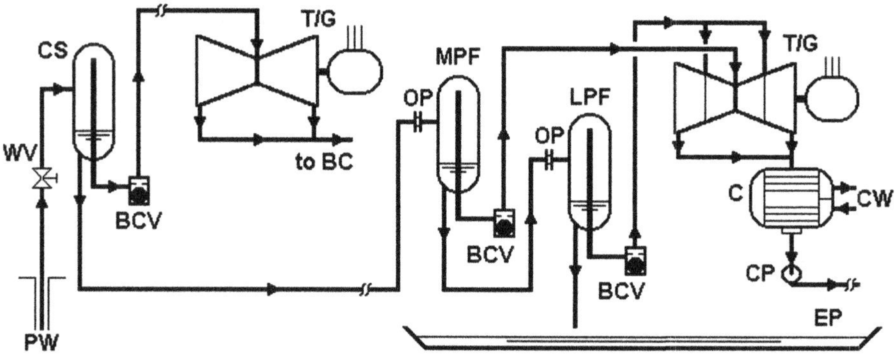

Fig. 13.9 Simplified flow diagram for Cerro Prieto I Units 1−5. The left side represents each of Units 1−4; the right side shows Unit 5.

Table 13.2 Technical specifications for Cerro Prieto I, Unit 5 [10].

Plant	Unit 5
Start-up year	1981
Type	single flash
Rating, MW	30.0
Maximum output, MW	35.3
Brine inlet temperature, °C	169.5
Brine outlet temperature, °C	122.0
Brine flow rate, kg/s	827.9
Turbine:	
cylinders	1
flows/turbine	2
stages/flow	5
MP inlet pressure, bar,a	4.15
MP inlet temperature, °C	145
MP steam mass flow, kg/s	39.7
LP inlet pressure, bar,a	2.05
LP inlet temperature, °C	121
LP steam mass flow, kg/s	37.8
last-stage blade height, mm	584.2
exhaust pressure, kPa	11.1
speed, rpm	3600
Condenser:	
type	low-level, spray-tray, direct-contact
CW flow, kg/s	2850
CW temperatures, °C:	
inlet	32.0
outlet	45.8
NCG system:	
steam ejector:	yes
stages	2
steam flow, kg/s	1.71
Plant performance	
SSC, (kg/s)/MW	2.58
utilization efficiency, %	31.7 (1)

(1)Based on a dead state at 25°C.

Fig. 13.10 Turbine rotor for each 110 MW unit at CP-II and CP-III at the Toshiba factory. Toshiba photo, used by permission [12].

The steam for these turbines comes from wellhead separators and flashers, one pair per well, as shown in the schematic Fig. 13.11. Thus there are pipe runs for high- and low-pressure steam from each well serving these units. At the yard of the power house, the steam is passed through large moisture removers before entering the turbines.

A unique feature of these units is the use of steam-driven turbo-compressors, T/C, to remove the noncondensable gases from the condensers. A side-stream of HP steam is used to power two turbo-compressor modules, one for each condenser, and the exhaust steam is fed back into the LP steam line for use in the LP stages of the turbines. This innovative design contributes to the high utilization efficiency of these units, namely, 49.3%.

Table 13.3 Technical specifications for Cerro Prieto II and III; data are per unit [12,13].

Plant	CP-II, Units 1 & 2, CP-III, Units 1 & 2
Start-up year	1982 (II), 1983 (III)
Type	double flash
Rating, MW	110.0
Turbine:	
cylinders	2
flows/turbine	2
stages/flow	3 (HP), 4 (LP)
HP inlet pressure, bar, a	10.75
HP inlet temperature, °C	182.2 (sat.)
HP steam mass flow, kg/s	177
LP inlet pressure, bar, a	3.16
LP inlet temperature, °C	134.6
LP steam mass flow, kg/s	25
last-stage blade height, mm	584.2
exhaust pressure, kPa	11.4
speed, rpm	3600
Condenser:	
type	low-level, spray-tray, direct-contact
No.	2
CW flow, kg/s	7012
CW temperatures, °C:	
inlet	34
outlet	46
NCG system:	
turbo-compressors:	
stages	1
sets	2
steam flow, kg/s	8.6
Plant performance	
SSC, (kg/s)/MW	1.84
utilization efficiency, %	49.3 [(1)]

[(1)]Based on geofluid temperature of 320°C, separation at 11.2 bar, a, flashing at 3.9 bar, a, and dead state at 25°C.

In 2003, plants CP-II and CP-III had capacity factors of 78.4% and 82.7%, respectively [1].

13.3.3 Cerro Prieto IV – Units 1–4

The last units to be added to the Cerro Prieto power station are designated Cerro Prieto IV (CP-IV) and comprise four identical 25 MW single-flash units. The power house is to the east of the earlier units and draws its geofluid from the deepest part of the reservoir; see Fig. 13.5.

The turbine-generator sets were supplied by Mitsubishi; all four units started up in the year 2000, one each in the months of March, April, June, and July. While each

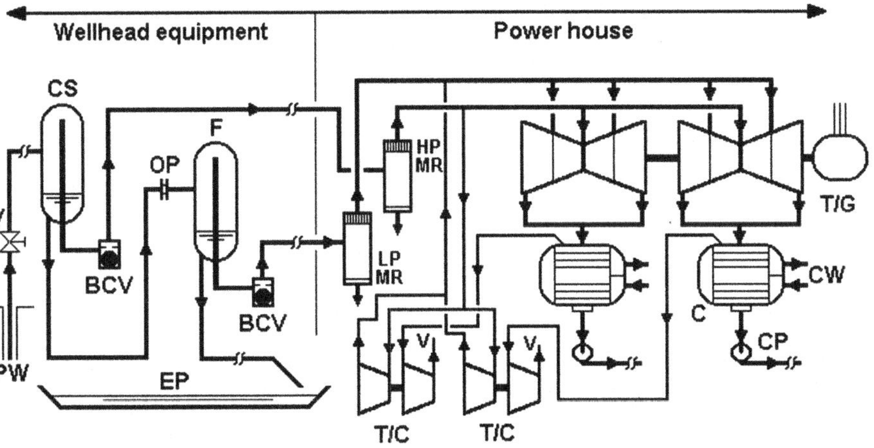

ig. 13.11 *Simplified flow diagram for Cerro Prieto Units II and III [13].*

unit is nominally rated at 25 MW net power, Mitsubishi rates each one at 26.95 MW, gross, with a maximum capability of 28.15 MW [10]. Table 13.4 gives the technical pecifications for one of the four identical units [10]. The entire CP-IV plant had a 3.9% capacity factor in 2003 [1].

Taken as a whole, all of the Cerro Prieto units have shown a decline in performance relative to the design specifications. Data from 2008 showed that the gross specific steam consumption (SSC) for the Cerro Prieto system had risen to 2.45 kg/s per MW [14] compared to the power-weighted average SSC of 1.94 kg/s per MW from Tables 13.1–13.4, roughly 26% poorer performance.

13.4 Expansion of Cerro Prieto and nearby prospects

For the time being, further expansion of the Cerro Prieto field is on hold; only make-up wells are now being drilled. The typical life of a well at Cerro Prieto is roughly ten years, with a work-over at around 4–5 years.

For the 35 years ending at the end of 2008, 369 wells had been drilled at Cerro Prieto for a total drilled depth of 868.4 km, giving an average well depth of 2353 m. The deepest well is 4400 m in depth. During 2008, 167 production wells were active, delivering 1,451 kg/s of steam to the 13 power units, for an average of 8.69 kg/s per well. This is 25% lower than the production rate five years earlier [1]. Thirteen injection wells were operating in 2008, returning 31% of the total separated brine (19.6×10^9 kg vs. 63.0×10^9 kg). The remaining 43.4×10^9 kg of waste brine was sent to the evaporation pond that had grown to 14.3 km^2 by the end of 2008; see Fig. 13.3. In 2011, in response to concerns expressed by nearby residents, CFE announced a US$72.6 million project to reduce the size of the pond and implement a monitoring program and environmental impact study of the possible effects of the pond [14].

Table 13.4 Technical specifications for Cerro Prieto IV; data are per unit [10].

Plant	CP-IV, Units 1–4
Start-up year	2000
Type	single flash
Rating, MW	25
Turbine:	
cylinders	1
flows/turbine	1
stages/flow	7
inlet pressure, bar, a	10.5
inlet temperature, °C	182 (sat.)
steam mass flow, kg/s	48.9
last-stage blade height, mm	584.2
exhaust pressure, kPa	11.5
speed, rpm	3600
Condenser:	
type	low-level, spray-tray, direct-contact
CW flow, kg/s	1681
CW temperatures, °C:	
inlet	32.9
outlet	47.4
NCG system:	
steam-jet ejector:	
stages	2
sets	2
steam flow, kg/s	2.16 (total)
vacuum pump:	
power requirement, kW	225
Cooling system:	
type	counterflow, mechanical-draft cooling tower
No. of cells	3
water flow rate, kg/s	1848
fan power, kW	337.5 (total)
Plant performance:	
SSC, (kg/s)/MW	1.96
utilization efficiency, %	39.7 (1)

(1)Based on geofluid temperature of 320°C, separation at 11 bar, a, and a dead state at 25°C.

Based on earlier studies, there are several other geothermal prospects in the general area of Cerro Prieto that have been explored to the level of prefeasibility. These studies recommended that drilling be carried out at the most promising sites, including Tulicheck, Riíto, Aeropuerto, Laguna Salada, Pescadores, Guadalupe Victoria, and Nayarit [15].

Tulecheck is about 27 km NW of Cerro Prieto, within the Mexicali Valley. Riíto is about 40 km from Cerro Prieto in the opposite direction, within the Colorado River delta. Guadalupe Victoria is about halfway between Cerro Prieto and

Riíto. Aeropuerto is roughly 27 km north of Cerro prieto, close to the Mexico-U.S. order and just west of the Imperial fault. Laguna Salada and Pescadores are ome 20 km to the WSW in the Sierra de los Cucapahs. All of these areas have een identified using gravity and magnetic surveys, followed by seismic reflection and refraction studies, in conjunction with high temperatures observed in irrigation wells [4].

Of course, the area lying east of the currently exploited Cerro Prieto field is also an xcellent prospect for extending the resource. Thus, the region of Baja California Norte holds the potential to add to the already installed 720 MW of geothermal electric power at Cerro Prieto.

References

[1] Gutiérrez-Negrín, L., C.A., R. Maya-González and J.L. Quijano-León, "Current Status of Geothermics in Mexico," *Proc. World Geothermal Congress 2010*, Paper 0101, Bali, Indonesia, 25–29, 2010.

[2] DiPippo, R., *Geothermal Energy as a Source of Electricity: A Worldwide Survey of the Design and Operation of Geothermal Power Plants*, U.S. Dept. of Energy, DOE/RA/28320-1, U.S. Gov. Printing Office, Washington, DC, 1980.

[3] Coordinadora Ejecutiva Cerro Prieto, *Geothermal Project Cerro Prieto, Mexico*, Comision Federal de Electricidad, March 1981.

[4] Fonseca L., H.L., A. de la Peña L., I. Puente C. and E. Díaz C., "Extension of the Cerro Prieto Field and Zones in the Mexicali Valley with Geothermal Possibilities in the Future," *Proc. Third Symposium on the Cerro Prieto Geothermal Field, Baja California, Mexico*, U.S. DOE in cooperation with Comisión Federal de Electricidad, March 1981, pp. 384–421.

[5] de la Peña L., A., I. Puente C. and E. Díaz C., "Geologic Model of the Cerro Prieto Field," *Proc. Second Symposium on the Cerro Prieto Geothermal Field, Baja California, Mexico*, Comisión Federal de Electricidad in cooperation with U.S. DOE, October 1979, pp. 29–56.

[6] Castillo B., F., F.J. Bermejo M., B. Dominguez A., C.A. Esquer P. and F.J. Navarro O., "Temperature Distribution in the Cerro Prieto Geothermal Field," *Proc. Third Symposium on the Cerro Prieto Geothermal Field, Baja California, Mexico*, U.S. DOE in cooperation with Comisión Federal de Electricidad, March 1981, pp. 474–483.

[7] *Proc. Second Symposium on the Cerro Prieto Geothermal Field, Baja California, Mexico*, Comisión Federal de Electricidad in cooperation with U.S. DOE, October 1979.

[8] *Proc. Third Symposium on the Cerro Prieto Geothermal Field, Baja California, Mexico*, U.S. DOE in cooperation with Comisión Federal de Electricidad, March 1981.

[9] Butler, S.J., S.J. Sanyal, R.C. Henneberger, C.W. Klein, H. Gutiérrez P. and J.S. de León V., "Numerical Modeling of the Cerro Prieto Geothermal Field, Mexico," *Proc. World Geothermal Congress 2000*, International Geothermal Association, 2000, pp. 2545–2550.

[10] Anon., "List of Geothermal Power Plant," Brochure No. H480-48GP01E2-H-0, Mitsubishi Heavy Industries, Ltd., Yokohama, Japan, 2000.

[11] I. Canchola F., "Rehabilitation and Modernization Project for Unit 5 of Cerro Prieto Power Plant," *Proc. World Geothermal Congress 2005*, Antalya, Turkey, 2005.

[12] Anon., "Geothermal Double Flash Turbine Generator," Brochure No. 7976, Toshiba Corporation, Tokyo, Japan, 1985.

[13] Ocampo, D., J.d.D., "Statistics of Electric Generation and Steam Production for 17 Years of Operation of the Cerro Prieto Geothermal Project," *Geotermia, Rev. Mex. Geoenergía*, V. 7, N. 1, 1991, pp. 69–93.

[14] Blodgett, L. and K. Gawell, "Mexico: CFE to Reduce Evaporation Reservoir Surface at Cerro Prieto Plant," *GEOTHERMAL ENERGY WEEKLY*, Geothermal Energy Association, August 1, 2011.

[15] Montiel, A.R., "Geothermal Exploration in Mexico," *Proc. Ninth Annual Geothermal Conf. and Second IIE-EPRI Workshop Geothermal Conf. and Workshop*, EPRI AP-4259-SR, Electric Power Research Institute, Palo Alto, CA, 1987, pp. 15-1–15-14.

Nomenclature for figures in Chapter 13

BC	Barometric condenser
BCV	Ball check valve
C	Condenser
CP	Condensate pump
CS	Cyclone separator
CW	Cooling water
EP	Evaporation pond
F	Flasher
HPMR	High-pressure moisture remover
LPF	Low-pressure flasher
LPMR	Low-pressure moisture remover
MPF	Medium-pressure flasher
MR	Moisture remover
OP	Orifice plate
PW	Production well
T/C	Turbo-compressor
T/G	Turbine/generator
V	Vent
WV	Wellhead valve

Chapter 14

Hatchobaru Power Station, Oita Prefecture, Kyushu, Japan

14.1 Overview of Japanese geothermal development

The islands of Japan lie directly along a section of the Pacific Ring of Fire. Large portions of the main islands of Hokkaido, Honshu, and Kyushu are occupied by impressive volcanoes. Although Japan comprises only about 0.3% of the earth's land surface, it is home to 10% of the world's active volcanoes.

There are abundant geothermal resources having a wide range of temperatures and applications. The enjoyment of thermal spas is a well-known part of Japanese culture. It should not be surprising that Japan was one of the earliest countries to experiment with the generation of electricity from geothermal resources. Extensive surveys were begun in 1918 to identify promising areas for geothermal development and the following year saw the successful drilling of a steam well – the Tsurumi well – in the Bozujigoku district of the famous hot spring city of Beppu on the island of Kyushu. In 1925, a tiny experimental electric generator produced some 1.12 kW from the Tsurumi well. Two years later a geothermal well was successfully drilled at Otake, south of Beppu, but the potential of the well was not captured owing to the primitive state of technology at the time for harnessing the power of liquid-dominated geothermal resources [1, 2].

Throughout the 1950s and early 1960s, scientific exploration and well drilling continued at the Otake area. In 1951, a 30 kW generator was installed at Minami Tateishi in Beppu – the Hakuryu Experimental Facility – using the steam from a newly-drilled well; see Fig. 14.1. All of this work culminated in 1967 with the construction and commissioning of the first commercial geothermal plant in Japan using a liquid-dominated resource – the 12.5 MW Otake power plant.

Geothermal Power Plants: Principles, Applications, Case Studies and Environmental Impact, Third Edition.

© 2012 Elsevier Ltd. All rights reserved.

Fig. 14.1 Hakuryu Experimental Facility at Beppu in 1951. Photo from Kyushu Electric Power Co., Ltd. Used with permission [2] [WWW].

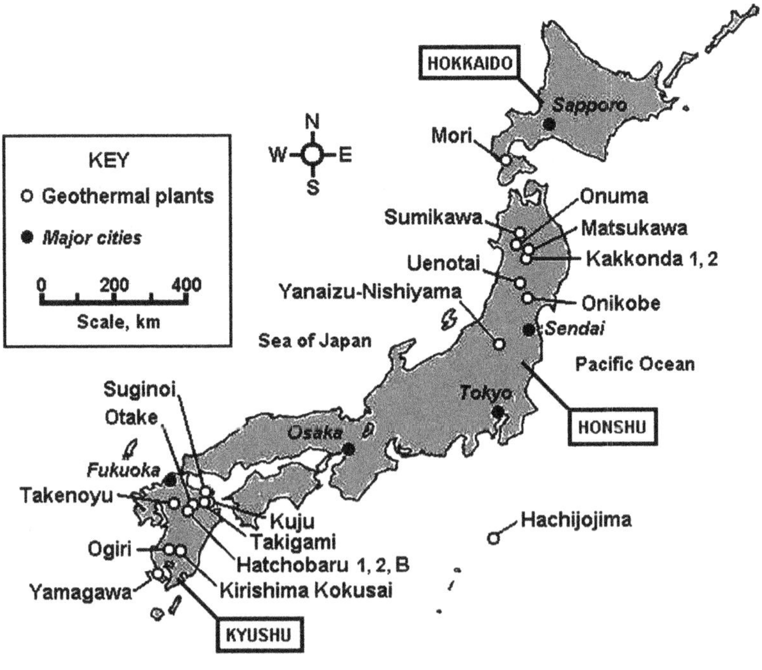

Fig. 14.2 Japanese geothermal power plants as of July 2004; compiled from various sources.

One year before the Otake plant began operating, Japan had commissioned its first ommercial geothermal plant, the 23.5 MW dry-steam plant at Matsukawa. At that ime in 1966, there were only two other sites in the world with dry-steam geothermal lants, Larderello in Italy and The Geysers in the USA.

Since then, Japan has pursued an active program of geothermal power development hat has resulted in 21 units at 18 different locations on the three largest islands of Ionshu, Hokkaido and Kyushu plus the small island of Hachijojima; see Fig. 14.2. he total installed capacity is 535.26 MW, good enough for eighth among the counries of the world; see Appendix A for more details. The Japanese geothermal power ystem recorded an overall capacity factor of 65.3% for the year 2007 [3].

14.2 Hatchobaru geothermal field

4.2.1 Geological setting

Iatchobaru is within the Aso National Park and as such is subject to stringent conrols on well drilling locations, atmospheric and surface discharges, noise, and visual listractions from the natural scenery. One of its neighbors is the 12.5 MW Otake geohermal power plant, only 2 km to the north. About halfway between the two plants ies the village of Sujiyu Hot Springs, a popular tourist attraction; see Fig. 14.3. As we

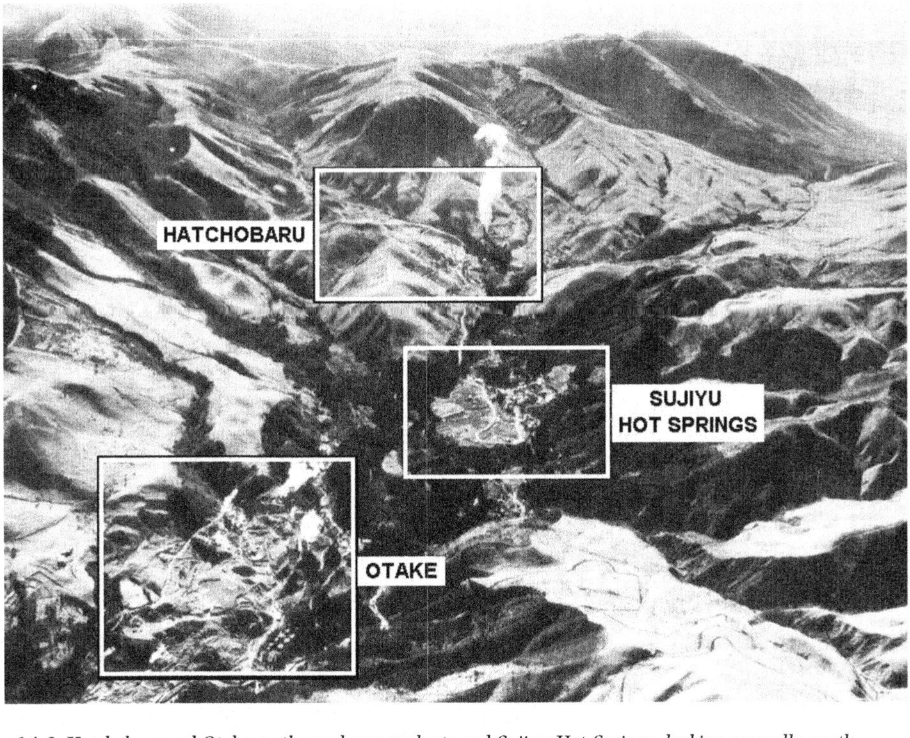

ig. 14.3 Hatchobaru and Otake geothermal power plants and Sujiyu Hot Springs, looking generally south. he distance between the two power plants is about 2 km. Modified from Yoshida and Aikawa [4].

will see, these conditions have played an important role in the development and maintenance of power from the Hatchobaru plants.

As the aerial site photo in Fig. 14.3 shows, the Hatchobaru field is rather narrowly confined within the rugged terrain that forms the Kuju mountain range. The region is marked by several major NNW-SSE-trending, steeply dipping faults that appear to bound the Hatchobaru reservoir. There are also orthogonal faults in the southeast sector that appear to play a major role in bringing the hot geofluid from its source toward the surface.

Figure 14.4 is a cross-section looking NW across the region [5]. In this plane, the principal fluid-controlling faults for the Hatchobaru field are the Hatchobaru fault and the Komatsuike sub-fault, both of which dip steeply to the southwest.

The Komatsuike fault, which gives rise to the prominent Komatsuike fumarole adjacent to the Hatchobaru plant site (see Fig. 14.8), actually dips to the northeast and forms a slip boundary across the Komatsuike sub-fault. As such, the hot fluid rising along the Komatsuike sub-fault probably completes its journey to the surface on the upper section of the Komatsuike fault.

Permeability and production have been found generally in the foot walls of the fluid-controlling faults, relatively close to the faults. This has guided the targeting of production wells. The permeability of the rocks constituting the formation is low whereas the reservoir exhibits very high horizontal connectivity along the aforementioned faults. The basement rocks lie about 1750 m below ground level at Hatchobaru where temperatures of about 300°C have been found.

14.2.2 Production and reinjection

The siting of wells for production and reinjection is severely constrained by the restricted size of the field. Figure 14.5 focuses in on the Hatchobaru field in a plan view, with the trace of the faults shown at sea level [6, 7]. Included in Fig. 14.5 are

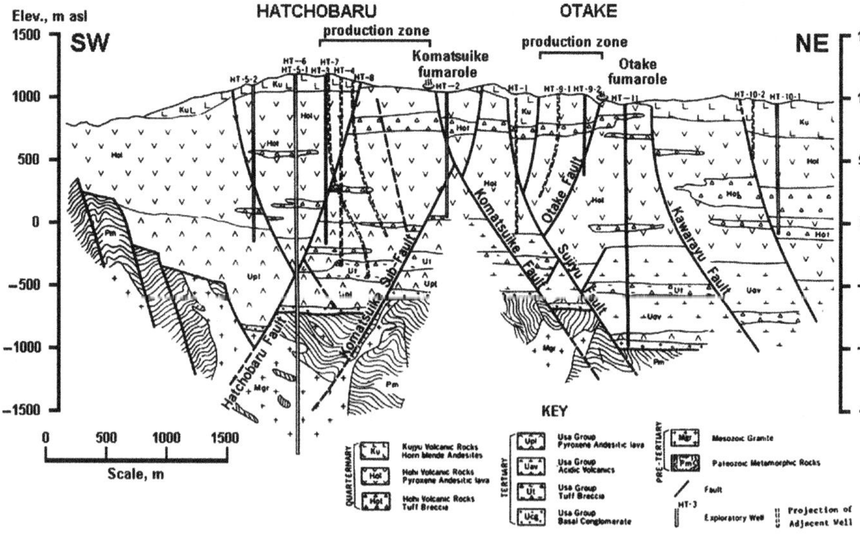

Fig. 14.4 Geologic cross-section of the Hatchobaru-Otake fields; after [5].

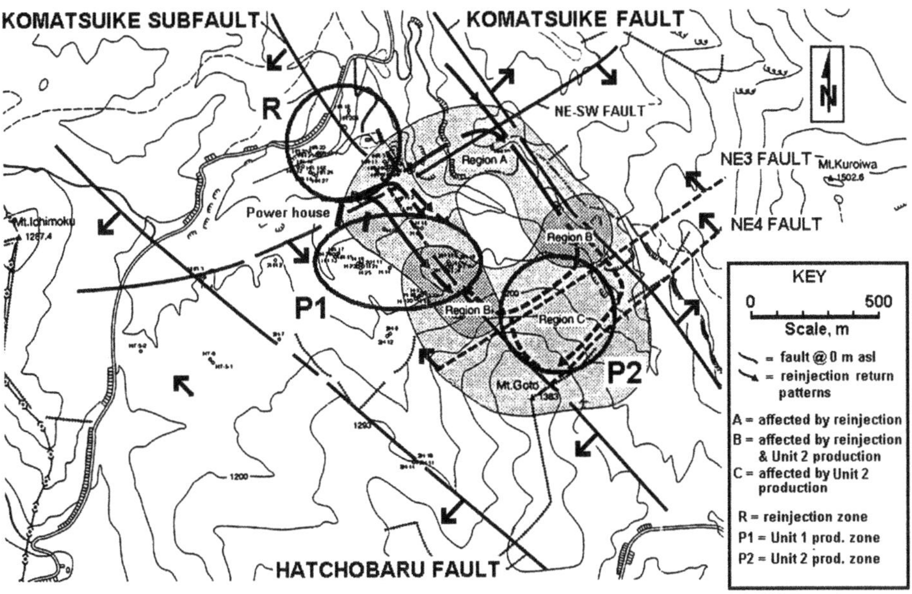

ig. 14.5 Apparent flow patterns caused by reinjection at the north side of the field in 1995. After Mimura, et al 6]; updated based on [7].

he zones that have been used for production and reinjection. In general, the best place or production is in the southeast portion of the field, while reinjection takes place as ar to the north as feasible to avoid interference with the production wells. Since the redominant natural reservoir flow is from the southeast to the north-northwest, the einjection wells are, therefore, in the outflow zone of the reservoir. However, the pres-ure difference that drives the fluid flow is disturbed when production occurs simulta-eously with reinjection. The former reduces the pressure in the hot portion of the eservoir while the latter increases the pressure in the cold zone. This pressure reversal as allowed "cold" injectate to find its way back into the production zone and has aused the demise of some previously excellent production wells, notably well H-4.

When well H-4 was tested upon completion of drilling in 1973 it was capable of enerating 18 MW, a truly impressive well. By the time it was connected to the plant n 1977, it was down to about 13 MW, still a very respectable amount. From 1977 to 1980 during the first three years of plant operation, albeit at half-load, the output leclined precipitously to 1 MW, rendering the well useless. Over this time, the temper-ature at the feedzone fell by some 50°C. It was clear that recirculation of injected fluid ad caused the death of the well [8, 9].

Well H-4 was located just to the east of the power plant, essentially at the inter-ection of the Komatsuike sub-fault and the NE-SW fault; see Fig. 14.5. It was within 240 m of the nearest reinjection well, measured on the surface, but it was only 140 m distant from the completion point of one directionally-drilled reinjection well, HR-14. Tracer tests showed velocities of injectate returns in parts of the field as high as 80 m/h [8].

As of 1996, there were 28 production wells supplying up to 710 kg/s of geofluid to two 55 MW power units (16 for Unit 1; 12 for Unit 2), and 16 reinjection wells (12 for Unit 1; 4 for Unit 2) accepting about 50% of the produced fluid.

14.3 Hatchobaru power units

The Hatchobaru geothermal power station comprises three generating units:

- Unit 1, 55 MW, double-flash, since June 24, 1977
- Unit 2, 55 MW, double-flash, since June 22, 1990
- Binary unit, 2 MW, since December 15, 2003.

14.3.1 Double-flash units

Hatchobaru Unit 1 was among the world's first geothermal power plants designed and built on the double-flash principle. Both it and the Krafla 30 MW unit came online in 1977. The multi-flash plant at Wairakei, New Zealand, had been operating since 1958, but it was not a true double-flash unit [11]. Unit 2 was a near replica of Unit 1 with a few technical improvements based on the dozen years of operating experience with Unit 1.

Figure 14.6 shows the flow diagram for Unit 2 [10]. Unit 1 does not have final moisture removers (demisters), MR, to protect the turbine from entrained moisture droplets in the main steam. Although only the radial blower NCG extraction system is shown, there is also a backup steam jet ejector system.

Figure 14.7 is the plan layout of the two units [12]. Another difference between the two units is that the cooling tower for Unit 2 has five, rather than four, cells to allow better performance during hot summer months. Since the plant is intended to operate unmanned from the control room at the nearby Otake plant, the generator for Unit 2

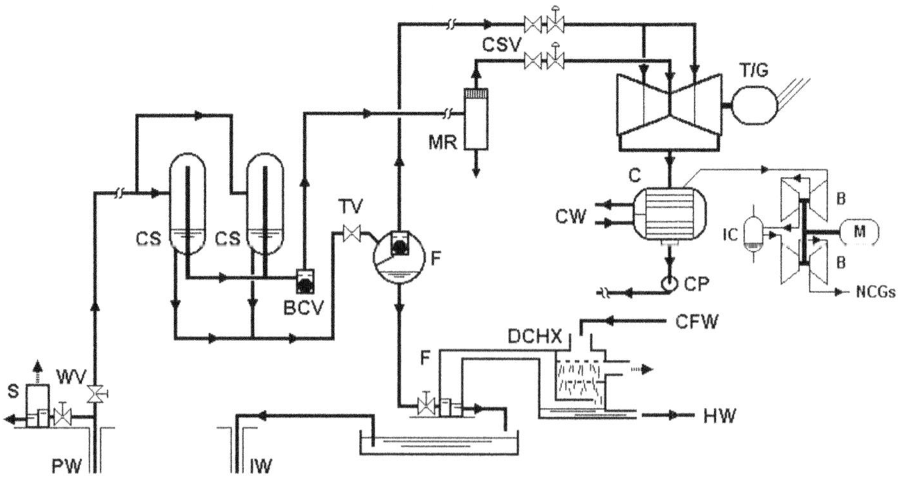

Fig. 14.6 Simplified flow diagram for 55 MW Hatchobaru Unit 2.

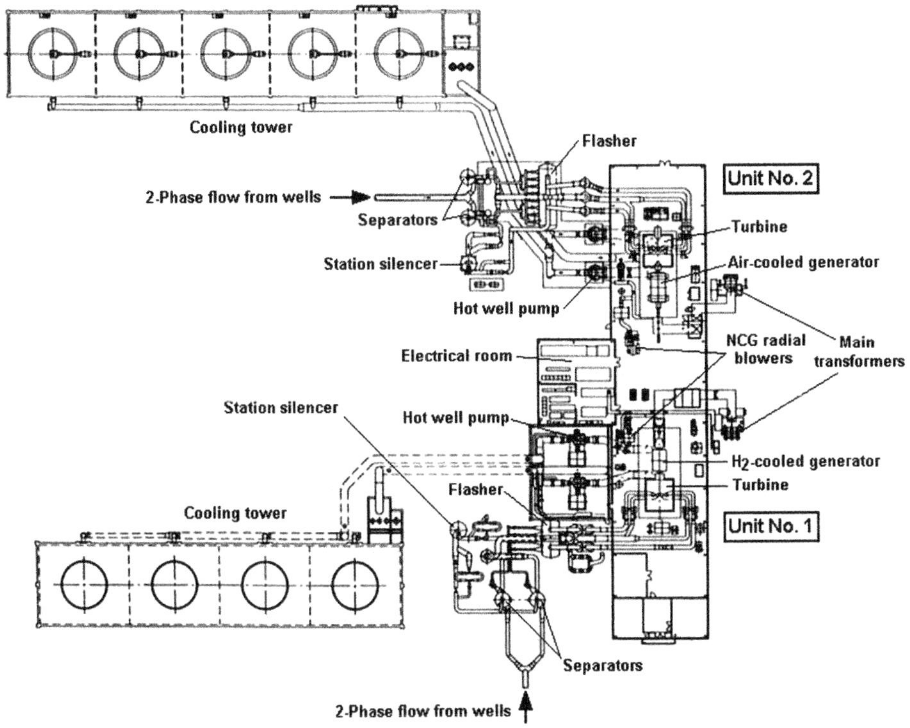

ig. 14.7 *Layout of Hatchobaru 2 × 55 MW geothermal power plant; after [12].*

s air-cooled, rather than hydrogen-cooled, for more reliable operation and less mainte-ıance. To save space at the power plant, Unit 2 uses vertical canned hot well pumps nstead of the horizontal ones in Unit 1. Finally, the turbine for Unit 2 is an advanced lesign incorporating twisted nozzles and an integral blade shroud that yields a 7% mprovement in turbine efficiency relative to Unit 1 [12].

An aerial view of the plant and part of the well field is presented in Fig. 14.8; the ayout of the production wells as of 1992 is given in Fig. 14.9. From the scale shown n Fig. 14.9, it is evident that the wells are drilled in a relatively small area, less than 1 km^2. The use of well pads with multiple wells drilled directionally saves the construc-ion of numerous well pads in rugged terrain. Taking into account the reach of the lirectional wells, the area encompassed by all the wells, production and reinjection, still only comes to about 1.3 km^2.

The reader may wish to compare the Hatchobaru field with the Miravalles field liscussed in Chapter 16; see Fig. 16.4. At Miravalles, Units 1 and 2 combine for 110 MW – the same as Hatchobaru – but the well field that supports them, both or production and reinjection, covers at least 15 km^2.

The difficulty in maintaining an output of 110 MW at Hatchobaru became evident soon after Unit 2 started up in 1990. Figure 14.10 shows the production history for Jnit 1 and Unit 2 from 1977 to 1995 [6]. It is clear that Unit 1 was able to reach and

Fig. 14.8 Aerial view of Hatchobaru facility, looking generally south. See Fig. 14.9 for identification of well sites. The Komatsuike fumarolic area is at the lower left side. Photo by Mitsubishi Heavy Industires, Ltd. Used with permission [10] [WWW].

maintain its power rating of 55 MW about three years after commissioning, once sufficient steam was won from the drilling program. But soon after Unit 2 came online, trouble began.

After the initial spike to 110 MW, the combined output began a steady decline of about 7 MW per year, a serious cause for alarm. Since most of the decline was with Unit 1 that was fed by wells in the middle of the field, rather than with Unit 2 that was fed from the southernmost wells, it was again clear that the reinjection wells at the north end of the field were causing the problem.

Since then a drilling program was undertaken to target the NE4 and Komatsuike faults for new production wells, and to drill wells farther north for reinjection. About one-third of the waste brine from Hatchobaru has for many years been sent via pipeline to the Otake field to be reinjected there.

One of the features of Hatchobaru, and of many Japanese geothermal power plants, is the use of the waste brine to heat water for the local communities. A direct-contact heat exchanger, DCHX in Fig. 14.6, receives cold fresh water from a nearby river and produces hot water by mixing it with steam flashed from the waste brine at atmospheric pressure; see the photo in Fig. 14.11 [10]. Direct use of brine in the

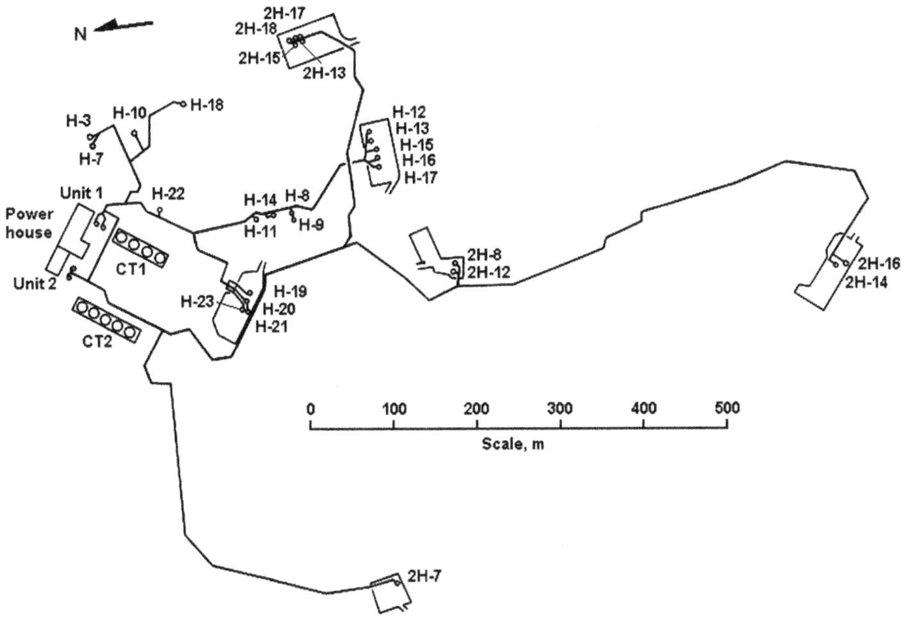

'ig. 14.9 Production wells for Hatchobaru power station in 1992; modified from [10].

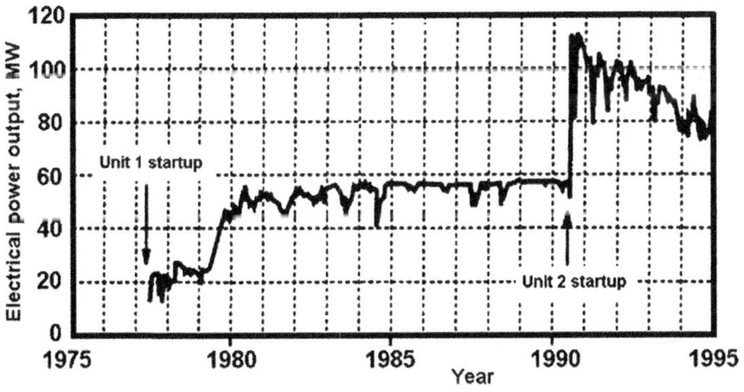

'ig. 14.10 Power production history for Unit 1 (1977–1990) and Units 1 and 2 (1990–1995) [6].

community is not feasible owing to high concentrations of arsenic, 3 ppm. Since the steam is practically free of minerals that are dissolved in the brine, the hot water that s sent to the communities is of good quality.

A negative consequence for the power plant of this community service is that the brine that is sent to reinjection, after passing through the silica settling pond, is at about 90°C [13]. In comparison, the waste brine for a single-flash plant is typically

Fig. 14.11 District hot water heating facility at Hatchobaru. Direct-contact water heater is seen in the foreground, adjacent to settling pond. Photo from Mitsubishi Heavy Industries, Ltd. Used with permission [10] [WWW].

between 140–165°C. Injecting relatively cool water into a reservoir with strong coupling along the major faults has exacerbated the problem with reinjection. Additionally, some of the injection wells have become clogged from silica that carries over with the brine and deposits in the formation, reducing permeability [14]. Owing to the high temperature of the geofluid, there is a high concentration of silica in the brine that is increased significantly during the two flash processes; see Fig. 6.15. As a result, the waste brine is greatly supersaturated with respect to amorphous silica and precipitation is guaranteed. The settling pond is one way to mitigate the problem, but may not eliminate it altogether.

The primary beneficiary of the free hot water is the town of Sujiyu Hot Springs mentioned earlier. A preliminary assessment in 1983 of the effect of plant operations on the hot springs indicated only seasonal fluctuations in the output of the springs [9]. However, it was recently reported [16] that the springs stopped flowing sometime after the startup of the geothermal power plants. However, the hot water received by the town from the power plant has allowed the spas to continue to operate and to provide the hot baths expected by the visitors to Sujiyu.

14.3.2 Binary unit

The Japanese have been exploring the use of binary plants at their geothermal fields since the 1970s. Two prototype 1 MW units were built and tested from 1977–1979, one at Nigorikawa near Mori on Hokkaido [11] and one at Otake, 2 km north of

Iatchobaru [15]. Although both plants performed well from a technical standpoint, either plant was pursued commercially for economic reasons.

In 2004, a 2 MW binary unit was commissioned at Hatchobaru. This unit is esigned to use a subpar well that cannot be connected to the main gathering system. 'he two-phase fluid is separated at the wellhead; the steam is sent to the vaporizer of he binary unit, while the separated brine and the steam condensate go to the reheater. Because the unit is air-cooled, all of the geofluid from the well is reinjected vith minimal environmental impact. Figure 14.12 shows the unit and Table 14.1 ives some of the technical specifications. The unit was supplied by Ormat, Inc., and, s with the double-flash units, is owned by Kyushu Electric Power Co., Ltd.

The plant will be closely monitored for a 2-year period and if successful could be eplicated at other noncommercial geothermal wells.

'ig. 14.12 Hatchobaru 2 MW binary unit. Photo courtesy of Ormat [17] [WWW].

Table 14.1 Technical specifications for Hatchobaru binary unit [17].

Gross power output	2,000 kW
Heating media	steam and brine separated at wellhead
Steam inlet pressure	4.0 bar,a
Steam inlet temperature	143.6°C, saturated
Steam flow rate	4.96 kg/s
Brine inlet temperature	143.1°C
Brine flow rate	17.82 kg/s
Cooling system	6-cell, air-cooled condenser
Generator type	brushless, synchronous, 1800 rpm
Generator output	3-phase, 2230 kVA, 6.6 kV, 60 Hz
Utilization efficiency, gross	35.0% (25°C dead state temp.)

14.4 Conclusion and forecast

The Hatchobaru geothermal resource is a high-temperature system that is currently host to three power units having a total installed capacity of 112 MW. It has experienced difficulty in avoiding interference between its reinjection and production wells owing to the close coupling of the wells along the fluid-controlling faults. Given the small areal size of the field and its proximity to the Otake field, conceptual models must include Otake to be of most value. A new model that does just that predicts a 4% annual decline in power output, amounting to about 20 MW over five years, without any new wells. However, the simulation indicated that by drilling one new production well per year aimed at the NE4 fault the power could be stepped back up to full capacity in five years. It is predicted that the cooling of the reservoir will nevertheless continue unabated in the interim periods between the new well additions [7].

As a postscript, there were nine geothermal power plants, housing ten units, located in northern Honshu and Hokkaido close to the epicenter of the 9.0 magnitude earthquake that struck Japan at 2:46 PM on March 11, 2011. One unit was out of service for maintenance, six automatically tripped off-line instantly, while three continued to operate normally. All units that shut down were back online by 7 PM on March 15, generating the same power as before the quake [18].

References

[1] Masukawa, T., "Geothermal Development of Otake and Hatchobaru," Internal Report, Research Dept., Kyushu Electric Power Co., Ltd., October 1977.

[2] Anon., "Hatchobaru Geothermal Power Station," undated brochure, Kyushu Electric Power Co., Ltd., Fukuoka City, Japan.

[3] Sugino, H. and T. Akeno, "2010 Country Update for Japan," *Proc. World Geothermal Congress 2010*, Paper 0142, Bali, Indonesia, April 25–29, 2010.

[4] Yoshida, K. and K. Aikawa, "Development and Operation Results of Double-Flash Cycle Geothermal Power Generation in Hatchobaru Power Plant," *Thermal Nuclear Power*, V. 31, N. 9 (year unknown).

[5] Fujino, T. and T. Yamasaki, "Geologic and Geothermal Structure of the Hatchobaru Field, Central Kyushu, Japan," *Geothermal Resources Council TRANSACTIONS*, V. 8, 1984, pp. 425–430.

[6] Mimura, T., K. Oishi, Y. Ogata, H. Tokita, Y. Tsuru and K. Matsuda, "Changes of Well Characteristics in the Hatchobaru Geothermal Field (Japan) by Exploitation of Unit No. 2," *Proc. 17th New Zealand Geothermal Workshop*, The U. of Auckland, 1995, pp. 179–184.

[7] Tokita, H., E. Lima and K. Hashimoto, "A Middle-Term Power Output Prediction at the Hatchobaru Field by Coupling Multifeed Wellbore Simulator and Fluid-Gathering Pipeline Simulator to Reservoir Simulator," *Proc. World Geothermal Congress 2005*, Antalya, Turkey, 2005.

[8] Horne, R.N., "Geothermal Reinjection Experience in Japan," *J. Pet. Tech.*, V. 34, 1982, pp. 495–503.

[9] Yoshida, K., K. Tanaka and K. Kusunoki, "Operating Experience of Double-Flash Geothermal Power Plant (Hatchobaru)," EPRI AP-3271, *Proceedings Seventh Annual Geothermal Conf. and Workshop*, Electric Power Research Institute, Palo Alto, CA, 1983, pp. 6-42–6-51.

[10] Anon., "Hatchobaru Geothermal Power Plant: Mitsubishi 2 × 55,000 kW Double Flash Cycle Plant," Brochure H480-48GP02E1-B-0, Mitsubishi Heavy Industries, Ltd., Tokyo, Japan, 1992.

[11] DiPippo, R., *Geothermal Energy as a Source of Electricity: A Worldwide Survey of the Design and Operation of Geothermal Power Plants*, U.S. Dept. of Energy, DOE/RA/28320-1, U.S. Gov. Printing Office, Washington, DC, 1980.

[12] Fujikawa, T. and M. Ikegami, "110 MW Geothermal Power Station in Japan – Completion of Hatchobaru No. 2 Unit," *Geothermal Resources Council TRANSACTIONS*, V. 16, 1992, pp. 561–565.

[13] Tokita, H., T. Yahara and I. Kitakoga, "Cooling Effect and Fluid Flow Behavior Due to Reinjected Hot Water in the Hatchobaru Geothermal Field, Japan," *Proc. World Geothermal Congress 2000*, International Geothermal Association, 2000, pp. 1869–1874.

[14] Inoue, K. and K. Shimada, "Reinjection Experiences in the Otake and Hatchobaru Geothermal Fields," *Proc. 7th New Zealand Geothermal Workshop*, The U. of Auckland, 1955, pp. 69–74.

[15] DiPippo, R., "Second Law Assessment of Binary Plants for Power Generation from Low-Temperature Geothermal Fluids," *Geothermics*, V. 33, 2004, pp. 565–586.

[16] Editorial, "Fallacy of Hot Springs," *The Weekly Post*, Pub. Shogakukan, Inc., Tokyo, Aug. 20–29, 2004: http://www.weeklypost.com/040820/040820b.htm.

[17] Sapiro, G., "First ORMAT Geothermal Plant in Japan," *IGA News*, Quarterly No. 55, Jan–Mar. 2004, pp. 10–11.

[18] Yasukawa, K., "All Geothermal Power Plants in Northeastern Japan Survived the M9.0 Earthquake," *IGA News*, No. 84, April–June 2011, pp. 8–10: www.geothermal-energy.org.

Nomenclature for figures in Chapter 14

B	Blower
BCV	Ball check valve
C	Condenser
CP	Condensate pump
CS	Cyclone separator
CSV	Control & stop valves
CW	Cooling water
CFW	Cold fresh water
DCHX	Direct-contact heat exchanger
F	Flasher
HW	Heated water
IC	Inter-condenser
IW	Injection well
M	Motor
MR	Moisture remover
NCG	Noncondensable gases
PW	Production well
S	Silencer
T/G	Turbine/generator
TV	Throttle valve
WV	Wellhead valve

Chapter 15

Mutnovsky Flash-Steam Power Plant, Kamchatka Peninsula, Russia

"It appears to be a well established opinion that flows of natural steam at workable pressures, when found, can be depended upon as a constant source of power, due to the belief that the steam production process is continuous and everlasting."

H.N. Siegfried – July 1, 1925

15.1 Setting, exploration, and early developments

The Russian city of Petropavlovsk-Kamchatksy on the Kamchakta Peninsula is 7840 km (4870 mi) and nine time zones east of Moscow. It is not connected to the Russian electricity grid. Imported fuel is a very expensive commodity for the 385,000 residents of the city and its neighboring communities. But it is only 70 km (44 mi) north of the geothermal resource at Mutnovsky; see Fig. 15.1. Indeed, the site of the world's first geothermal binary plant, Paratunka, is only about 30 km (20 mi) away "as the crow flies" across Avachinsky Bay from the city.

The whole length of the peninsula, about the size of the state of California, is studded with gigantic volcanoes, part of the Pacific Ring of Fire. Kamchatka is home to one of the world's premier geyser fields along the Geysernaya River in the Uzon caldera, so remote that it was not discovered until 1941 [2]. It is so spectacular that the area is now part of the Kronotsky State Biosphere Reserve. There are some 22 major geysers and about 100 lesser ones active along a 5 km (3 mi) length of the river, with some playing on a regular basis to a height of as much as 40 m (130 ft).

Despite its isolated location, Kamchatka was recognized as having great geothermal potential. Exploration began in the late 1950s, culminating in the construction of a

Geothermal Power Plants: Principles, Applications, Case Studies and Environmental Impact, Third Edition.
© 2012 Elsevier Ltd. All rights reserved.

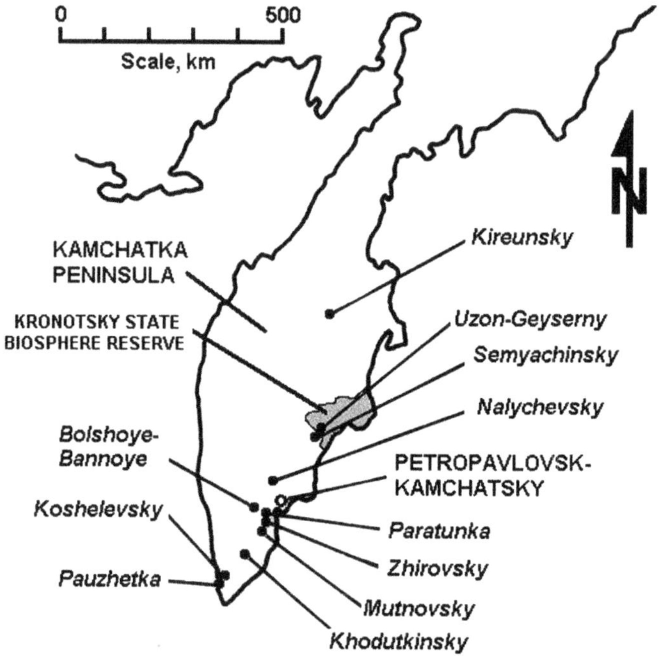

Fig. 15.1 Kamchatka peninsula, Russia. Geothermal areas are shown as solid circles; main city is an open circle. Mutnovsky geothermal field lies between the Gorely and Mutnovsky volcanoes at the southern end of the peninsula [1].

5 MW (2 × 2.5 MW) flash-steam plant at Pauzhetka in 1967 [3]. In 1982, a 6 MW unit was added, raising the output to 11 MW; see Fig. 15.2. The plant continues to operate and produces the least expensive electricity in Kamchatka [4].

The pioneering binary plant at Paratunka also started up in 1967, but generated only about 500 kW using low-temperature geofluids [3], and was eventually decommissioned and abandoned [1]. A view of the turbine hall is shown in Fig. 15.3, a vintage photograph from a 1970 article by Moskvicheva and Popov [5].

Further development of the geothermal resources of Kamchatka did not continue until the price of fossil fuels rose in the 1980s, thereby enhancing the value of the indigenous energy resources [4]. Renewed interest has resulted in scores of wells being drilled, in numerous geoscientific studies, and technical articles being published and presented to the geothermal community [1,4,6–13].

15.2 Conceptual model of Mutnovsky geothermal field

A detailed study of the reservoir at Mutnovsky has been carried out by Vereina using the TOUGH2 reservoir simulator [10]. This section is based on the conclusions reached in that work.

The region is marked by numerous thermal manifestations that appear to cluster near the intersections of the major faults and fractures. The fumarolic activity in the

'ig. 15.2 Pauzhetka flash-steam power plant. Key: 1 = power house; 2 = steam inlet pipe; 3 = offices and ıboratory; 4 = geofluid inlet. Photo from O.A. Povarov [4] [WWW].

'ig. 15.3 Paratunka 750 kW Ref-12 binary plant. Turbine-generator is at left; heater and evaporators are in 'enter; desuperheater-condenser is to the right. Photo from Moskvicheva and Popov [5].

:rater of the Mutnovsky volcano, just 8 km to the southwest of the center of the field, s the strongest of all volcanoes on Kamchatka.

As of 2003, there had been about 100 wells drilled at the Mutnovsky geothermal 'ield. The measured temperature and pressure data from 22 of those wells was gathered and combined with estimates for temperature and pressure in another 12 wells. This information was used in the simulator along with reservoir properties inferred from the geological studies of the region to develop the natural state model of the field. The model was based on a 5-layer structure with 160 irregularly-shaped cells per layer.

The best fit to the observed temperature and pressure distribution was obtained by assuming there are three heat sources for the system: one beneath the North Mutnovksy hot springs on the northern foothills of the Mutnovsky volcano to the south of the field; one beneath the Dachny hot springs in the center of the field; and one beneath the Upper-Mutnovsky area in the northeast of the field; see Fig. 15.4. The model correctly depicts the flow of 300°C (572°F) geofluid upwelling from a zone 3 km to the south of the center of the field, moving northward along a dominant N-S trending graben, then branching to the northeast at the Dachny hot springs area following the regional SW-NE trending fault system. It also predicts the observed steam cap in the center of the field beneath the Dachny hot springs where the two fault systems intersect. A variable pattern of permeability was imposed on the cells of the

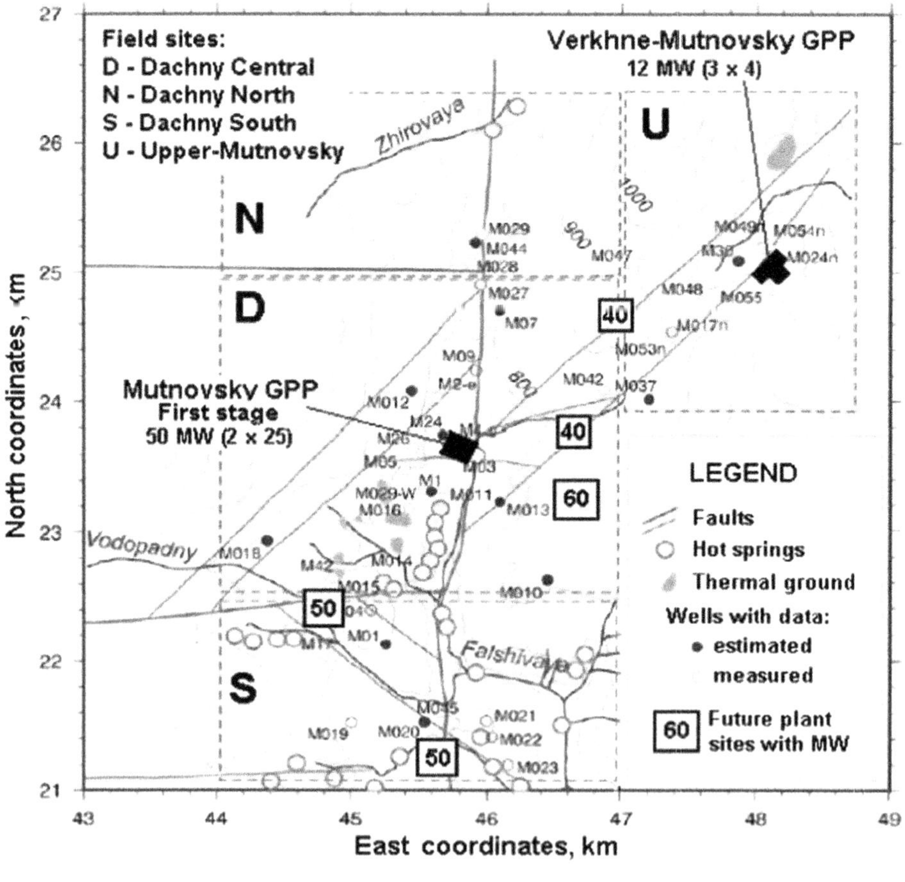

Fig. 15.4 Mutnovsky wellfield and power plant locations; after [6,10,11] [WWW].

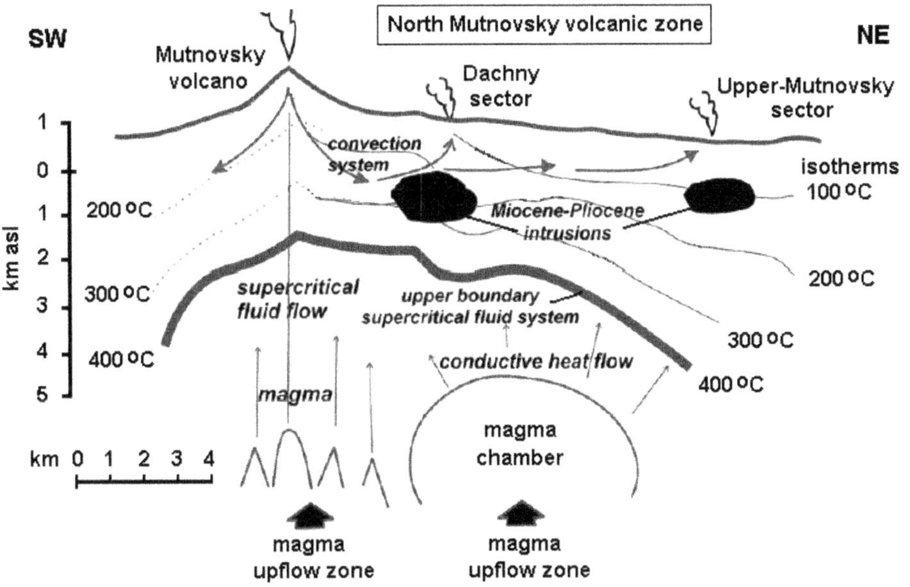

Fig. 15.5 Simplified conceptual model of Mutnovsky geothermal field, looking NW across the regional SW-NE trending fault zone; modified from [7,10].

model to match the observed conditions. Within the reservoir, the permeability varies from 30–50 mD, while at the margins it falls to less than 1 mD.

If the reservoir is defined as that portion of the field limited by the 240°C isotherm at sea level, which is generally at about 800 m below the surface, then the field covers about 10 km^2. The model's predictive capability should allow good siting of new wells and a forecast of the life of the field as more power units are constructed. The schematic conceptual diagram is shown in Fig. 15.5.

15.3 Verkhne-Mutnovsky 12 MW power plant

The first power generating station at the Mutnovsky field was commissioned in 1999 in the northeast sector of the field—the Verkhne-Mutnovsky power plant; see Fig. 15.4 for location. The plant comprises three 4 MW single-flash, air-cooled, condensing units. The turbine steam is obtained from a flash vessel followed by another separator designed to scrub the steam and remove any mineral-laden moisture from the steam. The quality of the steam entering the turbine is greater than 99.98% [8]. Interestingly, the separated liquid is flashed, similar to a double-flash plant, but the low-pressure steam is sent only to the steam jet ejectors to remove the noncondensable gases from the condenser.

A 3-D model of the plant is shown in Fig. 15.6; a site photograph is shown in Fig. 15.7. A schematic flow diagram is given in Fig. 15.8.

This plant is unique among flash-steam plants in its use of air-cooled condensers. The noncondensable gases (NCG) leaving the turbine are removed by a 2-stage steam jet ejector system, with motive steam obtained from the flash vessel. The NCGs, which

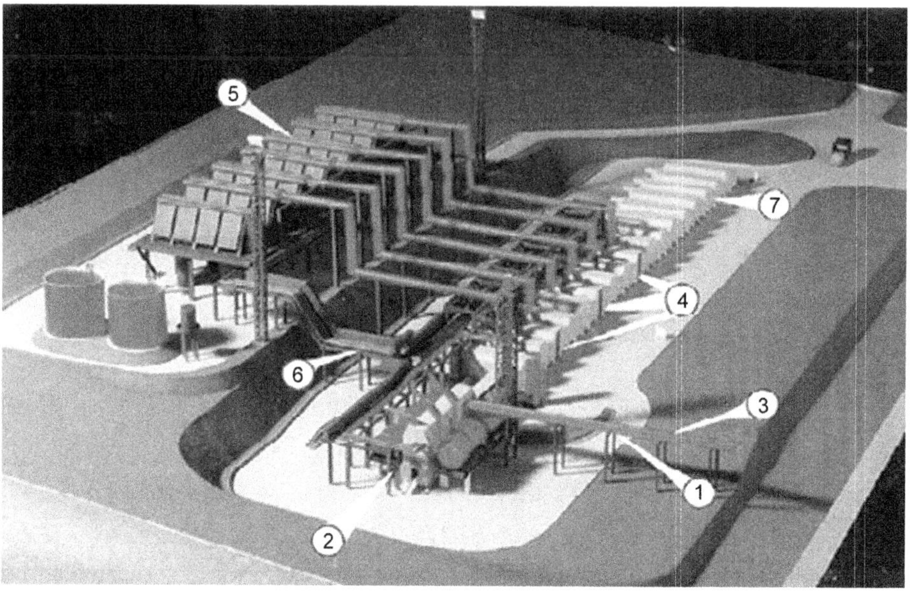

Fig. 15.6 3-D model of the Verkhne-Mutnovsky modular flash-steam 12 MW plant. Key: 1 = inlet two-phase geofluid pipeline; 2 = steam processing facility; 3 = brine return line; 4 = three 4 MW turbine modules; 5 = air-cooled condensers; 6 = steam condensate return lines; 7 = offices and domestic facilities [4] [WWW].

Fig. 15.7 Verkhne-Mutnovsky 12 MW plant; see Fig. 15.6 [9]. Photo by K. Povarov [WWW].

contain hydrogen sulfide, are captured, compressed, redissolved into the steam condensate in the absorber AB, and then reinjected into the formation via the condensate injection well CIW. Thus, the plant has minimal impact on the environment in that all of the geofluid—brine, steam condensate, and NCGs—are returned to the formation.

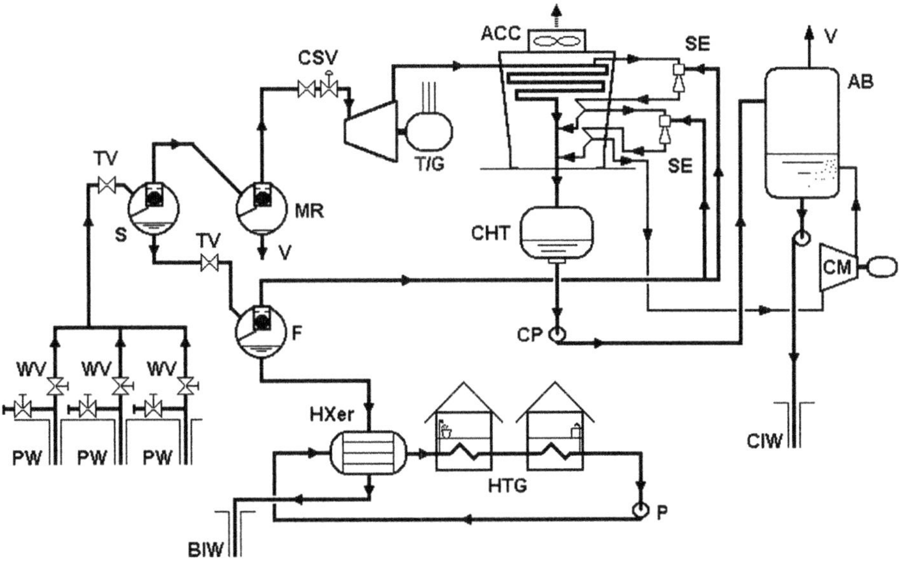

'ig. 15.8 *Simplified schematic flow diagram for Verkhne-Mutnovsky 12 MW plant. There are six 2.5 MW_{gross} urbines arranged 2 per module, with 3 modules; each module produces 4 MW_{net}.*

Since the plant site experiences extremely harsh winters, with snow drifts as high s 12 m (40 ft), winds of 40–50 m/s (90–110 mi/h), and average winter temperaures from October to May of −5°C (23°F), the waste brine from the flash vessel is put o use heating buildings and melting snow around the plant via a heat exchanger IXer. This plant is intended to demonstrate certain unique features that will eventully be incorporated into future plants at the site, and elsewhere in Kamchatka. Iowever, relatively little detailed technical information is available on the plant.

The turbine inlet steam pressure is 800 kPa and the temperature is 170°C, saturated vapor. The temperature of the condensate at the injection well is 50°C; this low emperature does not create a problem with scaling because the condensate is relaively free of minerals. The turbine is directly coupled to the generator and both turn t 3000 rpm. The pressure of the flash steam for the ejectors is 400 kPa and the flow ate is 2.78 kg/s. The bank of air-cooled condensers is set 6 m (20 ft) higher than the urbine-generator level, and each of the six condensing modules comprises eight panels of tubes in a hip arrangement, as shown in Fig. 15.6.

Finally, as a consequence of the severe weather at the site and the short period available for outdoor construction, most of the plant was built at the factory, assembled as much as possible, tested, and brought to the site via cargo air-lift, leaving only he bolting up of the pieces to the site crew. For example, each of the separation stations that consist of the horizontal separator, the horizontal scrubber/moisture remover, and the horizontal flash vessel and all the related piping is built as a package for delivery to the site. It is then only necessary to connect it to the geofluid manifold from the production wells.

15.4 Mutnovsky first-stage 50 MW power plant

The first stage of the construction of the Mutnovsky flash-steam power plant was com pleted in 2002 when the 2 × 25 MW plant was commissioned; see Fig. 15.9. It i expected that the resource will eventually be able to support several more units, a many as 20, with a total power capacity of 300 MW; see Fig. 15.4 [8].

This plant is a conventional single-flash plant. Although it retains the horizontal type separators and scrubber/moisture removers as in the Verkhne-Mutnovsky plant it uses water cooling towers instead of air-cooled ones. The turbine features a standar low-level, direct-contact, jet condenser, the type found in many other geotherma plants around the world. The NCGs are removed from the condenser by a steam je ejector followed by a liquid ring vacuum pump, and vented through one of the fou cooling tower cells. The steam for the ejector is taken as a side-stream from the turbin main steam, not flashed off the separated brine.

To protect the plant equipment and personnel from the harsh winter weather, a the equipment except the geofluid pipelines is located inside buildings, which are con nected by enclosed elevated corridors. The main building houses the scrubber/moistur removers, the turbine-generators, the condensers, cooling water circulating pumps cooling water chemical treatment facility, the control room, 400 V and 10,000 V switchgear/transformers, and the NCG removal system [8].

Within the separator building (see Fig. 15.10), one finds the main separators, th brine injection pumps, silencers, vent valves, the DCS controls, and the 400 V switch gear [8]. There are two other smaller buildings for the rest of the equipment. The plan receives geofluid in the form of 2-phase flow from seven wells: Unit 1 from wells M5E A2, M29W, and M16; Unit 2 from M4E, A3, and M26. There is an interconnectio between the lines from M16 and M26 to allow balancing the flows to the two unit [8]. After being used in the plant heating system, the waste brine is pumped to injec tion wells M27, M28, and M44; condensate is pumped into M07 and M09 [8,9].

Each turbine requires 42.9 kg/s of steam at 620 kPa to produce 25 MW of power i.e., the SSC is only 1.72 (kg/s)/MW. This is an impressive figure and is due mainly t the extraordinarily low condenser pressure of 0.05 bar that corresponds to a condens ing temperature of 32.9°C (91.2°F). The NCG content of the steam is 0.4% by weight

Fig. 15.9 Wide-angle view of the first stage 2 × 25 MW Mutnovsky flash-steam plant. Power house at left contains turbine-generators and associated equipment to protect them and plant personnel during winter weather. The separator building is out of sight to the left of the power house. Photo from Povarov, et al [8] [WWW].

Fig. 15.10 Interior of the separator building of the Mutnovsky power plant, showing the main separators. Photo from Povarov, et al [8] [WWW].

Assuming a dead-state temperature of +5°C and a reservoir at 240°C, the plant has a net utilization efficiency of about 33.5%.

15.5 Future power units at Mutnovsky

There are two units either under construction or in advanced planning at Mutnovsky: Verkhne-Mutnovsky IV and Mutnovsky second stage.

15.5.1 Verkhne-Mutnovsky IV

This will be the fourth unit at the northeast sector of the field, and will be in essence a bottoming unit for the existing 3 × 4 MW plant. It will be a 9 MW total plant, with a 3 MW back-pressure steam turbine coupled to three 2 MW binary units, along the lines of the combined single-flash and basic binary plant described in Sect. 9.3.1 and shown in Fig. 9.7, except that this unit will have an air-cooled condenser and the Russian-style horizontal separators and scrubber/moisture removers. A highly simplified flow diagram is presented in Fig. 15.11. The unit will receive 2-phase fluid from currently unused wells in the northeast area. The separated brine will boil/superheat the binary cycle working fluid, while the steam leaving the turbine will preheat it. Both geofluids will then be reinjected, maintaining an unimpaired environment.

15.5.2 Mutnovsky second stage

The concept to be demonstrated in the V-M Unit IV will be expanded in the next Mutnovsky unit. The second stage Mutnovsky plant is envisioned as a 60 MW plant.

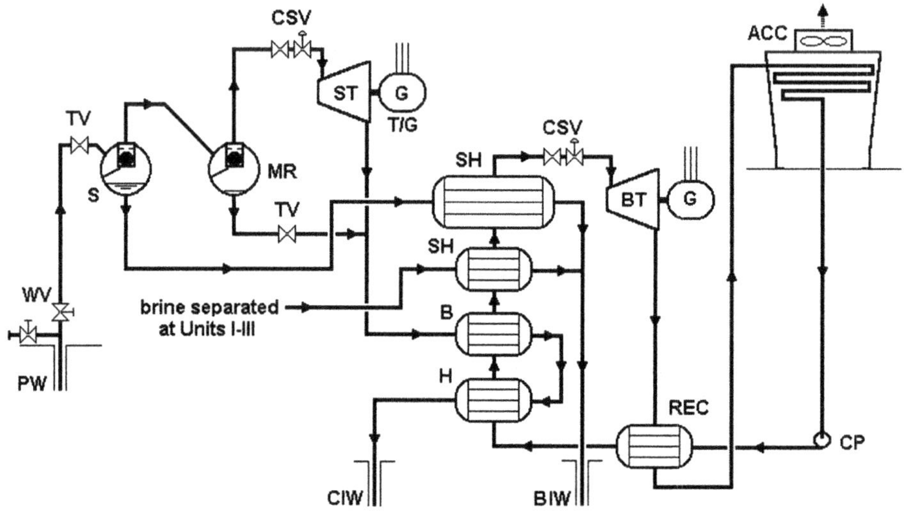

Fig. 15.11 Simplified flow diagram for the 9 MW Verkhne-Mutnovsky Unit IV. There will be three binary loops, each 2 MW, to accompany a single 3 MW steam turbine.

There will be two 30 MW plants combining to give the 60 MW total for the second stage development. The steam turbine will contribute 12 MW and three binary units will each contribute 6 MW, giving the total of 30 MW per plant [6].

It will be an integrated flash-binary cycle using a back-pressure steam turbine exhausting first to the evaporators and then the heaters of the binary cycles. The spent steam is delivered to an absorber, as in the first-stage plant (see Fig. 15.8), where it receives the extracted NCGs before being reinjected. The separated brine from the primary separators provides the heat to superheat the binary cycle working fluid, and is then reinjected. Each binary cycle will incorporate a recuperative heat exchanger wherein the working fluid turbine exhaust will preheat the working fluid returning from the air-cooled condensers prior to its being sent to the main heaters.

The preliminary estimates for the specific steam and brine consumption are 1.48 (kg/s)/MW and 2.96 (kg/s)/MW, respectively. These are very good performance figures and indicate that the design is well-matched to the environmental conditions found at the site. A binary cycle is better able to take economic advantage of the low ambient temperatures than a simple steam unit. The condensing temperature can be as low as 10–20°C (50–68°F). The volume occupied by steam at this uncommon condensing temperature is about eight times larger than at the usual conditions of about 50°C (122°F), which implies a very large turbine exit flow area. This in turn would require either very long last-stage blades or several double-flow turbines to allow the steam to pass. Furthermore, the removal of the NCGs from the condenser would consume much more work considering that the condenser pressure would be one order of magnitude lower than at the usual condensing temperature, i.e., 0.015 bar vs. 0.15 bar. By incorporating a back-pressure steam turbine, this difficulty is avoided.

By contrast, the binary cycle working fluid, typically a hydrocarbon or ammonia, is ot so limited and can operate effectively at these low condensing conditions. ecognizing the importance of having as low a sink temperature as possible for ottoming cycles, the binary cycles benefit greatly from this circumstance and can chieve gains of 20–40% compared to cycles that operate in warmer environments.)f course, such fluids have very low freezing temperatures, which are also appropriate or the Kamchatka climate.

With so many advantages for the integrated steam-binary plant, it is likely that iture developments at the geothermal fields of Kamchatka will be accomplished with his type of energy conversion system.

References

[1] Kononov, V., B. Polyak and B. Kozlov, "Geothermal Development in Russia: Country Update Report 1995–1999," *Proc. World Geothermal Congress 2000*, International Geothermal Association, 2000, pp. 261–266.

[2] Bryan, T.S., "*The Geysers of Yellowstone*," Revised Ed., Colorado Associated University Press, Boulder, CO, 1986.

[3] DiPippo, R., *Geothermal Energy as a Source of Electricity: A Worldwide Survey of the Design and Operation of Geothermal Power Plants*, U.S. Dept. of Energy, DOE/RA/28320-1, U.S. Gov. Printing Office, Washington, DC, 1980.

[4] Povorov, O.A., "Geothermal Power Engineering in Russia – Today," *Proc. World Geothermal Congress 2000*, International Geothermal Association, pp. 1587–1592.

[5] Moskvicheva, V.N. and A.E. Popov, "Geothermal Power Plant on the Paratunka River," *Geothermics – Special Issue 2, U.N. Symposium on the Development and Utilization of Geothermal Reseources*, Pisa, V. 2, Pt. 2, 1970, pp. 1567–1571.

[6] Britvin, O.V., O.A. Popov, E.F. Klochlov, G.V. Tomarov and V.E. Luzin, "Mutnovsky Geothermal Power Complex in Kamchatka," *Proc. World Geothermal Congress 2000*, International Geothermal Association, pp. 3139–3144.

[7] Fedotov, S.A., G.A. Karpov, I.F. Delemen, et al., "Development of Conceptual Model of Mutnovsky Geothermal Field (Dachny and South Sites) and Assessment of Heat Carrier Resources for Production," Internal Report, Inst. of Volcanology, Russian Academy of Sciences (Far East Div.), Petropavlovsk-Kamchaysky, Russia, 2002 (in Russian).

[8] Povarov, O.A., A.I. Nikolski and G.V. Tomarov, "Development of Geothermal Power Plants in Russia," *Proc. 2nd KenGen Geothermal Conference*, UNEP, Nairobi, Kenya, 2003, pp. 131–139.

[9] Povarov, O.A., V. Saakyan, A.I. Nikolski, V.E. Luzin, G.V. Tomarov and M. Sapozhnikov, "Experience of Creation and Operation of Geothermal Power Plants at Mutnovsky Geothermal Field, Kamchatka, Russia," Paper No. S01-052, *International Geothermal Conference*, Reykjavik, Iceland, 2003.

[10] Vereina, O., "Natural State Modelling of the Mutnovsky Geothermal Field, Kamchakta, Russia," Report No. 21, Geothermal Training Programme, U.N. University, Reykjvik, Iceland, 2003, pp. 505–526.

[11] Povarov, O.A., A.I. Postnikov and A.I. Nikolski, "Power Plants at the Mutnovsky Geothermal Field (Kamchatka)," *Geothermal Resources Council TRANSACTIONS*, V. 25, 2001, pp. 543–548.

[12] Belousov, V.I., S.N. Rychagov and S.P. Belousova, "A Conceptual Model of Mutnovsky Geothermal Area (Kamchatka)," *Geothermal Resources Council TRANSACTIONS*, V. 26, 2002, pp. 553–557.

[13] Boyarskiy, M., O.A. Povarov, A.I. Nikolskiy, and A. Shipkov, "Comparative Efficiency of Geothermal Vapor-Turbine Cycles," *Proc. World Geothermal Congress 2005*, Paper no. 1305, Int'l Geothermal Ass'n., Antalya, Turkey, 2005.

Chapter 16

Miravalles Power Station, Guanacaste Province, Costa Rica

"It can scarcely be denied that the supreme goal of all theory is to make the irreducible basic elements as simple and as few as possible without having to surrender the adequate representation of a single datum of experience." Paraphrased as: *"Everything should be made as simple as possible, but no simpler."*

Albert Einstein – 1933

16.1 Traveling to Miravalles

The drive from San Jose, Costa Rica's capital, to its geothermal power complex at Miravalles is best attempted during the dry season and in daylight. It is a spectacular trip.

It is about 220 kilometers from point to point and can take up to four hours under favorable conditions. As you leave San Jose in the Central Valley, the magnificent Irazu Volcano is in your rearview mirror as you head out along the Autopista General Cañas. On your right lie the two massive volcanoes, Poas and Barva, forming a wall to the north of the valley. There is another wall on your south formed by the non-volcanic Southern Cordillera.

Miravalles sits in the Guanacaste province in the northwest part of the country. To get there, you will take the Inter-American highway, the road that runs from Nuevo Laredo on the U.S.-Mexican border to Panama City, linking all the countries between that border on the Pacific Ocean. You must first climb out of the temperate

Geothermal Power Plants. Principles, Applications, Case Studies and Environmental Impact, Third Edition.

© 2012 Elsevier Ltd. All rights reserved.

Fig. 16.0 Miravalles Volcano at sunrise. Photo by author [WWW].

Central Valley, which is at 4000 ft (1200 m) elevation, reaching a peak of about 6000 ft (1800 m) before descending to sea level. It is this ascending and descending that can be difficult to negotiate since you share the two-lane, winding, mountain road with heavy commercial and agricultural vehicles. If you can glimpse the scenery, you will see splendid coffee plantations covering the steeply sloped terrain on all sides.

Eventually you reach the lowlands of the Pacific. As you continue on the now smooth and straight road toward Cañas some 70 kilometers ahead, the majestic volcanoes of the Guanacaste Cordillera begin to appear on your right. First is Arenal, unmistakable with its near-perfect symmetry. Then, as you approach Cañas, Tenorio comes into view. And soon after, there is Miravalles; see Fig. 16.0. When you arrive at the city of Bagaces, it is time to turn north and head for the geothermal field.

Once you leave the small community of Bagaces, you begin a steady, easy climb on a good road – one of the many benefits of the geothermal development. You are climbing the remnants of a pre-Miravalles volcano that erupted itself out of existence some 600,000 years ago. The current Miravalles Volcano grew up within a large caldera that was formed by the catastrophic eruption of the ancient pre-Miravalles volcano. Evidence of the massive eruption is all around you – basalt boulders scattered about the landscape and thick tuff formations evident whenever you pass through a road cut. The road continues up, in serpentine fashion, riding on the crests of ridges with deeply eroded canyons on either side, until you reach the top of the caldera rim.

From this impressive vantage point, the Miravalles Volcano and the geothermal power plants on its slopes present themselves to you; see Fig. 16.1. If the day is clear, there will probably be wispy clouds attached to the peak of Miravalles, obscuring the jagged multi-cratered crest. Thin vapor columns may be visible from the cooling towers of the six power plants. And it will be quiet.

'ig. 16.1 Miravalles Volcano and power plants; photo by author [WWW].

Before you are power plants generating 158 MW (net), about 9% of Costa Rica's otal electric power. They are running reliably, cleanly, quietly, economically, and fficiently, providing the country with roughly 15% of its electrical energy needs.

We will now learn how this important resource came into use. We will examine the letails of the various power plants at Miravalles, and see how the Costa Ricans overcame wo serious obstacles that could have blocked the development of the field.

16.2 History of geothermal development

This account is based on several articles by members of the technical staff of the nstituto Costarricense de Electricidad (ICE) [1–3] and by the author's own experience ıs a member of the Costa Rican Geothermal Advisory Panel since 1984.

Costa Rica, like many countries with limited or no indigenous fossil fuel resources, vas rocked by the oil shocks of 1973 and 1979. It caused the country to begin serious exploration for alternative energy sources within its own borders. In 1975, ICE began dentifying geothermal areas worthy of further exploration. They used their own funds ınd received the help of the Interamerican Development Bank (IDB), the Italian government, and the United Nations Development Program (UNDP). The best prospects vere in the volcanic cordillera of the Guanacaste province and included areas adjacent o the volcanoes Tenorio, Rincon de la Vieja, and Miravalles. Each of these had thermal manifestations such as fumaroles, boiling springs, mud pots, steaming ground, ınd areas of surface alteration. Miravalles was selected as the focus of the early studies owing primarily to its relative ease of access.

In 1977 funds were secured from IDB, and the first three deep exploratory wells vere drilled at Miravalles from 1979 to 1980. They were all successful, indicating that a liquid-dominated reservoir existed on the southwestern slope of the volcano. It was estimated that about 15 MW could be supported by these wells, and that the reservoir covered an area of some 15 km^2. From 1984 to 1986 six more wells were completed and flow tested. This data was analyzed and in 1987 the feasibility study for the first power unit was completed.

The feasibility report was compelling enough to convince the IDB to commit to a major development program that would involve the drilling of 20 more wells, the hiring of a consultant to conduct additional feasibility studies, to purchase equipment and construct the power station, and to connect the station to the main electrical transmission line via a new 50 km power line.

Even as construction on the first power unit was underway, ICE was looking for other good geothermal prospects. Between 1989 and 1991, seven other sites besides Miravalles were identified and characterized as to their potential for electric power generation. Starting from the westernmost site in the Guanacaste Cordillera and extending to the middle of the country along the Central Cordillera, they are: Rincon de la Vieja (very high potential), Tenorio (very high potential), Cerro Pelado (medium potential), Poco Sol (medium potential), Platanar-Poas (high to medium potential), Barva (medium to low potential), and Irazu-Turrialba (medium potential); see Fig. 16.2.

As of 2004, Tenorio has been drilled with two deep wells and found to be lacking sufficient temperature or permeability. Rincon de la Vieja is being explored with deep wells in two areas – Las Pailas and Borinquen – with excellent results at Las Pailas, where a total of five wells have been completed. They discovered a 245°C reservoir with good permeability. Three of the wells have a combined capability of 18 MW [4]. It seems probable that a new geothermal electric power plant will soon be in place at Miravalles' neighbor to the northwest.

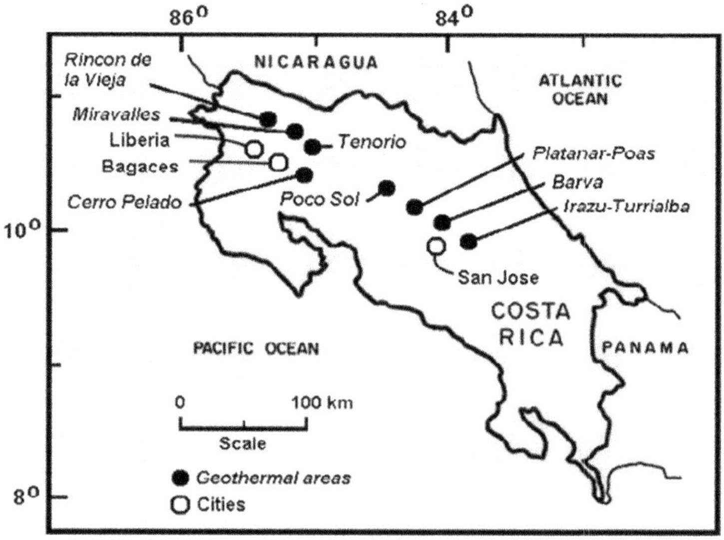

Fig. 16.2 Geothermal areas of Costa Rica.

6.3 Wells

n extensive well field, 24 km^2 in area, supports the six power generating units that col-ectively produce 158 MW. There are 53 wells including production, injection, and mon-oring wells. Figure 16.3 shows the wells that support these units. The wells lie within north-south graben marked roughly by limits 404–409 in Fig. 16.3. The graben cuts hrough the 15 km diameter caldera formed by the pre-Miravalles eruption 0.6 M years go. Succeeding eruptions by resurgent volcanoes have laid down various formations of umice, andesites, dacites, volcanic sediments, lavas, and ignimbrites.

The wells intercept these formations at various depths. While the major fluid conduits ppear to be related to faults associated with the graben and its sub-faults, the produc-ion comes mainly from the volcanic sediments at depths of between 500–1500 m, pproximately. The highest temperatures occur in the middle of the graben and fall off oward the edges, particularly steeply at the western margin. The highest temperatures how up in the northern wells, about 245°C, thought to be the main upflow zone,

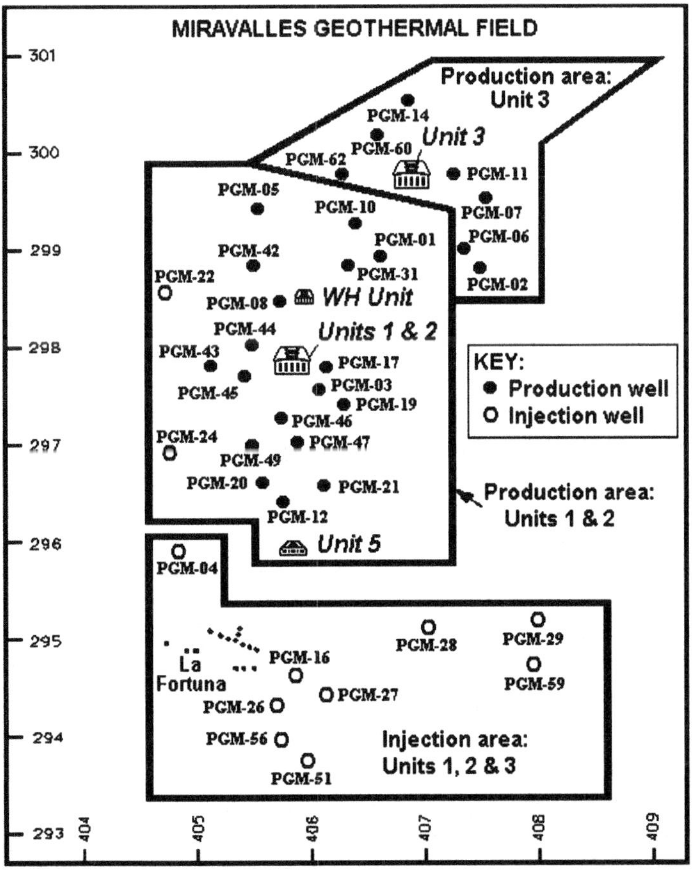

ig. 16.3 Wells in support of Miravalles power plants; after [1].

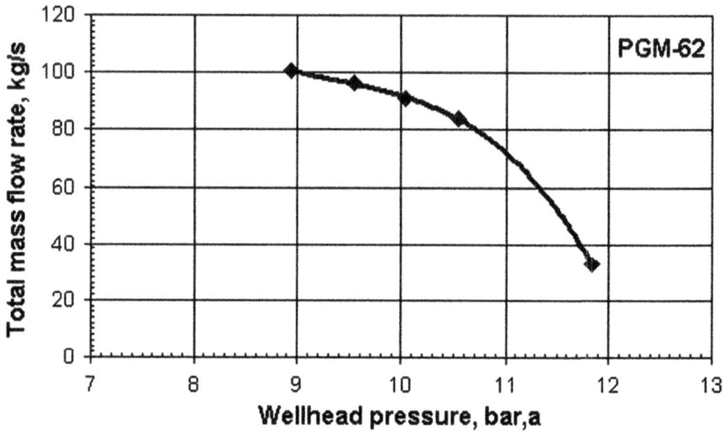

Fig. 16.4 Productivity curve for well PGM-62 [5].

whereas the temperatures in the central and southern areas are about 225–230°C. The reservoir flow is from the north-northeast to the south.

A productivity curve for a typical production well, PGM-62, is shown in Fig. 16.4 [5]. This well is located in the northern sector of the field and supplies fluid to Unit 3. The enthalpy of the flowing two-phase mixture was 1037 kJ/kg and the reservoir temperature was 241°C. These values correspond very closely to saturated liquid conditions in the reservoir. Thus, the steam fraction can be calculated from thermodynamic considerations (see eq. (5.7)); at a separation pressure of 8 bar, the total flow will be roughly 106 kg/s and the steam flow will be about 16.3 kg/s. Since the turbine is characterized by a specific steam consumption of 2.10 (kg/s)/MW, this well should produce about 7.8 MW when connected to Unit 3.

Well PGM-62 has a wellhead elevation of 692 masl; it was drilled to a depth of 1750 m, after experiencing total loss of circulation between 670–760 m depth. These LOC zones correspond to the first appearance of the hydrothermal mineral epidote in the rock cuttings. The 13–3/8 in casing was set at 702 m. The production comes from a zone of high permeability between 1025–1200 m depth. The drawdown coefficient is 0.0926 bar/(kg/s); alternatively, the productivity index is 10.8 (kg/s)/bar.

Whereas most of the early wells were sited close to the major thermal manifestation, Las Hornillas (see Fig. 16.5), the development wells cover a large area that is generally devoid of surface thermal features. The geoscientific studies together with data acquired by drilling have led to a fairly clear picture of the reservoir and a good working model of the field.

16.4 Power generation

Exactly 15 years from the completion of the discovery well PGM-01, the first power unit came online at Miravalles. Unit 1, a 55 MW single-flash plant, began operating in March 1994, and consists of a Toshiba double-flow turbine-generator set. It shares the

'ig. 16.5 A small portion of the Las Hornillas thermal manifestation area. Located just east of well PGM-7 see Fig. 16.3), this spectacular area displays boiling pools, fumaroles, bubbling mud pots, and highly altered, ιot ground. Photo by author [WWW].

urbine hall with another 55 MW single-flash unit that was installed in August of 1998. The turbine for Unit 2 was supplied by Ansaldo.

During the field development stage while many wells were being drilled, ICE installed hree 5 MW wellhead units to capitalize on the power available from newly-drilled wells. Two of the small units were rented from Mexico's Comisión Federal de Electricidad CFE); these operated from September 1996 to December 1998 (WH Unit 2) and from February 1997 to April 1998 (WH Unit 3). The other 5 MW unit has become a permanent part of the operation and utilizes excess steam from the separators serving Units 1 and 2. All three were back-pressure units that exhausted directly to the atmosphere. The chemical nature of the steam allows this type of operation because the hydrogen sulfide concentration is very low.

Unit 3 is a 27.5 MW single-flash unit that was privately built but is supplied with steam by ICE-owned wells, under a power-purchase agreement. This plant came online n March of 2000. The turbine-generator was supplied by Mitsubishi Heavy Industries.

The most recent plant, Unit 5, is a 2-unit binary cycle that started operating in March of 2004 [4]. This plant uses the waste brine from Units 1, 2, and 3 to capture energy that would otherwise be returned to the reservoir via injection wells. The 15.5 MW plant generates 7.75 MW from each unit. Each unit comprises two turbines driving a common generator. The cycle working fluid is normal-pentane, C_5H_{12}. Each unit uses two preheaters, two evaporators with their own separators, two heat recuperators, and four water-cooled condensers. The units were supplied by Ormat.

Figure 16.6 shows the wells that serve the 55 MW Unit 1. Each production well feeds two-phase geofluid to a satellite separator, except wells PGM-11, which has its own separator, and PGM-19, which produces practically all steam. There are three satellites serving Unit 2 and one for Unit 3. Note that a few injection wells can take fluid from more than one satellite. For example, well PGM-28 can receive waste liquid from either Satellite 1 or 2. Satellite 3 is shown in Fig. 16.7.

A view of the turbine hall for Units 1 and 2 is shown in Fig. 16.8 and the permanent wellhead unit is shown in Fig. 16.9.

Table 16.1 summarizes the technical specifications for the four steam power units. Table 16.2 gives the data for the two units comprising the bottoming binary Unit 5. The combined power installed at all six units is 158 MW.

Unit 5 deserves special attention. It was added to the existing plants in 2004 as a way of increasing the output of the field without drilling more wells. Unit 5 raises the power output by 15.5 MWe, about 11%, by extracting roughly 110 MW-thermal from the waste brine before it enters the injection wells. A highly efficient binary cycle, having about 14% thermal efficiency, it incorporates a heat recuperator to capture energy from the turbine exhaust for use in preheating the cycle working fluid, normal-pentane. A photograph of the plant is shown in Fig. 16.10 [7] and a simplified schematic flow diagram is given in Fig. 16.11.

ICE considered using a low-pressure flash system for Unit 5, but the binary plant had the lowest life-cycle cost compared with two bids for flash systems. The exit brine temperature was pegged at 136°C to avoid silica scaling in the heat exchangers or the

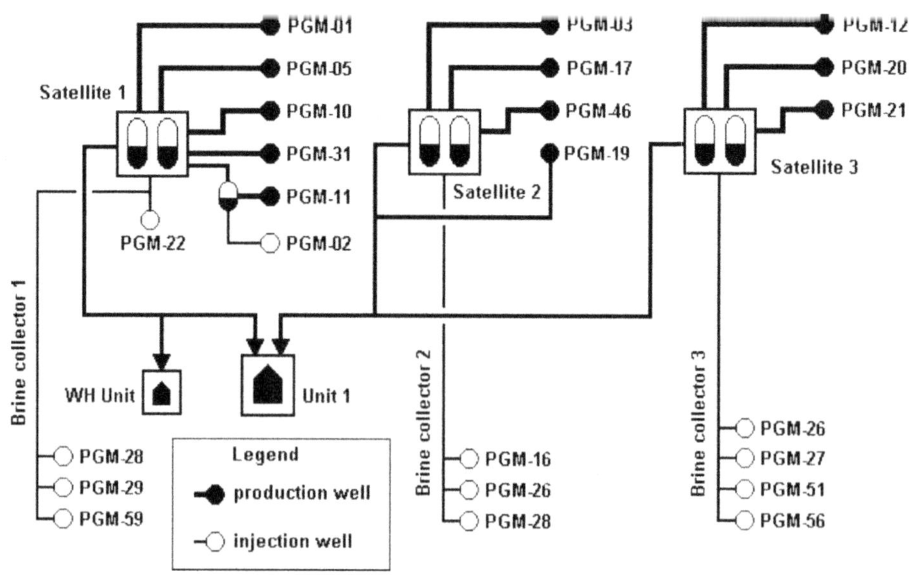

Fig. 16.6 Geofluid gathering system for Unit 1 and the Wellhead Unit [6].

'ig. 16.7 Satellite separator station 3. Twin cyclone separators receive two-phase fluid from wells PGM-12, 20, and -21 (upper pipes at right). Steam pipes emerge from bottom of vessels (lowest pipes), and liquid is directed rom the vessels into vertical surge tanks (one is visible at the far right). Photo by author [WWW].

'ig. 16.8 Unit 2 turbine-generator with Unit 1 in the background; photo by author [WWW].

Fig. 16.9 Permanent 5 MW wellhead unit; steam is tapped from the main steam line from Satellite 1, passes through a moisture remover (left) on the way to the turbine. Exhaust steam is discharged to the atmosphere through a roof-mounted silencer. Photo by author [WWW].

injection piping downstream from Unit 5. The utilization efficiency is slightly lower than the thermal efficiency (12% vs. 14%) owing to the heat-saving effects of the recuperator that raise the internal cycle efficiency. However, the functional Second Law efficiency based on the exergy drop of the brine is an impressive 45.7%. The recuperator also lowers the heat load on the condenser, allowing for a smaller cooling tower compared with a non-recuperative cycle. The input exergy is that of the separated brine at 165°C.

16.5 Calcite inhibition system

Shortly after the initial wells were flow tested it became clear that there was a severe calcium carbonate scaling problem in the wells; see Sect. 4.4. The flow rate began to decrease, gradually at first and later precipitously, owing to the buildup of calcite, $CaCO_3$, inside the casing of the well. The wells would require periodic cleaning in order to remain in production, a costly and time-consuming task that would involve moving a work-over rig to each well successively as it became clogged. Eventually, casings would be damaged from repeated reaming and need to be replaced, a very costly situation.

ICE studied the problem and came up with a method of continuously treating the geofluid by injecting a calcite scale inhibitor, downwell, below the flash horizon. In this way the fluid could be allowed to flash without having it become supersaturated

able 16.1 Technical specifications for Miravalles flash-steam plants.

lant	Unit 1	Wellhead Unit 1	Unit 2	Unit 3
tart-up year	1994	1995	1998	2000
ype	single flash	single flash – atmospheric exhaust	single flash	single flash
ating, MW	55	5	55	29
et MW	52	5	52	27.5
eofluid flow rate, kg/s	760	20.14 (steam only)	726	400
esource temp., °C	240	240	240	240
urbine:				
cylinders	1	1	1	1
flows/turbine	2	1	2	1
stages/flow	6	4	6	5
inlet pressure, bar,a	6.0	5.9	6.0	5.6
inlet temperature, °C	158.9 (sat)	158.1	159	156.8 (sat)
steam mass flow, kg/s	110.0	20.14	105	57.7
exhaust pressure, kPa	12.5	99.1	10.0	9.0
last-stage blade height, mm	584	116	584	635
speed, rpm	3600	3600	3600	3600
ondenser:		none		
type	DC, low-level		DC, low-level	DC
CW flow, kg/s	4,233.6		4150	2,143
CW temperatures, °C				
inlet	28.9		28.9	27.4
outlet	49.4		47	40.8
wet-bulb	21.7		21.7	21.7
CG system:		none		
steam ejector	yes		yes	yes
stages	2		2	2
steam flow, kg/s	4.057		5.145	1.517
compressor	yes		yes	no
stages	4		4	—
power, kW	450		600	—
vacuum pump	no		no	yes
power, kW	—		—	460
lant performance:				
net SGC, (kg/s)/MW	14.6	—	14.0	14.5
net SSC, (kg/s)/MW	2.19	4.03	1.95	2.10
tilization efficiency, %				
gross	29.6	32.6(1)	29.6	29.7
net	28.0	32.6(1)	28.0	28.1

(1) Based on the exergy of saturated steam at 158.1°C; dead state at 21.7°C.

vith calcite. The treatment stabilized the brine and eliminated the need for any mechancal cleaning. The method, however, is effective only when the flashing occurs in the vellbore.

It took a research program lasting about five years to perfect the system, both the nechanical design of the elements and the selection of the best chemical agent and

Table 16.2 Technical specifications for Miravalles Unit 5 binary plant.

Start-up year	2004
Type	recuperative cycle
No. of units	2
Rating, MW	19 (2 × 9.5)
Net power, MW	15.5 (2 × 7.75)
Geofluid flow rate, kg/s	885 (2 × 442.5)
Working fluid	n-pentane, C_5H_{12}
Preheaters:	
No. per unit	2
type	shell & tube
brine outlet temp., °C	136
Evaporator:	
No. per unit	2
type	shell & tube
brineinlet temp., °C	165
HXers. heat duty, MW-th	110.95
Turbine:	
type	axial flow
cylinders/unit	2
flows/turbine	1
stages/flow	3
inlet temperature, °C	147
pressure, bar,a	
inlet	15.5
outlet	1.35
mass flow/unit, kg/s	58.15
speed	1800
Recuperator:	
No. per unit	2
type	shell & tub
hot-side temps., °C	
inlet	88
outlet	65
cold-side temps., °C	
inlet	38
outlet	61
Condenser:	
No. per unit	4
type	shell & tub
CW temperatures, °C	
inlet	28
outlet	37
Feed pumps:	
No. per unit	2
power input, kW	260
Plant performance:	
net SGC, (kg/s)/MW	57.1
net utilization effic., %:	12.05[1]
net functional effic., %	45.73
net thermal effic., %:	13.97

[1]Based on the exergy of saturated liquid at 165°C; dead state at 22°C.

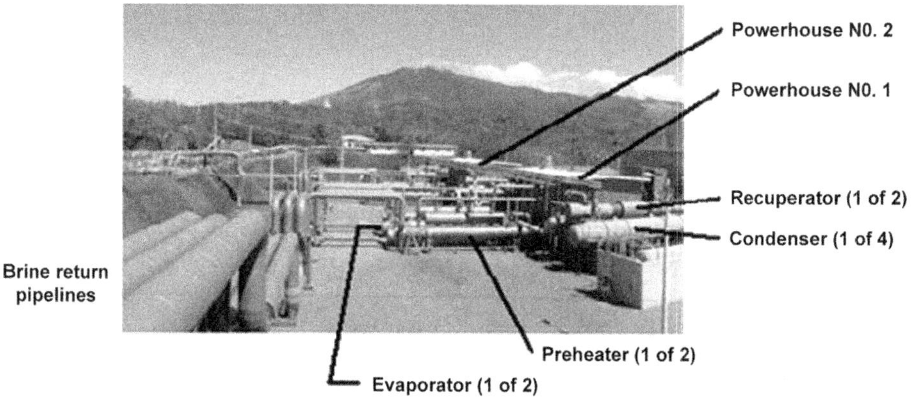

Fig. 16.10 Miravalles Unit 5 binary plant, looking northeast toward Miravalles Volcano. Photo courtesy of Ormat, Inc. [7] [WWW].

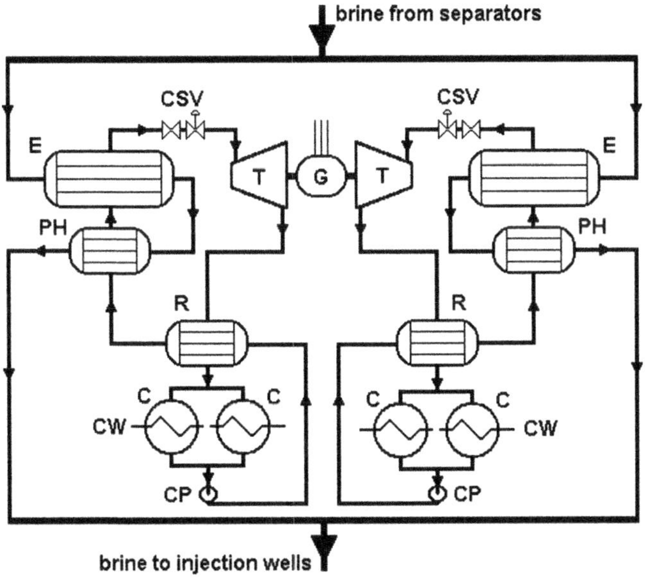

Fig. 16.11 Unit 5 flow diagram for one of two identical units.

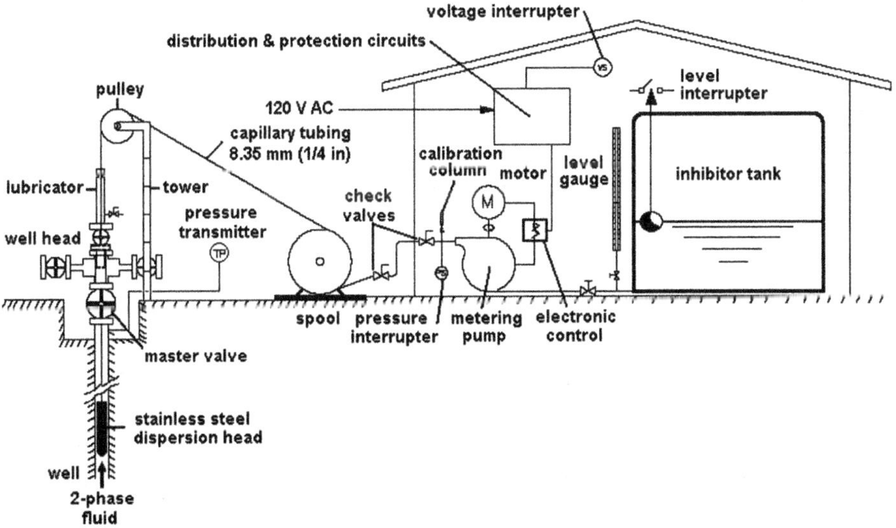

Fig. 16.12 Calcite inhibition equipment; each of the 21 production wells is served by a similar facility [8].

the correct dosages [8]. The inhibition facility is shown in Fig. 16.12 [8]. The inhibition systems operate continuously at each production well, but periodic sampling is necessary to monitor the effectiveness of the treatment. The dosage can be adjusted if the geofluid properties change during production. Given the characteristics of the Miravalles fluids, the best types of inhibitor are polymers with carboxylic functional groups. These are stable at the temperatures encountered in the reservoir, up to 245°C. Currently, the inhibitor of choice is Nalco 1340 HP, a polyacrylate, in dilute aqueous solution. During chemical testing from 1989 to 1994, the optimal dosage was 3 ppm, but under operating conditions it was increased to 4 ppm. In the following years, each well was adjusted to its own level and the range in dosage is from 0.5 to 2.5 ppm, with most wells using 0.75 ppm. It is crucial for all the components of the system to run reliably since a failure of the inhibition system will cause a well to scale up in a matter of a few days.

All 21 production wells now in operation are fitted with the inhibition system and all are performing satisfactorily, except for well PGM-3. The latter well receives two-phase fluid from the reservoir (i.e., flashing occurs in the formation) and the only way to remove the calcite that deposits in the fractures is periodically to treat the well with acid. To date, the system has performed smoothly, allowing the production wells to deliver a constant flow of geofluid to the plants for more than 17 years.

16.6 Acid neutralization system

The second problem encountered by ICE with the Miravalles wells is that the wells in the northeast sector of the field, closest to the volcano, produce highly acidic geofluid. The pH of these fluids is in the range of 2.3 to 3.2. The first well to exhibit this corrosive characteristic was PGM-02. Originally drilled to 1208 m in 1980, the well began producing acidic fluid after it was deepened in 1984 to 2000 m. Since then, wells PGM-06, -07, -19, and -64 all yield similar acidic fluids [9].

As can be seen from the well map in Fig. 16.3, PGM-07 is about 370 m from PGM-11, and PGM-19 is only 260 m from PGM-3, i.e., two acidic wells lie rather close to two neutral wells. Thus, it is thought this effect is a localized one related to upwelling magmatic gases along a lineament connecting the several craters and eruptive centers of the Miravalles Volcano. All acidic wells show high sulfate concentrations, but after a period of production, these concentrations are trending downward. Well PGM-07 has gone from 300 to 225 ppm SO_4^{-2}, PGM-19 has gone from 400 to 217 ppm SO_4^{-2}, and the natural pH of both wells has risen to 4.0. Whatever effect is giving rise to the acidic fluids seems to be abating as the wells are being produced.

Nevertheless, ICE has devised a method for neutralizing the acidic geofluid using a downwell injection scheme similar to the one used to inhibit calcite scaling. They inject a dilute sodium hydroxide, NaOH, solution into the geofluid deep inside the well, below the flash horizon and below the bottom of the deepest casing, allowing it to neutralize the H^+ groups, and thereby increase the pH. Thus, the casings will be safe from corrosive attack, and the geofluids will not cause damage to surface equipment. So far, two of the wells have been successfully connected to the fluid gathering system: PGM-19 since February 2000 and PGM-07 since October 2001 [9]. Well PGM-19 is seen in Fig. 16.13.

One unfortunate complication has been the deposition of anhydrite scale, $CaSO_4$. When the geofluid temperature is higher than 220°C and the pH is greater than 3.5, the geofluid becomes supersaturated with respect to $CaSO_4$. These conditions

'ig. 16.13 Acid neutralization system at well PGM-19. The tower supports the lubricator and pulley assembly)r the chrome-nickel capillary tubing. The control room is at the right and the cyclone separator is at the left. 'hoto by author [WWW].

ıre met once the neutralization process is initiated. Furthermore, ICE has seen ımorphous silica deposition along with anhydrite. As a result, it has been neces- ary to carry out a mechanical cleaning operation at both wells to remove the ›uildup every six months or whenever the flow rate has been curtailed to a large legree. ICE is working on developing a new chemical inhibition process to counter- ıct these adverse side effects of the neutralization process. Even with all these extra neasures that are needed to keep the wells in production, the neutralization cheme has proved itself to be economically viable, with each well returning the :ost of investment within 1.5 to 2 years [9].

16.7 Environmental protection and monitoring

Throughout the development of geothermal resources at Miravalles, and now at Rincon le la Vieja, ICE has been sensitive to preserving the environment from any harmful effects. As described in Sect. 5.7, there are several aspects of geothermal power development hat might adversely affect the environment, but at each step ICE took action to mitigate hese effects. For example while it is necessary to open wells for flow testing – a noisy)rocess – the duration of the tests was kept to the absolute minimum to obtain the equired data. No flow tests were conducted unless a sufficiently large holding pond vas built to capture the fluid being discharged; see Fig. 16.14. These fluids were then :onveyed to an injection well, PGM-4, dedicated to returning "cold" geofluid to the

Fig. 16.14 Flow testing the discovery well at Miravalles, PGM-01. Photo by author [WWW].

reservoir. Holding ponds were constructed at each drill pad to contain the drilling fluids and geofluids that might be produced in the process of drilling. All brine produced from the reservoir is returned to the reservoir, either through "hot" injection wells (136°C or 165°C), or the "cold" injection well (98–100°C). A seismic network is in place to monitor micro-earthquakes that might be generated by production or injection. Rain is sampled to detect any possible change in the acid content of the rain. An extensive network of ground-level measuring stations is in place to detect any subsidence or uplift over the entire Miravalles region and surrounding areas. No adverse effects of any kind have been detected to date after 25 years of development and more than 17 years of power generation.

16.8 Other geothermal power projects

A new power plant was inaugurated on July 24, 2011 at the Las Pailas geothermal field, which is part of the Rincon de la Vieja volcanic complex, about 19 km northwest of Miravalles. The nominal 42 MW plant consists of two 21 MW binary units that use both the brine and the steam separated from the wellheads to heat and boil, respectively, the n-pentane working fluid. Each unit has two turbines driving a centrally located generator, similar to the arrangement at Miravalles Unit 5; see Fig. 16.11. The net plant power output is 35 MW [10,11]. The power generating equipment cost US$65 million; the total project required an investment of US$221.8 million [12,13]. Borinquen, another site on the southwestern side of Rincon de la Vieja, is under development; it is 7 km further northwest from Las Pailas.

References

[1] Moya, P. and A. Yock, "First Seven Years of Exploitation at the Miravalles Geothermal Field," *Proc. Twenty-Sixth Workshop on Geothermal Reservoir Engineering*, Stanford University, Stanford, CA, 2001.
[2] Moya, P. and A. Mainieri, "Geothermal Energy Development in Costa Rica," *Geothermal Resources Council TRANSACTIONS*, V. 26, 2002, pp. 43–48.
[3] Moya, P. and D. Pérez, "Unit 5: A Binary Plant at the Miravalles Geothermal Field," *Geothermal Resources Council TRANSACTIONS*, V. 27, 2003, pp. 551–555.
[4] Mainieri, P.A., "Costa Rica Geothermal Update," *Geothermal Resources Council BULLETIN*, V. 32, No. 4, 2003, pp. 157–158.
[5] Chavarria R., L. and S. Castro Z., "New Wells Drilled from April 1999 to November 2000," Report to the Nineteenth Meeting of the Geothermal Advisory Panel, San Jose, Costa Rica, March 2001.
[6] Moya, P., "Development of the Miravalles Field (1994–2000)," Report to the Nineteenth Meeting of the Geothermal Advisory Panel, San Jose, Costa Rica, March 2001.
[7] Ormat, Inc., Sparks, NV, http://www.ormat.com/index_projects.htm.
[8] Moya, P. and A. Yock, "Calcium Carbonate Inhibition System for Production Wells at the Miravalles Geothermal Field," *Geothermal Resources Council TRANSACTIONS*, V. 25, 2001, pp. 501–505.
[9] Moya, P. and E. Sánchez, "Neutralization System for Production Wells at the Miravalles Geothermal Field," *Geothermal Resources Council TRANSACTIONS*, V. 26, 2002, pp. 667–672.
[10] Moya, P. and L.D. Pérez. "Las Pailas Geothermal Project: A 35 MW Plant," *Proc. World Geothermal Congress 2010*, Paper 0632, Bali, Indonesia, April 25–29, 2010.
[11] Richter, L.X., "Costa Rica inaugurates 42 MW geothermal plant by Ormat Technologies," *Think GeoEnergy*, August 7, 2011: http://thinkgeoenergy.com/archives/8280.
[12] Moresco, J., "Ormat Signs $65M Deal for Geothermal Plant," *Red Herring*. January 2009: http://redherring.com/Home/25754.
[13] Araya, A.M., "Chinchilla Will Open Las Pailas Geothermal Plant," *La Nacion*, July 17, 2011, http://translate.google.com/translate?js5n&prev5_t&hl5es&ie5UTF-8&layout52&eotf51&sl5es&tl5en&u5http%3A%2F%2Fwww.nacion.com%2F2011-07-18%2FElPais%2Fchinchilla-inaugurara-geotermica-las-pailas–.aspx.

Chapter 17

Heber Binary Plants, Imperial Valley, California, USA

"Hitch your engine to a volcano, not a live, active volcano, but one that has simmered down after a few hundred or a few thousand centuries and now contents itself with spouting up steam."

Warren Bishop – October 1921

17.1 Introduction

Heber is a town in Imperial County, California, located some 6 miles (10 km) north of the United States-Mexico border. Situated in the fertile Imperial Valley, it is the home to 3000 people who are mainly engaged in agriculture and raising livestock. The Imperial Valley is flat for miles in all directions, being bounded on the west by the Santa Rosa and Coyote Mountains and on the east by the Chocolate Mountains.

Heber is also home to two geothermal power stations, one a 47 MW double-flash plant and one a 33 MW binary plant; see Fig. 17.1. It is this latter plant that will be the focus of this chapter. Both plants in recent years have been augmented by small binary units to optimize the use of the resource. In addition, there used to be a 45 MW binary plant, just south of the current binary plant, and that will be an important part of the story.

17.2 Exploration and discovery

There are no surface thermal manifestations at Heber, but the Salton Trough is marked by thermal anomalies from the Salton Sea at its northern extremity, through

Geothermal Power Plants: Principles, Applications, Case Studies and Environmental Impact, Third Edition.
© 2012 Elsevier Ltd. All rights reserved.

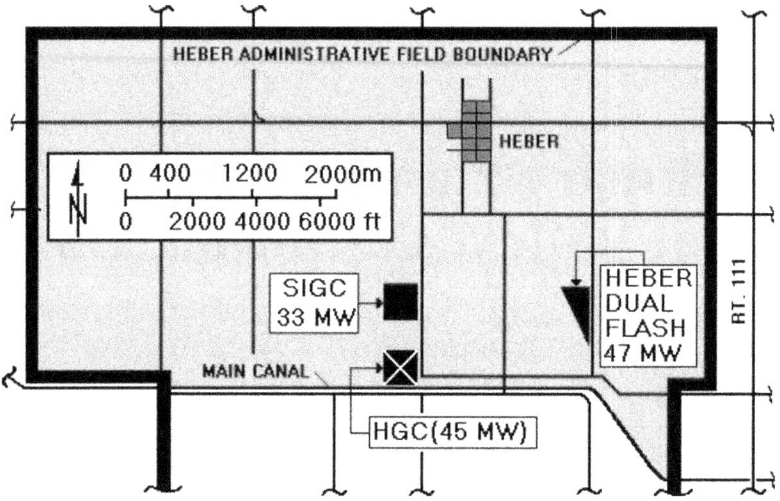

Fig. 17.1 Map of a portion of the Heber KGRA showing the town and its geothermal power stations; filled symbols represent operating plants; after [1]. SIGC is now called Heber 2; Heber Dual Flash is now Heber 1 [WWW].

the Imperial Valley, into Mexico, and all the way into the Gulf of California. The entire region in an active spreading zone characterized by higher than normal heat flow and recent volcanism.

In fact Heber was first explored as an oil and gas prospect by Chevron in the 1960s, based on a positive gravity anomaly just south of the town of Heber [2]. In the 1970s, spurred by the recent geothermal discovery at Cerro Prieto to the south, several shallow temperature gradient wells were drilled. It became clear that there coincided a thermal and a gravity anomaly that extended over an area of at least 6 mi^2 (15.5 km^2).

Two wells drilled in 1972 confirmed the existence of a liquid-dominated, moderate-temperature, low-salinity reservoir: the No. 1 Holtz well drilled by Magma Energy with financial support from Chevron, and the No. 1 Nowlin Partnership well by Chevron. The reservoir was seen to consist mainly of sandstones with porosities of 15–30% lying below a caprock of shale at about 2000 ft (600 m). The shales and sandstones are inter-bedded between 8000–10,000 ft (2400–3000 m), with the shales being thinner at the deeper depths [3]. The geofluid emanated from a source at about 380°F (195°C) some 2 mi to the SSE of the town, and spread toward the NNW as it rose through the formation, yielding fluids in the range of 350–370°F (175–188°C) at depths of about 4000 ft (1200 m) [2].

The early development planning was very optimistic. It was envisioned that seven power plants with a combined capacity of 500 MW would be built – 3×100 MW and 4×50 MW, all within an area of approximately 1.5 mi^2 (4 km^2). This plan failed to anticipate that the production wells to support such a large-scale development would most likely be drawing from the same geofluid source and therefore would interfere with each other. Indeed, interference was observed soon after the first two nominal 50 MW plants were put online at about the same time in 1985, and further power development beyond the 80–90 MW level has not taken place.

17.3 The first Heber binary plant

The U.S. Department of Energy (DOE) had as one of its missions in the early 1980s the development of geothermal resources, particularly moderate temperature ones, for the generation of electricity. The Heber field was chosen to host a joint government-industry-utility partnership to demonstrate the viability of this concept. In 1980, the DOE signed an agreement with the San Diego Gas & Electric Company (SDG&E) to build a commercial-scale binary plant at Heber. SDG&E, which owned 82.5% of the plant, provided 48.9% of the funding, DOE put in 34.8%, and eight other partners contributed the remaining 16.3%; the Electric Power Research Institute (EPRI) and the Imperial Irrigation District (IID) were the next largest at about 6% each [4].

At the same time, another binary power plant – the Magmamax plant – was going through its startup operations at East Mesa, some 17 mi to the east of Heber; see Chap. 18. That plant was rated at 12.5 MW, so the new demonstration plant would have to be larger in accordance with the principle of economy of scale. Since the early Heber plans had envisioned plants of either 50 or 100 MW, it was decided to make the demonstration plant a nominal 50 MW plant.

It had already become clear that the initial design of the Magmamax plant was overly complex and had led to many operational problems. By the end of 1982, the Magmamax plant had been rebuilt to correct the original design problems. The new Heber plant would be relatively simple and consist of a basic Rankine cycle with a single power loop and one working fluid for the cycle. Nevertheless, the state-of-the-art demonstration plant would showcase a mixed hydrocarbon working fluid, the properties of which would need to be determined by a separately funded research program with the National Bureau of Standards [5]. Furthermore, it would feature the largest hydrocarbon turbine ever built, one with a rated power capacity of 70 MW.

The flow diagram for the Heber Binary Demonstration plant (also known as the Heber Geothermal Corporation plant) is shown in Fig. 17.2. The plant required an estimated 15 production wells to provide the plant with 7,450,000 lbm/h (940 kg/s) of 360°F (182°C) brine, assuming each well produces 500,000 lbm/h (63 kg/s), along with 9 injection wells. The working fluid in the cycle was a mixture of 90% isobutane and 10% isopentane. The turbine inlet pressure was set at 580 lbf/in^2 (40 bar), supercritical. Eight massive heat exchangers were arranged in two trains; each exchanger was 80 ft long by 80 in in diameter (24 × 2 m) and weighed 319,000 lbm (145,000 kg). The turbine-generator assembly was 61 ft (18 m) long and weighed 460,000 lbm (209,000 kg). There were two condensers that received cooling water from a 9-cell, induced-draft, counterflow cooling tower. The total parasitic power requirements for the cooling water pumps, cooling tower fans, working fluid circulating pumps, etc. came to 18.5 MW. The brine production and injection pumps added another 4.9 MW to the station load. Thus the net power at full design conditions would be 46.6 MW [6].

Owing to the availability of only six production wells when the plant was ready for operation, it was decided to run just one train of heat exchangers and operate the plant at about half-load until all the production wells became available. However, the plant was plagued by numerous equipment failures during its first two years of operation [4], including excessive vibrations in many components, cracked valve seats,

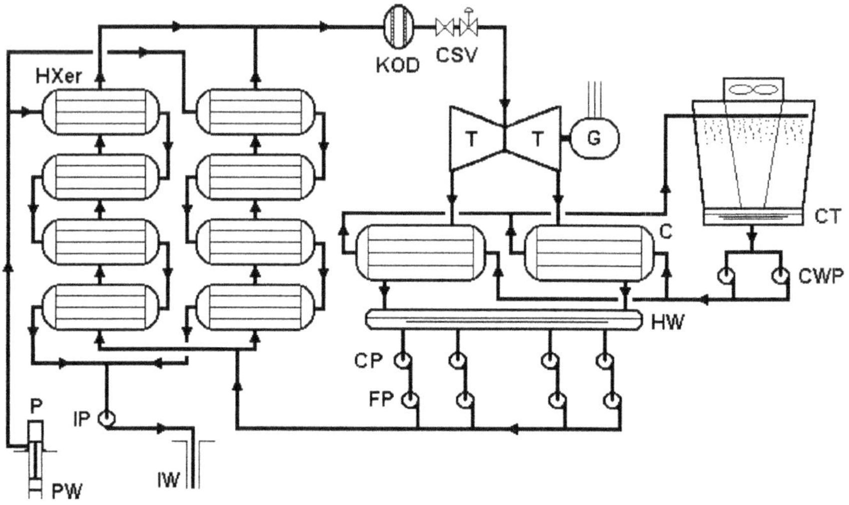

Fig. 17.2 Flow diagram for the first Heber binary plant; after [6].

valves that did not close, and working fluid pump breakdowns, all of which resulted in very sporadic operation. The downwell pumps did perform reliably but with so many plant equipment failures, this was small consolation.

The highest plant output occurred after a prolonged shutdown when eight wells returned to production and for a short time yielded about 20% more flow than under steady conditions. The plant generated 36 MW(gross) and 21 MW(net) from a brine flow of 4,870,000 lbm/h (614 kg/s) for that short period [4]. On another occasion in 1986, the flow rate to the binary plant increased by 30% over the previous steady value when the neighboring flash plant was taken out of service for maintenance [7], clear evidence of interference between the production wells for the two plants. An aerial view of the plant is shown in Fig. 17.3.

One of the weaknesses of the design was that the parasitic power requirements were not proportional to the load, and remained high even when the turbine was running at 40–50% of its rated capacity. The cooling tower, for example, had to run at full capacity even under 50% power load. Another problem was related to the use of a supercritical pressure for the working fluid that raised the cycle internal pumping power without raising the net power relative to a subcritical pressure cycle [8]. Finally, toward the end of the plant's operation, the pressure needed to force the waste brine into the injection wells began to increase, possibly owing to scaling in the reinjection wells. This in turn increased the power to drive the reinjection pumps and further exacerbated the already high parasitic power demands.

A Second Law analysis of the results of the part-load operation [9] based on an EPRI report [7] showed that the plant should have been able to meet its design objectives if it had been supplied with the design brine flow rate. Unfortunately, the needed production wells were not completed as the schedule called for and the owner of the plant, SDG&E, decided to halt operations in June 1987 [4]. The plant was decommissioned, placed on a long-term storage status, and eventually dismantled.

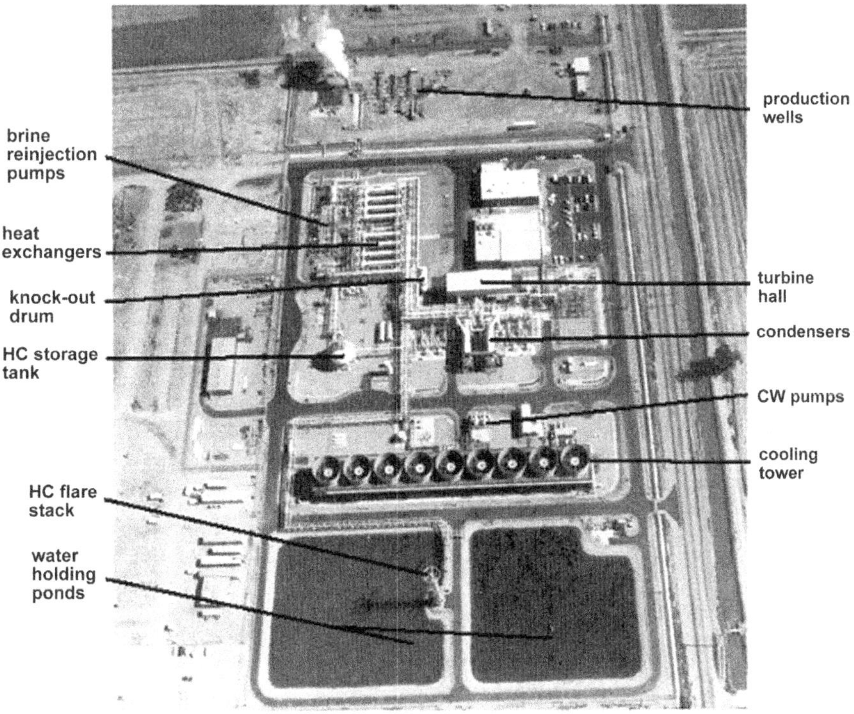

Fig. 17.3 Overview of the first Heber binary plant, looking north; after [4].

17.4 The second Heber binary plant

The second attempt to develop the Heber resource with binary technology began in 1992 when the Second Imperial Geothermal Company (SIGC) undertook to construct and operate a 48 MW(nameplate), 33 MW(net) modular-type binary plant [8]. The brine temperature had fallen to about 330°F (165°C) owing to the continued operation of the double-flash plant since July of 1985.

The development of binary plants for geothermal power generation saw two competing concepts: (1) the use of a large turbine-generator in a Rankine cycle, relying on the principle of economy of scale that applied in conventional fossil and nuclear power plants, vs. (2) the use of modular Rankine cycles wherein a number of small turbine-generator units would be connected to the brine in parallel to generate a reasonable amount of power. The failure of the HGC binary plant and several successful applications of the latter option led to the modular approach as the more effective one. By 1990, the Ormat companies had installed 68 small binary units in four power plants at the nearby East Mesa field that were generating about 55 MW. Furthermore, the technology of binary plants had advanced on a practical level to achieve higher efficiencies through more effective matching of brine cooling curves with working fluid heating curves. This experience was then applied to the Heber resource with the installation of the SIGC plant that was designed with six clusters of two-level power units.

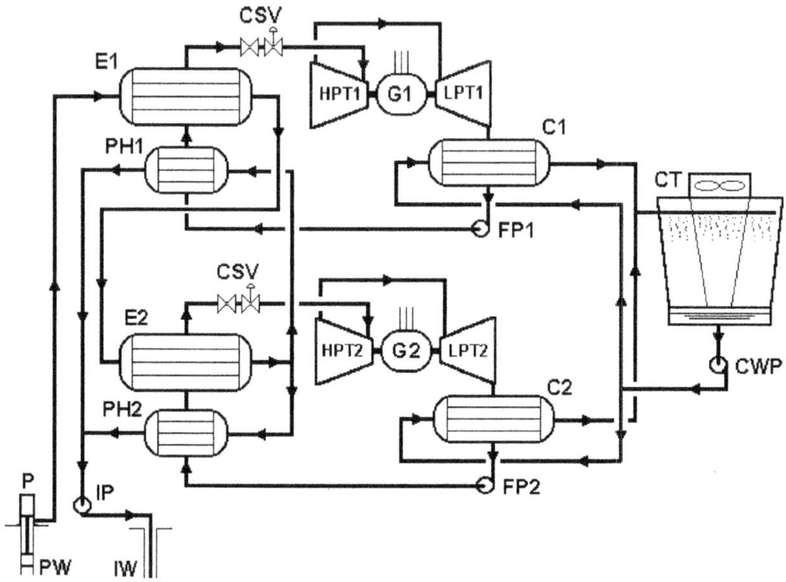

Fig. 17.4 Flow diagram for Ormat's Integrated Two-Level Unit.

The flow diagram for Ormat's Integrated Two-Level Unit (ITLU) is shown in Fig. 17.4. The Rankine cycle is designed with a subcritical boiler pressure using isopentane as the cycle working fluid. Two sets of turbine-generators are used for each modular cluster, the upper one, Level 1, being rated at 4.5 MW(gross) and the lower one, Level 2, at 3.5 MW(gross); the net power for the entire cluster is about 6.8 MW.

The SIGC Heber plant (now Heber 2) consists of six ITLUs and can deliver 33 MW of electricity, net. Figure 17.5 shows one of the six modular clusters [10].

Figure 17.6 is an aerial view of the configuration of the ITLUs, cooling towers, brine piping, and reinjection pumps.

The technical specifications and performance figures for the SIGC plant are given in Table 17.1. Note that the Second Law utilization efficiency is roughly the same as for a single flash plant, and that the First Law cycle thermal efficiency is the typical value for a simple binary plant. If this is compared to the cycle efficiency for Miravalles Unit 5 (see Chap. 16), it will be seen that the presence of the recuperator and the lack of reinjection pumps at Miravalles result in an increase in the net efficiency of about 3.5 percentage points, or a gain of over 30%.

The SIGC plant is served by 11 production wells having an average depth of 4000 ft (1219 m); each well is pumped by a motor-driven, line-shaft, centrifugal pump set at 1400 ft (427 m). The average flow rate from the wells is 1600 gpm or 722,650 lbm/h (91 kg/s). Only seven wells were needed to meet, and actually exceed by 13%, the design power output on startup in 1993 [8]. There are 13 reinjection wells, averaging 4500 ft (1370 m) in depth.

In the year 2000, after six years of continuous operation it was reported that the plant was online 99% of the time and averaged over 90% annual capacity factor [8].

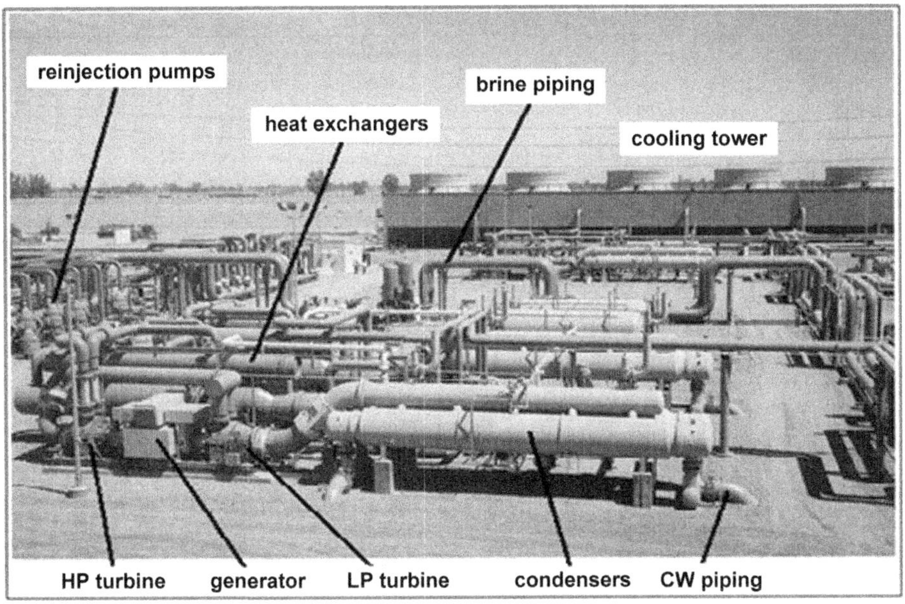

Fig. 17.5 A view of the SIGC binary plant at Heber. Photo courtesy of Ormat, Inc. [10] [WWW].

Fig. 17.6 SIGC Heber binary plant showing six clusters of ITLUs. Brine reinjection pumps are located in the upper center of the photo. The photograph in Fig. 17.5 was taken from the top of the left (southern) cooling tower. Photo courtesy of Ormat, Inc. [WWW].

This high reliability comes about partly through redundancy. If a component of some unit needs replacing, only that unit needs to be removed from service. The other units can receive some of the brine from the down unit, and can be over-driven for a short period of time to compensate for the lost power. Say 4 MW is lost from one Level 1 unit; each of the other 11 units needs to boost its output by only 365 kW to keep

Table 17.1 Technical specifications for SIGC Heber binary plant.

Start-up year	1993
Type	ITLU, dual-pressure cycle
No. of units(1)	12 (6 × 2)
Rating, MW	44
Net power, MW	33
Brine flow rate, lbm/h	1,000,000/ITLU
Working fluid	isopentane, $i\text{-}C_5H_{12}$
Preheaters:	
No. per ITLU	2
type	shell & tube
brine outlet temp., °F	154
heat duty, MW-th/ITLU	8.94 (L. 1), 9.44 (L. 2)
Evaporators:	
No. per ITLU	2
type	shell & tube
brine inlet temp., °F	330
heat duty, MW-th/ITLU	18.0 (L. 1), 15.7 (L. 2)
Condensers:	
No. per ITLU	4
type	shell & tube
heat duty, MW-th	22.7 (L. 1), 22.1 (L. 2)
CW temperatures., °F	
inlet	68.0
outlet	82.4

Turbine:	
type	axial inflow
cylinders/ITLU	2
flows/turbine	1
stages/turbine	2
inlet temperature, °C	n.a.
inlet pressure, psia:	225
speed, rpm	1800
Feed pumps:	
No. per ITLU	2
power input, kW	225 (L. 1), 120 (L. 2
Plant performance:	
net SGC, lbm/kW.h	181.8
utilization efficiency, %:	
gross	42.4
net	35.0
thermal efficiency, %:	
gross	12.7
net	10.5

(1)Throughout this book, a "unit" refers to a generator, whether it is driven by one or two turbines.

the whole plant at its rated output. In addition, the units are manufactured and largely assembled and tested in the factory, leaving only bolting and welding to be done on site. This means a high level of quality control can be maintained.

The current SIGC Heber binary plant (Heber 2) has achieved the success that was hoped for but not realized when the original Heber demonstration plant was built. The technical reasons for this outcome have been described in this chapter. One other important reason is that the power plants at both Heber sites and the Heber reservoir are now under ownership of a single company, which has led to a better understanding of the reservoir. The conceptual model and reservoir simulation have been improved, allowing for better placement of new production and injection wells that has greatly reduced the interference between the production wells for the binary plant and the double-flash plant.

This knowledge has also led to the construction of two 10 MW binary units, Gould II and Heber South, immediately to the west of the main plant; see Fig. 17.7.

To further optimize the performance of the entire Heber power complex, in 2006 a bottoming binary facility, Gould I, was added to the Heber 1 double-flash plant; see Fig. 17.8. The "full OEC" has two turbines driving a 5 MW generator; the "½ OEC" is a 2.5 MW unit; and the "TP unit" is a unique turbine-driven brine injection pump system; see Fig. 17.9. All four turbines operate in parallel using waste brine from the

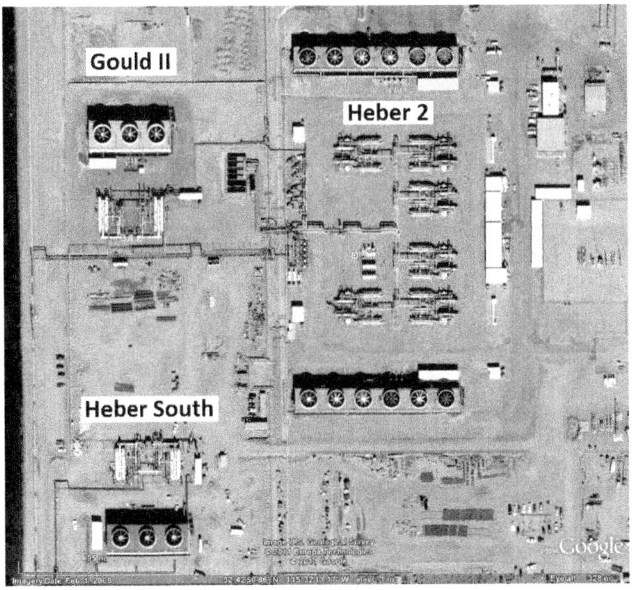

Fig. 17.7 *Heber 2 complex:* Google Earth *image, Feb. 1, 2008, labels by author [WWW].*

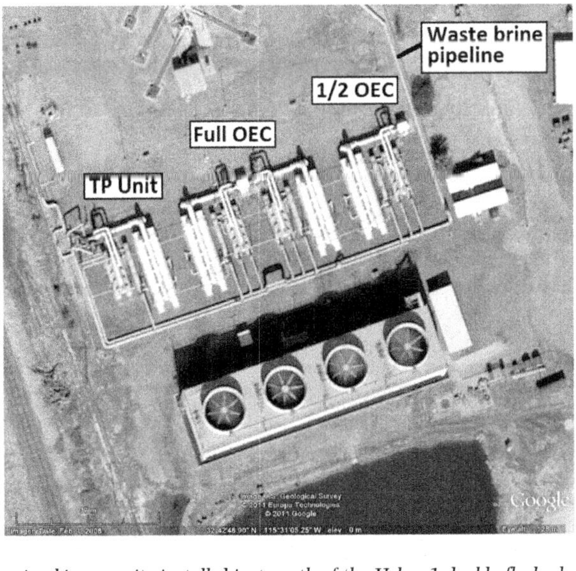

Fig. 17.8 *Gould I bottoming binary units installed just south of the Heber 1 double-flash plant:* Google Earth *image, Feb. 1, 2008, labels by author [WWW].*

Fig. 17.9 Turbine (center left) – injection pump (center right) unit at Gould I bottoming plant south of Heber 1 double-flash plant. Photo courtesy of Ormat.

flash vessels at Heber 1. As with the new OECs at Heber 2, all units employ water-cooled condensers in conjunction with cooling towers.

As of this writing (August 2011) the entire Heber complex has 110 MW installed using 19 units, separated by about one mile (1.6 km). The two plant sites occupy slightly over 0.4 km^2 with several outlying well pads each approximately 0.01 km^2 in area. The areal power density is roughly 275 MW/km^2 of surface area usage.

References

[1] DiPippo, R., "Geothermal Electric Power Production in the United States: A Survey and Update for 1990–1994," *Proc. 1995 World Geothermal Congress*, Int'l. Geothermal Assn., Vol. 1, 1995, pp. 353–362.

[2] Salveson, J.O. and A.M. Cooper, "Exploration and Development of the Heber Geothermal Field, Imperial Valley, California," *Geothermal Resources Council TRANSACTIONS*, V. 3, 1979, pp. 605–608.

[3] Tansev, E.O. and M.L. Wasserman, "Modeling the Heber Geothermal Reservoir," *Geothermal Resources Council TRANSACTIONS*, V. 2, 1978, pp. 645–648.

[4] Berning, J., J.E. Bigger and J. Fishbaugher, "Heber Binary-Cycle Geothermal Demonstration Power Plant: Half-Load Testing, Performance, and Thermodynamics," Special Report EPRI AP-5787-SR, Electric Power Research Institute, Palo Alto, CA, 1988.

[5] Gallagher, J.S., D. Linsky, G. Morrison and J.M.H. Levelt Sengers, *Thermodynamic Properties of a Geothermal Working Fluid; 90% Isobutane-10% Isopentane*, NBS Tech. Note 1234, National Bureau of Standards, U.S. Gov. Printing Office, Washington, DC, 1987.

[6] Nelson, T.T., "Heber Binary Project," *Geothermal Resources Council TRANSACTIONS*, V. 11, 1987, pp. 459–463.

[7] Berning, J., J.E. Bigger and J. Fishbaugher, "Heber Binary-Cycle Geothermal Demonstration Power Plant: Startup and Low-Power Testing," Special Report EPRI AP-5240-SR, Electric Power Research Institute, Palo Alto, CA, 1987.

[8] Sones, R. and Z. Krieger, "Case History of the Binary Power Plant Development at the Heber, California Geothermal Resource," *Proc. World Geothermal Congress 2000*, International Geothermal Association, pp. 2217–2219.

[9] DiPippo, R., "Heber Binary Demonstration Plant: A Second Law Assessment of Low-Power Tests," *Geothermal Hot Line*, V. 18, 1988, pp. 67–68.

[10] Anon., "Heber Second Imperial Geothermal, Imperial County, California, USA," Data Sheet, Ormat, Inc., Sparks, NV, undated.

Nomenclature for figures in Chapter 17

C	Condenser
CP	Condensate pump
CSV	Control & stop valves
CT	Cooling tower
CWP	Cooling water pump
E	Evaporator
FP	Feed pump
G	Generator
HPT	High-pressure turbine
HW	Hotwell
HXer	Heat exchanger
IP	Injection pump
IW	Injection well
KOD	Knock-out drum
LPT	Low-pressure turbine
P	Pump
PH	Preheater
PW	Production well
T	Turbine

Chapter 18

Magmamax Binary Power Plant, East Mesa, Imperial Valley, California, USA

"... *the greatest dividend I ever had in my life came the day I went into the geothermal steam business, not from the standpoint of money, but from the standpoint of mental interest.*"

Barkman C. "B.C." McCabe – 1969

18.1 Setting and exploration

The East Mesa geothermal field lies roughly 17 mi east of Heber in California's Imperial Valley, across the flat, irrigated fields of alfalfa. The East Mesa resource is about five miles east of the city of Holtville and one mile east of the East Highline Canal, on the unirrigated side in flat desert land. It stretches for four miles north-south, just north of Interstate 8, and covers an area of 6–8 mi^2.

Like the Heber field, East Mesa is a "hidden" geothermal resource, lacking any surface thermal manifestations; see Chap. 17. Interest in this area began in 1968 with the passage by the U.S. Congress of the Colorado River Basin Project Act (Public Law 90–537). The intent was to augment the water supply from the Colorado River that was under heavy pressure from agricultural interests for irrigation usage. The idea was to desalinate the geothermal brines that were known to exist in aquifers beneath sections of the Imperial Valley. The energy contained in the hot brines could be used to generate electricity that in turn could be used to pump water from the Salton Sea for reinjection into the geothermal reservoir and thereby provide replenishment of the extracted fluids and lower the level of the Salton Sea.

Geothermal Power Plants: Principles, Applications, Case Studies and Environmental Impact, Third Edition

© 2012 Elsevier Ltd. All rights reserved.

The U.S. Bureau of Reclamation (USBR) was given the task of making this a reality. It contracted for an exploration program with the University of California at Riverside that produced geophysical studies showing a well-defined thermal anomaly with temperature gradients exceeding 20°C/100 m, coinciding with a residual Bouguer gravity high of more than 5 mgal and a seismic ground-noise anomaly of 42 dB [1–3].

In the summer of 1972, the discovery well, Mesa 6–1, was drilled to 8030 ft (2450 m) into the center of the anomaly. The reservoir temperature was 399°F (204°C); the well could produce 222,000 lb/h (28 kg/s) at a wellhead pressure of 65 lbf/in^2 (4.5 bar), and the reservoir pressure was 3175 lbf/in^2 (219 bar). Four more wells were completed soon thereafter, and the USBR constructed a test facility at the site. In keeping with the aim of the law, the initial focus was on producing as much fluid as possible from the reservoir and desalinating it using two proposed processes: a multistage flash process and a vertical tube evaporation process [4]. Since the geofluid after desalination would be delivered off-site, the resource would become depleted unless make-up water could be found and reinjected. This led to tests of reinjection using the one of the five wells, Mesa 5–1, as the designated receiving well.

Concurrent with the desalination research, other tests were conducted on corrosion of materials exposed to the brines, on heat exchangers, and on a tiny mobile power plant – a helical rotor expander built by the Hydrothermal Power Company (see Sect. 9.5.3). The latter test produced 18 kW of electric power using part of the flow from well Mesa 6–1 [4]. All of this work was carried out in the late 1970s and would pave the way for the power plants that would be built in the coming years.

18.2 Magmamax binary power plant

The only geothermal binary power plant is existence in the 1970s was the tiny 670 kW unit at Paratunka on the Kamchatka peninsula in Russia that was put into operation in 1967 [5]. At that time, research into developing a practical binary plant that could use low-to-moderate-temperature resources was being carried out in the U.S., England, Poland, and Russia [6]. However up to 1970 there was little reported in the technical literature except for a few Russian articles cited in Ref. [6].

In March 1970, B.C. McCabe, Chairman and CEO of Magma Energy, Inc., a subsidiary of Magma Power Company, informed his stockholders of the company's intention to develop a new process for the utilization of liquid geothermal resources. He named it the "Magmamax" process and received U.S. Patent No. 3,757,516. Magma acquired Federal leases at East Mesa in 1974, conducted drilling and well testing, and in 1976 announced their intention to construct the Magmamax plant at a location about one mile SSW of the USBR Test Site. Two years later, construction got underway and the plant came online in September 1979 [7]. The flow diagram for the 12.5 MW(gross), 11 MW(net) power plant is shown in Fig. 18.1.

The unique features of the Magmamax design are as follows:

- Downwell pumps are used to prevent flashing.
- True counterflow heat exchangers are used to maximize heat transfer.
- Two Rankine cycles with different working fluids are used, one with isobutane and one with propane (shaded components in Fig. 18.1).

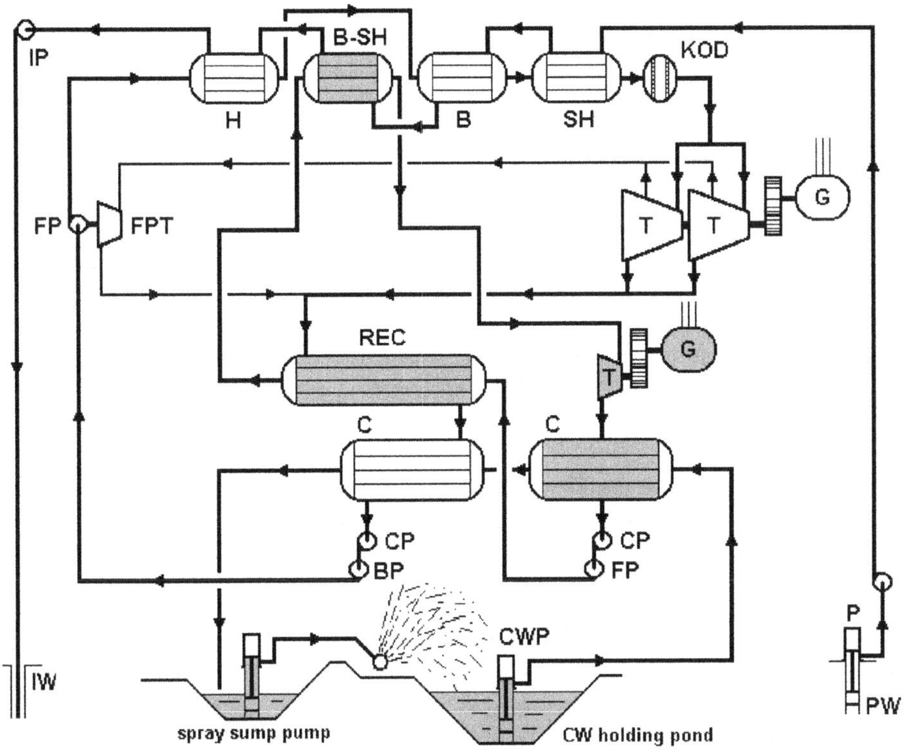

'ig. 18.1 Flow diagram for the original Magmamax binary plant.

- A heat recuperator couples the two cycles.
- 100% of the brine is reinjected.
- Cooling is achieved by spray evaporation.
- Ponds hold the cooling water for use at optimum times.

As a first-of-a-kind power plant, there was a great deal of engineering that needed o be done. The consulting firm of J. Hilbert Anderson was hired to perform the engi-ıeering for the plant [8]. Particular attention was paid to the heat exchangers that vould transfer the thermal energy of the geothermal brine to the cycle working fluid, ınd to the recuperator that coupled the two cycle working fluids. Since there were no ommercial hydrocarbon turbines for this type of service, they also had to be designed rom scratch. Furthermore, the novel cooling system was intended to take maximum ıdvantage of the desert climate to enhance the plant power output during times of ɔeak electrical demand [9,10].

The heat exchanger configuration was driven by the need to have easy access to he internals for purposes of cleaning the surfaces that might become fouled from scale leposition. The brine was placed on the shell side, with the hydrocarbons on the tube side. This somewhat unusual design was chosen because the pressure of the hydrocarbon working fluids was greater than that of the brine, allowing for lighter materials and lower cost. However, this meant that the tube bundles would have to be removed for cleaning the outside surfaces of scale deposited by the brine. Also since the isobutane was at a supercritical pressure, the heating and vaporization processes could best be achieved with the fluid confined within the tubes.

To achieve pure counterflow, the heat exchanger shells would be long, 80 ft (24 m) and slender, and there would be no baffles in the brine path. To enhance heat transfer, the tubes would have very thin walls, 35 mils or 0.035 in (0.0014 mm). To allow for thermal expansion, there would be a floating head at one end to allow the tube bundle to grow and shrink as the unit was subjected to varying temperatures. A floating balance piston, lubricated by oil, would prevent mixing of the brine and the hydrocarbon working fluids [11]. The units were specially engineered and constructed to specification, rather than "off-the-shelf." The heat exchangers are shown in Figs. 18.2 and 18.3.

The recuperator was a one-of-a-kind component that provided heating and boiling of the propane within a single tube pass. It supplied 75% of the total heat into the propane bottoming loop, the remaining 25% coming from a relatively small brine heat exchanger; see the shaded elements, B-SH and REC, in Fig. 18.1. The recuperator is shown in Fig. 18.4, along with the condensers.

The York Division of Borg-Warner Corporation designed and built the 10 MW isobutane turbine, a 2-cylinder, tandem-compound, 3-stage, radial-flow, single-extraction machine. This was essentially a compressor that had been modified for expansion duty. It was compact and had several advantages relative to steam turbines of the same rating. The extraction point was used to bleed off sufficient isobutane to drive an auxiliary turbine that was connected to the final boiler feed pump through a reduction

Fig. 18.2 Overview of heat exchangers, original Magmamax design. Brine-hydrocarbon heat exchangers are in center; recuperator is at upper left. June 1981, photo by author [WWW].

'ig. 18.3 Isobutane liquid heaters. Note vertical brine crossover pipes (shell side) and horizontal end-connectors arrying isobutane (tube side) from one heater to the next. June 1981, photo by author [WWW].

'ig. 18.4 Condensers and recuperator (upper vessel) at original plant. The two condensers on the near side are for he main isobutane cycle; the third one behind them served the propane cycle. June 1981, photo by author [WWW].

gear box. This was seen as another efficiency measure whereby internal power from the cycle would be used for this parasitic load, rather than using a motor-driven pump and purchasing electricity from the grid. In Fig. 18.5 it is seen awaiting installation at the East Mesa site; in Fig. 18.6 it is in operation.

As for the cooling system, the designers decided against a simple cooling tower and instead opted for a phased or batch cooling system with daytime storage and nighttime cooling via spray ponds. Daytime temperatures at East Mesa during the summer can reach 125°F (50°C), but clear nights can result in low temperatures due to radiational cooling. To capitalize on this, two daytime holding ponds were excavated into the desert soil to receive the spent cooling water from the condensers, and one deep stratified holding pond for the water that had been cooled overnight. The deepest layers in this pond held the coldest water, and the circulating pump was designed to draw water from different levels depending on where the coldest water happened to be. This approach was designed to reduce the parasitic power requirements associated with cooling water pumps and cooling tower fans.

Make-up water to account for the water lost to evaporation in the spray process was obtained from the nearby East Highline canal. The use of the heat recuperator had the added benefit of lowering the heat load on the condensers and therefore reduced the amount of heat that had to be rejected to the cooling water. The ponds are shown in Figs. 18.7 and 18.8.

Table 18.1 displays the design technical specifications for the Magmamax plant in its original configuration. Table 18.2 shows the cost of the plant by major category [10].

Fig. 18.5 Original turbine awaiting installation at East Mesa. The reduction gear is behind the turbine. March 1978, photo by author [WWW].

Fig. 18.6 Isobutane turbine in operation. June 1981, photo by author [WWW].

Fig. 18.7 Cooling water spray pond. October 1980, photo by author [WWW].

Fig. 18.8 Deep holding pond for cooling water with pump island. Note spray facility at right; cooled water drains by gravity to the holding pond. October 1980, photo by author [WWW].

18.3 Modified Magmamax binary power plant

After about two years of sporadic operation during which the propane bottoming cycle never was put into use, the plant was shut down for an extensive modification. The troubles began even before the plant went into full operation. Just around the time when the plant had been undergoing its original startup procedure, there was a severe earthquake centered in the Imperial Valley (October 19, 1979). Although there was no obvious external damage to the plant, the brine heat exchangers sustained internal damage to the joints between the tubes and the tubesheets. This only became clear after it was noticed that leaks were present that allowed brine to mix with the isobutane [10]. Upon inspection, it was seen that corrosion was present in the brine system and that the 35 mil-thick carbon steel tubes were undergoing corrosion at rates of between 5–25 mils/year [12]. The balance piston at the floating end of the heat exchangers had been damaged and also was leaking. There were also numerous problems related to excessive vibrations of various components, including the main turbine.

Following various attempts to patch the problems, it was decided to conduct a major overhaul. Magma hired Dow Engineering Company to carry out the modifications. What emerged in the fall of 1982 was a drastically different plant. Gone were the main turbine, the true counterflow heat exchangers, and the interconnected Rankine cycles. In their place, one found a simple, non-extraction, isobutane turbine from Mafi Trench, and nine larger brine heat exchangers (60 ft × 29 in) with the brine on the tube side and the hydrocarbons on the shell side. The isobutane needed to drive the boiler feedpump turbine came from a separate side-stream off the superheater.

able 18.1 Design specifications for the original Magmamax dual-fluid binary plant.

tart-up year	1979
ype	dual-fluid, recuperative
o. of units	2
ating, MW	12.5 (10 + 2.5)
et power, MW	11.2 (9 + 2.2)
rine flow rate, lbm/h	1,426,400
Vorking fluid 1	isobutane, $i-C_4H_{10}$
Vorking fluid 2	propane, C_3H_8
$-C_4H_{10}$ preheaters:	
No. per unit	6
type	shell/tube, counterflow
inlet pressure, psi,a	605
temperatures, °F:	
inlet	93
outlet	275
brine outlet temp., °F	180
$_4H_{10}$ evaporators:	
No. per unit	2
type	shell/tube, counterflow
$_4H_{10}$ superheaters:	
No. per unit	2
type	shell/tube, counterflow
$i-C_4H_{10}$ outlet temp., °F:	345
brine inlet temp., °F	360
$_3H_8$ superheater:	
No. per unit	1
inlet temperature, °F:	177
outlet temperature, °F:	205
$-C_4H_{10}$ turbine:	
type	radial inflow, extraction
cylinders/unit	2
flows/turbine	1
stages/flow	3
inlet temperature, °F:	345
outlet temperature, °F:	225
inlet pressure, psi, a:	500
outlet pressure, psi, a:	60
speed, rpm	6391
$_3H_8$ turbine:	
type	radial inflow
cylinders/unit	1
flows/turbine	1
stages/flow	3
inlet temperature, °F:	205
inlet pressure, psi, a:	460
outlet pressure, psi, a:	129
Recuperator:	
No. per unit	1
type	shell/tube, counterflow
hot-side temps., °F:	
$i-C_4H_{10}$ inlet	225
C_3H_8 outlet	177
cold-side temps., °F:	
C_3H_8 inlet	72.5
$i-C_4H_{10}$ outlet	150
$i-C_4H_{10}$ condensers:	
No. per unit	2
type	1-pass shell & 2-pass tube
inlet temperature, °F:	150
outlet temperature, °F:	87
C_3H_8 condensers:	
No. per unit	1
type	1-pass shell & 1-pass tube
inlet temperature, °F:	115
outlet temperature, °F:	72.5
Cooling water:	
temperatures, °F:	
inlet to C_3 cond.	62.0
outlet from C_3 cond.	65.5
inlet to $i-C_4$ cond.	65.5
outlet from $i-C_4$ cond.	79.5
flow rate, gal/min	25,000
$i-C_4H_{10}$ feed pumps:	
No. per unit	3
temperatures, °F:	
inlet	87
outlet	93
Plant performance:	
net SGC, (lbm/h)/kW	127.4
utilization efficiency, %:	
gross	46.1
net	41.3
thermal efficiency, %:	
gross	16.2
net	14.5

Table 18.2 Cost of the plant by major category [10].

Category	Total cost, 1980 $	Cost, $/net kW
Land improvements	531,700	47.50
Building	319,500	28.50
Cooling system	3,472,700	310.00
Power equipment	1,017,400	90.80
Isobutane system	2,382,800	212.75
Propane system	1,050,000	93.75
Brine system	2,026,100	180.90
General support system	1,565,100	139.75
Engineering	948,000	84.65
Overhead and A&G	1,413,900	126.25
TOTAL	**14,727,200**	**1314.90**

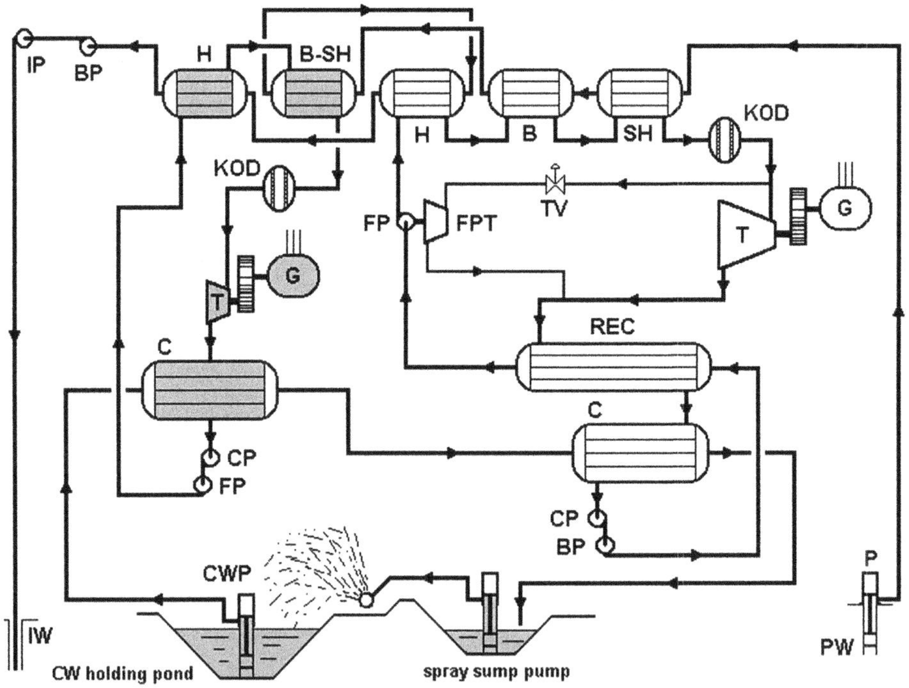

Fig. 18.9 Flow diagram for redesigned Magmamax binary plant.

A pressure-letdown (throttle) valve was inserted to lower the isobutane pressure to match the BFPT inlet pressure. The recuperator was still there but it now served only as a preheater for the isobutane cycle; the propane cycle was now independent of the isobutane cycle. Figure 18.9 is a flow diagram for the new Magmamax design; Figs. 18.10 and 18.11 show the reconfigured heat exchangers.

'ig. 18.10 Overview of redesigned Magmamax plant. Note new heat exchangers and the Y-connecting pipe (center) etween the single exhaust of the new turbine and the recuperator. October 1984, photo by author [WWW].

'ig. 18.11 Isobutane boiler-superheaters at redesigned plant. Note brine vertical connectors on the tube side; sobutane connectors (shell side) are seen between alternating sets of heat exchangers. October 1984, photo ɔy author [WWW].

Table 18.3 Specifications for redesigned Magmamax plant [7].

Item	Value
Mass flow rates:	
Brine	1,700,000 lbm/h
Isobutane	1,300,000 lbm/h
Propane	342,500 lbm/h
Brine temperatures:	
Inlet	350°F
Outlet	167°F
$i-C_4H_{10}$ turbine:	
Mass flow rate	1,150,000 lbm/h
Temperatures:	
Inlet	335°F
Outlet	209°F
Pressures:	
Inlet	565 psi, a
Outlet	70 psi, a
Generator output	10,500 kW
$i-C_4H_{10}$ boiler feed pump turbine:	
Mass flow rate	150,000 lbm/h
Temperatures:	
Inlet	285°F
Outlet	209°F
Pressures:	
Inlet	300 psi, a
Outlet	70 psi, a
Pumping power	1,120 kW
C_3H_8 turbine:	
Temperatures:	
Inlet	205°F
Outlet	120°F
Pressures:	
Inlet	480 psi, a
Outlet	155 psi, a
Generator output	2,000 kW
Cooling water:	
Volumetric flow rate	25,000 gal/min

A summary of the technical specifications for the redesigned plant is given in Table 18.3 [7]. Note that the brine inlet temperature is 10°F lower than the original plant. Also the brine is cooled to 167°F, 13°F lower than the original design. Together with the increased brine flow rate, the new plant receives 21% more heat from the brine. The new utilization efficiencies are: 41.1% gross, 36.8% net; the new thermal efficiencies are: 13.4% gross, 12.0% net. All of these values are significantly lower than the original design.

The new plant began its operating life in October 1982 and went through a 2-month period of adjustment and repair for relatively minor problems. Then the plant entered into a period of essentially continuous operation except for scheduled outages

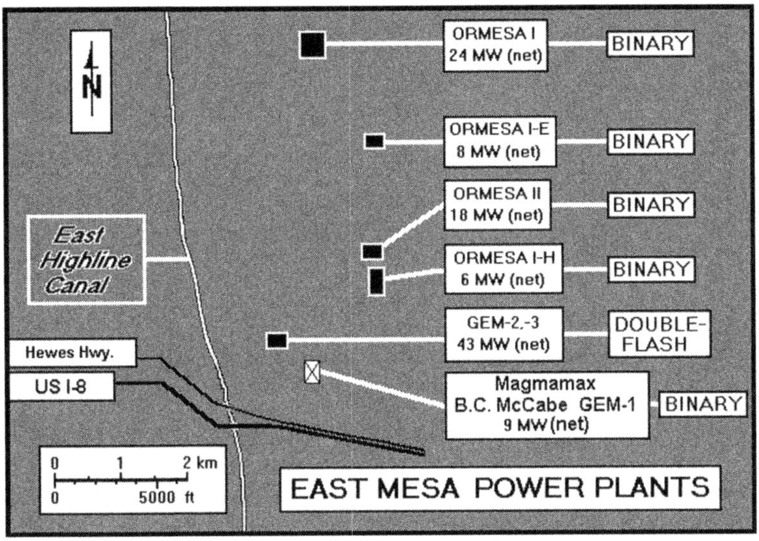

ig. 18.12 East Mesa geothermal power plants as of 2011; filled rectangles denote operating plants.

or maintenance and inspection. For the next 15 months, excluding scheduled utages, the plant averaged 96.8% online availability [7].

18.4 Conclusion

n 1983, Magma signed a 30-year power purchase agreement with Southern California Edison, whereby SCE would buy all the power that the Magmamax plant could generate. n honor of Magma's founder and geothermal pioneer, the plant was renamed the "B.C. McCabe Power Plant." In the early 1990s the plant was sold to GEO East Mesa L.P. and was renamed yet again to "GEM 1." Eventually the plant was taken over by Ormat as part of their other holdings in the East Mesa area and decommissioned and dismantled.

The downwell pumps that had been so problematic in the early years became very reliable and operated routinely for up to two years without trouble. This improvement n brine production was the key to further development of the moderate-temperature geothermal resources in East Mesa. Today, there are five geothermal power plants at East Mesa, all operated by Ormat; see Fig. 18.12. With the exception of the two double-flash units at GEM-2 and -3, the plants are modular binary units.

The technology that was brought to life by Magma in 1979 is now flourishing in he desert of East Mesa, generating about 90 MW in a reliable and environmentally riendly fashion.

References

[1] Combs, J., "Heat Flow and Geological Resource Estimates for the Imperial Valley," in *Cooperative Geological-Geophysical-Geochemical Investigations of Geothermal Resources in the Imperial Valley of California*, Educational Research Serv., Univ. of California at Riverside, 1971, pp. 5–27.

[2] Bielher, S., "Gravity Studies in the Imperial Valley," in *Cooperative Geological-Geophysical-Geochemica Investigations of Geothermal Resources in the Imperial Valley of California*, Educational Research Serv. Univ. of California at Riverside, 1971, pp. 29–41.

[3] Douze, E.J., "Seismic Ground-Noise Survey at the Mesa Anomaly," Teledyne-Geotech, Dallas, TX 1971.

[4] Anon., *Geothermal Resource Investigations: East Mesa Test Site, Imperial Valley, California*, Status Report U.S. Bureau of Reclamation, April 1977.

[5] DiPippo, R., *Geothermal Energy as a Source of Electricity: A Worldwide Survey of the Design and Operatio of Geothermal Power Plants*, U.S. Dept. of Energy, DOE/RA/28320–1. U.S. Gov. Printing Office Washington, DC, 1980.

[6] Moskvicheva, V.N. and A.E. Popov, "Geothermal Power Plant on the Paratunka River," *Geothermics – Special Issue 2, U.N. Symposium on the Development and Utilization of Geothermal Reseources*, Pisa, V. 2 Pt. 2, 1970, pp. 1567–1571.

[7] Hinrichs, T.C., "Magmamax Power Plant – Success at East Mesa," EPRI AP–3686, *Proceeding Eighth Annual Geothermal Conf. and Workshop*, Electric Power Research Institute, Palo Alto, CA, 1984 pp. 6-21–6-30.

[8] Anderson, J.H., "The Vapor-Turbine Cycle for Geothermal Power Generation," Chap. 8 in *Geotherma Energy: Resources, Production, Stimulation*, P. Kruger and C. Otte, Eds., Stanford University Press Stanford, CA, 1973.

[9] Dambly, B.W., "Heat Exchanger Design for Geothermal Power Plants," *Proc. 13th Intersociety Energy Conversion Engineering Conference*, V. 2, Society of Automotive Engineers, 1978, pp. 1102–1108.

[10] Hinrichs, T.C. and B.W. Dambly, "East Mesa Magmamax Power Process Geothermal Generating Plant: A Preliminary Analysis," EPRI TC–80–907, *Proceedings Fourth Annual Geothermal Conf. and Workshop*, Electric Power Research Institute, Palo Alto, CA, 1980, pp. 5-1–5-13.

[11] Pundyk, J.M., "Performance Evaluation of Magma 11.2 MWe Binary Plant," *Proceedings Third Annual Geothermal Conf. and Workshop*, Electric Power Research Institute, Palo Alto, CA, 1979 pp. 2-24–2-30.

[12] Shannon, D.W., "Monitoring the Chemistry and Materials of the Magma Geothermal Binary Cycle Generating Plant," *Geothermal Materials Review*, V. 4, No. 4, U.S. DOE/SF/11503–03, 1982.

Chapter 19

Nesjavellir and Hellisheidi Plants, Iceland

"What angered the gods, when the lava flowed, on which we are standing now?"
Snorri Godi, *Kristni Saga*

19.1 Introduction

Iceland, sitting astride the Mid-Atlantic Ridge with abundant active volcanism and a population of around 300,000, has the highest per capita concentration of geothermal energy of any country in the world. Furthermore most of the residents are clustered in the Reykjavik metropolitan area making it feasible to use geothermal energy to provide electricity and heat to nearly everyone living in Iceland.

Between geothermal and hydro, about 81% of Iceland's primary energy needs are being met with renewable sources. Geothermal contributes 62% of the total need. Space heating from geothermal hot water takes care of 89% of all the homes. In addition, the hot water is used in greenhouses, fish farming, snow melting, swimming pools, and various industries. The most recent estimate of direct heating (2009) is 7,300 GWh-th. The total installed geothermal electric power capacity stands at 573 MW with an annual generation of 4,400 GWh. The system has an overall capacity factor of 88%. Geothermally produced electricity accounts for 26% of the total in the country [1].

The full listing of the geothermal power plants in Iceland can be found in Appendix A.6.2. This chapter focuses on two power stations the lie close to Reykjavik and on opposite sides of the Hengill volcano, Nesjavellir and Hellisheidi.

Geothermal Power Plants: Principles, Applications, Case Studies and Environmental Impact, Third Edition.
© 2012 Elsevier Ltd. All rights reserved.

19.2 Geology and geosciences

Iceland is being pulled apart at an average rate of 2 cm per year as the American and European plates separate from each other. This places enormous tensile stress on the Earth's crust that underlies the island. The effects are manifest on the surface along wide bands of intense faulting that run continuously from the north to the south and then to the southwest. From Fig. 19.1 it may be seen that the major high-temperature geothermal fields lie within the youngest formations, but that low-temperature fields are abundant everywhere except along the eastern and southeastern coasts.

The Hengill volcano is about 30 km east of Reykjavik in the north branch of the Rift Zone. Three volcanic systems comprise the Hengill region; these have been active within the last 11,000 years, including the most recent eruption some 2000 years ago. The volcanic region covers an area of 112 km^2, one of the largest in all of Iceland.

The Nesjavellir and Hellisheidi power plants lie along the north-northeast trending fault zone, about 10.5 km apart; the Hengill volcano sits between them about 3.5 km from Nesjavellir. The surface expressions of many of the faults are clearly visible in the aerial image in Fig. 19.2.

A conceptual model of the Hengill geothermal field, Fig. 19.3, has been developed [3]. It shows that there are two apparent heat sources for the two power plants although both are related to the Hengill volcano. The model indicates a central upflow zone beneath the volcano and outflows to the southwest and northeast. On the Nesjavellir side, in some areas the 300°C isotherm rises to about 1000 m of the surface.

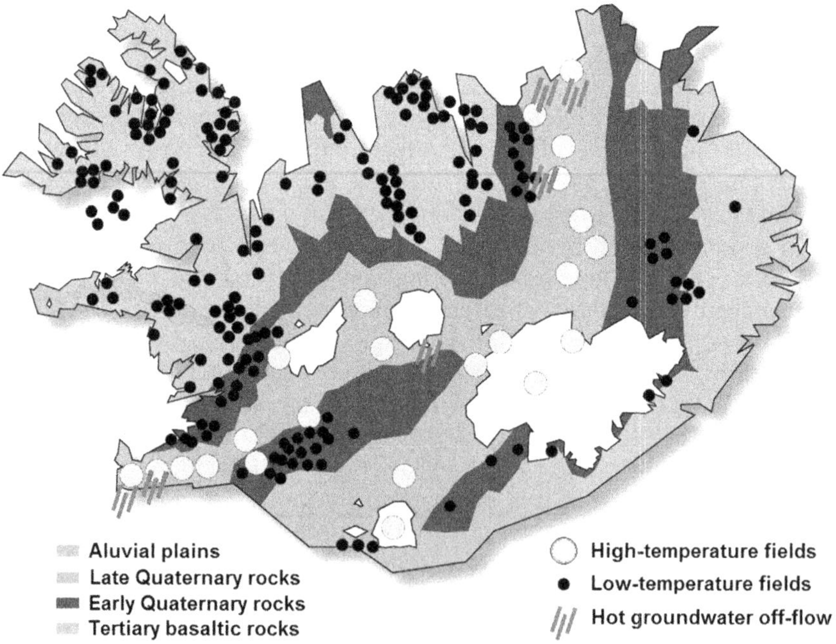

Fig. 19.1 Highly simplified geologic map of Iceland [2] [WWW].

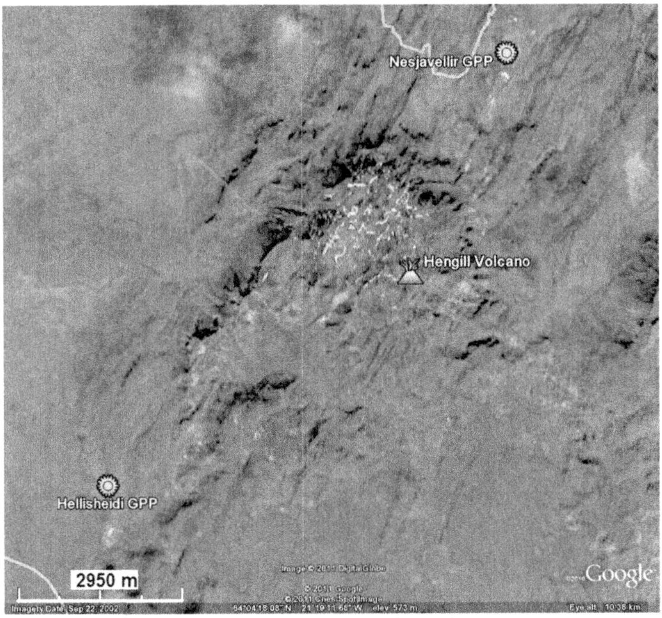

Fig. 19.2 *Aerial view of Hengill geothermal area,* Google Earth *image, Sept. 22, 2002 [WWW].*

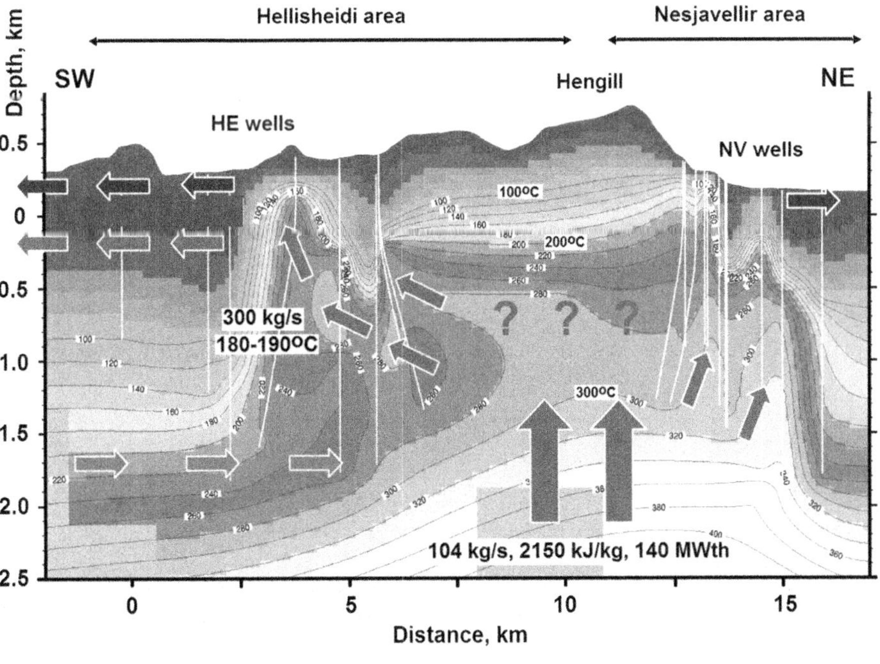

Fig. 19.3 *Conceptual model of Hengill geothermal formation, after [3] [WWW].*

Fig. 19.4 depicts a cross-section along the direction of the fault zone at Nesjavellir [3 showing a multilayered formation with alternating hyaloclastites and lavas consisting of basaltic dykes. The Hengill system has an estimated age of 300,000–400,000 years The basaltic intrusions and shallow-dipping diorite intrusions contribute to the permeability of the reservoir. There is a similar pattern in the Hellisheidi area but the volcanics are found at a somewhat deeper depth.

The faults are nearly vertical throughout the area, necessitating directionally-drilled wells to intercept as many faults as possible. The wellhead locations at the Hellisheidi and Nesjavellir fields are shown in Fig. 19.5, along with wells in two other nearby

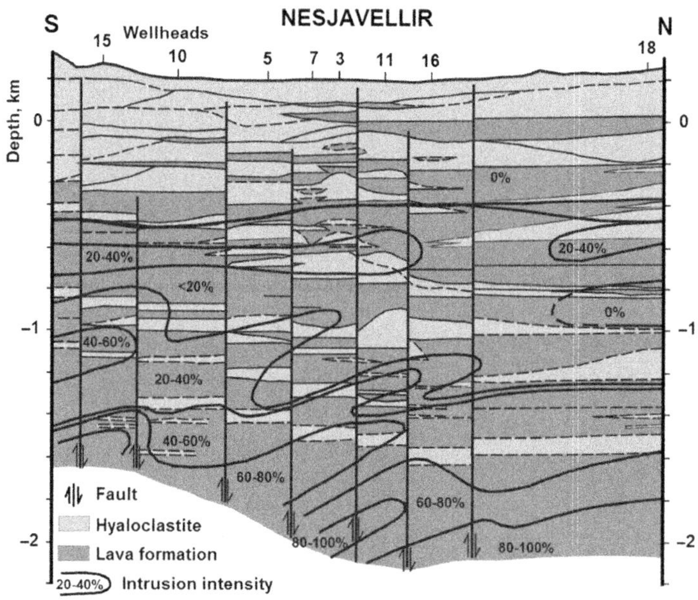

Fig. 19.4 Lithology of the Nesjavellir reservoir, after [3].

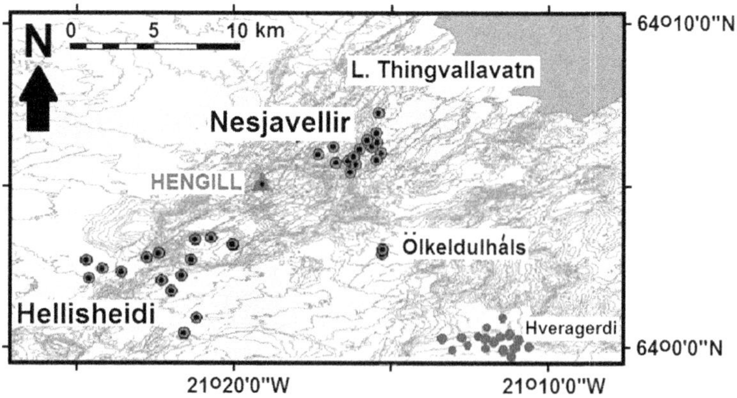

Fig. 19.5 Wellhead locations in Hengill area, after [4] [WWW].

reas. There is no power plant at Hveragerdi but there are numerous wells that are ısed for other purposes in the town. This area is no longer volcanically active but hermal manifestations are present, and there could be sufficient stored thermal energy o support a new power plant.

9.3 Nesjavellir power plant

)riginally the Nesjavellir power plant was built to supply hot water to Reykjavik. ;everal stages of development followed that included incremental installations of ›ower turbines and an increase in the hot water flow to Reykjavik. Studies began in 947 but drilling started at Nesjavellir in 1965 along with a series of geologic, geo-hemical, and geophysical surveys of the wide area surrounding Hengill. The plant ›egan supplying 100 MW-th of hot water to Reykjavik in 1990 [5]. The route of the ›ipeline from Nesjavellir to Reykjavik is shown in Fig. 19.6.

The first generation of electricity began in 1998 with the installation of two 0 MW turbines. The plant's power capacity was raised to 90 MW in 2001 and to 20 MW in 2005 by adding 30 MW turbines to reach the current state of the plant. 'he hot water rate also increased in increments over this period, as shown in 'able 19.1.

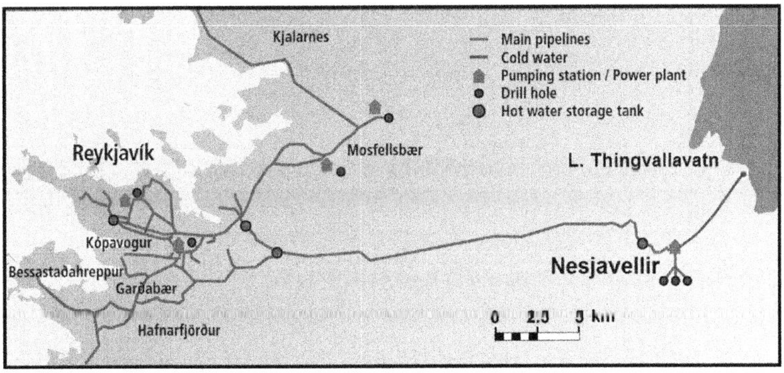

'ig. 19.6 Pipeline route from Nesjavellir to Reykjavik, after [2] [WWW].

'able 19.1 History of development at Nesjavellir [6].

'ear	Hot water		Electric power
	L/s	MW-th	MW
1990	560	100	—
1991	840	150	—
1998	1120	200	60 (2 × 30)
2001	1120	200	90 (3 × 30)
2003	1640	290	90 (3 × 30)
2005	1640	290	120 (4 × 30)

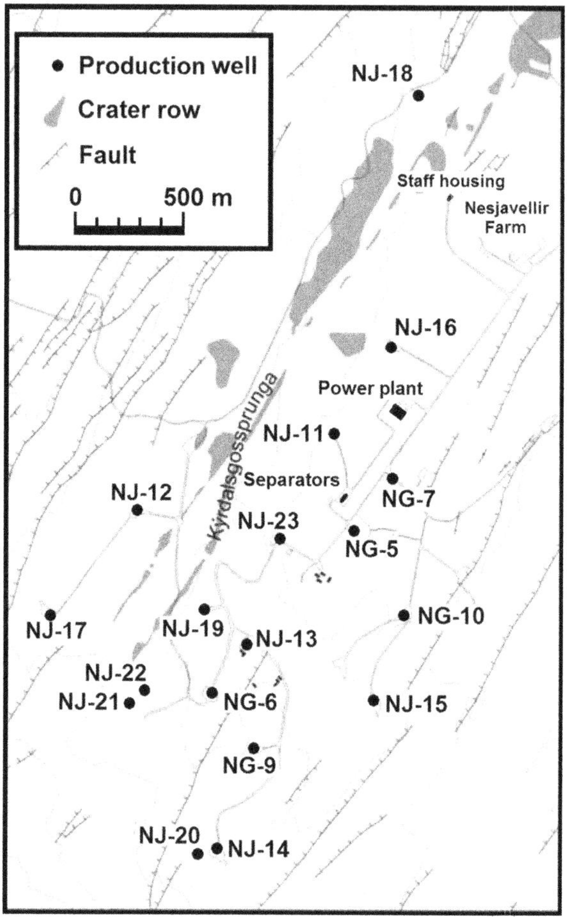

Fig. 19.7 Location map of wellheads at Nesjavellir, after [7].

Since the heating demand varies by season, the plant is designed to be flexible to accommodate these seasonal swings in demand. When excess steam is available, the plant can generate more electricity, and if the heating demand exceeds the normal value, steam can be diverted from the turbines to the heating plant.

Figure 19.7 shows the locations of the wells in the Nesjavellir area. The combined heat-power plant operates as follows; see Fig. 19.8. Geothermal fluid is produced from wells and fresh, cold water is pumped from wells near Lake Thingvallavatn. The geofluid is sent to the plant via 2-phase, liquid-vapor pipelines. A geofluid header receives the fluid and passes it to a set of horizontal separators, the inlet ends being outside and the outlet ends being inside the separator building. The steam is sent to the turbine hall, passing through a set of horizontal moisture removers before entering the power house. The steam drives the turbine-generators and discharges axially into the steam condenser. The latter is an element in the heating plant to raise the fresh water temperature to a sufficient level to make the trip to Reykjavik and to serve the heating

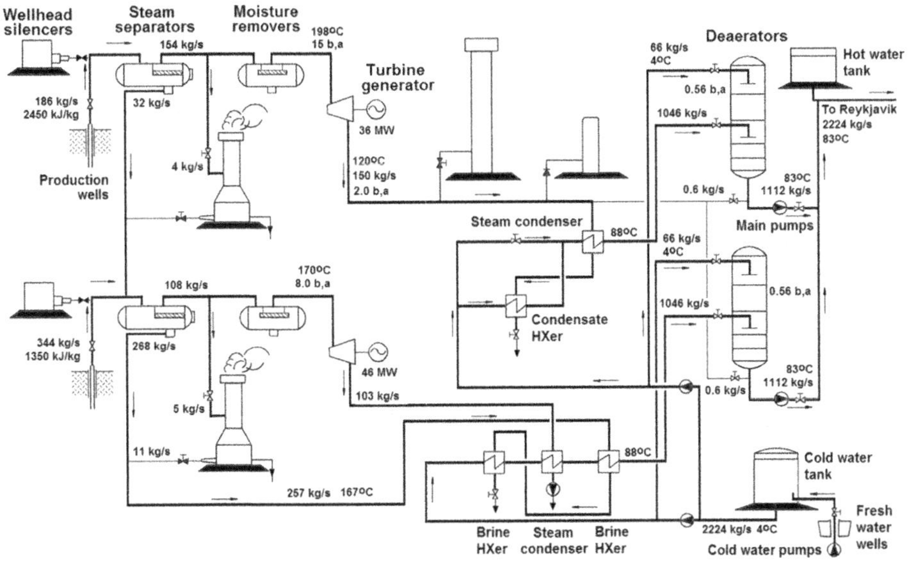

'ig. 19.8 *Simplified process flow diagram for Nesjavellir cogeneration Units 1 and 2, after [6].*

ıeeds of the city. There is another heat exchanger downstream of the condenser to ubcool the condensate and preheat the fresh water. Then the steam condensate is ›ither reinjected or sent to a drainage channel.

The cold water is pumped into a storage tank from whence it is divided into two low paths; one goes to the aforementioned steam heat exchangers and one goes to a et of brine heat exchangers. The brine that is separated from the steam in the separa-ors is sent to a set of heat exchangers that raise the cold water temperature to the lesired level, with the help of the steam condenser from the turbine of the second unit. ›ince the geofluid characteristics differ from well to well, the two units are not identi-:al in term of flow rates and temperatures, as can be seen from Fig. 19.8, and thus nust be carefully controlled to maintain proper and stable conditions. The pressure egulation steam stacks are essential to keeping the plant running smoothly, even hough it may appear to be wasteful of steam.

Once the cold water is heated to the required 88°C, it must be processed to remove all lissolved oxygen which is highly corrosive in hot water. This is accomplished in deaera-ors by means of a vacuum flash process that produces a small amount of steam and eleases the oxygen and other NCGs. These are vented (not shown in Fig. 19.8) to the ıtmosphere while the steam is condensed by a very small flow rate of cold water that is ›prayed into the top of the deaerator. The last step occurs at the bottom of the deaerator vhere a tiny flow of steam bled from the turbine exhaust is fed into the hot water just ›efore it leaves the plant for the hot water storage tank. This removes any residual oxy-gen that might have survived the deaeration process by reacting it with the hydrogen ›ulfide (H_2S) in the steam, adjusts the pH of the hot water to about 8.5, and imparts a 'hot spring" smell to the water, which is evident to anyone who takes a shower in Reykjavik or any other place that receives geothermally heated water for domestic uses.

It may be noted from Fig. 19.8 that the turbines are shown with power ratings higher than their nominal 30 MW ratings. By oversizing the turbine-generators, the plant can take advantage of those times when the heating demand is low and produce excess electricity which is always in demand. Also there are no cooling towers shown because the cold water drawn from fresh water wells serves the purpose of cooling water in the steam condensers. The plant does include a 4-cell cooling tower as a backup system; this is much smaller than a 120 MW geothermal plant normally would require.

The technical specifications for the turbine, condenser, NCG extractor, and cooling tower may be found in Tables 19.2–19.5, respectively [8]. A site photo with captions identifying the major items is given in Fig. 19.9; a scale layout of the plant is shown in Fig. 19.10 [9] and the layout of one turbine-generator in the power house is shown in Fig. 19.11 [9].

Table 19.2 Turbine specifications for Nesjavellir Units 1–4 [8].

Type	single-cylinder, single-flow, reaction blading axial exhaust
Manufacturer	Mitsubishi Heavy Industries (MHI)
Rated capacity	30.0 MW
Maximum capacity	31.5 MW
No. of stages	8
Speed	3,000 rev/min
Steam inlet pressure	1197.5 kPa (173.6 lbf/in^2,a)
Steam inlet temperature	188°C (370°F)
Exhaust pressure	21.5 kPa (3.12 lbf/in^2,a)
Steam mass flow rate	57.25 kg/s (454,390 lbm/h)
Last stage blade height	609.6 mm (24 in)

Table 19.3 Condenser specifications for Nesjavellir Units 1–4 [8].

	Units 1–3	Unit 4
Type	surface, shell-and-tube	surface, shell-and-tube
Manufacturer	MHI	MHI
Shell pressure	20.0 kPa (2.90 lbf/in^2,a)	10.0 kPa (1.45 lbf/in^2,a)
Cooling water inlet temp.	4°C (39.2°F)	20°C (68°F) + 10°C (50°F)
Cooling water outlet temp.	57°C (134.6°F)	39°C (102.2°F)
Cooling water flow rate	574.4 kg/s	1,268 kg/s + 99.26 kg/s

Table 19.4 Gas extractor specifications for Nesjavellir Units 1–4 [8].

	Units 1–3	Unit 4
Type	vacuum pumps	vacuum pumps
Manufacturer	MHI	MHI
No. of sets	1 set (each 50%)	1 set (each 33%)
Suction pressure	19.0 kPa (2.76 lbf/in^2,a)	9.32 kPa (1.35 lbf/in^2,a)
Discharge pressure	99.14 kPa (14.4 lbf/in^2,a)	99.14 kPa (14.4 lbf/in^2,a)
NCG capacity	18,530 m^3/h (10,902 ft^3/min)	27,795 m^3/h (16,353 ft^3/min)
Power consumption	300 kW (150 kW × 2)	450 kW (150 kW x 3)

able 19.5 Cooling tower specifications for Nesjavellir plant [8].

ype	mechanical draft, counterflow
Ianufacturer	MHI
'o. of cells	4
Varm water temp.	46°C (114.8°F)
old water temp.	20°C (68°F)
esign wet-bulb temp.	12°C (53.6°F)
an type	vertical axial

'ig. 19.9 Nesjavellir plant with captions, modified from: http://en.wikipedia.org/wiki/ 'ile:NesjavellirPowerPlant_edit2.jpg [WWW].

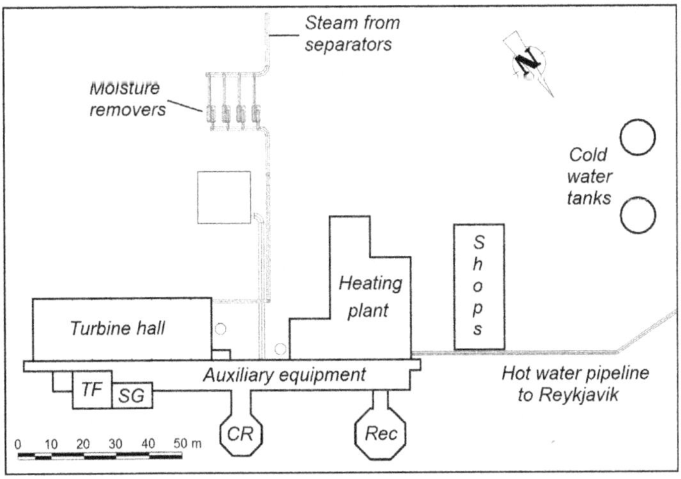

'ig. 19.10 Nesjavellir power station layout, simplified from [9].

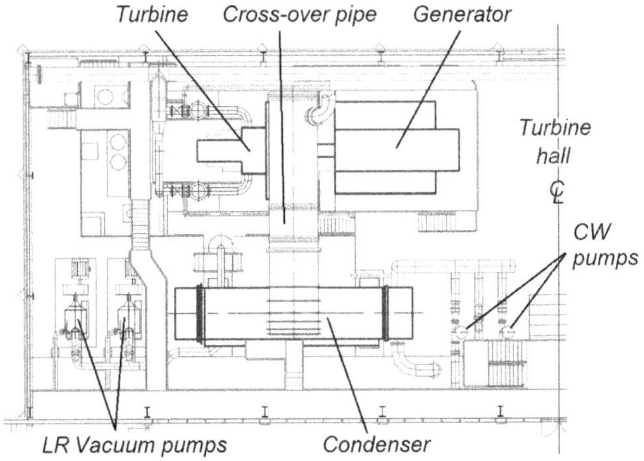

Fig. 19.11 Nesjavellir turbine hall arrangement (one unit), simplified from [9] [WWW].

19.4 Hellisheidi power plant

Once the Nesjavellir plant was brought to its full level of exploitation in 2005, attention turned to the other side of the Hengill volcano, the Hellisheidi field. As it happened, this field proved to be even larger in capacity than Nesjavellir.

As can be seen from Fig. 19.3, the two reservoirs share similar characteristics. The wells at Hellisheidi produce geofluids that are somewhat lower in temperature but still high enough to support several moderate sized power plants. Unlike Nesjavellir, Hellisheidi was first developed as an electric generating station and then later became a cogeneration, heat and power plant. Table 19.6 shows the timeline of development.

A total of 57 wells have been drilled to depths of 2,000–3,000 m in support of the fully developed plant [11]. All the units are owned and operated by Orkuveita Reykjavikur (Reykjavik Energy). Whereas the hot water is sent to Reykjavik, the electricity is mainly sent to nearby aluminum refineries.

The first two power plants are single-flash type; see Table 19.7 for details on the turbine. Given the large volume of hot water that is available from the separators, a bottoming flash plant was built to capitalize on this source of energy or exergy. The process flow diagram and some state-point properties for this unit are given in Fig. 19.12.

The processes for this plant are shown in Fig. 19.13; the state points are identified in the plant flow diagram. State A represents the high-pressure steam from the inlet to the turbines for Units 1 and 2 that is throttled down to the LP turbine inlet pressure of 2 bar,a. This slightly superheated steam (state 4) is mixed with the steam from the flash separator (state 2) to form the inlet steam condition for the LP turbine (state 5). State 6s is the ideal isentropic turbine outlet point and state g represents the saturated vapor at the condenser pressure.

The results are shown in Table 19.8 where the calculated values were obtained using REFPROP and the specifications cited in Ref. [12]. Note the slight differences

able 19.6 History of development at Hellisheidi [10].

hase	Year	Hot water MW-th (total)	Electric power MW (total)
	2006	—	90 (2 × 45)
	2007	—	123 (2 × 45 + 1 × 33)
	2008	—	213 (4 × 45 + 1 × 33)
	2010	133	213 (4 × 45 + 1 × 33)
	2011	400	303 (6 × 45 + 1 × 33)

able 19.7 Turbine specifications for Hellisheidi Units 1-2 [8].

ype	single-cylinder, single-flow, impulse blading, axial exhaust
anufacturer	Mitsubishi Heavy Industries (MHI)
ated capacity	40.0 MW
aximum capacity	47.0 MW
o. of stages	6
peed	3,000 rev/min
eam inlet pressure	750.3 kPa (108.8 lbf/in^2,a)
eam inlet temperature	167.8°C (334°F)
xhaust pressure	9.8-21.6 kPa (1.42-3.13 lbf/in^2,a)
eam mass flow rate	75.0 kg/s (595,242 lbm/h)
ast stage blade height	762 mm (30 in)

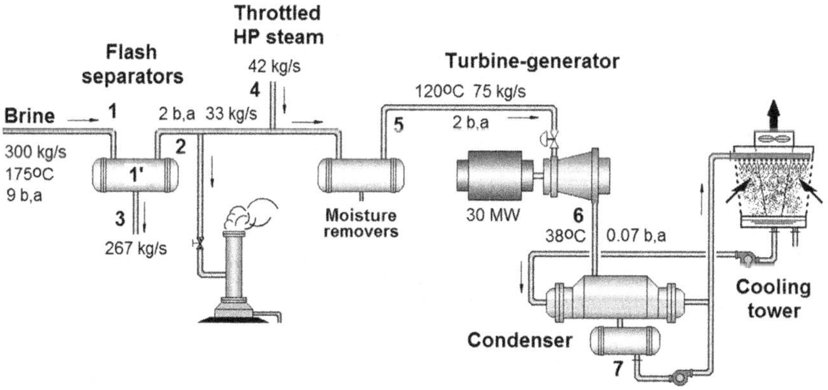

g. 19.12 *Process flow diagram for Hellisheidi LP-flash plant, after [12].*

etween some of the quantities. The Baumann rule was used to find the actual turbine utlet state assuming a dry expansion efficiency of 85%. The actual turbine efficiency is bout 81%, the outlet quality is 90%, and the calculated turbine power is 30,082 kW.

The Second Law utilization efficiency of the LP unit is simply the ratio of the power utput to the input exergy of the brine and steam. The steam condition should be aken as the high-pressure state before throttling since that steam could have been sed in a HP turbine. The dead state was taken at the design wet-bulb temperature of

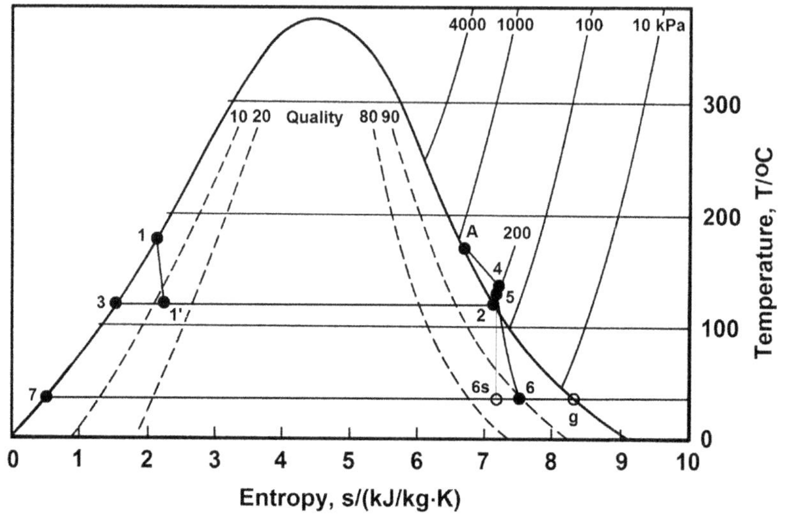

Fig. 19.13 Temperature-entropy process diagram for the Hellisheidi LP-flash plant.

Table 19.8 State-point analysis for the Hellisheidi LP-flash plant.

State	Pressure kPa	Temperature °C	Entropy kJ/kg · K	Enthalpy kJ/kg	Quality —	Mass flow kg/s
1	900	175	2.0940	742.56	0	300
1'	200	120	2.1349	742.56	0.1080	300
2	200	120	7.1269	2706.2	1	32.4
3	200	120	1.5302	504.70	0	267.6
A	750	168	6.6836	2765.6	1	42
4	200	148	7.2727	2765.6	SH	42
5	200	136	7.2104	2739.7	SH	74.4
6s	7		7.2104	2239.6	0.8621	
6	7	39	7.5177	2335.5	0.9019	74.4
7	7	39	0.5590	163.35	0	74.4
g	7	39	8.2745	2571.7	1	

the neighboring Nesjavellir plant, namely, 12°C or 285.15 K and the local atmosphere pressure was taken as 98.2 kPa, corresponding to the plant elevation of 260 masl. The inlet brine exergy rate is 43,934 kW, the inlet steam exergy rate is 36,152 kW, with a total inlet exergy of 80,086 kW; thus the utilization efficiency comes to 37.6%.

Lastly, in a 6−12 month experiment to demonstrate carbon sequestration, the Hellisheidi reservoir will receive aqueous CO_2 (a kind of seltzer water) via injection wells to learn whether the CO_2 can be reacted with the basalt to form solid limestone as a permanent means of disposing of unwanted CO_2 [13]. A similar effect was observed at the EGS test site at Ogachi in Japan when CO_2 was used as the heat transfer fluid in the reservoir; see Sect. 22.3.2.

References

[1] Ragnarsson, Á., "Geothermal Development in Iceland 2005–2009," *Proc. World Geothermal Congress 2010*, Paper 0124, Bali, Indonesia, April 25–29, 2010.

[2] "Nesjavellir Power Plant," Orkuveita Reykjavíkur (Reykjavik Energy), 2006.

[3] Franzson, H., E. Gunnlaugsson, K. Árnason, K. Sæmundsson, B. Steingrímsson and B.S. Harðarson, "The Hengill Geothermal System, Conceptual Model and Thermal Evolution," *Proc. World Geothermal Congress 2010*, Paper 1177, Bali, Indonesia, April 25–29, 2010.

[4] Mutonga, M.W., A. Sveinbjornsdottir, G. Gislason and H. Amannsson, "The Isotopic and Chemical Characteristics of Geothermal Fluids in Hengill Area, SW-Iceland (Hellisheidi, Hveragerdi and Nesjavellir Fields)," *Proc. World Geothermal Congress 2010*, Paper 1434, Bali, Indonesia, April 25–29, 2010.

[5] Zarandi, S.S.M.M. and G. Ivarsson, "A Review on Waste Water Disposal at the Nesjavellir Geothermal Power Plant," *Proc. World Geothermal Congress 2010*, Paper 0242, Bali, Indonesia, April 25–29, 2010.

[6] Gunnarsson, A., B.S. Steingrimsson, E. Gunnlaugsson, J. Magnusson and R. Maack, "Nesjavellir geothermal co-generation power plant," *Geothermics*, V. 21, 1992, pp. 559–583.

[7] Gíslason, G., G. Ívarsson, E. Gunnlaugsson, A. Hjartarson, G. Björnsson and B. Steingrímsson, "Production Monitoring as a Tool for Field Development: A Case History from the Nesjavellir Field, Iceland," *Proc. World Geothermal Congress 2005*, Paper 2415, Antalya, Turkey, April 24–29, 2005.

[8] Mitsubishi Heavy Industries, Ltd., "List of Geothermal Power Plants," August 2005.

[9] Ballzus, C., H. Frimannson, G.I. Gunnarsson and I. Hrolfsson, "The Geothermal Power Plant at Nesjavellir, Iceland," *Proc. World Geothermal Congress 2000*, Kyushu – Tohoku, Japan, May 28–June 10, 2000, pp. 3109–3114.

[10] Bertani, R., "Geothermal Power Generation in the World 2005–2010 Update Report," *Proc. World Geothermal Congress 2010*, Paper 0008, Bali, Indonesia, April 25–29, 2010.

[11] Blog at WordPress, "Iceland: Hellisheidarvirkjun: Hellisheidi Geothermal Power Plant," February 15, 2010: http://systemsthatseep.wordpress.com/2010/02/15/iceland-hellisheidarvirkjun-hellisheidi-geothermal-power-plant/.

[12] Kjartansson, G., "Low Pressure Flash-Steam Cycle at Hellisheidi –Selection Based on Comparison Study of Power Cycles, Utilizing Geothermal Brine," *Proc. World Geothermal Congress 2010*, Paper 2643 , Bali, Indonesia, April 25–29, 2010.

[13] *IceNews – News from the Nordics*, "Icelandic Carbon Storage Experiment About to Begin," 29 August 2011: http://www.icenews.is/index.php/2011/08/29/icelandic-carbon-storage-experiment-about-to-begin/#ixzz1Wdo484Xr.

Chapter 20

Raft River Plants, Idaho, USA

"A pessimist sees the difficulty in every opportunity; an optimist sees the opportunity in every difficulty."

Sir Winston Churchill

0.1 Introduction

he Raft River geothermal field in southeastern Idaho has been the site of two power lant developments. The original nominal 5 MW project was supported by the United tates Department of Energy (USDOE), and the other, a nominal 10 MW plant, is a rivate project of U.S. Geothermal, Inc. The contrast between these two efforts is illumi-ating. They represent two approaches to exploitation of the same geothermal esource.

The chapter begins with a description of the geologic setting that is common to the vo projects, and includes descriptions of the means of geofluid production, energy onversion systems, cooling systems, power plant performance, environmental impact, nd economics for both plants.

2012 Elsevier Ltd. All rights reserved.

20.2 Geology and geosciences

The Raft River of southern Idaho flows directly through the Raft River geothermal field in a generally northeast direction. Figure 20.1 shows the location of the field and includes an aerial view of the site from *Google Earth* as of June 24, 2009. The reservoir lies in a basin bounded on the west by low-angle faults and on the east by secondary faulting associated with the Black Pine Mountains; see Fig. 20.2. The basement rocks

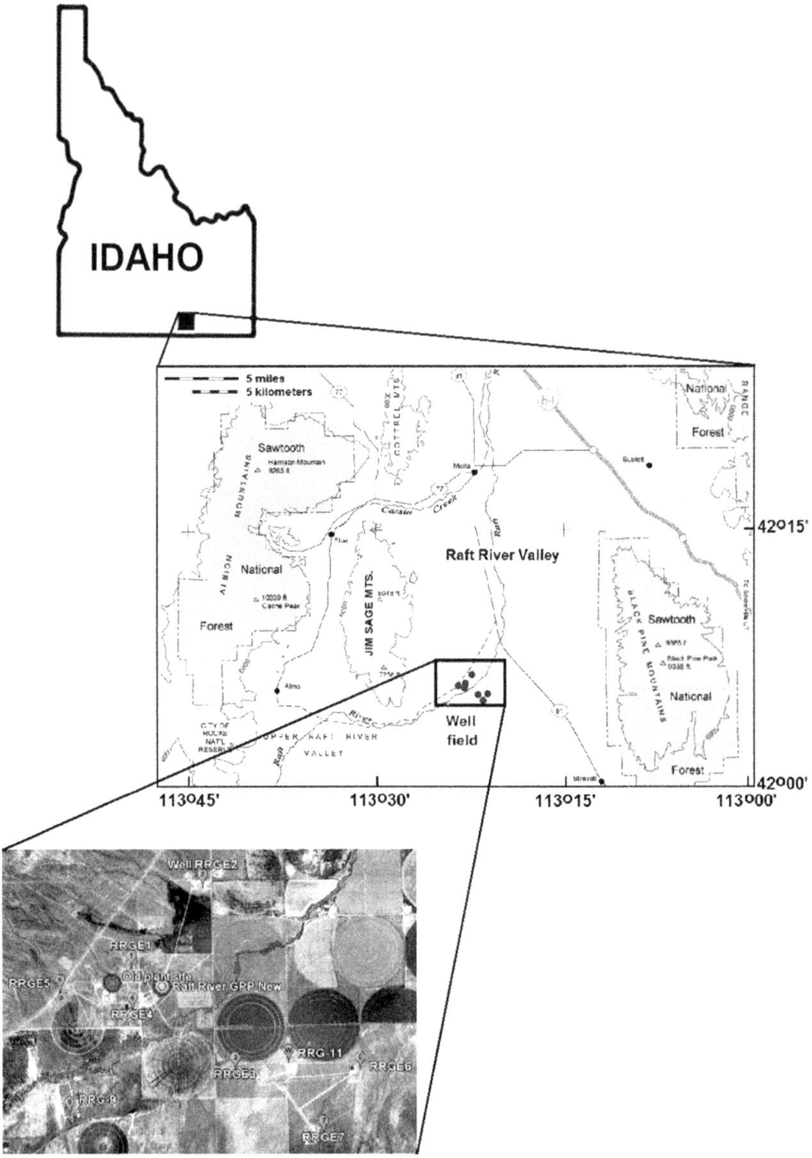

Fig. 20.1 Location of the Raft River geothermal field [1, 2, 3] [WWW].

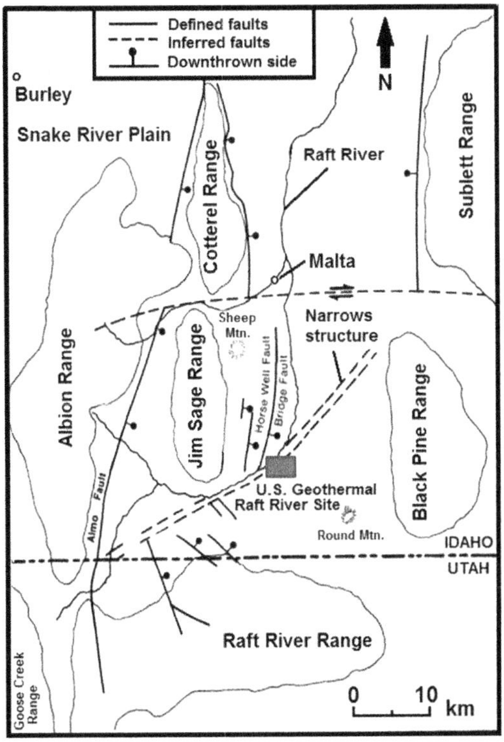

'ig. 20.2 Mountain ranges and fault mapping surrounding the Raft River area, after [4].

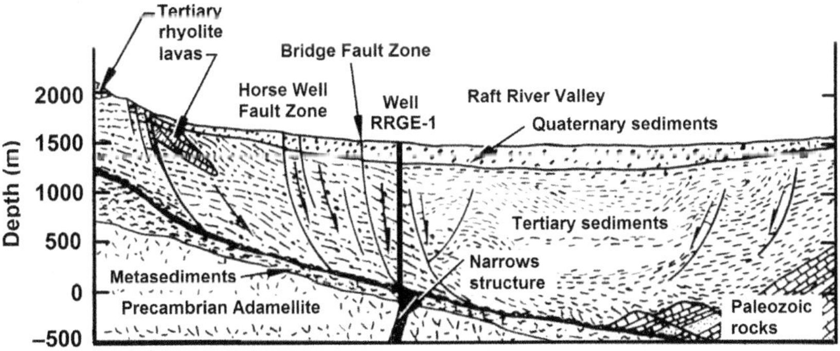

'ig. 20.3 East-west cross-section through the formation, from 1982, after [5].

ie about 1600 m (5250 ft) deep and are traversed at depth by a northeast-trending ineament known as the Narrows Structure, thought to be caused by basement shear.

Figures 20.3 and 20.4 illustrate the reservoir concept; the former was presented in 1982 and the latter in 2011. Geofluid production occurs at the intersection of the 3ridge Fault zone and the Narrows Structure. The hot fluid was originally thought to 'ise along the Narrows Structure and then spread through the sedimentary formations

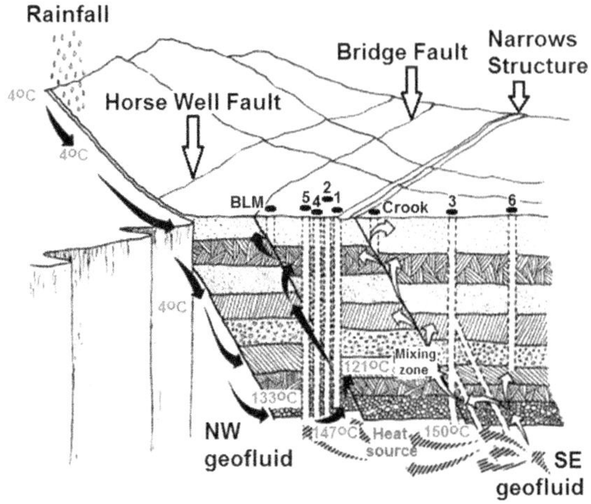

Fig. 20.4 Conceptual model of the reservoir in 2011, after [4]. Wells 1–6 are geothermal wells; BLM and Crook are shallow wells.

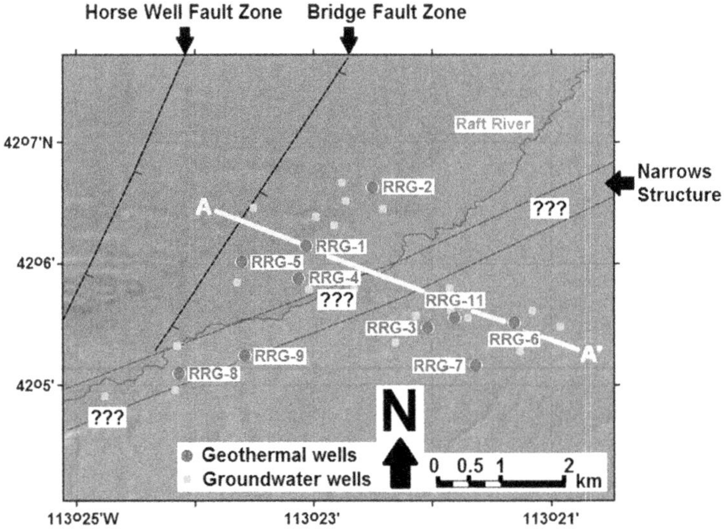

Fig. 20.5 Well field layout including major controlling features, after [7] [WWW].

along other faults and fractures. More recent thinking posits two feed zones [6], one from the northwest along the Horse Well Fault zone and one from the southeast, perhaps associated with the deep Narrows Structure.

Figures 20.5 and 20.6 show, respectively, the well field layout and the main fault zones that control the geothermal and groundwater flows, and the presumed flows of these fluids within the formation [7]. Numerous shallow water and monitoring wells exist across the field, denoted by small circles and thin lines in Figs. 20.5 and 20.6, respectively.

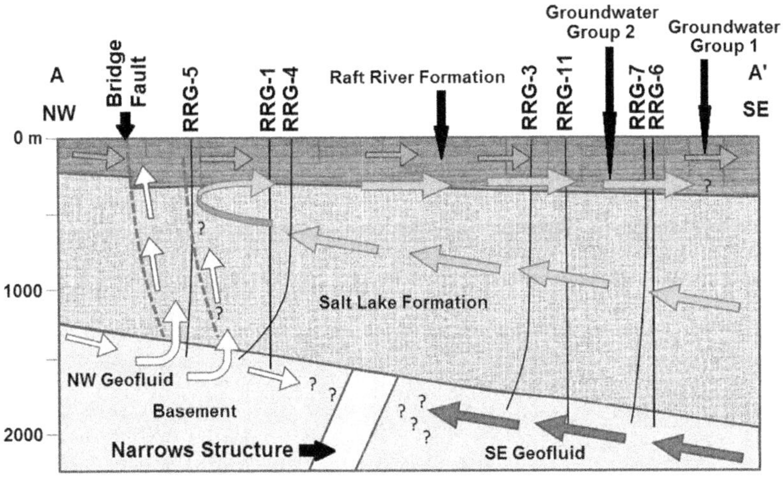

Fig. 20.6 *Speculative model of fluid flow in the Raft River system, after [7]. Section taken across A-A' shown in Fig. 20.5 [WWW].*

The Group 1 groundwater appears to be of recent meteoric origins, whereas the other three fluids share a common parent believed to be of glacial origins [7]. The difference in their chemical characteristics is attributable to the different paths they have taken to arrive at the Raft River field. The SE geofluid has taken a deeper path, is hotter, and has been in contact with evaporites, accounting for its higher dissolved solids relative to the NW geofluid.

20.3 Original development – DOE pilot plant

20.3.1 Original wells and power plant

The early development from 1974 to 1978 involved drilling seven deep wells, RRGE-1 to RRGE-7; see Fig. 20.7. The discovery well RRGE-1 was targeted to intercept the Bridge Fault toward the bottom of its planned total depth. It was completed in April 1975 at a total depth of 1,521 m (4,989 ft). It penetrated 35 m (116 ft) of the metasedimentary formation and bottomed in the basement rock. The primary production horizon was near the bottom of the Tertiary sediments from 1,128–1,372 m (3,700–4,500 ft). The well was capable of artesian flow. Although the development strategy called for production wells to be sited to the northwest and injection wells to the southeast, well RRGE-3 was used for production since RRGE-4 was incapable of sufficient flow. Thus only two wells, RRGE-6 and -7, served as injectors. Table 20.1 lists some properties of the seven wells [8].

All wells were fitted with pumps to increase the flow rates. For example, RRGE-1 was able to flow only 300 GPM under artesian conditions but 1120 GPM when pumped. Originally electric submersible pumps were planned to be installed but were abandoned after problems with cable and motor failures. Peerless line-shaft multistage

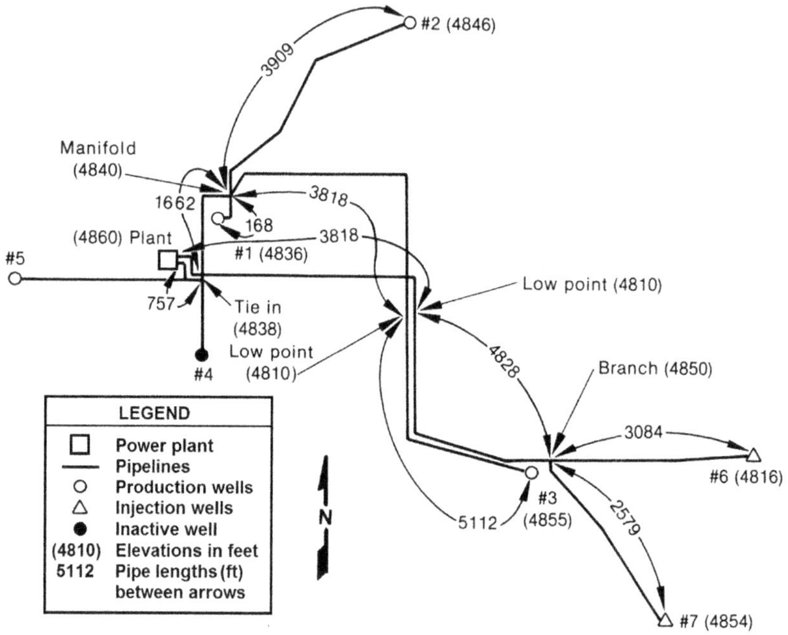

Fig. 20.7 Original Raft River deep wells; all wells are designated RRGE, after [8].

Table 20.1 Well characteristics – original wells [8].

Well	Type	Total depth ft	Casing size Diameter in	Casing size Depth ft	Bottom hole temp. °F	Flowing temp. °F	Wellhead pressure psig	Max. artesian flow GPM
RRGE-1	Production	4989	30	40	294	279	134	300
			20	901				
			13-5/8	3623				
RRGE-2	Production	6543	30	40	296	280	121	250
			20	901				
			13-5/8	4227				
RRGE-3	Production	Note a	20	120	298	296	102	200
			13-5/8	1383				
			9-5/8	3565				
RRGE-4	Unusable	Note b	20	150	273	248	126	55
			13-5/8	1915				
			9-5/8	3408				
RRGE-5	Production	Note c	20	150	168	257	110	200
			13-5/8	1510				
			9-5/8	3408				
RRGE-6	Injection	3888	13-5/8	1698	253	NA	38	NA
RRGE-7	Injection	3444	13-5/8	2044	162	NA	4	NA

Note a. Forked well with three legs: A-5853, B-5532, C-5917.
Note b. Forked well with two legs: A-4911, B-5115.
Note c. Forked well with two legs: A-4911, B-4925.

Table 20.2 Characteristics of production well pumps [2].

Well	Set depth ft	Flow rate GPM	TDH[(a)] ft	Motor horsepower
RRGE-1	1000	1120	1340	500
RRGE-2	1000	680	1340	350
RRGE-3	998	375	1425	250

[(a)]Total Developed Head.

entrifugal pumps were more successful at each production well. Table 20.2 gives ome information on the production well pumps. The two injection wells received pent brine into holding ponds at each well that served as a fluid capacitor to allow mooth operation of the entire flow system. The brine was pumped out of the ponds by neans of vertical-turbine pumps and delivered to each injection well.

The geofluid was delivered from the well pumps to the plant via buried concrete-sbestos pipes. Owing to thermal expansion and contraction during plant operations, hese pipes suffered cracks and led to loss of hot brine. The location of the cracks was vident from surface pools of hot brine. The operators were forced to adopt a practice f making gradual changes in all operating procedures to reduce the effect of tempera-ure changes on the buried pipes.

Given the low temperature of the geofluid, a binary plant was designed for the Raft River site. In the late 1970s binary plants were novelties with only two tiny plants nstalled in the world, one in Russia at Paratunka [9] and one at Kiabukwa in the Democratic Republic of the Congo [10]. Whereas the former plant had been reported n the literature, the latter was essentially unknown to the wider technical community ntil recently. Thus, the U.S. Department of Energy was breaking new ground with he Raft River demonstration binary plant. Approximately $13 million was spent on he project.

The cycle chosen was a dual-pressure (dual-boiling) Rankine cycle, similar, but not dentical, to the cycle discussed in Sect. 8.4.2. Fig. 20.8 is a simplified flow diagram. The cycle working fluid was isobutane, i-C_4H_{10}.

The pressurized geofluid passed sequentially through the high-pressure evaporator (HPE), the high-pressure preheater (HPH), the low-pressure evaporator (LPE), and inally the low-temperature preheater (LTH). Notice that there is only one isobutane irculating pump (CP); it produces the high-pressure fluid for all the high-pressure ele-nents. The LTH operates with the maximum isobutane flow and at the high pressure. The fluid is divided as it leaves the LTH into two streams; one continues on to the HPH and the rest is throttled down to the low-pressure portion of the cycle by a con-rol valve (TV) and enters the LPE. The turbine was a radial-inflow machine within a ingle cylinder but containing both the high- and low-pressure wheels. The turbine drove a generator through a gear box. There was a common exhaust (state 6) that dis-charged into the water-cooled condenser (C). Fig. 20.9 shows the processes followed by the isobutane as it passes through the plant components. The reader may wish to compare this to Fig. 8.11.

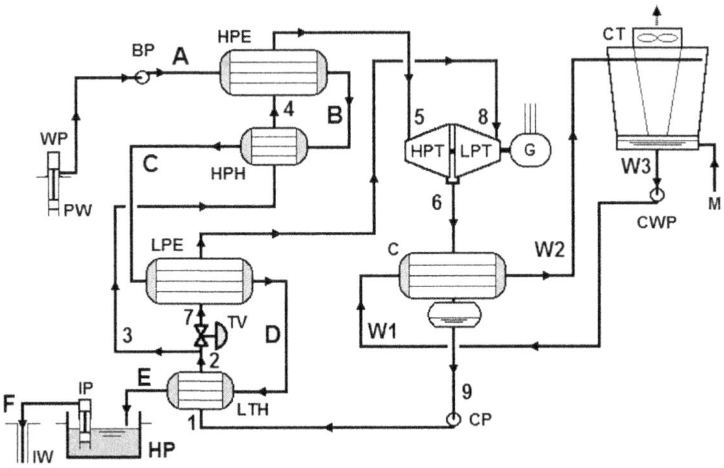

Fig. 20.8 Simplified flow diagram for the original power plant, after [8].

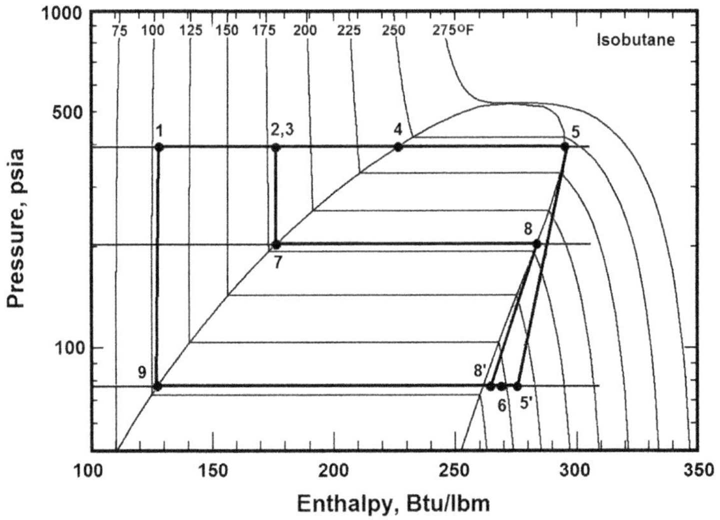

Fig. 20.9 Pressure-enthalpy process diagram for the original plant.

20.3.2 Design performance assessment

The performance of the plant can be determined for the design conditions, which did not actually materialize, as well as for the few test runs that were carried out. Table 20.3 lists the cycle properties for the nominal design case [11]; the state-point labels correspond to those shown in Fig. 20.8. Table 20.4 includes the enthalpy and entropy values, in U.S. Customary units, taken from the NIST REFPROP Database, and serves as the basis for the cycle and plant analysis.

able 20.3 Nominal design specifications for original plant, data from [8].

ate	Temperature		Pressure		Mass flow rate	
	°C	°F	MPa	psia	kg/s	10^6 lbm/h
eothermal fluid						
	143	290	91.1	628	131.0	1.04
	121	250	NA	NA	131.0	1.04
	106	222	NA	NA	131.0	1.04
	88	190	NA	NA	131.0	1.04
	62	144	NA	NA	131.0	1.04
	62	144	NA	NA	131.0	1.04
obutane						
	41	105	2.63	381.6	117.7	0.934
	82	180	2.63	381.6	117.7	0.934
	82	180	2.63	381.6	77.24	0.613
	116	240	2.63	381.6	77.24	0.613
	116	240	2.63	381.6	77.24	0.613
	53	128	0.534	77.5	117.7	0.934
	82	180	1.40	203.0	40.45	0.321
	82	180	1.40	203.0	40.45	0.321
	38	101	0.534	77.5	117.7	0.934
ooling water						
V1	24	75	1.99	289	948.8	7.53
V2	35	95	NA	NA	948.8	7.53
V3	35	75	0.087	12.6	948.8	7.53

able 20.4 Nominal design properties for original plant.

tate	Temp °F	Pressure psia	Entropy Btu/lbm · °F	Enthalpy Btu/lbm	Mass flow rate 10^6 lbm/h
eothermal fluid					
	290	628	0.42391	259.6	1.04
	250	NA	0.36804	218.8	1.04
	222	NA	0.32733	190.4	1.04
	190	NA	0.27893	158.2	1.04
	144	NA	0.20533	112.1	1.04
	144	NA	0.20533	112.1	1.04
sobutane					
	105	381.6	0.31569	128.42	0.934
	180	381.6	0.39539	176.43	0.934
	180	381.6	0.39539	176.43	0.613
	240	381.6	0.46563	223.55	0.613
	240	381.6	0.56790	295.12	0.613
	128	77.5	0.56773	269.01	0.934
	180	203.0	0.39746	176.63	0.321
	180	203.0	0.56500	283.72	0.321
	101	77.5	0.31376	125.65	0.934
ooling water					
V1	75	289	0.08404	43.1	7.53
V2	95	NA	0.12073	63.1	7.53
V3	75	12.6	0.08404	43.1	7.53

lotes: All geothermal fluid and cooling water properties are for saturated liquid. Isobutane properties at tates 4, 5, 7, 8, and 9 are for saturated conditions.

The property values shown in Table 20.4 were used to calculate the design performance of the cycle. Pressure losses in piping and heat exchangers were neglected and therefore the results will be somewhat optimistic.

The cycle power terms may be calculated from the following equations, with reference to Figs. 20.8 and 20.9:

$$\dot{Q}_{LTH} = \dot{m}_1(h_2 - h_1) \quad (20.1)$$

$$\dot{Q}_{LPE} = \dot{m}_7(h_8 - h_7) \quad (20.2)$$

$$\dot{Q}_{HPH} = \dot{m}_4(h_4 - h_3) \quad (20.3)$$

$$\dot{Q}_{HPE} = \dot{m}_4(h_5 - h_4) \quad (20.4)$$

$$\dot{Q}_C = \dot{m}_1(h_6 - h_9) \quad (20.5)$$

$$\dot{W}_{HPT} = \dot{m}_4(h_5 - h_{5'}) \quad (20.6)$$

$$\dot{W}_{LPT} = \dot{m}_7(h_8 - h_{8'}) \quad (20.7)$$

$$\dot{W}_{CP} = \dot{m}_1(h_1 - h_9) \quad (20.8)$$

The equations for the turbine power contain the terms $h_{5'}$ and $h_{8'}$ which represent the actual exhaust states for the high- and low-pressure turbines, respectively, prior to merging in the single pipe to the condenser. These were found from the isentropic efficiency equations with an assumed value of 85% for each machine:

$$h_{5'} = h_5 - 0.85\ (h_5 - h_{5s}) \quad (20.9)$$

$$h_{8'} = h_8 - 0.85\ (h_8 - h_{8s}) \quad (20.10)$$

The terms h_{5s} and h_{8s} represent the ideal isentropic turbine exhaust states. These may be calculated in the manner shown in eqn. (5.14). Table 20.5 presents the findings. The efficiency values were found from:

$$\eta_{CYC} = \dot{W}_{NET}/\dot{Q}_{IN} \quad (20.11)$$

$$\eta_{PLANT} = \dot{W}_{PLANT}/\dot{Q}_{IN} \quad (20.12)$$

The plant design auxiliary power requirements were taken from [8]; these account for the cooling tower fans, cooling water circulating pump, and the geofluid boost pump, but not the production well pumps or the injection pumps.

Utilization efficiencies based on the Second Law can also be calculated using as the dead-state conditions the design wet-bulb temperature of 65°F and the local ambient

Table 20.5 Results of power analysis for original Raft River plant.

Quantity	Value	Units
$\dot{Q}_{LTH}$	13.1	MW-th
$\dot{Q}_{LPE}$	10.1	MW-th
$\dot{Q}_{HPH}$	8.47	MW-th
$\dot{Q}_{HPE}$	12.9	MW-th
$\dot{Q}_{C}$	39.2	MW-th
$\dot{Q}_{IN}$	44.5	MW-th
$\dot{Q}_{NET}$	5.30	MW-th
$\dot{W}_{HPT}$	3.97	MW
$\dot{W}_{LPT}$	2.10	MW
$\dot{W}_{T-TOT}$	6.08	MW
$\dot{W}_{CP}$	0.758	MW
$\dot{W}_{NET}$	5.32	MW
η_{CYC}	11.9	%
$\dot{W}_{AUX}$	0.7	MW
$\dot{W}_{PLANT}$	4.62	MW
η_{PLANT}	10.4	%

ressure of 12.1 psia at 4,800 ft elevation. The reference enthalpy h_0 is 33.13 Btu/lbm nd reference entropy s_0 is 0.06516 Btu/lbm · °F. Thus the exergy of the geofluid at eservoir conditions relative to the dead state may be found from:

$$\dot{E}_{GEO} = \dot{m}_A[h_R - h_0 - T_0(s_R - s_0)] \quad (20.13)$$

vhere $h_R = 259.62$ Btu/lbm and $s_R = 0.4239$ Btu/lbm · °F. This comes to 11.63 MW. 'hus, the design cycle utilization efficiency is 45.7% (5.32*100/11.63) and the overall lesign plant utilization efficiency is 39.7% (4.62*100/11.63).

The resulting thermal and utilization efficiencies are typical of binary plants operating on lower temperature geothermal resources. However, several simplifying and optimistic assumptions went into these calculations.

20.3.3 Actual performance assessment

t is very difficult to obtain actual plant operating data from the final report for the Raft River project [8]. The turbine failed to perform for about half of the small number of ests that were conducted, and the inconsistencies in the data from various instruments nade it nearly impossible to obtain a clear assessment of the performance. Nevertheless he final report does list the results from one test that may shed some light even though he test parameters were not identical with the design values. The observed power erms and calculated thermal efficiencies are shown in Table 20.6. Adopting the dead-state conditions for the test, the geofluid exergy at plant inlet is 13.69 MW; this means he cycle utilization efficiency was 24.7% and the plant efficiency was 10.7%.

Based on this single test result, the Raft River plant performed well below the design xpectations for many reasons, including lower geofluid temperatures, malfunctions of arious components, poor design of the geofluid gathering system, and inconsistencies among measurements.

Table 20.6 Performance results from Test 1A as reported in [8].

Quantity	Value	Units
$\dot{Q}_{LTH}$	11.7	MW-th
$\dot{Q}_{LPE}$	8.81	MW-th
$\dot{Q}_{HPH}$	9.78	MW-th
$\dot{Q}_{HPE}$	9.96	MW-th
$\dot{Q}_{C}$	35.6	MW-th
$\dot{Q}_{IN}$	40.25	MW-th
$\dot{Q}_{NET}$	4.62	MW-th
$\dot{W}_{HPT}$	NA	
$\dot{W}_{LPT}$	NA	
$\dot{W}_{GEN}$	4.010	MW
$\dot{W}_{CP}$	0.633	MW
$\dot{W}_{NET}$	3.377	MW
η_{CYC}	8.39	%
$\dot{W}_{BP}$	0.115	MW
$\dot{W}_{CT-FANS}$	0.134	MW
$\dot{W}_{CT-PUMPS}$	0.388	MW
$\dot{W}_{PW-PUMPS}$	0.821	MW
$\dot{W}_{IW-PUMPS}$	0.448	MW
$\dot{W}_{AUX}$	1.906	MW
$\dot{W}_{PLANT}$	1.471	MW
η_{PLANT}	3.65	%

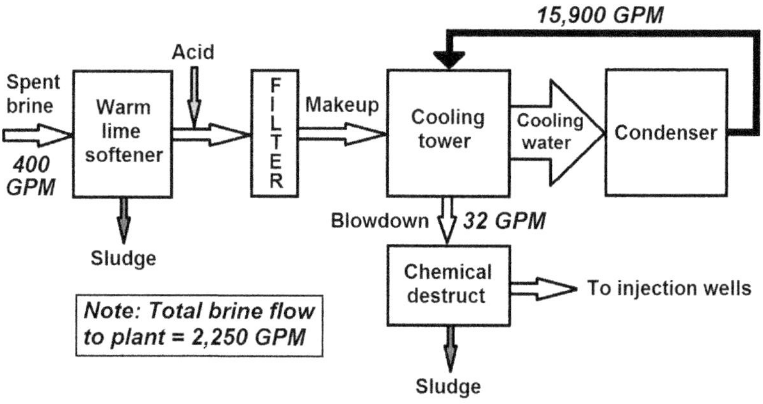

Fig. 20.10 Make-up water treatment system.

20.3.4 Cooling tower make-up water treatment

The plant used a water cooling tower that requires a continuous flow of make-up water to compensate for the water that evaporates to create the cooling effect plus blowdown and drift. The available surface waters in the area are neither plentiful nor of very high quality, which led to an innovative means of providing for the make-up. Approximately 18% of the total spent brine was side-streamed to a chemical treatment facility where it was made suitable for use as make-up. Fig. 20.10 is a flow diagram of ıe processes. The system was plagued by numerous mechanical failures mainly ıvolving chemical feed pumps [2].

0.3.5 Plant photographs

'he photos shown in Figs. 20.11 to 20.14 were taken by the author in March 1981.

'ig. 20.11 Heat exchangers. LTH is at lower right, LPE is at upper right, HPH is at lower center, HPE is above the IPH. The condenser is the large horizontal vessel at the upper left rear [WWW].

'ig. 20.12 Preheaters and evaporators: lower row (L-R) – HPH and LTH; upper row (L-R) – HPE and LPE WWW].

Fig. 20.13 Turbine-generator on elevated pedestal [WWW].

Fig. 20.14 Turbine showing dual inlet pipes, control valves, and position indicators; large horizontal pipe at left center is the exhaust pipe [WWW].

20.3.6 Lessons learned

Although the pilot plant never completed the full five years of operation and testing that was anticipated at the outset, several important conclusions could be drawn from the limited test program. These include:

- It is technically viable to operate an organic Rankine cycle (ORC) at a moderate-temperature geothermal resource.
- In spite of the complex nature of the dual-pressure (dual-boiling) Rankine cycle, there were no plant operational problems and the performance was predictable.
- The gathering system pipes should not be made of concrete-asbestos since this places severe restrictions on plant operations owing to thermal expansion problems.
- Line-shaft mechanical well pumps were preferred over electric submersible pumps, at the time of the project.
- The use of waste brine for cooling tower make-up, while technically feasible, was chemically complex and problems were encountered owing to the use of carbon steel condenser tubes. Appropriate coatings on the tubes would alleviate these problems.
- Feed pumps should be oversized to allow sufficient working fluid flow rate to overcome pressure losses in the flow circuit.
- Kettle-type evaporators should be operated with relatively low liquid levels and fitted with moisture separators to eliminate liquid carryover to the turbine.
- To fully exploit a demonstration project, adequate funding should be allocated at the outset to ensure the project will complete its full series of tests.

20.3.7 Concluding remarks

After nine years of research and development and the expenditure of $13 million dollars, the Raft River power plant project fell victim to budget cuts and a philosophical change in direction for the USDOE from demonstration to research-oriented projects. As a result it operated only briefly during the fall of 1981 and spring of 1982; it was permanently shut down on June 15, 1982. Eventually, the equipment was bought by a private company, dismantled, and partially reassembled at the Bradys Hot Spring geothermal field, but it never operated again.

20.4 New development – U.S. Geothermal plant

20.4.1 Site renewal process and well modifications

Following the termination of the Raft River demonstration project by the USDOE, although the equipment was dismantled and removed, the wells remained in a closed condition, perhaps in anticipation of another attempt to develop the field. That opportunity came 20 years later when in 2002 U.S. Geothermal (USGeo) set about renewing the Raft River prospect. USGeo is a Canadian company headquartered in Vancouver, BC with an office in Boise, Idaho.

Some important events in the chronology of the project are summarized below [12]:

Jun. 2002	USGeo acquires old DOE Raft River parcel, including five production wells, two injection wells, and seven monitoring wells; see Fig. 20.7.
Apr. 2004	Flow tests conducted on the production wells with a grant from DOE.
Jan. 5, 2005	USGeo signs a Power Purchase Agreement (PPA) with Idaho Power Company (Idaho Power) for 10 MW from Phase 1 Raft River plant.
Dec. 7, 2005	USGeo enters into a \$20.3M Engineering, Procurement and Construction (EPC) contract with Ormat Nevada for Phase 1; contract duration is 21 months.
Jan. 10, 2006	Raft River Energy 1 (RRE1), a subsidiary of USGeo, seeks \$35M to construct a 13.8 MW(gross), 10 MW(net) power plant; expects to be online early in 2007. Phase 2 envisioned to cost \$70M for two 10 MW plants.
Jan. 31, 2006	Air quality application for Permit To Construct submitted to Idaho Department of Environmental Quality (IDEQ).
Feb. 8, 2006	Request for authorization for non-contact cooling water reuse filed with the IDEQ.
Mar. 9, 2006	Interconnection agreement signed with Raft River Rural Electric Co-Op (RRREC) to wheel power from Phase 1 power plant to Bridge, Idaho substation on a new 3.2 mile, 34.5 kV transmission line to be constructed by RRREC, paid for by RRE1, and owned by RRREC.
Apr. 4, 2006	Sale completed of \$Cn25M in shares in USGeo.
Apr. 27, 2006	USGeo issues Notice to Proceed to Ormat Nevada to start work on the EPC contract that calls for 10 MW average monthly power for 20 years delivered to Idaho Power Company in accord with USGeo's PPA.
May 26, 2006	IDEQ issues the Permit to Construct.
Aug. 10, 2006	Project financing completed: \$34M for Phase 1.
Aug. 15, 2006	Union Drilling mobilizes its rig, under a May 25, 2006 contract, to begin the well improvement program by deepening the two existing injection wells, RRG-6 and -7. Then it will drill an additional production leg in each of two existing production wells. Total time to complete job expected to be 90 days.
Aug. 17, 2006	Geofluid gathering and distribution system 30% complete under a \$2.6M contract with Industrial Builders of

	Ontario, Oregon; production piping is 10" and 12" diameter; injection piping is 16" and 24" diameter.
Oct. 19, 2006	Upon completion of deepening RRG-6 and -7 and progress reports from major contractors, project on schedule to be online in September 2007. RRG-6 and -7 turn out to be excellent producers, with flowing wellhead temperatures of 240°F and 270°F, respectively. RRG-7 has a 300°F feed zone at the bottom, indicating cool water inflow somewhere along the open hole. Both wells flow more than 1000 gallons per minute (GPM).
Oct. 19, 2006	Redrilling of RRG-4 is 20% complete; deviated sidetrack being drilled through the casing at 2800 ft toward the Bridge Fault zone and planned to reach a depth of 5400 ft (see Fig. 20.4).
Dec. 14, 2006	RRG-4 completed; several permeable zones encountered, indicative of passing through the Bridge Fault zone. The existing leg in RRG-3 is deepened from 5937 ft to 6185 ft and a new second leg is drilled to 5735 ft. Several highly permeable sand layers, fractures, and lost circulation zones are encountered in the new leg. Piping gathering system 95% complete.
Feb. 2007	Mechanical construction work is awarded to Industrial Builders of Ontario, Oregon by Ormat Nevada.
Mar. 21, 2007	PPA with Idaho Power amended to 45.5 MW annual average for 25 years, including 13 MW from Phase 1
Apr. 23, 2007	RRREC completes construction of 3.2-mi power transmission line. Well improvement program nearly done: 4 wells enhanced by deepening or adding extra legs; 2 new wells drilled for Phase 2; well RRG-7 to become a producer and RRG-3 an injector.
May 2007	Electrical construction work awarded to Merit Electric.
Jun. 22, 2007	USGeo leases 1,685 acres (2.6 mi^2) from U.S. Bureau of Land Management (BLM), increasing its Raft River holdings area by 31.7% to 6,933 acres (10.8 mi^2).
Jul. 25, 2007	A 29-stage, 900 hp production pump installed in producer RRG-2; a 33-stage, 1000 hp pump to be installed in RRG-1. Both pumps are from Goulds Pumps/ITT. Two more production well pumps scheduled for delivery to site within 60 days. Seven technicians being trained to operate the plant. Plant expected online in 4th qtr. 2007.
Sep. 26, 2007	25-year 13 MW PPA for Phase 1 officially signed with Idaho Power.

Oct. 2007	Test runs of power plant conducted from Oct. 18–23; plant ran for 108 hrs and total gross generation was 1,022 MWh, i.e., 9.5 MW average power; a few mechanical problems were found and are being repaired. Commercial operation set for the end of the 4th qtr. 2007.
Nov. 22, 2007	Test runs resume after repairs to two injection pumps and the high-pressure turbine; testing and training will continue for 60–90 days.
Dec. 3, 2007	Plant capacity test begins at 6:15 P.M.
Dec. 4, 2007	Plant capacity test ends at 6:15 A.M. Plant meets contractual net output.
Dec. 28, 2007	Construction is substantially complete; 4 production wells in operation; net output is 8–9 MW; 14.4 and 9.5 MW, max. and min. gross outputs; 9.4 and 7.1 MW, max. and min. net outputs.
Jan. 3, 2008	Raft River 1 officially achieves commercial power generation status.
Jun. 19, 2008	With 4 production and 4 injection wells in operation, net power output is between 10.5–11.5 MW; the maximum net output achieved during March, April, and May were 11.2, 12.0, and 11.7 MW, respectively; plant had a 99% availability during that period. Reservoir studies continue with the aim of determining how best to use the existing wells.
Aug. 27, 2008	For the past three months, plant had net output of 9.5–10.5 MW and 99.9% operating availability. New reservoir model completed and being used to plan new well drilling to increase geofluid flow to raise annual average power up to the PPA rating of 13 MW. Options include: adding more legs to existing production wells, drilling a new production well, and perhaps a new injection well to support increased production.
Oct. 14, 2008	DOE announces grant of $9M to USGeo to demonstrate the viability of Enhanced Geothermal Systems (EGS) at Raft River. DOE will provide up to $6M while USGeo will contribute the rest in "in kind" contributions. Program to perform a monitored thermal stimulation of an existing injection well to improve permeability.
Oct. 14, 2008	Raft River 1 operated at 99.9% availability for the past 6 months. Power generation is increasing, currently producing 11.0–11.5 MW(net). Reverse osmosis filtering unit being installed to significantly reduce chemical treatment costs for cooling tower and reduce high

	levels of dissolved chloride from the cooling tower feed water. Discussions underway with several parties to examine the viability of cascaded use of the geothermal energy, including year-round heating of greenhouses, etc.
Sep. 15, 2009	DOE officially awards \$10.21M to USGeo, wherein DOE to provide up to \$7.39M with rest coming from USGeo in "in kind" contributions. This is the same award that was announced 11 months earlier.
Feb. 23, 2010	Raft River EGS project gets underway. Program team consists of the Energy & Geoscience Institute (EGI), The University of Utah (program lead), USGeo, APEX Petroleum Engineering Services, and HiPoint Reservoir Imaging. USGeo contributed an existing production well (for Phase 2) for testing thermal fracturing of reservoir rock some 6,000 feet below the surface. Halliburton Energy Services placed bridge plug in well RRG-9 at 2235 ft. Work in this portion of the program includes: borehole imaging, fracture analysis, development of a geologic-structural reservoir model, interference testing, construction of injection pipelines, and installation of a liner in the wellbore to the stimulation target horizon.
Jun. 2010	Production well RRG-2 shut down after a pump failure.
May 16, 2011	Pump rig mobilized to the site and begins work on well RRG-2.
May 17, 2011	USGeo signs \$1.65M Repair Services Agreement (RSA) between its wholly owned subsidiary US Geothermal Services and RRE1 for the repair of two production wells, RRG-2 and RRG-7. Production well RRG-7 has a leak in a cement seal that failed where two steel casing sections overlap, allowing cooler geothermal fluid to enter the well bore. Thus, over the last two years, wellhead temperature declined from 299°F to 240°F. Successful workovers to these wells should raise net power from 8 to 10 MW, annual average. Flow stimulation technique called deflagration (controlled, rapid combustion that produces jets of high-temperature, high-pressure gas that impact the wellbore in selected areas) will also be applied to wells RRG-2 and RRG-7 to increase fluid flow.
Jul. 2, 2011	RRG-7 successfully repaired and put back into service after a cement squeeze job to fix the casing leak; the production pump was refurbished and the well is now flowing 1300 GPM of geofluid at 297°F. RRG-2 is now under repair.

Fig. 20.15 USGeo Raft River, aerial view from Google Earth *image, June 24, 2009 [WWW].*

20.4.2 Power plant design

The new USGeo Raft River power plant is located about 600 m SSE of the site of the original plant, see Fig. 20.15.

The energy conversion system is a partially recuperated, dual-level, double-pressure binary cycle. There are two separate units each with two axial-flow turbines driving a common central generator [13]. A 4-cell water cooling tower provides chilled water for the shell-and-tube condensers. Make-up water is obtained from water wells and requires treatment to maintain water quality in the towers. About 3,150,000 lbm/h (397 kg/s) of 280°F (138°C) brine is pumped to evaporators and preheaters where the heat is absorbed by the cycle working fluid, isopentane, i-C_5H_{12} [14]. The recuperator is used only in the high-pressure (HP) cycle. The process flow diagram is shown in Fig. 20.16 [13]; a site photo is given in Fig. 20.17 [15].

20.4.3 Design performance assessment

With the arrangement of the components shown in the flow diagram, Fig. 20.16, together with key parameters obtained from the manufacturer's (Ormat) heat balance diagram (HBD) and the governing equations from thermodynamics, all the state-point properties can be found, and the heat and work terms computed. Table 20.7 shows temperatures, pressures, and mass flow rates from the HBD, keyed to the state labels used in Fig. 20.16 The design wet-bulb temperature (assumed as the dead-state temperature) is 39.8°F.

Next, these data were analyzed and the values of temperature, pressure, entropy, and enthalpy were found with the aid of REFPROP; the results are given in Table 20.8.

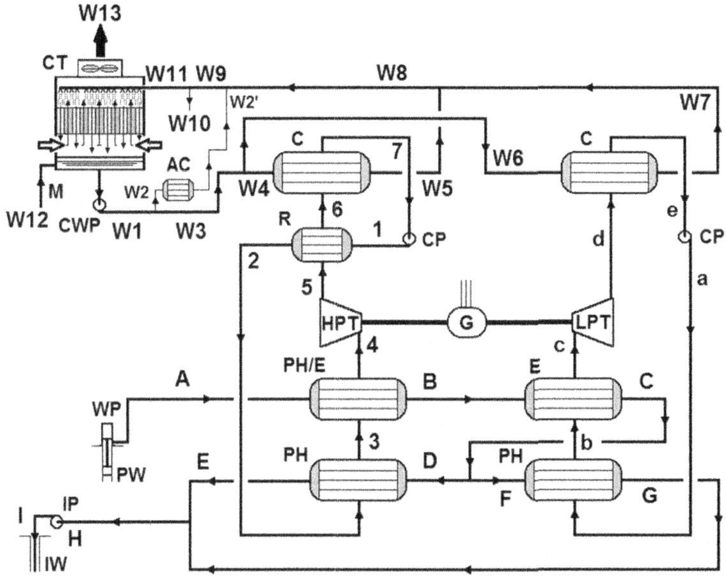

Fig. 20.16 Simplified process flow diagram for USGeo Raft River power plant, after [13].

Fig. 20.17 USGeo Raft River plant: Equipment is arrayed symmetrically on the left and right of generator except for the recuperator, which is above the condensers on the HP side of the plant only; after [15] [WWW].

Table 20.7 Nominal design specifications for USGeo plant, data from [13].

State	Temperature		Pressure		Mass flow rate	
	°C	°F	MPa	psia	kg/s	10^6 lbm/h
Geothermal fluid						
A	137.8	280	0.772	112	396.9	3.15
B	110.0	230	NA	NA	396.9	3.15
C	87.4	189.4	NA	NA	396.9	3.15
D	87.4	189.4	NA	NA	198.4	1.575
E	68.2	154.8	0.634	92	198.4	1.575
F	87.4	189.4	NA	NA	198.4	1.575
G	63.0	145.4	0.634	92	198.4	1.575
H	65.6	150.1	0.634	92	396.9	3.15
I	65.8	150.5	NA	NA	396.9	3.15
Isopentane – HP cycle						
1	NA	NA	0.9566	138.7	139.3	1.1057
2	38.1	100.6	0.9566	138.7	139.3	1.1057
3	113.1	235.6	0.9566	138.7	139.3	1.1057
4	113.1	235.6	0.9566	138.7	139.3	1.1057
5	62.4	144.3	0.0897	13.0	139.3	1.1057
6	40.3	104.5	0.0897	13.0	139.3	1.1057
7	19.9	67.8	0.0897	13.0	139.3	1.1057
Isopentane – LP cycle						
a	NA	NA	0.5124	74.3	128.8	1.0226
b	84.3	183.7	0.5124	74.3	128.8	1.0226
c	84.3	183.7	0.5124	74.3	128.8	1.0226
d	48.2	118.7	0.0876	12.7	128.8	1.0226
e	19.4	66.9	0.0876	12.7	128.8	1.0226
Cooling water						
W1	11.9	53.5	0.2828	41	3,010.9	23.896
W2	11.9	53.5	0.2828	41	78.8	0.6256
W2'	14.4	58.0	0.1655	24	78.8	0.6256
W3	11.9	53.5	0.2828	41	2,932.1	23.270
W4	11.9	53.5	0.2828	41	1,466.0	11.635
W5	20.7	69.2	0.1655	24	1,466.0	11.635
W6	11.9	53.5	0.2828	41	1,466.0	11.635
W7	20.3	68.6	0.1655	24	1,466.0	11.635
W8	20.5	68.9	0.1655	24	2,932.1	23.270
W9	20.3	68.6	0.1655	24	3,010.9	23.896
W10	20.3	68.6	0.1655	24	17.63	0.1399
W11	20.3	68.6	0.1655	24	2,993.3	23.756
W12	11.9	53.5	NA	NA	52.97	0.4204
W13	NA	NA	NA	NA	35.31	0.2802

The working fluid processes are shown in the pressure-enthalpy diagram, Fig. 20.18. Note that state d, the exhaust from the LP turbine, contains some superheat, 44.1°F, but the designers chose not to employ a recuperator in the LP cycle, whereas they did so in the HP cycle where the exhaust superheat (state 5) was significantly greater, 68.5°F. Given the extra cost for the recuperator and the limited value it

Table 20.8 Nominal design state-point properties for USGeo plant.

State	Temp °F	Pressure psia	Entropy Btu/lbm·°F	Enthalpy Btu/lbm	Mass flow rate 10^6 lbm/h
Geothermal fluid					
A	280	112	0.4101	249.49	3.15
B	230	105	0.3390	198.69	3.15
C	189.4	99	0.2779	157.79	3.15
D	189.4	99	0.2779	157.79	1.575
E	154.8	92	0.2230	123.09	1.575
F	189.4	99	0.2779	157.79	1.575
G	145.4	92	0.2076	113.68	1.575
H	150.1	92	0.2153	118.38	3.15
	150.5	NA	NA	NA	3.15
Isopentane – HP cycle					
1	68.8	137.55	−0.0140	−6.8982	1.1057
2	100.6	137.55	0.0180	10.524	1.1057
3	235.6	137.55	0.1504	93.613	1.1057
4	235.6	137.55	0.3101	204.65	1.1057
5	144.3	13.00	0.3225	174.40	1.1057
6	104.5	13.00	0.2931	157.24	1.1057
7	67.8	13.00	−0.0145	−7.7488	1.1057
Isopentane – LP cycle					
a	67.4	73.49	−0.0152	−7.8181	1.0226
b	183.7	73.49	0.1002	59.620	1.0226
c	183.7	73.49	0.2958	185.48	1.0226
d	118.7	12.70	0.3044	163.30	1.0226
e	66.9	12.70	−0.0154	−8.2327	1.0226
Cooling water					
W1	53.5	41	0.04296	21.704	23.896
W2	53.5	41	0.04296	21.704	0.6256
W2'	58.0	24	0.05171	26.162	0.6256
W3	53.5	41	0.04296	21.704	23.270
W4	53.5	41	0.04296	21.704	11.635
W5	69.2	24	0.07313	37.367	11.635
W6	53.5	41	0.04296	21.704	11.635
W7	68.6	24	0.07199	36.767	11.635
W8	68.9	24	0.07256	37.067	23.270
W9	68.6	24	0.07199	36.767	23.896
W10	68.6	24	0.07199	36.767	0.1399
W11	68.6	24	0.07199	36.767	23.756
W12	53.5	NA	NA	NA	0.4204
W13	NA	NA	NA	NA	0.2802

could contribute with only 44°F of superheat, it was not cost-effective for the LP cycle. To make a fair comparison with the original plant (see Fig. 20.9), the pressure losses in piping and heat exchangers were ignored.

Using the thermodynamic governing equations (similar to eqs. 20.1–20.13), all the heat and work transfer terms were calculated for the design case. These results appear in Table 20.9 for each of the two cycle loops; Table 20.10 shows the results for the

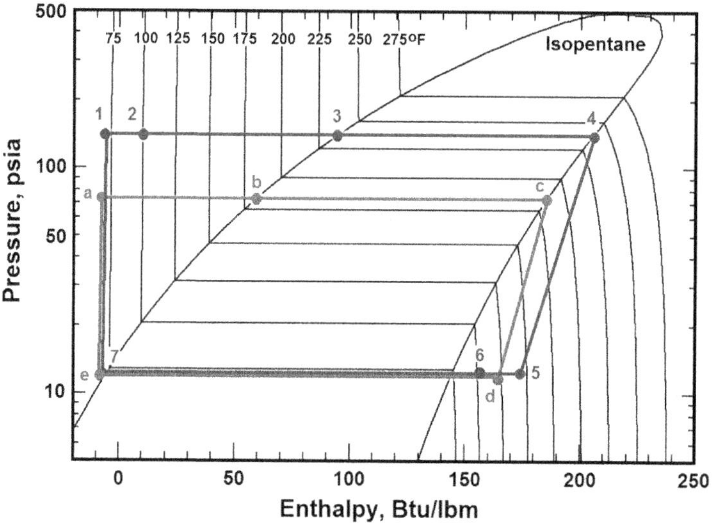

Fig. 20.18 Pressure-enthalpy process diagram for USGeo Raft River plant.

Table 20.9 USGeo Raft River cycle analysis: LP and HP cycles.

Low-pressure cycle			High-pressure cycle		
$\dot{Q}_{LPH}$	20.21	MW-th	$\dot{Q}_{HPH}$	26.93	MW-th
$\dot{Q}_{LPE}$	37.72	MW-th	$\dot{Q}_{HPE}$	35.98	MW-th
$\dot{W}_{LPT}$	6.65	MW	$\dot{W}_{HPT}$	9.80	MW
$\dot{Q}_{R}$		(none)	$\dot{Q}_{R}$	17.29	MW-th
$\dot{Q}_{LPC}$	51.41	MW-th	$\dot{Q}_{HPC}$	53.47	MW-th
$\dot{W}_{LPP}$	0.134	MW	$\dot{W}_{HPP}$	0.276	MW
$\dot{Q}_{IN,LP}$	57.93	MW-th	$\dot{Q}_{IN,HP}$	62.91	MW-th
$\dot{W}_{CT,FANS}$	0.2875	MW	$\dot{W}_{CT,FANS}$	0.2875	MW
$\dot{W}_{CWP}$	0.375	MW	$\dot{W}_{CWP}$	0.375	MW
$\dot{W}_{NET,LP}$	5.86	MW	$\dot{W}_{NET,HP}$	8.86	MW
$\eta_{TH,LP}$	10.1%		$\eta_{TH,HP}$	14.1%	

Note: CT fan and CW pump power for each cycle are assumed half the total for the plant.

Table 20.10 Design performance for new Raft River plant, excluding field pumps.

$\dot{W}_{T,TOT}$	16.45	MW
$\dot{W}_{P,TOT}$	0.410	MW
$\dot{W}_{CT,FANS}$	0.575	MW
$\dot{W}_{CWP}$	0.750	MW
$\dot{W}_{NET,OEC}$	14.72	MW
$\dot{Q}_{IN,TOT}$	120.84	MW-th
$\eta_{TH,OEC}$	12.2%	
$\dot{E}_{IN,BRINE}$	39.42	MW
$\eta_{U,OEC}$	37.3%	

Table 20.11 Performance for plant during capacity test conditions.

$\dot{W}_{OEC,GROSS}$	14.602	MW
$\dot{W}_{PLANT,NET}$	9.609	MW
$\dot{E}_{IN,BRINE}$	35.452	MW
$\eta_{U,GROSS}$	41.2	%
$\eta_{U,NET}$	27.1	%

Table 20.12 DOE and USGeo Raft River power plants compared.

	DOE	USGeo
T_{GEO}, °F	290	280
$\dot{m}_{GEO}$, 10^6 lbm/h	1.04	3.15
$\dot{W}_{NET,OEC}$, MW	5.32	14.72
$\eta_{TH,OEC}$, %	11.9	12.2
T_0, °F	65	39.8 (65)
$\eta_{U,OEC}$, %	45.7	37.3 (45.4)
$\dot{W}_{NET,OEC}/\dot{m}_{GEO}$, Wh/lbm	5.12	4.67

plant as a whole. The heat transfer in the recuperator, $\dot{Q}_R$, is the average of the values obtained from using the heat released by the vapor (state 5 to 6) and the heat picked up by the liquid (state 1 to 2); the difference is very small, about 1.5%.

The HP cycle is more thermally efficient, 14.1% vs. 10.1%. The plant as a whole has a thermal efficiency of 12.2% and converts 37.3% of the incoming brine exergy to useful power. This accounting includes all cycle parasitic loads but does not account for the power to run the production well pumps and injection pumps. And it should be emphasized that these results are for the design case.

20.4.4 Actual performance assessment

The actual plant performance has been limited by the available flow rate from the production wells. Instead of the design flow rate of 3,150,000 lbm/h, the actual mass flow rate during the capacity tests for performance validation averaged only 2,510,234 lbm/h or about 80% of the design value. The brine temperature was slightly higher than design, 282.7°F vs. 280°F. The tests were conducted on December 3–4, 2007 from 6:15 P.M. to midnight on the 3rd and continuing from midnight to 6:15 A.M on the 4th. Besides the geofluid flow rate and temperature, there were several other off-design conditions that needed to be corrected for using appropriate factors. Under the off-design conditions, the average plant performance is shown in Table 20.11; these were calculated using the actual data recorded during the test and provided by Ormat [13].

From these data, since the brine return temperature is not given, the heat input to the isopentane could not be determined and so the thermal efficiency could not be calculated. However, using a dead-state temperature of 33°F (the actual wet-bulb temperature during the test was below freezing), the incoming brine exergy can be found and the utilization efficiency calculated. The utilization efficiency for the actual conditions fell ten percentage points or 27% below the design value.

When the actual performance results were corrected using factors derived from the variation between the design and actual site conditions, the corrected net power was 13.906 MW which exceeded the guaranteed net power at the design point of 13.735 MW. Thus the plant passed its acceptance test.

20.4.5 Conclusion

The original (DOE) and the new (USGeo) Raft River plants have very similar design performance, despite using quite different energy conversion systems. The DOE plant was based on a geofluid temperature of 290°F and a 65°F wet-bulb temperature (assumed dead-state temperature); the USGeo plant uses 280°F brine temperature and a quite low 39.8°F wet-bulb temperature. Table 20.12 compares some important results for the plants. Note that the design performance for the USGeo plant has been calculated at its own wet-bulb temperature and at that for the DOE plant (values in parentheses) to put them on a common footing for utilization efficiency. The design thermal efficiency is roughly the same, with the USGeo plant holding a 2.5% advantage. The DOE plant has a 9.6% advantage in specific output, Wh/lbm, mainly due to the higher geofluid temperature. When the plants are designed for the same wet-bulb (dead-state) temperature of 65°F, they are essentially identical in terms of utilization efficiency, each plant converting about 45.5% of the geofluid exergy into useful output.

The DOE plant suffered from some serious design flaws and from a lack of reliable geothermal well pumps. The USGeo plant is well designed, constructed, and uses very reliable downhole pumps, but is lacking sufficient geofluid flow to meet the design output, a fundamental problem that is shared by some other recent geothermal plants. The only solution is to drill more wells, both producers and injectors, to raise the flow rate. However, the reservoir itself is capable of yielding and accepting only so much fluid within the leasehold. Since 100% of the produced fluid must eventually be returned to the formation, the lack of injection capacity is often the real limitation on plant output. For example, the newly refurbished well, RRG-7, which was switched from being an injector to a producer, will increase the hot brine flow, but unless all of that fluid can be effectively reinjected, the gain may not be fully realized.

References

[1] *Google Earth*, 5.2.1.1588, Build date: Sept. 1, 2010, Imagery date June 24, 2009.

[2] Toth, W.J., "Raft River Colloquy: A Series of Papers on Geothermal Development at Raft River, Idaho," *Geothermal Resources Council BULLETIN*, V. 11, No. 2, 1982, pp. 4–5.

[3] U.S. Geothermal Inc., http://www.usgeothermal.com/RaftRiverProject.aspx, accessed May 6, 2011.

[4] Mattson, E., M. Plummer, C. Palmer, L. Hull, S. Miller and R. Nye, "Comparison of Three Tracer Tests at the Raft River Geothermal Site," *Proc. Thirty-Sixth Workshop on Geothermal Reservoir Engineering*, Stanford University, Stanford, California, January 31–February 2, 2011, Rep. No. SGP-TR-191.

[5] Tullis, J.A. and M.R. Dolenc, "Geoscience Interpretations of the Raft River Resource," *Geothermal Resources Council BULLETIN*, V. 11, No. 2, 1982, pp. 6–9.

[6] Holt, R.J., "Numerical Model Development and Results: Raft River Geothermal Field, Cassia, County, Idaho," Geothermal Science, Inc., Technical Report, prepared for U.S. Geothermal, Inc., 2008.
[7] Ayling, B., P. Molling, R. Nye and J. Moore, "Fluid Geochemistry at the Raft River Geothermal Field, Idaho: New Data and Hydrogeological Implications," *Proc. Thirty-Sixth Workshop on Geothermal Reservoir Engineering*, Stanford University, Stanford, California, January 31–February 2, 2011, Rep. No. SGP-TR-191.
[8] Bliem, C.J. and L.F. Walrath, "Raft River Binary-Cycle Geothermal Pilot Power Plant Final Report," Idaho National Engineering Laboratory, EG&G Idaho, Inc., Rep. EGG-2208, Idaho Falls, ID, April 1983.
[9] Moskvicheva, V.N. and A.E. Popov, "Geothermal Power Plant on the Paratunka River," *Geothermics – Special Issue 2, U.N. Symposium on the Development and Utilization of Geothermal Reseources*, Pisa, V. 2, Pt. 2, 1970, pp. 1567–1571.
10] Kraml, M., K. Kessels, U. Kalberkamp, N. Ochmann and C. Stadtler, 2006. "The GEOTHERM Programme of BGR, Hannover, Germany: Focus on Support of the East African Region," The 1st African Geothermal Conference, Addis Ababa, Ethiopia: http://www.bgr.de/geotherm/ArGeoC1/pdf/50%20%20Kraml,%20M.%20GEOTHERM%20programme.pdf.
11] Ingvarsson, I.J. and W.W. Madsen, "Determination of the 5 MW Gross Nominal Design Case Binary Cycle for Power Generation at Raft River, Idaho," Idaho National Engineering Laboratory, EG&G Idaho, Rep. TREE-1039, Idaho Falls, ID, December 1976.
12] U.S. Geothermal, Inc., Boise, Idaho, *News Releases*, various dates: http://www.usgeothermal.com/NewsReleases.
13] Personal communication, L. Bronicki, Ormat, Inc., July 14, 2011.
14] Peltier, R., "Raft River Geothermal Project, Malta, Idaho," *POWER*, December 15, 2007: http://www.powermag.com/issues/cover_stories/Raft-River-Geothermal-Project-Malta-Idaho_231.html.
15] Richter, L.X., "Geothermal Could Be Significant Source of Power in Idaho," *ThinkGeoEnergy*, July 24, 2011: http://thinkgeoenergy.com/archives/8161.

Chapter 21

Geothermal Power Plants in Turkey

"Everything we see in the world is the creative work of women."

Mustafa Kemal Ataturk

21.1 Geologic setting

Turkey lies in a highly active tectonic region, the Alpine-Himalayan orogenic belt. There are almost 200 geothermal prospects, but nearly all of them are suitable only for low-temperature applications. The one area that has high-temperature resources is associated with the Büyük Menderes Graben in western Anatolia.

All three of Turkey's operating power plants lie in this region. In fact they are roughly aligned in an east-west direction approximately along the 38° North parallel, together with the famous Pamukkale UNESCO World Heritage Site. The latter is a popular tourist attraction consisting of numerous travertine terraces formed from hot springs that issue from the northeast side of the graben. Because the geofluid is

Geothermal Power Plants: Principles, Applications, Case Studies and Environmental Impact, Third Edition.
© 2012 Elsevier Ltd. All rights reserved.

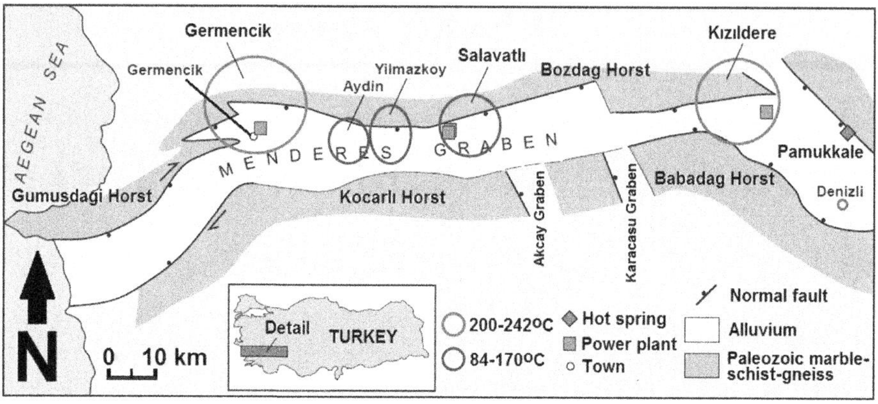

Fig. 21.1 Menderes Graben and locations of Turkey's geothermal power plants, after [1] and [2] [WWW].

supersaturated with calcium carbonate, when it reaches the surface and the dissolved carbon dioxide is released to the atmosphere, calcium carbonate precipitates and eventually hardens into spectacular white terraces. Unfortunately, the geothermal fluids produced at the fields supporting power plants also suffer from calcium carbonate scaling in the production wells, forcing the operators to resort to various means to remedy the situation.

Starting from Pamukkale, Turkey's oldest geothermal power plant, Kızıldere, is about 25 km west, followed by Salavatlı which is about 66 km west of Kızıldere, and ending at the newest power plant, Germencik, about 40 km west of Salavatlı. Fig. 21.1 shows the geologic setting for Turkey's geothermal power plants. The prominent graben lies between two parallel horsts in an alternating pattern of subsidence and uplift. To date high-temperature geofluids have been found only in wells drilled into the slip-step faults that form the northern boundary of the graben. These faults appear to be the main fluid controlling structures all across the east-west extent of the graben from Germencik to Kızıldere.

Fig. 21.2 presents a cross-section of the region with the location of the Kızıldere power plant and the hot spring Tekke Hamam thermal spa, which are a few kilometers apart on opposite sides of the Menderes River. Hot springs are found on both the north and south sides of the graben. The basement rock is gneiss and numerous layers of various formations constitute the lithology of the reservoir.

At the eastern end of the Büyük Menderes Graben in the vicinity of Kızıldere, three distinct production zones have been identified by well drilling. The first and shallowest is the Sazak Formation between Pliocene units; it extends to 706 m and has a temperature of 190–200°C. The next one in depth is the Igdecik Formation of Menderes metamorphics, which reaches to 1261 m and has a temperature of 200–212°C. The deepest one consists of gneiss and quartzite under the micaschists. There may even be a fourth, very deep (~3000 m) reservoir with a possible temperature in the 250–260°C range [5]. The upper layer, the cap rock, is comprised of Pliocene-aged clay, marl, and altered sandstones.

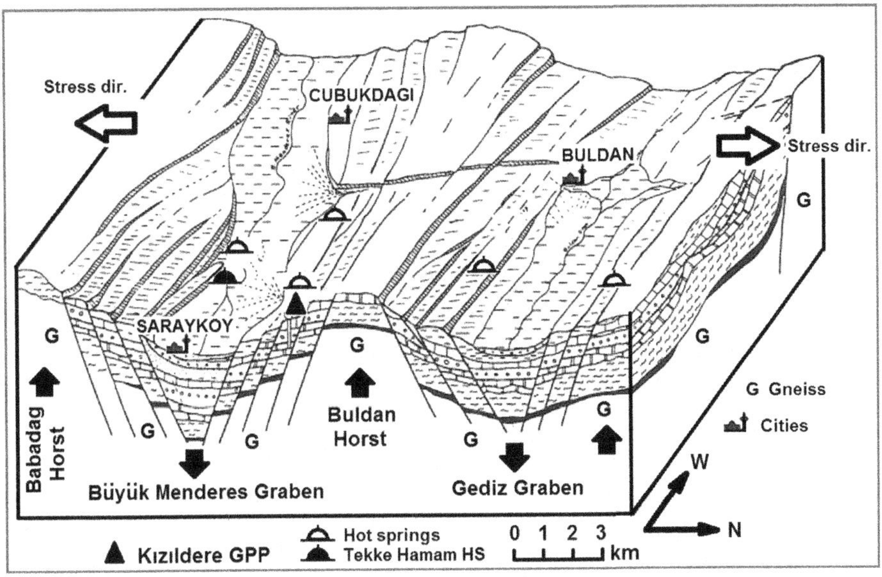

Fig. 21.2 Geologic cross-section showing Büyük Menderes Graben and associated horsts, modified from [3] as shown in [4].

21.2 Kızıldere single-flash plant

21.2.1 Early history and pilot plant

Discovered in the 1960s, the Kızıldere resource is a liquid-dominated reservoir at a temperature of between 200–242°C, with low total dissolved solids (TDS ~2500–3200 ppm), but very high noncondensable gases (NCG ~ 2.5% by weight of brine in the reservoir, 5% by volume of steam, 10–21% by weight of steam, and an average 13% by weight of steam at the turbine inlet). The NCG also contain 96–99% (by weight) carbon dioxide (CO_2) and 100–200 ppm of hydrogen sulfide (H_2S). The steam fraction (by weight) at the wellhead is 10–12% [6].

The field was first exploited by a small pilot plant under the direction of the General Directorate of Mineral Research and Exploration (MTA). The 0.5 MW noncondensing turbine went into operation in 1974 and ran for six years while providing free electricity to three neighboring communities. The turbine and associated piping and equipment were fabricated by MTA and served to demonstrate that the resource could be used for practical power generation purposes. Figures 21.3 and 21.4, provided to the author by Mr. Orhan Mertoğlu in 1981, show the pilot plant attached to well KD-13 and the turbine-generator hall, respectively.

The wellhead unit was designed to the specifications given in Table 21.1.

21.2.2 Wells

The wells supplying the plant were drilled starting in the 1970s, as shown in Table 21.2. The location of the wells is presented in Fig. 21.5. For more than 25 years the plant operated with no reinjection, disposing of the waste liquid via runoff

Fig. 21.3 Kızıldere 0.5 MW pilot plant [7]; photo courtesy of O. Mertoğlu.

Fig. 21.4 Steam turbine and generator for Kızıldere 0.5 MW pilot plant [7]; photo courtesy of O. Mertoğlu.

Table 21.1 Technical specifications for Kızıldere pilot wellhead unit [7].

Item	Value
Turbine type	single-cylinder, Curtis stage, back-pressure, geared
Rated capacity	500 kW
Speed, turbine/generator	4,500/1,500 rev/min
Maximum pressure	786 kPa (114 lbf/in^2,a)
Steam pressure	486 kPa (70.5 lbf/in^2,a)
Steam temperature	150°C (302°F)
Exhaust pressure	115 kPa (16.7 lbf/in^2,a)
Steam mass flow rate	3.61 kg/s (28,660 lbm/h)
NCG	17% by wt. of steam
Blade height	76 mm (3 in)

Table 21.2 Well information [4, 5, 6].

Well	Date	Depth m	T_{res} °C	T_{sep} °C	P_{sep} kPa	$\dot{m}_{tot}$ kg/s	$\dot{m}_{st}$ kg/s	$\dot{m}_{liq}$ kg/s	% CO_2 wt. of steam
KD-1	1968	540	198			NA			
KD-2	1968	769	174			18.6$^{(*)}$			
TH-1	1968	615.5	116			NA			
KD-1A	1969	573.1	195			37.5$^{(*)}$			
KD-3	1969	370	172			NA			
KD-4	1969	486	178			NA			
KD-III	1969	505.9	152			NA			
KD-6 (P)	1970	851	194	147	367	23.32	2.80	20.52	20.0
KD-7	1970	667.5	208			70.0$^{(*)}$			
KD-8	1970	576.5	190			NA			
KD-9	1970	1241	172			42.8$^{(*)}$			
KD-12	1970	404.7	148			NA			
KD-14 (P)	1970	597	210	148	387	29.13	3.50	25.63	10.0
KD-13 (P)	1971	760	198	145	377	26.31	3.16	23.15	17.4
KD-15 (P)	1971	510	208	147	377	31.59	3.79	27.80	17.5
KD-16 (P)	1975	666.5	212	148	387	45.45	5.45	40.00	12.0
KD-17	1975	365.2	144			NA			
KD-21 (P)	1985	898	204	147	387	32.03	3.84	28.19	10.6
KD-22 (P)	1985	888	205	147	367	28.24	3.39	24.85	14.0
KD-20 (P)	1986	810	205	147	377	30.48	3.66	26.82	13.7
TH-2 (A)	1996	2001	168			11.67			
R-1 (P)	1997	2261	242	148	369	44.44	5.33	39.10	21.0
R-2 (I)	2002	1371	197			55.55			

Notes: $^{(*)}$ = maximum flow, (P) = production, (I) = injection, (A) = abandoned.

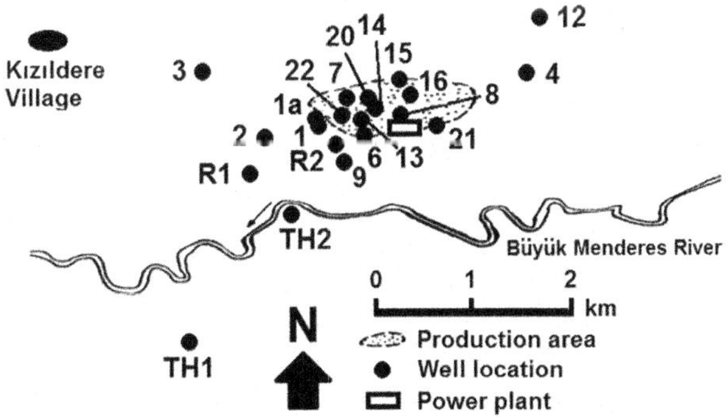

Fig. 21.5 Well field and power plant location, after [6].

channels to the Büyük Menderes River. In 1997 R-1 was drilled specifically as an injection well, being some 1.8 km southwest of the center of the production area. However, it was found that R-1, the deepest well, was also the hottest and very productive, having a power capability of about 6 MW. Thus, in 2001 it was incorporated into the plant as a producer and the output of the plant increased by 6.5%. The wells

TH-1 and TH-2 (Fig. 21.5) were sited on the south side of the river not far from the Tekke Hamam hot spring spa as potential injection wells. These wells showed little permeability and were abandoned. Reinjection is complicated by the high reservoir pressure, at least when the plant first began operating, that requires high pressure pumping to overcome the reservoir pressure. After some 20 years of operation, the reservoir pressure has declined by about 10 bar alleviating the reinjection problem [8].

Given the present understanding of a three-layered reservoir, a methodology has been proposed [9] to produce from the deepest layer and reinject the hot separated water into the shallow layer within the productive area. Since the separated water is at about 147°C and the limit temperature to avoid silica scale is 115°C, there should be no plugging of the injection wells. The cooler wastewater would then be sent to injection wells outside the production area to the south and west of the field.

The calcite scaling in the production wells required frequent work-overs to remove the deposits, reducing plant output and increasing the operating and maintenance costs. Recently, following the 2008 sale of the plants to a private company, the Zorlu Group, a system of downhole injection of scale inhibitor was initiated to prevent the formation of calcite. Such a system is used in several fields that are subject to this problem; see Section 16.5 for a discussion of the system as used at Miravalles in Costa Rica.

21.2.3 Kızıldere Unit 1

The main challenge for the exploitation of the resource is the very high percentage of noncondensable gases (NCG) in the steam [10]. The central station that was constructed at Kızıldere followed the practice developed at Larderello, Italy, to cope with the high levels of NCG, namely, the use of turbocompressors connected to the turbine shaft. Kızıldere Unit 1 (see Figs. 21.6 and 21.7) came online in 1984 and had a rated

Fig. 21.6 Kızıldere Unit 1; photo by author April 2005 [WWW].

ower of 20.4 MW but currently operates at 12–15 MW [11]. Figure 21.8 shows the lant flow diagram in highly simplified form. The flow diagram shows the NCG being ented, but since 1985 an adjacent CO_2 plant has been in operation taking all of he NCG from the plant and producing 120×10^6 kg of liquefied CO_2 and dry ice per ear [2].

The high parasitic power needed to operate the turbocompressors – they take 17% f the gross power – motivated a study to remove the NCG from the steam upstream of he turbine. The compressors alone consume 2.38 MW and the total plant parasitic load s 3.0 MW. A reboiler system was studied, and a pilot-scale unit was built and tested on a ide-stream from one well [12, 13] but has not been implemented in full scale.

Tables 21.3 and 21.4 show the technical particulars for the turbine-generator and he NCG compressors. Based on these data and neglecting the power generated by the xpanding NCG, the turbine has an isentropic efficiency of 71.5%; modern geothermal team turbines have efficiencies about 15 percentage points higher.

Table 21.5 lists the plant parasitics. The total, 2,851.6 kW, amounts to 16% of the ated capacity of the plant. Cooling water is produced by a 4-cell, mechanical-draft, ounterflow cooling tower. Make-up water is taken from the Büyük Menderes River.

ig. 21.7 Turbocompressors at Kızıldere Unit 1; photo by author April 2005 [WWW].

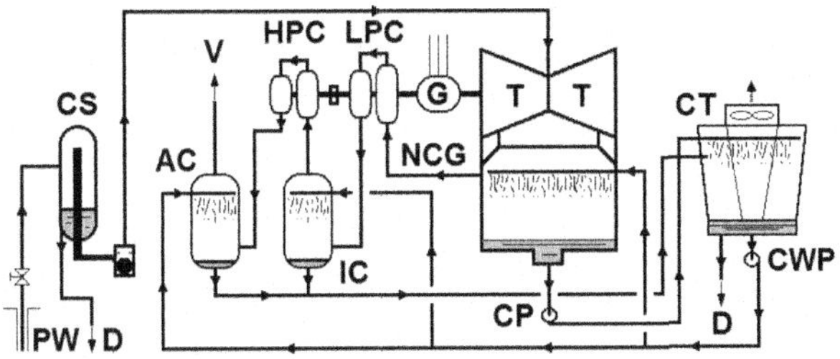

Fig. 21.8 Simplified flow diagram for Kızıldere Unit 1.

Table 21.3 Turbine-generator specifications for Kızıldere Unit 1 [7].

Turbine type	single-cylinder, double-flow, reaction blading
Turbine manufacturer	Ansaldo
Rated capacity	17.8 MW
No. of stages	7 per flow
Speed	3,000 rev/min
Steam inlet pressure	378 kPa (54.8 lbf/in^2,a)
Steam inlet temperature	147°C (297°F)
Exhaust pressure	10.19 kPa (1.467 lbf/in^2,a)
Steam mass flow rate(1)	42.42 kg/s (336,670 lbm/h)
Generator rating	20.6 MVA @ 10.5 V
General manufacturer	Ansaldo

(1)At maximum power output.

Table 21.4 Turbocompressor specifications for Kızıldere Unit 1 [7].

Compressor type	centrifugal
No. of stages	2 with inter-condenser & after-condenser
Speed	3,000 rev/min, LP; 3,900 rev/min, HP
Rated power input	2.38 MW
Inlet gas mass flow rate(1)	8.91 kg/s (70,715 lbm/h)
LP inlet pressure	80 kPa (11.6 lbf/in^2,a)
LP inlet temperature	53°C (127°F)
LP outlet pressure	101.3 kPa (14.7 lbf/in^2,a)
LP outlet temperature	51°C (124°F)
HP inlet pressure	93 kPa (13.5 lbf/in^2,a)
HP inlet temperature	51°C (124°F)
HP outlet pressure	340 kPa (49.3 lbf/in^2,a)
HP outlet temperature	37.5°C (99.5°F)
Manufacturer	Franco-Tosi

(1)Includes 6.44 kg/s NCG and 2.47 kg/s steam at LP inlet.

Table 21.5 Auxiliary power requirements for Kızıldere Unit 1 [7].

Item	No. of units	Total power (kW)
Compressor	1	2,380.0
Cooling tower fan motor	4	149.14
Chlorinated water pump	2	74.0
Service air compressor	2	74.0
Water pump	2	74.0
Main lube oil pump	1	37.3
Emergency bearing lube pump	1	37.3
Sand filters water pump	2	15.0
Sand filters back-wash pump	2	6.0
Miscellaneous	7	4.86
Total	24	2,851.6

21.2.4 Future plans for Kızıldere

The Zorlu Energy Group, the new owner of Kızıldere since September 2008, has indicated its intent to develop a new power unit with a 60 MW rating. The new plant would utilize 20 new wells, or deepened existing wells, to tap the hotter reservoir at temperatures of at least 240°C and at depths of up to 3,000 m [14, 15]. Zorlu also plans to rehabilitate Unit 1 and has already restored the output to 17.4 MW by cleaning out the calcite from the production wells, thereby increasing the flow rate to the plant. They anticipate an expenditure of $250 million for the entire project, which includes about $60 million for the new wells.

21.3 Salavatlı binary plants

There are two binary power plants in operation near the village of Salavatlı about 56 km west of Kızıldere. The geologic setting is similar to that found at Kızıldere since they lie in the same Büyük Menderes Graben. Reservoir temperatures reached so far are somewhat lower than those in the deep reservoir at Kızıldere, but the NGC contents are about the same.

21.3.1 Dora I

The first geothermal power plant to be built by a private company came online in April 2006 on the outskirts of the Salavatlı village and has been producing power continuously since May 2006. It is a 6.5 MW(net) integrated-two-level binary power station that was developed by the Menderes Geothermal Elektrik Uretim A.ş. (MEGE). MEGE also drilled the wells, constructed the power plant, and has maintained it since it commenced operation. The power block equipment was supplied by Ormat [16].

There are two production wells, AS-1 and ASR-2, and one reinjection well, AS-2. Another reinjection well was drilled, ASR-1, but it proved to be such a good producer, capable of 97.2 kg/s total flow, that it was reserved for the second power unit. The reservoir temperature is about 170°C and the combined flow from the two producers is 157.9 kg/s [17]. The geofluid is separated at the wellheads by means of vertical cyclone separators, and steam and brine are conveyed to the plant. Well AS-2 is immediately adjacent to the plant whereas AS-1 requires 900 m of steam and water pipelines and ASR-2 100 m of pipelines. The plant is at an elevation 13 m lower than the more distant producer, AS-1. At the plant there is a moisture remover to dry the steam and an accumulator to collect the brine. Figure 21.9 shows the locations of all wells to date that will serve both Dora units at Salavatlı. Figure 21.10 is an aerial view of the field. Note the close proximity of the Linde Gaz CO_2 facility to the power plant.

The geofluid at the wellhead consists of mainly liquid with a small weight percentage of steam and NCG, about 2.2%, however the NCG make up more than 50% of the vapor phase. The vapor fraction passes to the Level 2 vaporizer where it gives up its latent heat and condenses. The condensate is used no further and is pumped to the reinjection line. The liquid fraction (brine) passes first to the Level 1 vaporizer, then to the Level 2 vaporizer where it travels through a set of tubes separate from those used

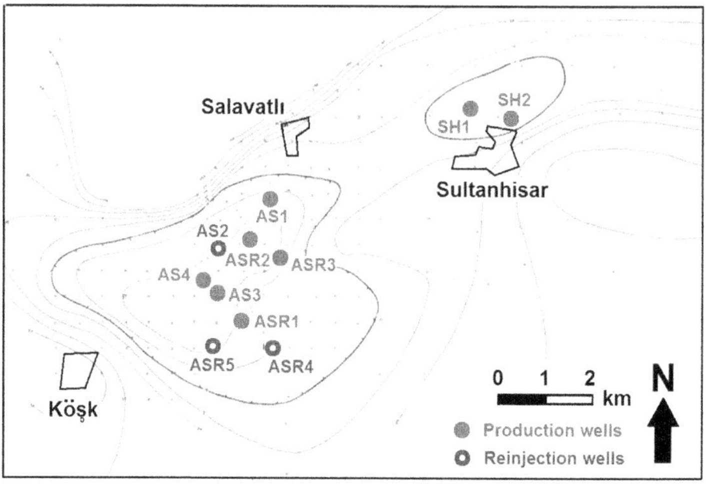

Fig. 21.9 Salavatlı wells drilled for Dora I and Dora II power plants, after [8].

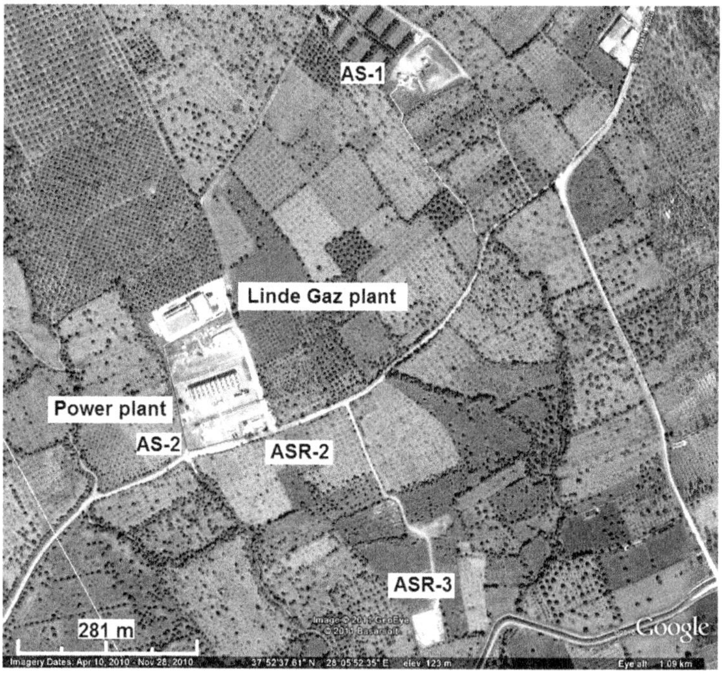

Fig. 21.10 Salavatlı Dora I plant and wells, Google Earth, *April 10, 2010 [WWW].*

by the steam, and then is split into two streams for use in the preheaters of each level. The cooled brine is then at a temperature of about 78–80°C.

The process flow diagram is given in Fig. 21.11. Components labeled with "1" and "2" are for Levels 1 and 2, respectively. Notice that half of the air-cooled condenser

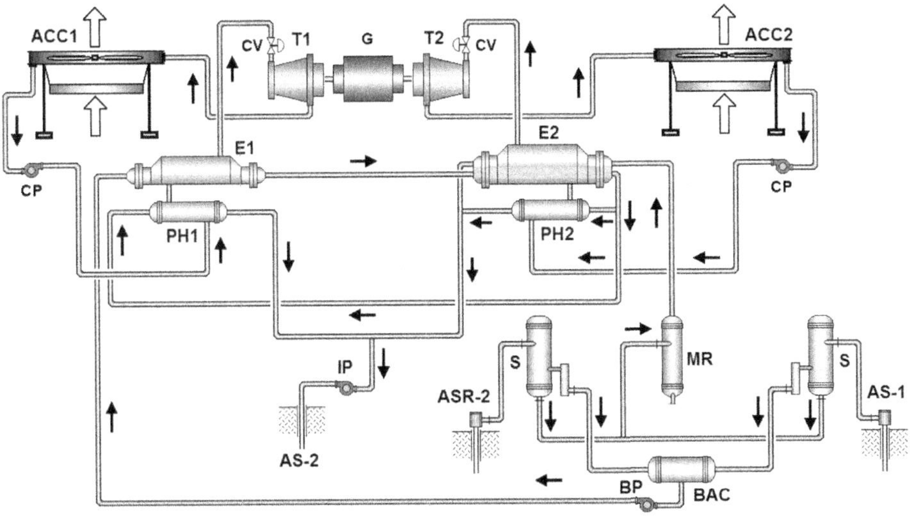

Fig. 21.11 *Salavatlı Dora I flow diagram. The organic Rankine cycle (ORC) working fluid is n-pentane.*

Fig. 21.12 *Salavatlı Dora I power plant [16], before construction of Linde Gaz plant. Reinjection well AS-2 is in the front [WWW].*

wells serve each level separately. Furthermore, after the steam fraction passes through the Level 1 vaporizer, no further use is made of it and its condensate is reinjected. The wells are designated in accordance with current (June 2011) operation. Figures 21.12 and 21.13 are site photographs [16].

The mass flows and geofluid conditions into and out of the power plant are shown in Fig. 21.14. Additionally but not shown, there is a vent stream that discharges the NCG and a small amount of steam: NCG flow rate is 2.10 kg/s and steam flow is

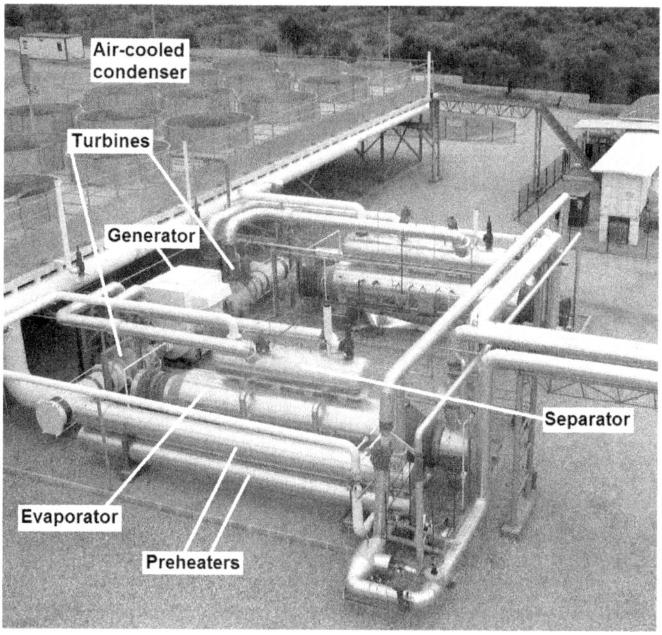

Fig. 21.13 Turbines, generator and heat exchangers at Salavatlı Dora I plant; Level 1 is in the foreground, Level 2 is at the rear [16] [WWW].

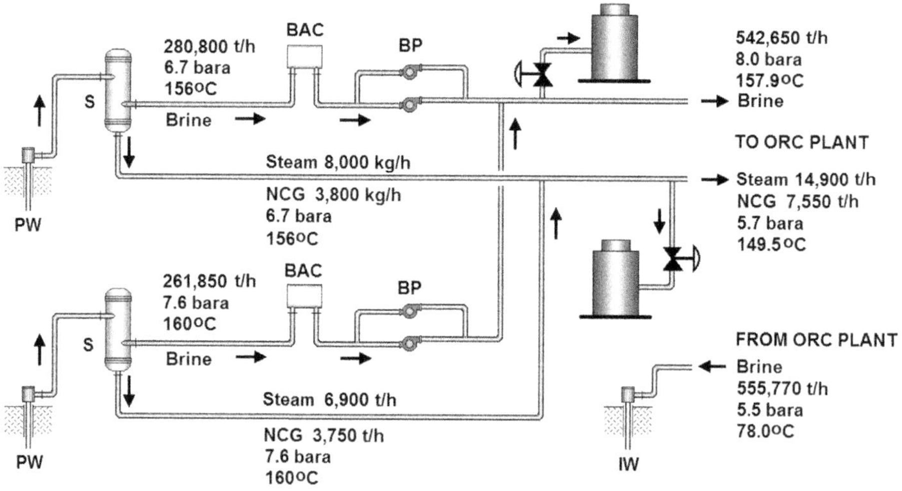

Fig. 21.14 Production steam and brine flows at Salavatlı Dora I plant, original design; NCG vent not shown.

0.49 kg/s. Both of these are sent via pipeline to a CO_2 facility operated by Linde Gaz, which is adjacent to and just north of the power plant.

Leaving aside the internal details of the ORC, the overall plant performance can be calculated from the external flows and the power generation. Table 21.6 lists the flows, their enthalpies, and exergies for the fluids as they enter the ORC.

Table 21.6 Properties of flows into and out of ORC.

Item	Mass flow rate kg/s	Enthalpy kJ/kg	Exergy[(1)] kJ/kg	Exergy rate kW
Brine in	150.7	666.36	110.75	16,693.8
Steam in	4.139	2745.3	761.63	3,152.3
NCG in	2.097	618.81	20.30	42.57
Brine out	154.4	327.0	23.96	3,699.4
Steam out	0.494	2625.8	491.4	242.7
NCG out	2.097	572.3	8.917	18.70

[(1)]Relative to a dead state at 17.1°C and 1 bar,a.

From these values the thermal and exergetic (utilization) efficiencies can be found. Thermal efficiency:

$$\eta_{th} = \frac{\dot{W}}{\dot{Q}_{in}} \quad (21.1)$$

$$\dot{Q}_{in} = \dot{Q}_{in,brine} + \dot{Q}_{in,steam} + \dot{Q}_{in,NCG} \quad (21.2)$$

$$\dot{Q}_i = \dot{m}_i(\Delta h_i) \quad (21.3)$$

Utilization efficiency:

$$\eta_u = \frac{\dot{W}}{\dot{E}_{in}} \quad (21.4)$$

$$\dot{E}_{in} = \dot{E}_{in,brine} + \dot{E}_{in,steam} + \dot{E}_{in,NCG} \quad (21.5)$$

$$\dot{E}_{in,i} = \dot{m}_i[h_{i,in} - h_{i,0} - T_0(s_{i,in} - s_{i,0})] \quad (21.6)$$

The gross power for Dora I is 7,350 kW and the net power is 6,500 kW. Thus, the thermal efficiency comes to 12.2% (gross) and 10.8% (net); the exergetic efficiency is 37.0% (gross) and 32.7% (net).

Owing to the displacement of fossil-fuel-generated electricity, the Dora I plant meets the requirements as a greenhouse gas reduction facility under the Voluntary Carbon Standard 2007.1 (VCS 2007.1). Over the period of inspection to verify compliance, i.e., July 1, 2008 to July 31, 2009, Dora I generated 54,560 MWh(net) of electricity. Each MWh in Turkey offsets 0.591 tCO_2,e and thus the plant constituted a reduction of 32,245 tCO_2,e [21]. Including any planned or forced shutdowns over that period, the average net power output was 5.74 MW and its capacity factor was 88.3% based on the design net power.

21.3.2 Dora II

The second power plant is located about 2 km south of Dora I. It was dedicated on March 25, 2010. The geofluid conditions are very nearly the same and they share essentially the same design of the power conversion equipment. The geofluid mass

Fig. 21.15 Heat exchangers at Salavatlı Dora II plant; Level 1 is at left, Level 2 is at right [19, 20] [WWW].

Fig. 21.16 Dedication of Salavatlı Dora II plant on March 25, 2010; background: Level 1 (L) and Level 2 (R) turbines and generator (C) [20] [WWW].

flow rate is higher allowing a higher power output, namely, 11.5 MW (gross), 9.5 MW (net). The capital cost was $45 million [18].

The photo in Fig. 21.15 shows the heat exchangers; notice that there is only one shell for the preheater per level (left side) whereas Dora I has two (cf. Fig. 21.13). Figure 21.16 shows the turbines and generator at the plant dedication.

The layout of the Dora II site is shown in Fig. 21.17. The northern wells are producers and the two wells at the south are injectors.

Fig. 21.17 Salavatlı Dora II plant and wells, Google Earth, *April 10, 2010 [WWW].*

21.4 Germencik double-flash plant

21.4.1 Overview

The first double-flash plant in Turkey came online on February 28, 2009 and began feeding electricity into the grid in March. It has a rated capacity of 47.4 MW and uses geofluid from eight wells drilled to depths between 965 and 2,432 m. The maximum reservoir temperature so far observed is 232°C. The total produced fluid is about 2,500 t/h with a temperature of 228°C at the wellhead. The plant owner is Gürmat, a subsidiary of the Güriş Group, making the plant the first of its type in Turkey to be privately financed. The new renewable energy law that provides $0.105/kWh has motivated this and other geothermal projects that may have been previously marginal financially [22, 23].

21.4.2 Geologic setting

Germencik is the last of the important, high-temperature geothermal resources along the Büyük Menderes Graben, moving from east to west. The field lies just south of the northern fault boundary of the graben and is host to hot springs and fumaroles. The boundary fault appears to be fractionated into several parallel faults at depth, creating what has been called a "horse-tail fault zone"; see Fig. 21.18 [24].

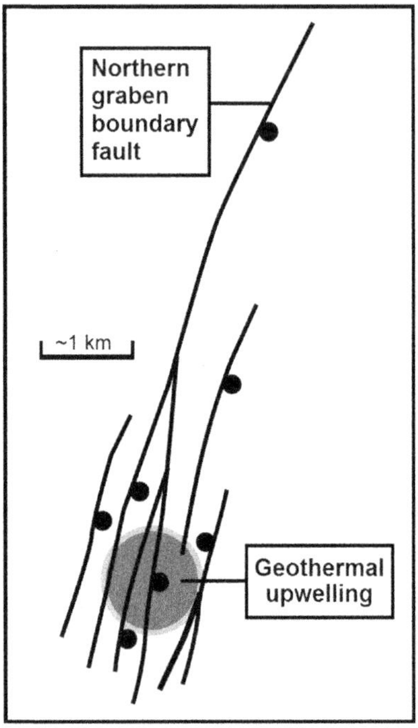

Fig. 21.18 Possible "horse-tail" faulting effect at Germencik, after [24].

Also, the Germencik area is subject to much earthquake activity with numerous quakes in the 3–5 magnitude range and a few in the range 5–6 [25]. This activity keeps the fault zones alive and ever-changing and promotes permeability in the reservoir. It is noteworthy that the highest levels of earthquake activity are focused at the west and east ends of the graben where Germencik and Kızıldere, the most productive Turkish fields, happen to be located.

21.4.3 Germencik double-flash power plant

The design of the power plant was settled upon after a study of a wide range of possible types of energy conversion systems and specific equipment. Capital and operating costs as well as practicality played crucial roles in the process. The plant engineering was performed by Power Engineers (Boise, Idaho); steam and brine piping design was done by Veizades & Associates (San Francisco); reservoir engineering was performed by Geologica (Berkeley, California); and the construction was done by the Turkish company the Güriş Group that owns the plant owner Gürmat. In selecting equipment for the plant, priority was given to materials and equipment that could be sourced locally as an in-country benefit of the project [26].

Among the designs considered were single-flash (rejected due to less efficient utilization of the resource), simple binary (rejected for the same reason, owing to the high temperature of the geofluid); and hybrid flash-binary (rejected due to complexity with only a small improvement in efficiency). Double-flash plants have been installed successfully at resources of similar temperature, e.g., Ahuachapan, El Salvador, exhibit

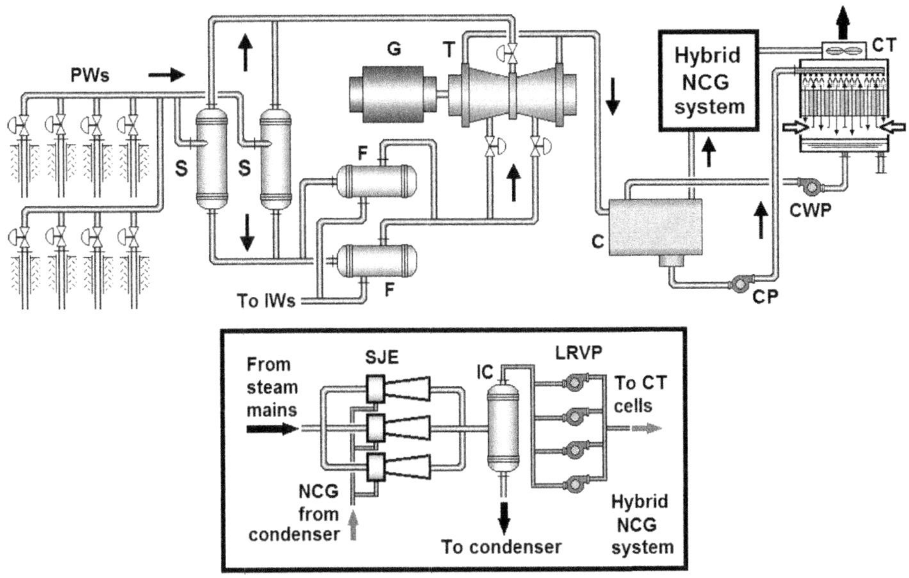

ʼig. 21.19 *Simplified process flow diagram for 47.4 MW Germencik double-flash plant. Note: There is a backup* *JE in parallel with the LRVP to allow maintenance without interruption of operations.*

ɡood efficiency, and are relatively straightforward to operate – all advantages in a ituation where the plant will be locally owned and operated.

The main challenges to be overcome were the high potential for wellbore scaling rom calcium carbonate and the high percentage of NCG in the geofluid, both common roblems for all Turkish geothermal power plants. After a study of various calcite inhiitors, one was selected for injection downhole below the flash point. The anti-scalant ised is called Geosperse (PowerChem Technology, Minden, Nevada). The NCG are handled by a hybrid extraction system with first stage steam-jet ejectors followed y liquid-ring vacuum pumps. The steam-jet ejectors are comprised of three ejectors n parallel, having capacities of 25%, 40%, and 60%, to allow flexibility in dealing with variations in NCG over time. Turbocompressors were considered but were not cost-effective either in capital or operating costs for this set of geofluid conditions.

A simplified process flow diagram is given in Fig. 21.19. There are seven production wells located generally to the north and southeast of the plant, seven hot injection wells located to the west, southwest, and south of the plant, and one cold injection well for the cooling tower overflow to the east of the plant (R. Kaderli, Güriş, personal communication June 23, 2011); see Fig. 21.20. Table 21.7 gives some data for the first nine wells. For the Germencik geofluid conditions, a wellhead mass flow of 100 kg/s corresponds roughly to 7 MW of power. The plant incorporates two high-pressure (HP) vertical cyclone separators, two low-pressure (LP) flash vessels, a demister for each steam flow (not shown in Fig. 21.19), a dual-pressure, pass-in steam turbine with vertical exhaust to a compact advanced direct-contact condenser (ADCC) that is located outside the turbine building, vertical can hot-well pumps, a hybrid NCG extraction system, and a 7-cell, counterflow, low-clog film-fill water cooling tower.

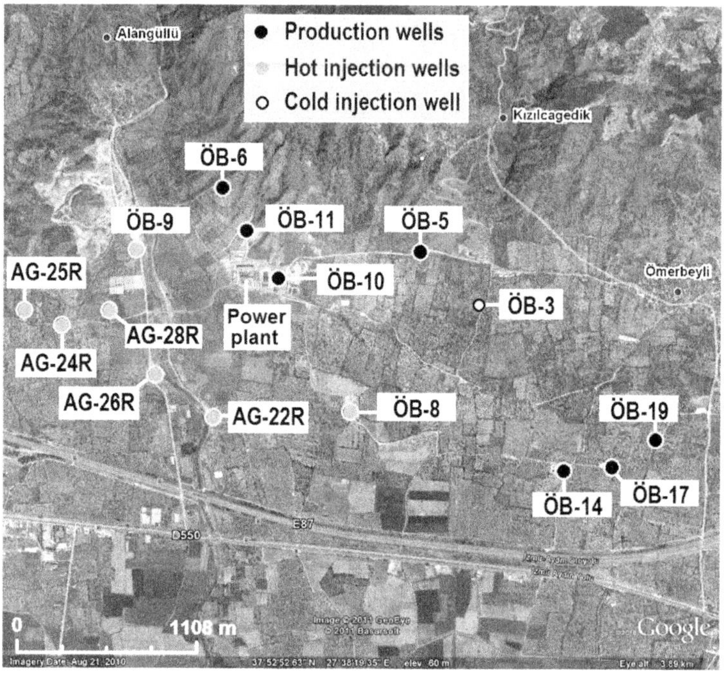

Fig. 21.20 Aerial view of the Germencik field showing wells and power plant locations, Google Earth, *August 21, 2010 [WWW].*

Table 21.7 Properties of the first nine deep wells drilled at Germencik [27].

Well No.	Date	Depth, m	Temp., °C	Mass flow, kg/s
OB-1	1982	1000	203	NA[(1)]
OB-2	1982	975.5	231	25
OB-3	1983	1196.7	230	65
OB-4	1984	285	213	180–100
OB-5	1984	1302	221	65
OB-6	1984	1100	221	140
OB-7	1985	2398	203	65
OB-8	1986	200	219.87	120
OB-9	1986	1464.7	223.8	145
Average			218.4	

[(1)]Geysering flow.

A close-up aerial view of the power house and its associated equipment is given in Fig. 21.21 and a site photo is presented in Fig. 21.22. The power house is the large building in the left center, and the geofluid processing equipment is to the right of the power house adjacent to the holding pond. The NCG extraction system and the external condenser lie to the north of the power house between it and the cooling tower. In Fig. 21.22, the insulated pipelines in the background carry the waste brine to the reinjection wells.

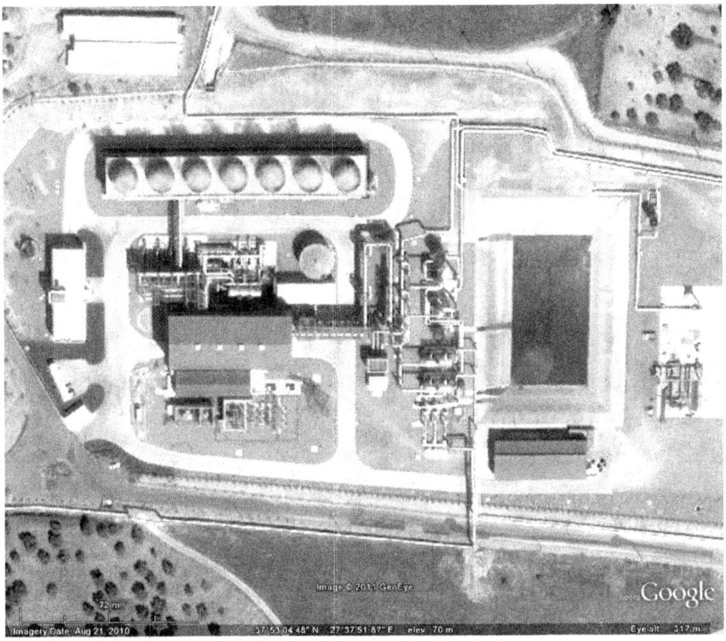

ig. 21.21 Aerial view of Germencik plant layout, Google Earth, *Aug. 21, 2010. The pipeline at the bottom ›reground is the brine reinjection line carrying the fluid to the left [WWW].*

Fig. 21.22 Germencik double-flash power plant, Google Earth, *image 21329604, by E. Atilir [WWW].*

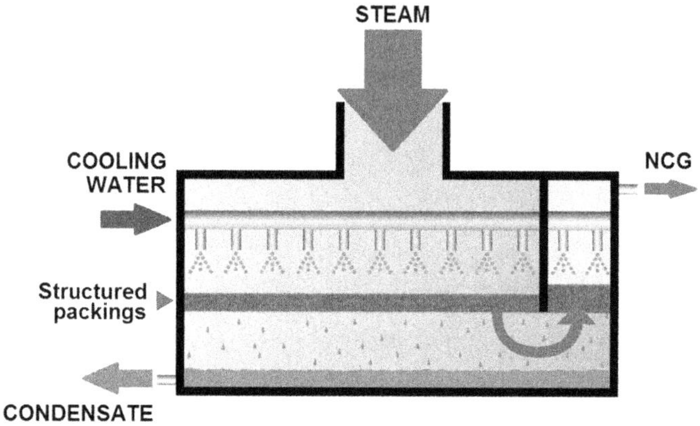

Fig. 21.23 Advanced direct-contact condenser (ADCC) [WWW].

Early trial deployments of the innovative ADCC were done as retrofits at The Geysers in California, but this was the first use in a new power plant as large as Germencik. The technology was developed at the U.S. National Renewable Energy Laboratory (NREL) and commercialized by Ecolaire (Bethlehem, Pennsylvania), specifically for renewable energy applications such as ocean thermal and geothermal power plants; see Fig. 21.23.

The internal surfaces are made of stainless steel to resist corrosion. There is a large area of packing (fill) that enhances heat and mass transfer by means of a very large surface area of contact between the steam and cooling water. In this way the unit mimics the behavior of a wet film fill cooling tower. It also results in less subcooling of the condensate and lower NCG exit temperatures leaving the gas cooler section (upper right corner in Fig. 21.23). The benefits of using this innovative condenser design include [26]:

- Lower cooling water flow, and lower cooling water system capital cost and parasitic loads
- Lower NCG cooler outlet temperature and vapor flow rate, lower NCG system capital costs, and lower ejector steam and vacuum pump electrical consumption

An estimate of the performance of the plant can be made from the reported mass flow rate of geofluid, 2,500 t/h = 694.4 kg/s [22], and the assumed reservoir temperature (at saturated liquid) of 218°C, obtained as the average temperature of the first nine deep wells drilled at Germencik (see Table 21.7). The specific exergy of the geofluid in the reservoir is 205.4 kJ/kg; the rate of exergy produced is 142.6 MW. Thus the utilization efficiency is 33.2% (gross) and 31.6% (net).

The performance record from start-up through May 2011 shows that the plant has operated highly successfully (R. Kadirli, Güriş, personal communication June 20, 2011). The data is summarized in Table 21.8. Since June 2009 through May 2011, the plant has been in operation a total of 17,345.2 hours out of a possible total of 17,520 hours, for a remarkable availability of 99.0%. The parasitic power consumption is 16.9% of the gross power, essentially the same as for the Kızıldere plant.

Table 21.9 lists the cumulative production and reinjection since February 2009. About 76% of the produced geofluid is being reinjected back into the formation.

Table 21.8 Performance of the 47.4 MW Germencik double-flash plant.

	2009(1)	2010	2011(2)
Avg. monthly maximum load, MW	48.97	48.31	48.52
Avg. monthly gross generation, MWh	33,382	33,481	33,115
Total gross generation, MWh	233,675	401,770	165,574
Avg. monthly net generation, MWh	27,530	27,963	27,620
Total net generation, MWh	192,707	335,559	138,099
Avg. monthly hours of operation	722.6	722.9	722.4
Avg. monthly capacity factor, %	95.98	96.54	96.41
Avg. monthly load factor, %	93.42	93.78	93.26
Avg. monthly availability factor, %	98.44	99.01	99.68

(1)7 months from June; (2)5 months through May.

Table 21.9 Cumulative geofluid production and reinjection in 10^6 kg.

From February 2009 through	Production	Reinjection	Net extraction
December 2010	36,800	28,212	8,588
January 2011	38,933	29,483	9,450
February 2011	40,565	30,766	9,799
March 2011	42,377	32,196	10,181
April 2011	44,124	33,568	10,557
May 2011	45,931	34,947	10,984

A more refined calculation can be made of the plant utilization efficiency from the most recent month for which data are available, May 2011: the plant operated full time for 744 hours, received $1{,}807 \times 10^6$ kg of geofluid from the reservoir (674.7 kg/s) and generated an average of 44.38 MW(gross) and 36.91 MW(net). Assuming a reservoir at 218°C, the specific exergy of the geofluid is 205.4 kJ/kg or a rate of 138.6 MW. Thus the plant has a utilization efficiency of 32.0% (gross) and 26.6% (net).

21.5 Environmental impact

The only plant that has been in operation long enough to assess the environmental impact is the Kızıldere plant. The two units at Salavatlı are closed-loop binary plants that practice full reinjection, which renders their impact on the environment negligible. As a flash plant, Germencik does have some impact on the environment owing to releases of steam and NCG to the atmosphere, but it is too early to assess its full impact. Furthermore, if a carbon dioxide capture plant is built, as has been done at the other three power plants, the impact will be minimal.

The impact of the Kızıldere plant centers on the disposal of the waste brine. Since in its early days, the plant did not reinject any brine, all of it was sent to the Menderes River via drainage channels. As of 2005, roughly 278 kg/s was being discharged with no treatment [4]. The brine temperature is about 140°C and the main problem constituent is boron with a concentration of 25 ppm. Since this amount of boron is excessive

Table 21.10 Installed geothermal plants in Turkey as of June 2011 [17, 28].

Plant name	Start-up Year	Type	No. of units	Installed capacity, MW(g)/MW(n)
Kızıldere	1984	Single-flash	1	20.4/17.0
Salavatlı: Dora I	2006	Binary	1	7.95/6.5
Saraköy: Bereket	2007	Binary	1	6.35/5.5
Germencik	2009	Double-flash	1	47.4/45
Tuzla-Çanakkale	2009	Binary	1	7.5/6
Salavatlı: Dora II	2010	Binary	1	11.5/9.5
Denizli: Jeoden	2011	Binary	3	0.78/0.75
Totals			9	101.88/90.25

relative to its use in agricultural irrigation, three possible solutions have been considered: (1) reinjection, (2) removal of boron, and (3) disposal of brine into the Aegean Sea The latter, although theoretically feasible, was ruled out on the basis of cost, practicality, and effectiveness. The second option is promising but as yet unproven on the scale needed for Kızıldere. Thus the only realistic solution is reinjection.

Given the three-layer nature of the Kızıldere reservoir, it has been suggested that one way to carry out this program would be to produce from the deepest, high-temperature but low-permeability reservoir and reinject into the middle-depth formation that has a moderate temperature but high permeability [9]. There are clear advantages to a properly planned and executed reinjection program but its success hinges on a reliable reservoir model.

One potential major environmental impact is avoided by the CO_2 capture facility adjacent to the Kızıldere plant. About 120,000 t/yr (33,333 kg/s) of CO_2 is recovered from the plant, purified, and sold as liquefied CO_2. In fact this one plant supplies over 90% of the demand of the carbonated beverage industry in Turkey.

The last of the impacts to be discussed is the impact on naturally occurring hot springs. The nearby Kızıldere spa lost its natural source of hot water as a result of the lowering of the water table by the producing wells. However, the plant sends cooled waste brine to the spa to make up for this loss. After some 27 years of operation, no impact from Kızıldere has been observed at the famous Pamukkale hot springs and terraces that lie about 25 km to the east.

21.6 Current state and future prospects of geothermal power

Turkey operates nine geothermal power units at five different locations; see Table 21.10.

The Bereket binary unit was designed to receive the 145°C waste brine from the Kızıldere power plant some 2 km to the west. Besides generating electricity the plant was also supposed to provide district heating to the town of Saraköy, about 8 km away from the plant, but several problems arose that rendered the plant of no use [17]. The make-up water for the wet cooling tower was not available for most of the time, being sourced from an irrigation canal. The water quality was also deficient and led to scaling problems. Eventually an air-cooled condenser was installed. There was a temperature

Table 21.11 Planned geothermal plants in Turkey as of June 2011 [28].

Plant location	Planned capacity, MW	Geofluid max. temp., °C
Saraköy-Kızıldere-Denizli	60	242
Salavatlı-Sultanhisar	34	171
Salihli-Manisa: Caferbeyli-1	15	168
Salihli-Manisa: Caferbeyli-2	15	168
Germencik-Aydın	9.5	143
Aydın: Umurlu-1	5	131
Aydın: Umurlu-2	5	131
Atça, Nazilli, Sultanhisar-Aydın	9.5	124
Total	153	

mismatch between the brine leaving the binary unit and the requirements of the district heating system that led to a failure to meet the heating needs in the winter of 2008. Finally, once the Kızıldere plant was privatized, the new owners decided that the waste brine would be better reinjected and the source of fluid for the Bereket plant was no longer available (U. Serpen, personal communication June 6, 2011).

The Tuzla-Çanakkale binary plant is located in northwest Turkey about 50 km south of the western entrance to the Dardanelles strait. The flow rates and temperatures (max. 174°C) observed in the first two deep wells were very encouraging, but later wells drilled to support the 7.5 MW plant have been disappointing.

In 2011, three small PureCycle binary units have been delivered to a private company, Jeoden Geothermal, to be installed in the city of Denizli to operate on low-temperature geofluids at 120°C.

In light of the favorable support for geothermally generated electricity under the new Turkish law for renewable energy, many new projects are now in the planning stage; see Table 21.11. However, the new 2007 Geothermal Code is not without its drawbacks [17]. First, all geothermal activities were frozen for one year at the beginning. Second, there is still a formidable bureaucratic structure to navigate before a project can be approved. And last, it was decided to divide the geothermal land leases into smallish parcels that could lead to overexploitation of resources by neighboring developers. This last difficulty was instrumental in the rapid decline of The Geysers in California when competing developers tried to extract as much fluid as they could without regard to the overall benefit of the resource.

Nevertheless, there are currently in planning enough megawatts to increase the installed geothermal capacity by 150%. The top listed project is the extension and modernization of the Kızıldere plant and resource, and the others should be amenable to exploitation by binary technology.

References

[1] Faulds, J.E., V. Bouchot, I. Moeck and K. Oğuz, "Structural Controls on Geothermal Systems in Western Turkey: A Preliminary Report," *Geothermal Resources Council TRANSACTIONS*, V. 33, 2009, pp. 375–381.

[2] Şimşek, Ş., "Present Status and Future Development Possibilities of Aydın-Denizli Geothermal Province," International Geothermal Conf., Reykjavik, Session 5, Paper 034, Sept. 2003.

[3] Şimşek, Ş., "Geothermal Model of Denizli-Buldan Area," *Geothermics*, V. 14, Is. 2–3, 1985 pp. 155–169.

[4] Şimşek, Ş., N. Yıldırım and A. Gülgör, "Developmental and Environmental Effects of the Kızılder Geothermal Power Project, Turkey," *Geothermics*, V. 34, 2005, pp. 234–251.

[5] Kaya, T. and A. Kindap, "Kızıldere – New Geothermal Power Plant in Turkey," *Proc. Internationa Conf. on National Development of Geothermal Energy Use and International Course/EGEC Business Semina on Organization of Successful Development of a Geothermal Project*, K. Popovski, A. Vranovska an S.P. Vasilevska, Eds., Slovakia 2009.

[6] Koyun, A., "Energy Efficiency and Renewable Energy: Turkey – National Study," Mediterranean an National Strategies for Sustainable Development, Priority Field of Action 2: Energy and Climat Change, Plan Bleu, Regional Activity Centre, Sophia Antipolis, March 2007.

[7] DiPippo, R., "Progress in Geothermal Power Development in The Azores, The People's Republic o China, Costa Rica, El Salvador, Indonesia, Kenya, Turkey, and The U.S.S.R.," *Proc. Fifth Annua Geothermal Conf. and Workshop*, EPRI Rep. No. AP-2098, November 1981, pp. 7-66–7-71.

[8] Serpen, U. and N. Aksoy, "Reassessment of Reinjection in Salavatli-Sultanhisar Field of Turkey," *Proc World Geothermal Congress 2010*, Paper 2302, Bali, Indonesia, April 25–29, 2010.

[9] Serpen, U. and A. Satman, "Reassessment of the Kızıldere Geothermal Reservoir," *Proc. Worl Geothermal Congress 2000*, Kyushu-Tohoku, Japan, May 28–June 10, 2000, pp. 2869–2874.

[10] Gökgöz, A., "Geochemistry of the Kızıldere-Tekkehamam-Buldan-Pamukkale Geothermal Fields Turkey," U.N. University, Geothermal Training Programme, Reykjavik, Iceland, Rep. No. 5, 1998.

[11] Mertoğlu, O., Ş. Şimşek, H. Dagistan, N. Bakir and N. Dogdu, "Geothermal Country Update Report fo Turkey (2005–2010)," *Proc. World Geothermal Congress 2010*, Paper 0119, Bali, Indonesia, Apri 25–29, 2010.

[12] Gökçen, G., and G.E. Coury, "Upstream Reboiler Design and Testing for Removal of Noncondensabl Gases from Geothermal Steam at Kizildere Geothermal Power Plant, Turkey," *Proc. World Geotherma Congress 2000*, Kyushu-Tohoku, Japan, May 28–June 10, 2000, pp. 3173–3178.

[13] Gökçen, G. and Özcan, N.Y., "Performance Analysis of Single-Flash Geothermal Power Plants: Gas Removal Systems Point of View," *Proc. World Geothermal Congress 2010*, Paper 2613, Bali, Indonesia April 25–29, 2010.

[14] Tayman, E., "Zorlu Discovers New Geothermal Field," *Hürriyet Daily News & Economic Review*, October 5, 2009: http://www.hurriyetdailynews.com/n.php?n=zorlu-finds-new-geothermal-field-2009-10-05.

[15] Kindap, A., T. Kaya, F.S.T. Hakhdır and A.A. Bükülmez, "Privatization of Kizildere Geothermal Powe Plant and New Approaches for Field and Plant," *Proc. World Geothermal Congress 2010*, Paper 0708, Bali, Indonesia, April 25–29, 2010.

[16] Shoshan, G. and U. Serpen, "Turkey–The First Private Geothermal Power Plant," *Geotherma Resources Council TRANSACTIONS*, V. 33, 2009, pp. 69–72.

[17] Serpen, U., N. Aksoy, T. Öngür and E.D. Korkmaz, "Geothermal Energy in Turkey: 2008 Update," *Geothermics*, V. 38, 2009, pp. 227–237.

[18] European Bank for Reconstruction and Development (EBRD), Renewable Development Initiative: http://ws2-23.myloadspring.com/sites/renew/geothermal.aspx (accessed June 9, 2011).

[19] Richter, L.X., "Dora-2 Geothermal Plant Goes Online in Turkey," Think Geoenergy, March 27, 2010: http://thinkgeoenergy.com/archives/4245.

[20] Original source (in Turkish) for [19]: http://www.mucadele.com.tr/haber/aydin/dora–2-jeotermal-enerji-santrali-elektrik-uretimine-basladi/19592.

[21] "Verification Report of the Dora I Geothermal Power Plant Project in Turkey, Voluntary Carbon Standard 2007.1," Report No. V001, rev. 03, Germanischer Lloyd Certification GmbH, Hamburg, Germany, Sept. 30, 2009.

[22] Richter, L.X., "Geothermal Plant Goes Online in Turkey," Think GeoEnergy, March 3, 2009: http://thinkgeoenergy.com/archives/1069.

[23] Original source (in German) for [22]: http://www.tiefegeothermie.de/index.php?id=49&tx_ttnews%5bpointer%5d=1&tx_ttnews%5btt_news%5d=200&tx_ttnews%5bbackPid%5d=48&cHash=073bf4e0e0.

[24] Faulds, J., M. Coolbaugh, V. Bouchot, I. Moeck and K. Oğuz, "Characterizing Structural Controls of Geothermal Reservoirs in the Great Basin, USA, and Western Turkey: Developing Successful

Exploration Strategies in Extended Terranes," *Proc. World Geothermal Congress 2010*, Paper 1163, Bali, Indonesia, April 25–29, 2010.

[25] Kumsar, H., Ö. Aydan, H. Tano, R. Ulusay, S.B. Celik, M. Kaya and M. Karaman, "An On-Line Monitoring System of Multi-Parameter Changes of Geothermal Systems Related to Earthquake Activity in Western Anatolia in Turkey," *Proc. World Geothermal Congress 2010*, Paper 1389, Bali, Indonesia, April 25–29, 2010.

[26] Wallace, K., T. Dunford, M. Ralph and W. Harvey, "Aegean Steam: The Germencik Dual Flash Plant," GeoFund-IGA Geothermal Workshop "Turkey 2009," Geothermal Energy in ECA Region Countries, Istanbul, Feb. 16–19, 2009: http://pangea.stanford.edu/ERE/pdf/IGAstandard/GeoFund/Turkey2009/6._wallace.pdf.

[27] Filiz, S., G. Tarcanand U. Gemici, "Geochemistry of the Germencik Geothermal Fields, Turkey," *Proc. World Geothermal Congress 2000*, Kyushu-Tohoku, Japan, May 28–June 10, 2000, pp. 1115–1120.

[28] Özcan, N.Y., "Modeling, Simulation and Optimization of Flashed-Steam Geothermal Power Plants from the Point of View of Noncondensable Gas Removal Systems," PhD Thesis in Mechanical Engineering, The Graduate School of Engineering and Sciences, Izmir Institute of Technology, Izmir, Turkey, June 2010.

Nomenclature for figures in Chapter 21

AC	After-condenser
ACC	Air-cooled condenser
BAC	Brine accumulator
BP	Brine pump
C	Condenser
CP	Condensate pump
CS	Cyclone separator
CT	Cooling tower
CV	Control valve
CWP	Cooling water pump
D	Drain
E	Evaporator
F	Flash vessel
G	Generator
HPC, LPC	High- and low-pressure compressor
IC	Inter-condenser
IP	Injection pump
IW	Injection well
LRVP	Liquid-ring vacuum pump
MR	Moisture remover
NCG	Noncondensable gases
PH	Preheater
PW	Production well
S	Separator
SJE	Steam-jet ejector
T	Turbine
V	Vent

Chapter 22

Enhanced Geothermal Systems – Projects and Plants

"Any time you try to do something innovative, you should expect that there's always going to be people who doubt it, who suggest that perhaps you'd be better off doing something else."

Irwin Jacobs, Qualcomm founder – 2011

22.1 Definitions

In Sect. 1.5.1 we introduced the concept of Hot Dry Rock (HDR) as a potential system for exploiting certain types of geothermal resources. The basic premise is to create a

Geothermal Power Plants: Principles, Applications, Case Studies and Environmental Impact, Third Edition.

© 2012 Elsevier Ltd. All rights reserved.

commercial-grade geothermal system that mimics a natural hydrothermal reservoir. Originally geothermal regions having high temperature but lacking sufficient permeability or fluid in the formation were seen as candidates for this approach. However, as drilling technology has advanced and as the urgency for finding alternative sources of energy has intensified, other potential sites have become targets for this type of development. New language has come into use and definitions are appropriate.

22.1.1 Hot Dry Rock

Formations consisting primarily of granites that have high temperature but very low permeability and lack of stored fluid are candidates for Hot Dry Rock (HDR) development. The first site for this work was the Valles Caldera in New Mexico at the Fenton Hill project. In some cases where the permeability was somewhat higher, the term Hot Fractured Rock (HFR) has been used. Fracture creation and stimulation by means of hydraulic fracturing ("hydrofracking") techniques are essential to produce a commercial grade resource.

22.1.2 Enhanced Geothermal Systems

Originally called Engineered Geothermal Systems, Enhanced Geothermal Systems (EGS) are formations that may possess some desirable characteristics of commercial hydrothermal reservoirs but are lacking in others. This is similar to HDR in that a heat source is inferred but fluid in-place or adequate permeability is lacking. Portions of known hydrothermal reservoirs that lie on the periphery of the developed area are good candidates for EGS. Other locations believed to lack sufficient permeability but which might yield attractive temperatures at deeper depths are also seen as appropriate for EGS development. As with HDR, hydrofracking is needed to create a productive reservoir.

22.1.3 Deep Hydrothermal Systems

This new category of geothermal resource came into existence upon the discovery at several recent EGS projects where permeable and fluid-saturated formations were encountered at depths of 2500–5000 m. In Deep Hydrothermal Systems (DHS), the fluids were under sufficient pressure to allow them to flow to the surface without the need for pumps. Since these regions are not associated with elevated geothermal temperature gradients, the fluid temperature was low to moderate, say 120–140°C, but high enough to be used in a suitable binary power plant. Downwell pumps could be deployed in some cases to enhance flow rates and guarantee that the geofluids remain in the liquid state. Also while some stimulation of the formation may be performed to enhance permeability, it is typically nothing more than is routinely done at normal hydrothermal systems. DHS should be distinguished from very deep wells drilled at existing hydrothermal plants that aim to reach parent fluids that feed the shallower layers of the formation. For example, at Larderello in Italy, wells are now being drilled to 4000 m and beyond to reach steam both beneath the existing reservoir and in areas surrounding the known steam fields.

A useful way to view the EGS/HDR/DHS spectrum is by examining the relationship between the geothermal temperature gradient and the natural formation permeability; see Fig. 22.1 [1]. Bear in mind that the average geothermal gradient is about 33°C/km.

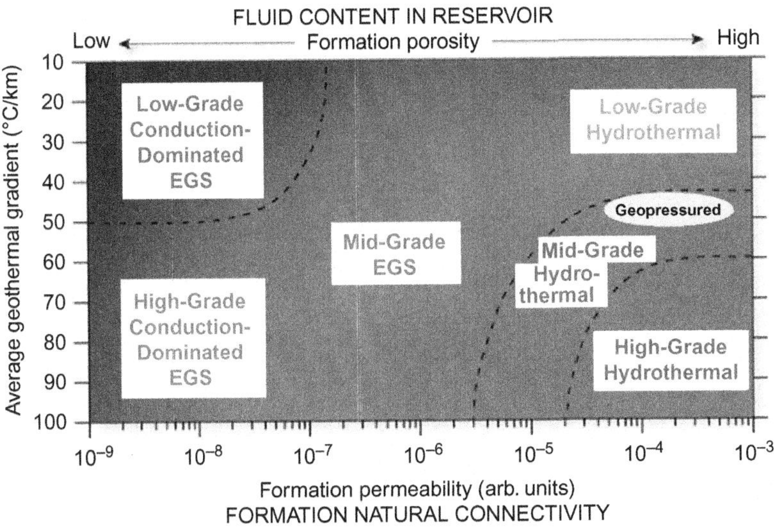

Figure 22.1 The EGS spectrum: temperature gradient vs. formation permeability, after [1] [WWW].

Thus even at that gradient, if the formation is highly permeable (upper right corner), one has the potential for a low-grade hydrothermal system, but this will require very deep wells to achieve temperatures suitable for power generation. This would be a DHS. In the diagonally opposite corner, one finds very little permeability (lack of natural fractures) but a very high temperature gradient. Thus it might be possible with relatively shallow wells to achieve commercial grade production by means of hydrofracking. This is the classic EGS system.

22.2 Early projects

We will describe two programs that began in the 1970s and 1980s, which aimed to develop the HDR concept to the commercial stage: the Fenton Hill and the Rosemanowes projects.

22.2.1 Fenton Hill HDR project

The U.S. Dept. of Energy sponsored a major HDR effort at the Fenton Hill site within the Valles Caldera in New Mexico from 1973–1996 [2]. In the first phase of the project, from 1973–1979, two wells were successfully drilled to a depth of about 2600 m (8500 ft) in hot, fractured crystalline basement rock [3]. The wells communicated through a fracture field in the 185°C (365°F) artificially-created reservoir and were capable of producing pressurized liquid at 135–140°C (275–285°F) at flow rates ranging from 7–16 kg/s (55,500–127,000 lbm/h). A 60 kW binary plant was installed as a means of generating power from the hot water as it circulated through the loop.

During the second phase, starting in 1979, two new wells were drilled about 50 m apart on the surface with the intention of achieving the reservoir conditions shown in

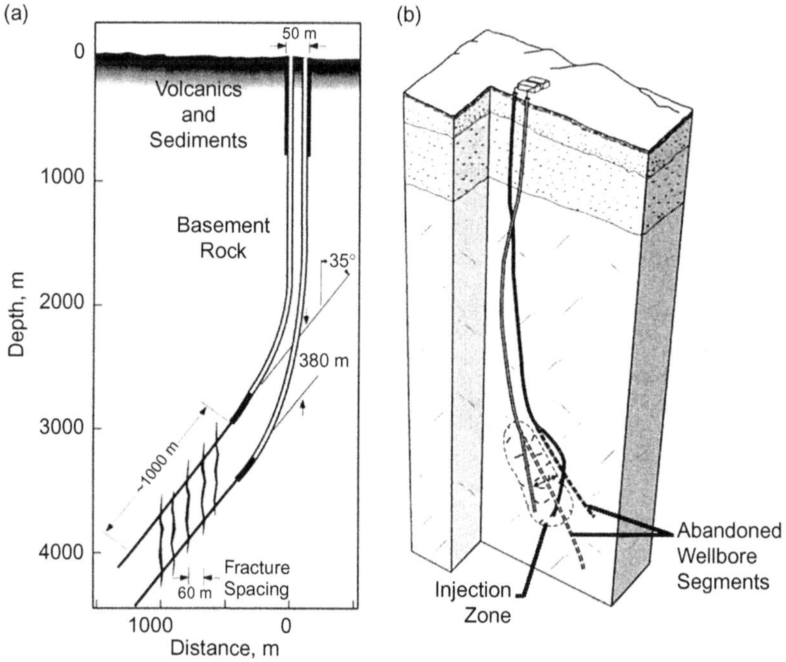

Figure 22.2 Fenton Hill HDR well configuration: (a) conceptual design, (b) actual [2].

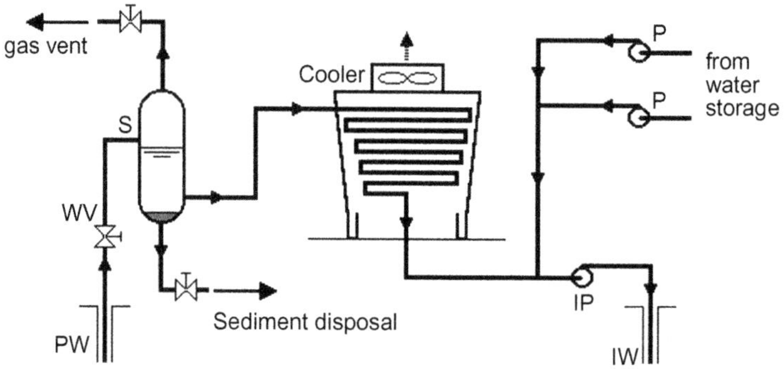

Figure 22.3 Fenton Hill HDR surface testing facility; after [2].

Fig. 22.2(a). The deeper one extended to 4390 m (14,400 ft) into 327°C (620°F) rock. Unfortunately the two new wells did not communicate; the fracture patterns did not conform to the expected ones based on the results of the first phase. After several periods of massive hydraulic fracturing and redrilling, the final configuration shown in Fig. 22.2(b) was achieved.

A simplified schematic of the flow test facility is shown in Fig. 22.3 and some of the results of flow tests are summarized in Table 22.1. The reader is referred to Refs. [2–4] for more details.

able 22.1 Fenton Hill HDR flow test results [2–4].

em	Phase I 1977	Phase II 1986	Phase II 5/1993	Phase II 5/1995	Phase II 6/1995
eservoir depth, m	2700	3550	4204	4204	4204
eservoir temp., °C[(1)]	185	232	228	227	227
ıjection flow rate, kg/s	NA	NA	8.14	7.98	7.82
ıjection pressure, MPa	NA	NA	26.54	27.3	27.3
ıjection temp., °C	NA	NA	22.3	22 (est.)	22 (est.)
roduction flow rate, kg/s	7–16	12–14	6.85	5.82	5.52
roduction pressure, MPa	NA	NA	9.74	9.65	15.2
roduction temp., °C	NA	NA	190.3	185	183
hermal power, MW	3	10	5.50	4.7	5.5
Vater "loss", kg/s	NA	NA	1.29	2.16	2.3
Vater "loss", %	7–1	35–19	16	27	29

)All Phase II temperatures were taken at 3280 m depth.

All of the thermal power developed by the circulating water during the Phase II xperiments was dissipated by an air-cooled heat exchanger; no attempt was made to enerate any power as had been accomplished in Phase I. The apparent water "loss" ıeeds some explanation. The term "loss" may be a misnomer because the fractures viden under increasing pressure, retract with decreasing pressure, and the rock may ontract while being cooled by the circulating water. Thus some injected water could e viewed as stored within the reservoir or outlying fractures. In fact, some tests howed a higher production rate than injection rate depending on the history of injecion pressures. In any case, under steady operating conditions, a fairly large percentge of make-up water would be required to keep a power plant running steadily.

Although the Fenton Hill experiments have ended and the facility has been dismanled, the basic idea of HDR is being applied to some hydrothermal systems that have een partially depleted owing to exploitation. Since roughly 80% of the thermal energy f a hydrothermal reservoir resides in the hot rocks, the life of a reservoir can be xtended by external fluid injection. The case of The Geysers mentioned earlier is such n example.

22.2.2 Rosemanowes, England

The Rosemanowes quarry was the site of England's first experimental HDR project. It s located in the far southwest corner of the country about 4.5 km from Falmouth; see Fig. 22.4. Because of extensive mining in the area, records were available on the type f rocks that would be encountered down to 1 km. The temperature gradient was reasonably favorable at 30–40°C/km. The purpose of this experiment was to study the nechanisms appropriate to create artificial fractures in granite, and not to generate ower. The maximum temperatures in the reservoir were purposely kept below 100°C 5]. The project was undertaken in 1977 by the Camborne School of Mines [6].

Initially several shallow 300 m wells were drilled but the granite at that depth was ıot representative of the deep formation. Then in 1980 two deep wells were drilled, RH11 and RH12; see Fig. 22.5. The bottom-hole temperature in the injection well

Figure 22.4 Location of the Rosemanowes HDR project, Google Earth, *May 11, 2009 [WWW].*

RH12 was 79°C. Stimulation was carried out by injecting up to 100 kg/s of cold water into RH12 using a wellhead pressure of 14 MPa. It was expected that the fracturing of the near-vertically jointed granite would progress upward from RH12 to RH11, but the opposite happened, which led to poor connection between the two wells.

Since a highly fractured formation had thus been created below RH12, in 1983 a third well, RH15, was drilled into the lower formation; see Fig. 22.5. RH15 went to a total vertical depth (TVD) of 2652 m and encountered granite at about 100°C. Following stimulation, RH15 was connected to the original injection well RH12 and a series of circulation flow tests were conducted from 1985 through 1990 under various flow rates and wellhead pressures. The long-term tests were run over the last four years and showed a cooling of the formation from 80.5°C to 70.5°C. Flow rates varied from 5–24 kg/s: when 5 kg/s was injected in RH12, 4 kg/s was returned in RH15 at a wellhead pressure of 40 bar,a; however, when 24 kg/s was injected, only 15 kg/s was returned at 10.5 MPa. At the higher flow rate, a short-circuit was created (low flow resistance, short residence time) that allowed cold water to travel rapidly through the formation without heating sufficiently. Since flow rates of at least 25 kg/s are essential for a commercial power plant, this result was discouraging.

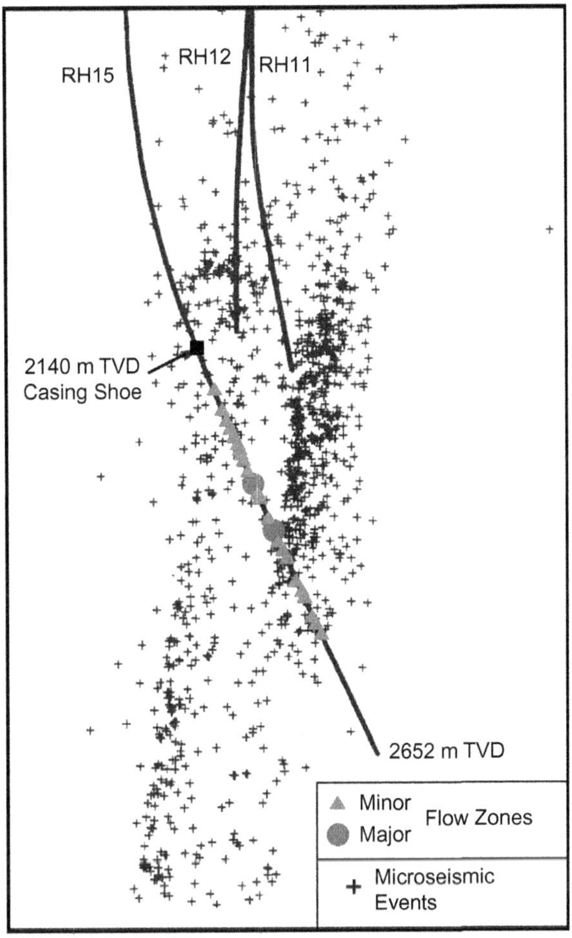

'igure 22.5 Flow zones and induced microseismicity in well RH-15 [6].

Although the Rosemanowes project did not achieve all of its goals, it demonstrated hat hydraulic fracturing (hydrofracking) alone could be used to create a fracture net- vork in hot, naturally-fractured granitic rocks. The main findings included:

1. Regardless of the details of the hydrofracking method, the natural fractures in the granite will control the creation of the fracture network [7].
2. Natural fractures are widely found in crystalline rocks at all depths and all locations that were studied. Thus, an artificial fracture will quickly intersect the natural system within meters of the well, and the natural system will dominate the spread of the fracture system from there.
3. The use of excessive wellhead pressure at the injection well can result in overstimulation of the fracture system and create too direct a connection from the injector to the producer, resulting in lower production temperatures.

4. Finally, it must be recognized that everything one does to a reservoir during hydrofracking is irreversible and can be detrimental to establishing a productive flow loop through the formation. High pressures can move rocks creating permanent short circuits or open pathways to the far field resulting in excessive water loss to the formation [6].

The expertise and knowledge obtained there was put to use later in France at the Soultz-sous-Forêts project (see Sect. 22.4.1). The research facilities and the project staff were transferred to the Camborne School of Mines Associates Ltd. (CSMA) in 1992. Asea Brown-Boveri in 1997 acquired from CSMA the Rosemanowes quarry and all the assets of the project including its intellectual property. In 2004 Schlumberger bought the intellectual property and hired members of the technical staff with expertise in microseismic monitoring. They then sold the quarry. Most recently, in 2006 the site was acquired by 3 K Facilities, a company that offers deep borehole test facilities in one of the most well logged sites in the world. Researchers can gain access to the three deep boreholes for a range of downhole testing. Additionally there are four uncased 300 m holes and one 150 m 30°-deviated hole, as well as a microseismic test station and a wind turbine test facility on the 81,000 m^2 site [8].

22.3 Later projects

22.3.1 Hijiori, Japan

Japan has invested in fundamental and applied HDR research at two sites in the northern part of Honshu, Japan: at Hijiori [9], Yamagata Prefecture (Figs. 22.6, 22.7, and 22.8) and at Ogachi, Akita Prefecture [10]. Hijiori will be covered in this section, Ogachi in the next one.

Figure 22.6 HDR research facility at Hijiori, Japan. Silencers at the production well during a 2-year circulation test, 2000–2002 are shown. The reservoir lies at a depth of 2300 m, with a temperature of 270°C [11].

Figure 22.7 Hijiori HDR test facility showing the make-up water holding pond, production well HDR-2a (rear left), production well HDR-3 (center), and separator-silencer station (right) [11].

'igure 22.8 Experimental binary-type power plant at Hijiori HDR test facility [12].

The Hijiori HDR project (1985–2002) was sponsored by Japan's New Energy and ndustrial Technology Development Organization (NEDO). NEDO was a partner, along vith West Germany, in the Fenton Hill project from 1981–1986.

As can be seen from the schematic, Fig. 22.9, the NEDO approach differed in several vays from the Fenton Hill experiment. First, the reservoir is considerably shallower, eaching only to 1800–2200 m. Next, the induced fractures lie in a generally horiontal orientation, separated by some 400 m, and follow the natural fractures in the ormation. There are two injection wells and two production wells, with each injection vell dedicated to one of the fracture zones: SKG-2 to the shallow zone and HDR-1 to he deep zone. Also there is a river adjacent to the site, the Nigamizu River, that suplied water for the injection. Even with a river at hand, there was a backup holding ond with a capacity of 450,000 kg. And lastly, a 130 kW binary power plant was nstalled in the last stage to convert some of the thermal power into electrical power.

Hijiori is located about 1 km southwest of the Hijiori Hot Springs at the southern oundary of a small, 2 km diameter caldera, in northern Honshu; see Fig. 22.10. The ite is about 50 km west-southwest of the Onikobe power plant.

Work began at the site in 1985 under the auspices of NEDO. There were two phases of the project. In the first stage from 1985–1991 only the shallow reservoir was xplored using SKG-2 as the injector and HDR-1, -2 and -3 as producers. All the wells vere drilled to about 1,800–1,900 m for this phase. The distances from the injector to he producers were intentionally short, i.e., 38 m to HDR-1, 33 m to HDR-2, and 63 m o HDR-3. The wells were drilled successively and the formation was stimulated by circuation among the wells as they were completed. Table 22.2 shows the steps involved.

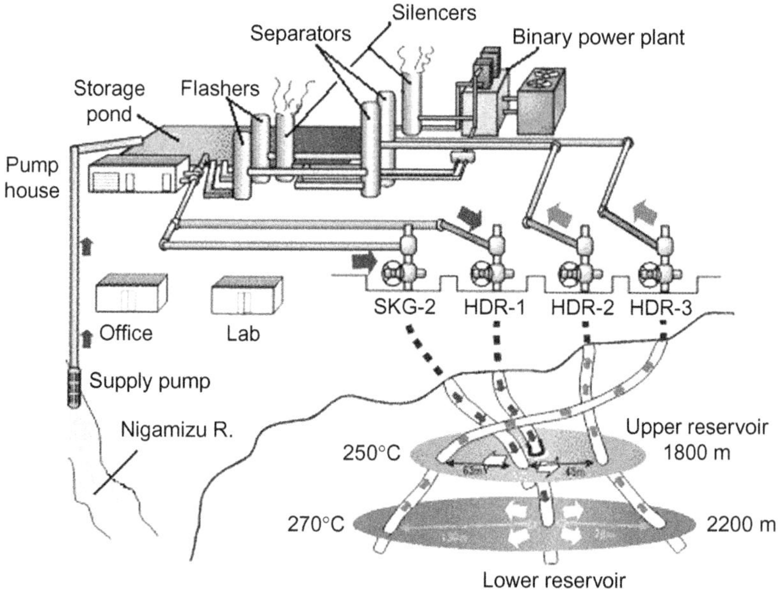

Figure 22.9 Schematic of Hijiori HDR reservoirs, test facility, and power plant; after [9] [WWW].

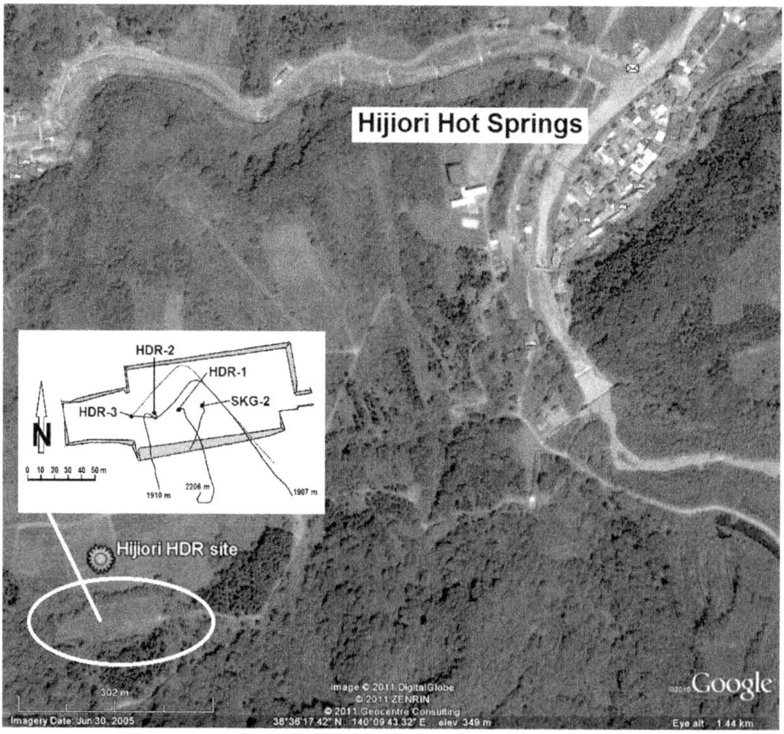

Figure 22.10 Former site of Hijiori HDR project, Google Earth, *June 30, 2005 and [13] [WWW].*

Table 22.2 Stages in well drilling and stimulation for Phase 1 at Hijiori [13].

Year	Events
1986	Hydrofracking at 1800 m via well SKG-2; 1,000 m^3
1987	Well HDR-1 drilled
1988	2-week circulation from SKG-2 to HDR-1
1989	Well HDR-2 drilled; 1-month circulation from SKG-2 to HDR-1 and -2
1990	Well HDR-3 drilled
1991	90-day circulation from SKG-2 to HDR-1, -2, and -3

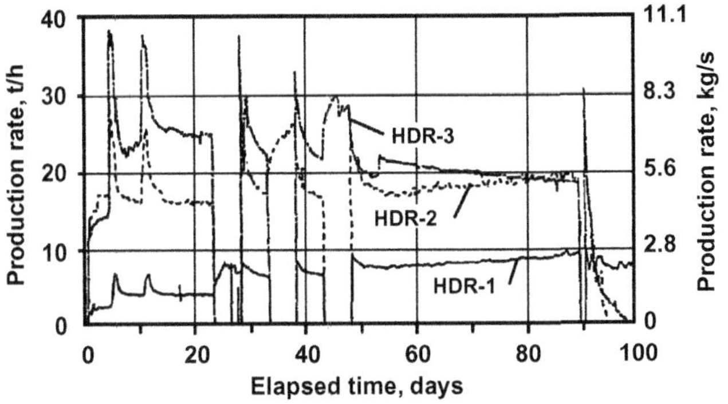

Figure 22.11 Flow rate results of 90-day circulation flow test in the shallow reservoir at Hijiori in 1991 [13].

The 90-day flow test conducted in 1991 had the injection point in SKG-2 at a vertical depth of 1,801 m; fluid entered all three producers at the following main depths: HDR-1 at 1,784 m; HDR-2 at 1,753 and 1,761 m; and HDR-3 at 1,754 m; there were several other secondary fluid entry points in the wells. The flow rates for each of the production wells are shown in Fig. 22.11. Under stable conditions achieved after about 50 days, the flow rates from HDR-2 and HDR-3 were both about double that from HDR-1. The retaining pond was used to hold cold water pumped from the river prior to injection along with the fluids produced from the wells.

Fluid temperatures issuing from each producer are shown in Fig. 22.12. The temperature of the injectate was roughly 50°C, having been obtained from the retention pond. Well HDR-1 was the hottest well at about 190°C and was continuing to heat up as the test came to an end; HDR-3 was the next hottest at 165°C, and HDR-2 appeared to be stabilizing at about 150°C at the end. Note that the natural reservoir temperature at the shallow zone is about 250°C. At the conclusion of the test, the total mass flow rate was slightly less than 50 t/h (13.9 kg/s) and the thermal power extracted from the reservoir was about 8 MW-th. Had a binary plant been installed, this would have supported about 0.8–1 MW of electrical generation.

The second phase began in 1992 and involved the deepening of the three HDR wells into the deeper fracture zone. Prior to deepening, well HDR-2 was plugged back to about 1,600 m and side-tracked; to avoid confusion, it was then renumbered as HDR-2a; see Fig. 22.13 [14].

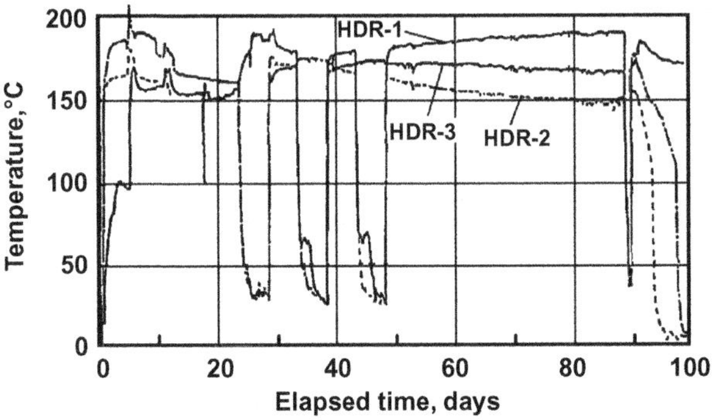

Figure 22.12 Temperature results of 90-day circulation flow test in the shallow reservoir at Hijiori in 1991 [13].

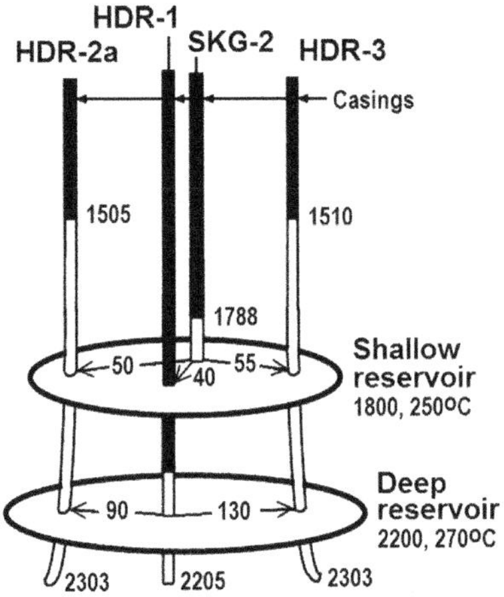

Figure 22.13 Configuration of Hijiori wells after deepening [14].

The deeper reservoir was stimulated in 1992, followed by flow tests in 1994, 199 and 1996; an analysis of these tests can be found in [15]. From December 2000 t August 2002, a 19-month long-term circulation test (LTCT) experiment was carrie out. The first year was dedicated to evaluating just the deep reservoir with HDR- serving as the injector and HDR-2 and -3 as the producers. The last seven month involved injection into both reservoirs with the addition of SKG-2 as an injector. Th binary plant was in operation for the last three months of the testing program [9].

The mass flow rates achieved for the first year of tests are shown in Fig. 22.14; th temperatures recorded for the last three months are shown in Fig. 22.15.

One of the most serious, and unexpected, problems encountered during the LTC was the precipitation of anhydrous calcium sulfate (anhydrite), $CaSO_4$, in the two pro duction wells, HDR-2a and -3 [16]. It was explained by the dissolution of anhydrite b the cold injectate as the water from the holding pond was delivered to the deep rese voir since calcium exhibits retrograde solubility, i.e., it is more soluble in water at lo temperature and becomes supersaturated at high temperatures (opposite of silica

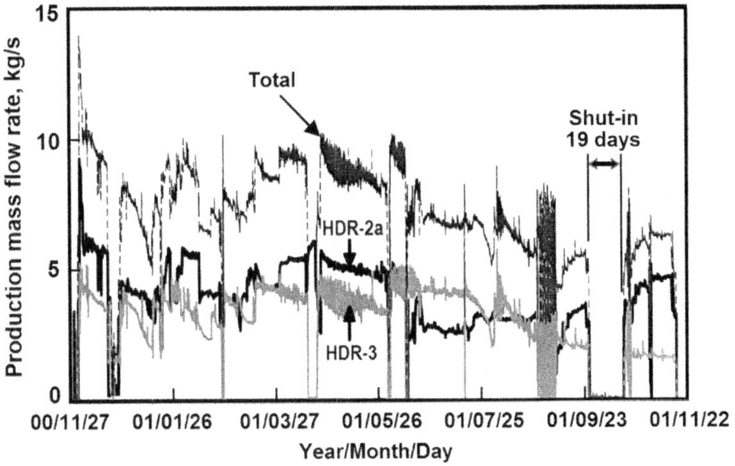

igure 22.14 Long-term flow tests mass flow rates for Hijiori [16].

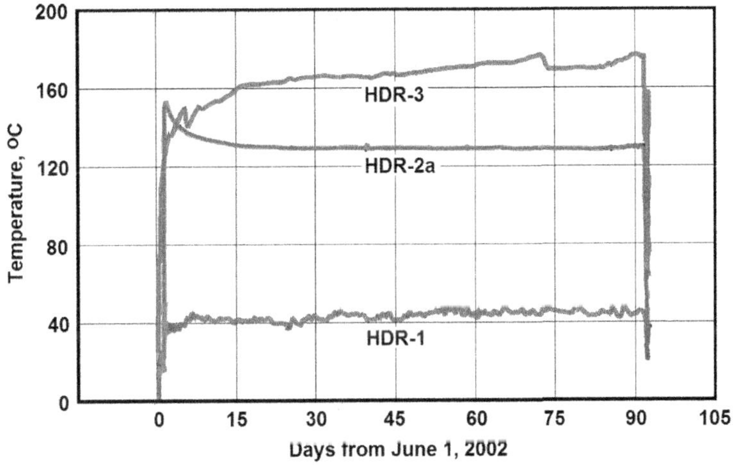

'igure 22.15 Temperature of production wells (HDR-2a and HDR-3) and injection well (HDR-1) [17] [WWW].

Then as the water makes its way through the hot formation at some point it reaches upersaturation and precipitation becomes possible; see Fig. 22.16.

However, the problem was further complicated by calcite and amorphous silica scale in both wells and surface pipelines, depending on the temperature of the fluids. Since the fluid was collected in the pond and recirculated during the LTCT, the dissolved minerals became concentrated and exacerbated the situation. This problem had not appeared for the previous 1-month flow tests presumably because insufficient time had elapsed to reach supersaturation conditions. The problem first appeared after 4–5 months of the LTCT [16].

At the end of the project, a committee reviewed the results focusing on five aspects: 1) overall system design, (2) field characterization, (3) reservoir creation, (4)

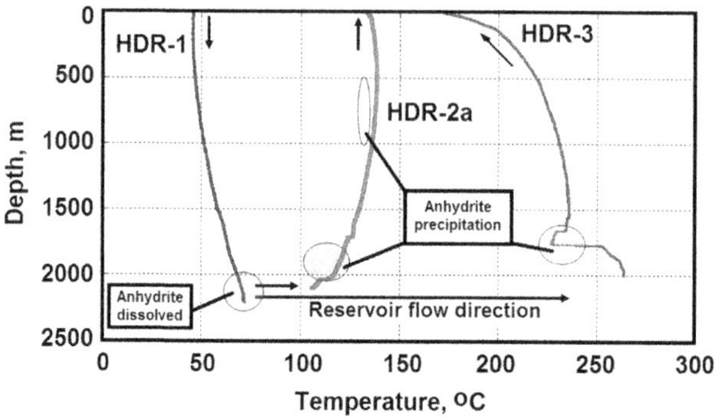

Figure 22.16 Anhydrite precipitation model for Hijiori project [17] [WWW].

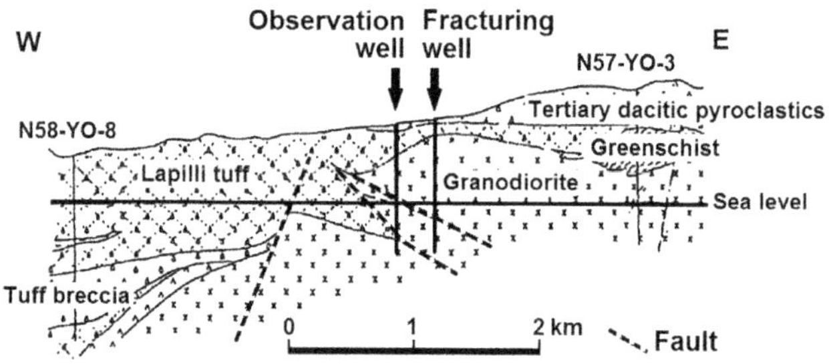

Figure 22.17 Geologic cross-section at Ogachi [18].

circulation/heat extraction, and (5) monitoring [9]. A search of the literature turned up no mention of the fate of the project, but an aerial view of the site from *Google Earth* dated June 30, 2005 (see Fig. 22.10) clearly shows the site restored to its undisturbed condition. The insert plot map is from [13].

22.3.2 Ogachi, Japan

The Ogachi HDR project was conducted during the same time period as the one at Hijiori. Its success as a HDR project was limited owing to lack of support for a multi-well/multi-reservoir system, such as the Hijiori project. In 1990 the original well OGC-1 was drilled to 1,000 m depth into a 228°C uplifted mylonitized granodiorite basement characterized by two major faults; see Fig. 22.17. This well was used to stimulate both fault zones. In 1992 a production well OGC-2 was drilled directionally to intercept both reservoirs; its true length was 1,100 m. A circulation test resulted in only 3% recovery of injected water in OGC-2. After further hydrofracks, first to OGC-2 and then to both wells simultaneously, the rate of recovery improved in steps to 10% to 25%.

The acoustic emissions (AE) from the stimulations were analyzed to determine the extent of the newly fractured region as well as the local stress field. Then in 1999 a new well OGC-3 was drilled to about 1,300 m into the new fractures. A flow test from

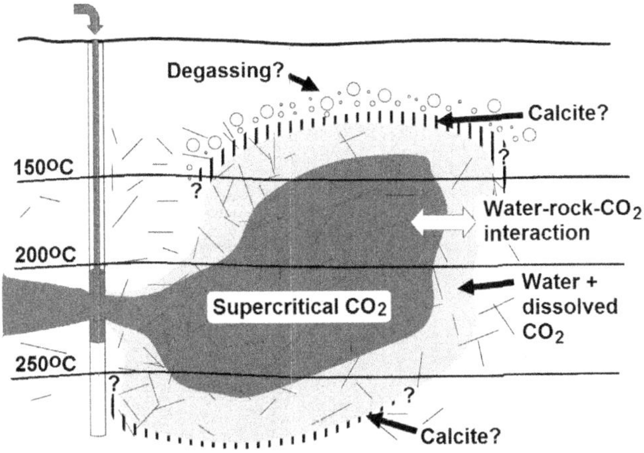

Figure 22.18 Conceptual model of CO_2 sequestration in geothermal reservoirs [20] [WWW].

OGC-1 to OGC-3 indicated connections at three locations at 770, 950, and 970 m depths, as predicted by the AE analysis [19].

In the course of this program a great deal of geoscience was performed that increased understanding of how the local stress field and natural fractures affect the creation of fluid flow paths between wells. A planned long-term flow test using the three Ogachi wells was never carried out. However, the wells were later used to study the concept of carbon dioxide sequestration in geothermal formations [10]. The basic concept is illustrated in Fig. 22.18 where a mixture of water saturated with dissolved CO_2 is forced down a well into a porous formation. Under the reservoir pressures and temperatures, the CO_2 exists in a supercritical state with enhanced ability to dissolve minerals from the formation. As the fluid migrates through the reservoir, chemical processes including degassing and precipitation occur in various parts of the formation.

At Ogachi, experiments were carried out using the existing wells. The first test was run in 2006 with 6.3 kg/s of neutralized river water being injected at a pressure of 15 MPa into OGC-1 and 0.67 kg/s produced via OGC-2; the produced fluid was at 127.5°C. For these experiments, OGC-3 was an observation well. Upon reaching steady conditions, CO_2 was introduced at the injection well as centimeter-size cubes of dry ice. Since the produced fluids were a mixture of the injectate and natural geofluids, no determination could be made about the fixation of carbon in the formation. In 2007 the next test involved injecting the dry ice/water mixture into OGC-2; see Fig. 22.19. The results indicated that the injected CO_2 precipitated as carbonates within a few days [20].

The importance of the Ogachi experiments lies in improving the understanding of CO_2-rock interactions in a non-aqueous, hot formation. It has been proposed that CO_2 may be a useful working fluid in an EGS system, instead of water [21]. Figure 22.20 is a schematic of how a CO_2-EGS reservoir might behave. The three zones will have different modes of interaction with the formation. Zone 1 is devoid of any natural water since it is assumed to be both displaced by and dissolved into the CO_2 stream. As such, the CO_2 would not be a super-solvent as it would be in a supercritical ionic aqueous mixture. This would eliminate the removal of minerals from the rock that would later

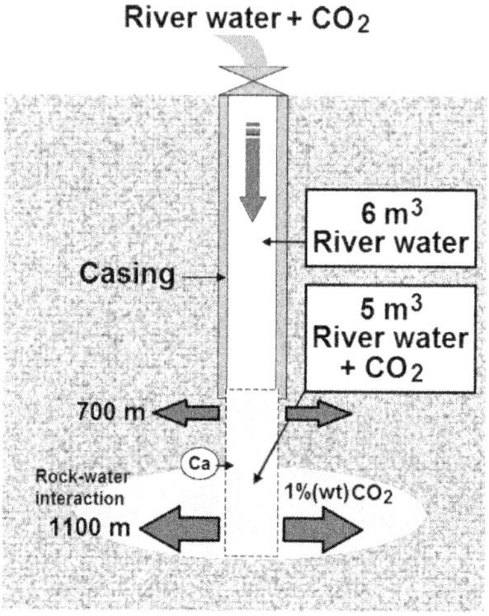

Figure 22.19 Carbon dioxide sequestration experiment at Ogachi HDR project [20] [WWW].

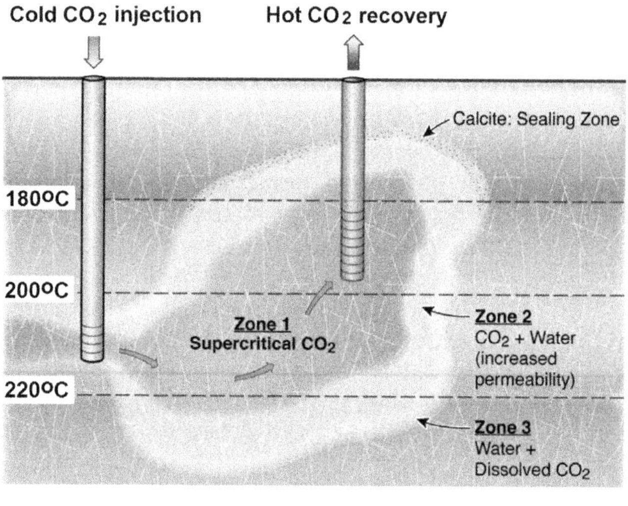

Figure 22.20 Conceptual model of CO_2 injection/production in an EGS reservoir [9] [WWW].

precipitate and cause trouble in the flow paths. Moreover, loosely bound water molecules in some minerals could be removed and lead to improved permeability [9]. All of these hypotheses must await further experiments for verification.

22.3.3 Basel, Switzerland

The city of Basel was selected for a Hot Dry Rock combined heat and power project partly because there was a demand for electric power and a district heating system was already in place. The geology was also favorable since Basel lies at the southernmost end of the Rhine Graben [22] and has a higher than normal heat flux. The fact

able 22.3 Environmental issues of concern to the DHMB [23].

npact area	Possible relevant aspects
'ater – ground & surface	Contamination, change of ground water level and/or flow direction
oil	Contamination, deposition, and treatment
ir	Emission of pollutants, smoke, and odor
oise	Emissions generated during construction, operation, and related transports
ibrations	Ground vibrations caused by construction, operation, and related transports
raffic	Change of traffic volume
iduced seismicity	Change of seismicity through activities in deep rock formations

nat the wells would be sited within Switzerland's second largest metropolitan center vith 830,000 area residents that include communities in France and Germany ttested to the confidence that the developers had in the safety of their technology. 'he project was known as the Deep Heat Mining Basel (DHMB) project and was conucted by Geopower-Basel, a consortium of 11 public and private entities.

In a 2002 report to the International Energy Agency (IEA) the subject of environnental impact was discussed briefly. The issues to be studied are shown in Table 22.3.

It is obvious that these are generic issues for any large construction project except or the last item. Apparently no serious risk assessment was carried out beforehand on he important matter for any EGS project.

The project officially got started in 1996 with support from the Swiss Federal Office of nergy, although theoretical research had been going on for 20 years prior. For the next en years, feasibility studies were performed and sites examined as possible locations for he first deep heat mining project in Switzerland. As a private project, the federal governnent would not be an owner of the facility. The city of Basel expressed an interest and eing situated favorably as mentioned above, it was selected as the project site.

The plant was designed to produce 3 MW electric and 20 MW thermal using one njection well and two production wells drilled to about 5,000 m depth into 200°C ock. The produced fluid was to be 170°C and the reinjection temperature would be '0°C. The power would be generated in a binary plant with the cooled discharged luid being used in a heat exchanger array to heat water for the district heating system efore being returned to the reservoir. The original design mass flow rate was an ambiious 70 kg/s, nearly three times greater than had ever been achieved in any EGS xperiments [22], but was later scaled back to 50 kg/s [24].

The first well (OT-1) was spudded in June 1999 in the Otterbach area. After reachng a depth of 1537 m, drilling problems forced the abandonment of the well in anuary 2001. The measured shallow temperature gradient, down to 537 m, was xtrapolated to 42°C/km. The well was converted into a seismic monitoring station by he Swiss Seismological Service.

The second well (OT-2) was started in March 2001 using the same site for OT-1 out with a higher capacity drill rig. It was completed to a total depth of 2755 m in une 2001. The well was left as an open hole from 2030 to 2755 m with a diameter of 5-7/8". The measured temperature gradient was 38°C/km. The deep formation was ight, fractured granite. This well provided data to allow for planning of the 5000 m vells and was converted to another seismic monitoring station, one of six to monitor

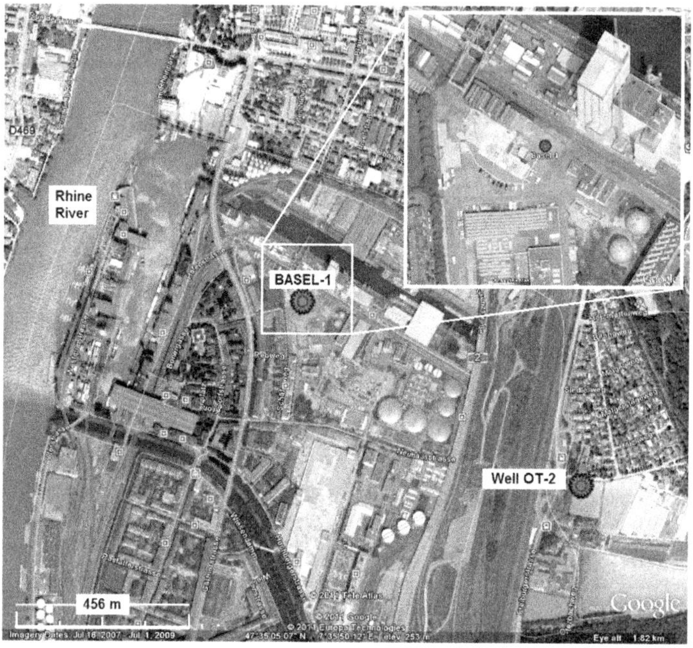

Figure 22.21 Location map for wells Basel-1 and OT-2, Google Earth, *July 1, 2009 [WWW].*

the deep system. By early 2006, all monitoring wells were ready at depths of around 2,700 m, just into the top of the granitic basement.

Prior to the start of drilling of the first deep well, the developers, R.J. Hopkirk and M.O. Häring, wrote these prescient words:

> *"Nature does not obey laws set by mankind and surprises are to be expected. Only if project planning and management ... can meet and deal with such surprises can any serious projections for hot fractured rock EGS be made."* [25]

Basel-1, the first deep well in the DHMB program, was drilled from May to October of 2006 in the industrial section called Kleinhüningen; see Fig. 22.21. It reached a depth of 5000 m, the lower 2.6 km being in the granitic basement; see Fig. 22.22 for the detailed casing and temperature profiles [26]. The thermal perturbations at about 2700 m and just below the final casing shoe are noteworthy in that they indicate interceptions with fractures or faults. The zone at 2700 m was a source of hot water at 120–125°C and possibly might be able to support a small binary plant depending on the flow rate. The obstruction shown near the bottom of the well could not be removed as the cutting tool itself got stuck. The locations for Basel-1 and one of the monitoring wells, OT-2, are shown in Fig. 22.23 against the background of the geologic formation.

Prior to the main stimulation procedure, a brief injection test was carried out in November 2006 under low-pressure and low-flow conditions. The entry to the formation was at the major, highly-altered fractures identified at 4700 and 4835 m depths.

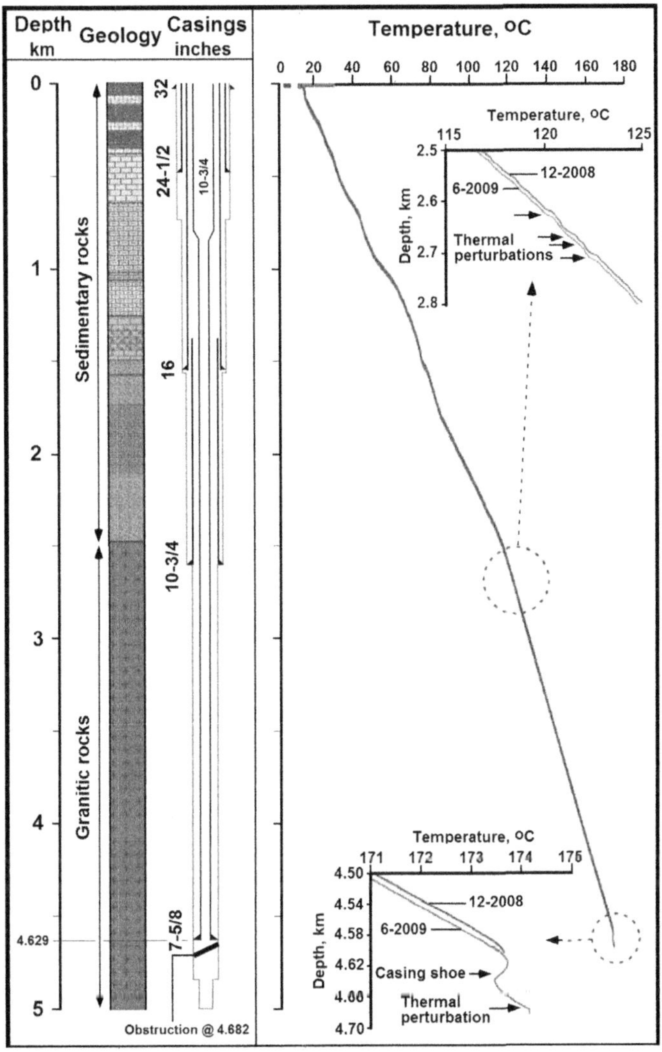

Figure 22.22 Well profile for Basel-1 including temperature measurements [26] [WWW].

It became clear that although the formation contained fluids the well would not flow under artesian conditions.

The massive stimulation began on December 2 using water pumped from the Rhine River basin adjacent to the well. The flow rate was raised in steps up to 1.67 kg/s over the next 16 hours, at which time the wellhead pressure reached 110 bar. Further injection over the next several days eventually reached 55 kg/s with the wellhead pressure at 296 bar. There was a continuing high frequency of microseismic events observed during this period, and when a seismic reading of magnitude 2.7 was noted, the injection rate was lowered in the morning of December 8. But the seismic activity continued, so five hours later, the injection was stopped. While plans were being made to bleed off

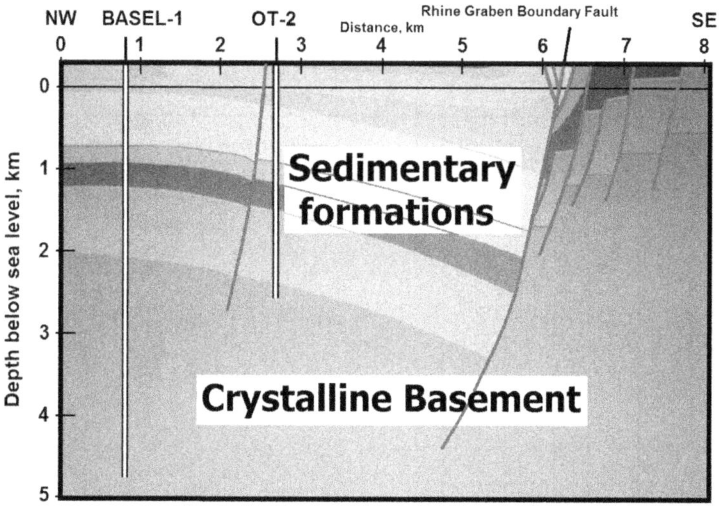

Figure 22.23 Geologic cross-section through the eastern end of the Rhine Graben showing wells Basel-1 and OT-2, modified from [27] [WWW].

the well, a magnitude 3.4 event occurred. This was felt in the community and caused great fear among the citizens who were well aware of the historic magnitude 6.5–6.9 earthquake that destroyed Basel in 1356. While a 3.4 event is not expected to cause damage, apparently some homes suffered cracked plaster walls and similar problems.

The stimulation procedure was highly instrumented with six borehole seismic observation sites arrayed around the Basel-1 well; see Fig. 22.24. The area was fitted with predefined triggers for several color-alerts, a so-called "traffic light" arrangement, set to deploy when seismic events of various magnitudes were recorded, starting with magnitude 2.9 and ground motion of 5 mm/s.

In the 4-month aftermath of the stimulation, the seismic monitoring network recorded over 13,500 potential events, of which the strongest 200 were between 0.7 and 3.4 magnitude with nine being 2.5 or greater [28]. A post-event risk analysis was ordered by the government and the results of that study effectively put an end to the DHMB project. The study showed that while the risk of a major earthquake from reservoir stimulation was low, the risk of similar low-magnitude, but damage-causing events was very high. The estimated cost to repair damage to homes was put at around $7–9 million per year. The study also did not rule out the possibility that other EGS projects might be acceptable in other locations Switzerland and elsewhere, provided a careful technical plan was worked out ahead of time including a thorough risk assessment [29].

A detailed technical analysis of the fluid-rock mechanics involved in the stimulation can be found in [30]. The main conclusions are: (1) the natural fracture system dictated the growth plane induced by the stimulation; (2) the created reservoir was essentially planar in the vertical direction; (3) the displacement achieved was irreversible, and (4) the hypothesis that a highly fractured, interconnected reservoir network could be formed by a massive stimulation was false, at least at Basel. The project developers speculated that by using a more gradual stepwise stimulation process using

Figure 22.24 Well field at Basel: the circles are monitoring wells showing the depth of sondes [24] [WWW].

lower flow rates and wellhead pressures together with a waiting period between pulses ("nudge and let it grow"), it might be possible to avoid the type of seismic events that led to the project's demise in December 2009.

22.4 EGS power plants

22.4.1 Soultz-sous-Forêts, France

The EGS effort at Soultz-sous-Forêts, France got started in 1987 with a research program aimed at developing a fractured reservoir using three deep wells, roughly 5000 m in depth, and ultimately producing both heat for the local district heating system and electric power. In mid-June 2008, the first electricity flowed from the 1.5 MW organic Rankine cycle (ORC) pilot plant. A timetable of major events is shown in Table 22.4

An enormous amount of scientific work has been carried out over the last 25 years and a recent Special Issue of the journal *Geothermics* is a good starting point for further study [33]. The plant is situated on the French side of the Rhine Graben near the border with Germany, about 25 km west of Karlsruhr; see Fig. 22.25.

There are three 5000 m wells that serve as one injector and two producers for the plant: GPK-2, -3, and -4. Several other wells have been either drilled from scratch or deepened (see Table 22.4 and Fig. 22.26) and an extensive array of seismic monitoring stations are in place.

Soultz is characterized by an exceptionally high near-surface gradient, about 110°C/km, about three times higher than normal. Each of the three deep wells revealed an interesting temperature profile, shown in Fig. 22.27 in which the high

Table 22.4 Summary of Soultz project [31, 32].

Year	Event
1987	GPK-1 drilled to 2002 m
1989	3 existing oil wells deepened as monitoring wells for microseismicity
1990	Existing oil well EPS-1 deepened to 2227 m
1991	GPK-1 stimulated from 1420-2002 m
1992	GPK-1 deepened to 3590 m
1993	GPK-1 stimulated from 2850-3590 m
1994	GPK-1 goes into production
1995	GPK-2 drilled to 3876 m
	GPK-2 stimulated from 3211 to 3876 m
	2-week circulation test between GPK-1 and GPK-2
1996	GPK-2 re-stimulated from 3211 to 3876 m
1997	Highly successful 4-month circulation test between GPK-1 and GPK-2
1999	GPK-2 deepened to 5084 m
2000	Seismic monitoring well OPS-4 drilled to 1537 m
	GPK-2 stimulated from 4431 to 5084 m
2002	GPK-3 drilled to 5031 m
2003	Injection test of GPK-2 while observing GPK-3 shows strong connection
	2-week test: injection in GPK-2, production from GPK-3 (as yet unstimulated)
	GPK-3 stimulated
	16-day circulation test between GPK-3 (I) and GPK-2 (P)
	GPK-4 (P) drilling started
2004	GPK-4 completed to 5105 m and stimulated
2005	6-month circulation test among GPK-2 (P), GPK-3 (I), and GPK-4 (P)
2008	1.5 MW pilot power plant installed and 2 circulation tests
2008–09	Tests of line-shaft and electric submersible pumps in GPK-2 and GPK-4
2010-	Continuing tests of reservoir, pumps, and power plant

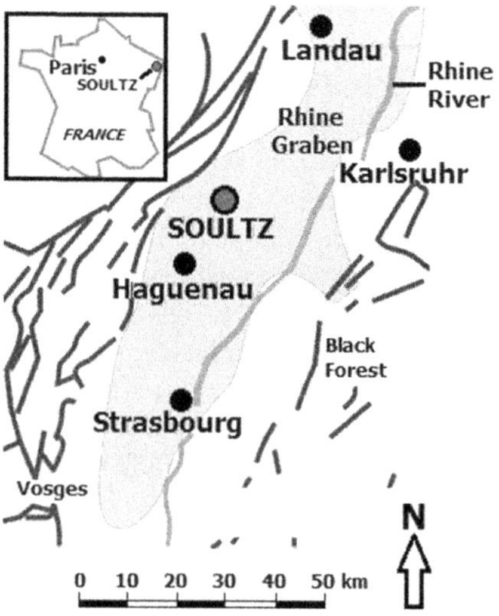

Figure 22.25 Soultz location map; shaded area has high potential for EGS [WWW].

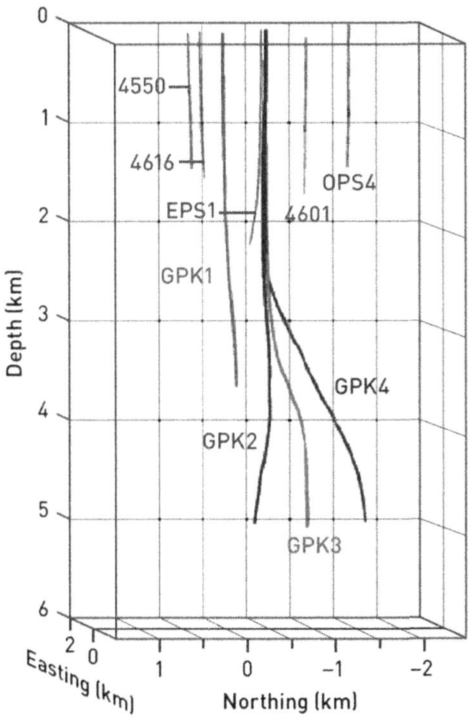

Figure 22.26 Well profiles at Soultz [6].

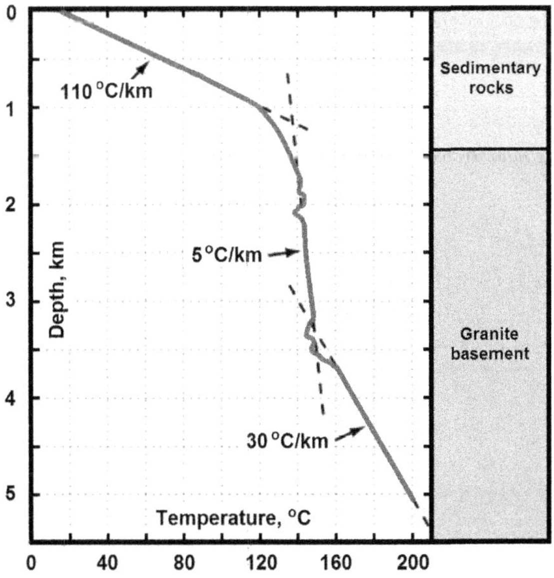

Figure 22.27 Temperature profile in Soultz reservoir, after [34].

near-surface gradient lies above 2.5 km of nearly isothermal rock where the gradient is only 5°C/km, while the lowest 1.5 km is marked by the expected gradient of about 30°C/km. Evidently there is a large section of the formation in which convection plays an important role, lifting high temperature fluids close to the surface and influencing the near-surface gradient. The perturbations seen in the temperature curve in Fig. 22.27 indicate fluid entry or exit points.

The geofluid is a sodium-chloride brine with about 100,000 ppm total dissolved solids (TDS), a pH of 4.9, and a reservoir temperature of 200°C, in equilibrium with the rock. By the time it reaches the plant, the temperature falls to 175°C.

The power plant was designed and built in 2007–2008 and has functioned as a test station while producing power from time to time. A site photograph (Fig. 22.28) shows a compact arrangement made possible because the three wells lie in a line with 6 m separation between them. A 3-D schematic diagram (Fig. 22.29) shows the locations of the major pieces of equipment in the power plant.

Figure 22.30 is the simplified flow diagram with labeled state points for ease of analysis. The geofluid is produced from GPK-2 with a line-shaft pump (LSP) and from GPK-4 with an electric submersible pump (ESP). After it passes through the heat exchangers of the power cycle, it is filtered and pumped back into the reservoir via GPK-3. The turbine is a 1.5 MW radial-flow machine supplied by Cryostar-Turboden turning at 13,000 RPM and driving a 1500 rpm, 11 kV asynchronous generator through a gear box. The power is stepped up and fed into the local 20 kV electrical grid. The plant operates with an air-cooled condenser (ACC) due to the lack of available fresh water at the site.

The thermodynamic cycle properties are shown in Table 22.5 in which the pressures, temperatures, and mass flow rates were taken from [36] but all the other properties were calculated using REFPROP. The pressure-enthalpy diagram for the cycle is given in Fig. 22.31; the shaded band represents the heat transfer in the recuperator.

Figure 22.28 Soultz 1.5 MW pilot power plant [35] [WWW].

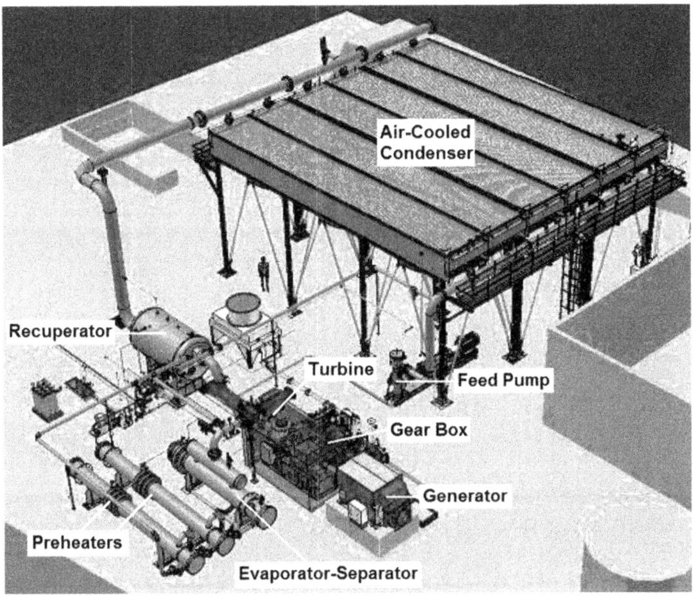

Figure 22.29 Schematic 3-D visualization of Soultz power plant, after [35] [WWW].

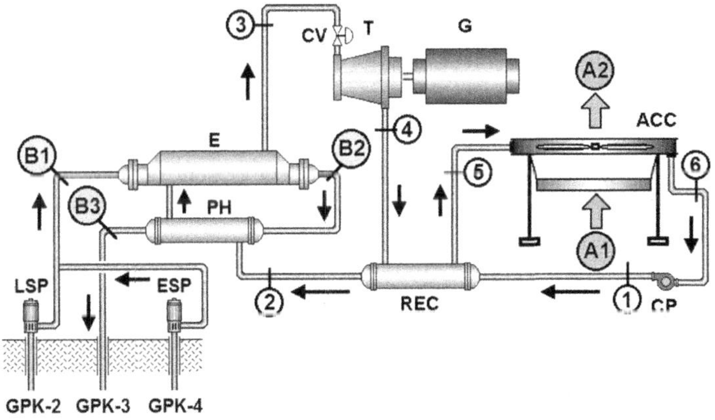

Figure 22.30 Soultz simplified flow diagram.

It is obvious that the original specifications were taken for an assumed ideal, isentropic feed pump since the entropy at states 1 and 6 are essentially identical. In determining the performance, it was here assumed that the feed pump is 75% efficient; the turbine specifications from [36] indicate that an 80.8% efficiency was used, a reasonable value.

The cycle analysis reveals that under these conditions, the power cycle would have a thermal efficiency of 13.9% and a brine utilization efficiency of 40.0% based on a dead-state temperature of 20°C. The presence of the recuperator adds about 1.5 percentage points to the thermal efficiency of a non-recuperated cycle or a gain of about 11.6%.

Table 22.5 Soultz state-point information [36].

State	Temperature, °C	Pressure, bar, a	Entropy, kJ/kg · K	Enthalpy, kJ/kg	Mass flow, kg/
Isobutane					
1	32.3	30.97	1.2528	278.32	34.76
2	50.5	30.57	1.3980	323.91	34.76
3	127.7	30.50	2.4064	702.58	34.76
4s	50.004	4.65	2.4064	629.59	
4	57.3	4.65	2.4493	643.60	34.76
5	32.4	4.25	2.3177	598.10	34.76
6	30.9	4.15	1.2530	273.45	34.76
Geofluid					
B1	175		2.0906	741.02	31.23
B2					31.23
B3	70		0.9551	293.07	31.23
Air					
A1	20.0	3.00	3.5513	419.03	
A2	29.6	2.90	3.5937	428.75	

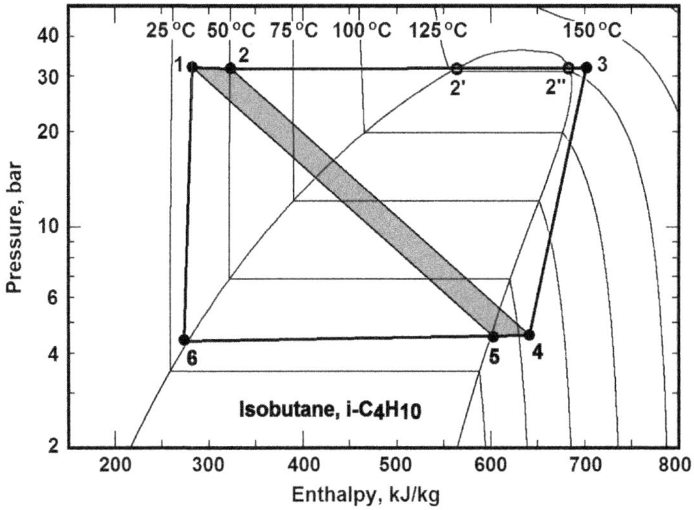

Figure 22.31 Pressure-enthalpy process diagram for design conditions at Soultz.

These values do not account for the external power requirements such as the fan motors in the ACC, geofluid production pumps, and other system pumps. Taking all parasitic power into account is expected to lower the thermal efficiency to about 11.4%.

22.4.2 Neustadt-Glewe, Germany

Neustadt-Glewe, a town of about 8000 residents, is in northern Germany about 70 km south of the Baltic Sea. It is the site of a combined heat-power (CHP) plant that uses both geothermal and fossil energy to provide district heating and electricity. The eating plant went into operation in 1995 to be followed by the electric power plant ome eight years later. It is served by two wells, one producer and one injector, located 350 apart. Depending on demand, the flow rate can vary between 12.8–38.2 kg/s. he geofluid is a sodium-chloride brine containing roughly 220,000 ppm TDS and has density of 1147 kg/m^3 [37]. The production zone lies at 2216 to 2248 m within a 7 m thick sandstone formation.

The EGS power plant at Neustadt-Glewe, the first one in Germany, went into com-nercial production in November 2003, five years before Soultz [38]. However, this lant belongs more in the category of Deep Hydrothermal Systems than Enhanced Geothermal Systems. It tapped into the water-saturated sedimentary formations that re prevalent in some parts of Germany and other European countries. The sole produc-ion well is 2250 m deep and yields highly mineralized water at 98°C. This fluid is used rimarily in a district heating system with excess hot water feeding a binary power plant ORC). In summer when heating demand is very low, essentially all of the hot water goes to the power plant. The thermal power is 11 MW-th of which about 6 MW-th is upplied by hot geofluid (the rest is provided in a peak mode by natural gas) whereas he small binary power unit is rated at 230 kW; see Fig. 22.32 [39]. The annual elec-ricity generation has been about 1.2 million kWh for the last few years [40].

The power plant is operated by Erdwärme-Kraft GbR. The unit receives 32 kg/s of geofluid at 98°C, which transfers heat to perfluoropentane (C_5F_{12}), the cycle working luid, and is discharged at 72°C. C_5F_{12} has a normal boiling point of about 30°C and a riple-point at 147°C and 2 MPa, making it well suited to this geofluid. The turbine is single-stage axial-flow machine with three inlet nozzles. Owing to the corrosive nature of the brine, the heat exchangers are made of titanium [41, 42]. It uses a wet ooling tower with a dedicated fresh water well to provide make-up.

Although the plant was commissioned in November 2003, it underwent several nodifications as problems became apparent. In 2004, the condenser was replaced; see Fig. 22.33. In August 2004 it began continuous operation only to be shut down for

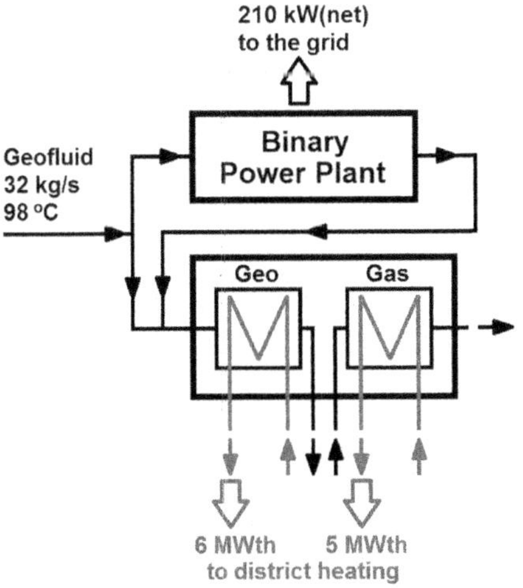

Figure 22.32 Simplified schematic of the Neustadt-Glewe CHP plant.

Figure 22.33 Neustadt-Glewe skid-mounted power unit with new condenser installed in 2004 [41] [WWW].

the winter from November 2004 to March 2005. From April to May 2005, a burned generator bearing caused an outage, immediately followed by another outage from June to August 2005 because of a leaky valve in the ORC. Then a brine pump failed causing another outage from October to November 2005. The total investment cost as of 2006 was 950,000 € (about $1.2 million) [42].

The cycle has a thermal efficiency of 6.6%, allowing for a 5% pressure loss in the preheater and assuming a 70% isentropic turbine efficiency and a 65% isentropic pump efficiency. The utilization efficiency based on the exergy of the incoming brine and a dead-state temperature of 20°C is 17.7%. Since the turbine exhaust is superheated, it is possible to incorporate a recuperator into the cycle (at additional cost) and the thermal efficiency would then rise to 8.3%. This analysis is based on the technical specifications for the plant; for an analysis of some early performance data, see [43].

22.4.3 Landau, Germany

Located about 34 km northeast of Soultz-sous-Forêts in the Upper Rhine Graben, the Landau power plant is the first geothermal CHP plant to be connected to the grid; see Fig. 22.34. The electric power part was the first to go online; the district heating was added later.

A timeline of progress is given in Table 22.6. The Landau project was supported by the German Federal Ministry for the Environment, Nature Conservation and Nuclear Safety (BMU). In addition, the German state of Rhineland-Palatinate assisted with the geological exploration, and mitigated the financial risk by providing an exploration subsidy and a state loan guarantee [45].

A pair of wells, a doublet, constitutes the geothermal connection to the reservoir. One well, Landau 1 with a deviation of 33°, produces fluids at a temperature of about 155°C with the aid of a pump from a highly permeable formation in the Malm limestone. This well is so productive that no stimulation was needed. The other, Landau 2 with a deviation of 25°, is the reinjection well. This well was not so productive and

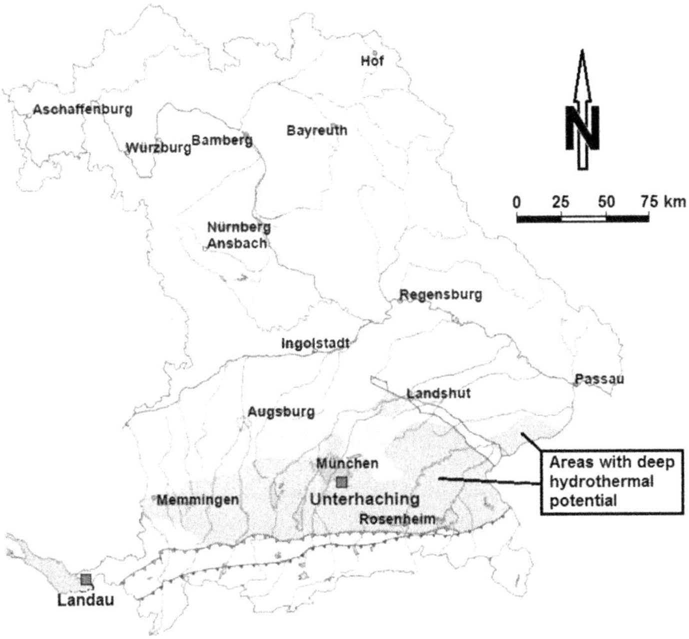

'igure 22.34 Southern Germany and the Molasse Basin showing the locations of the Landau and Unterhaching eothermal power plants, modified from [44] [WWW].

'able 22.6 Chronology of milestones for Landau power plant [45, 47].

)ate	Event
Iay 2003	Beginning of project
ugust–November 2005	Drilling of production well Landau 1
anuary–April 2006	Drilling of injection well Landau 2
'ebruary 2006	Beginning of power plant design
Iarch–April 2006	Stimulation and acidizing of Landau 2
Iarch–May 2006	Running of circulation flow tests
)ecember 2006	Beginning of power plant construction
April–May 2007	Circulation test of Landau 2
Iay 2007	Installation of power plant cooling system
August 2007	Installation of power plant turbine-generator
Vovember 21, 2007	Power plant commissioned
anuary 2008	Start of continuous operation
August 15, 2009	Microearthquake ML = 2.7; continuous operation curtailed, but research operation allowed
September 2009	Plant required to reduce output and reinjection pressure
Vovember 2009	40 residents report cracks in walls of homes
)ecember 2010	Official report: Landau operations very likely the cause for microearthquakes; requires monitoring wells be emplaced around the plant and €50 million in liability insurance be purchased by the plant owners

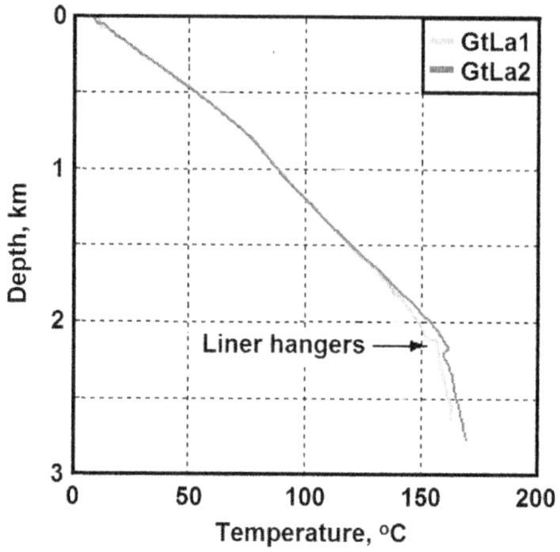

Figure 22.35 Temperature profiles in Landau wells, after [46].

required both massive hydraulic stimulation and acidizing to improve the injectivity [46]. They are separated by about 1200 m in the reservoir. After these efforts circulation tests proved that the doublet could support the 3 MW binary power plant.

From the temperature profiles in Fig. 22.35, it can be seen that there is a convective part of the formation beginning just below 2200 m depth; cf. Fig. 22.27 for the Soultz wells. In this case, the drilling was stopped just below 3000 m depth and production was taken from this permeable layer.

The power plant is a conventional, air-cooled, basic binary cycle that uses isopentane as the working fluid most appropriate for a geofluid temperature of 160°C. The brine discharge temperature from the unit is 72°C. This fluid then passes to the district heating plant and, after providing heat to the town's hot water system, is returned to the reservoir under pressure at 50°C via the 3170 m Landau 2 well. The average net power generated over a typical year is about 2.9 MW.

On August 15, 2009, the residents of Landau were jolted by a loud noise that many likened to a sonic boom. A magnitude 2.9 microearthquake had occurred. It was determined that the epicenter lay 500 m from the well at the plant and at a depth that corresponded to the reservoir. A special investigation was conducted and in 2010 concluded that the plant operation was very likely the cause of the event. Immediately after the microquake, the plant was forced to curtail power and reduce the reinjection pressure. After the final report, the plant was allowed to continue operating but with added seismic monitoring around the plant, which is located on the southern limit of the town. The plant was also forced to buy a €50 million per annum liability insurance policy to cover any future damage to property caused by its operations [47].

22.4.4 Unterhaching, Germany

With reference to Fig. 22.34, Unterhaching lies in southern Germany's Molasse Basin, roughly 290 km southeast of Soultz or Landau. In this area it is possible to reach eservoir temperatures of 120–140°C at depths of 3000–3500 m. Fig. 22.36 shows a orth-south schematic cross-section of the geology in the area.

Like Landau, Unterhaching operates with a doublet of wells: the injector reaches a epth of 3350 m and a temperature of 123.5°C; the producer goes to a depth of 3590 m nd yields fluid at 133.7°C. Both wells are capable of flowing at least 150 kg/s. Both vells experience good inflows from about 3100–3340 m in well Gt Uha1a and from 350–3700 m in well Gt Uha2. Several faults or fault zones are responsible for the ood flow conditions. The wellheads are about 3 km apart. The geofluid is a bicarbonate vater with low TDS of 600–1000 ppm, of which 300–400 ppm is bicarbonate; the pH s 8.8 [49].

The plant receives 150 kg/s of geofluid, but 25 kg/s is sent to the district heating lant directly. After the 125 kg/s is used to drive the 3.38 MW Kalina cycle, it rejoins he discharge brine from the heating plant and is reinjected at 60°C. The Kalina cycle ses a water-ammonia mixture as its working fluid: 11% H_2O + 89% NH_3. The thermal efficiency is reported to be 12%, but the electricity generated is compensated at a ate of €0.15/kWh according to the feed-in tariff of Germany's Renewable Energy Act. 'his allows the plant to be profitable. As of 2007 the total investment cost for the own of Unterhaching in the CHP plant was €60 million (about $75 million) [50].

The history of the project is summarized in Table 22.7.

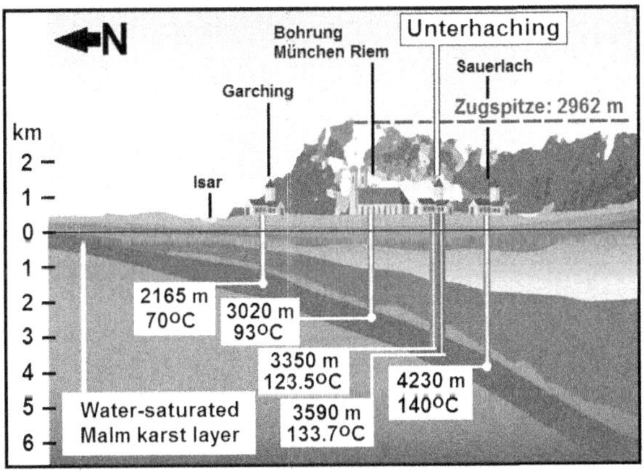

Figure 22.36 North-south schematic section through Molasse Basin [48] [WWW].

Table 22.7 Short history of events for Unterhaching project [49].

Date	Event
January–September 2004	Drilling of production well Gt Uha1a to 3350 m depth
September 2004	Successful acidizing of Gt Uhaa
June–November 2006	Drilling of injection well Gt Uha2 to 3331 m depth
January 2007	Deepening of Gt Uha2 to 3590 m and successful acidizing of Gt Uha2
January 2007	Completion of plant building
October 2007	Commissioning of 27 MW-th heating plant
June 2009	Commissioning of 3.38 MW Kalina cycle power plant

The Unterhaching project achieved several "firsts" for Germany:

- It marked the first time that the Molasse Basin in southern Germany wa exploited for geothermal electricity generation.
- It was the first time that a Kalina cycle was used to produce power on an indus trial scale.
- It was the first time in the Molasse Basin that a geothermal well was fitted wit a submersible pump and produced a mass flow rate of up to 150 kg/s at a tem perature of 123.5°C.
- Lastly, Gt Uha2 was the deepest geothermal well ever drilled in the Molass Basin, reaching a total vertical depth of approximately 3590 m.

22.4.5 Bruchsal, Germany

Located about 35 km east of Landau on the eastern border of the Upper Rhine Graben Bruchsal hosts a small, 0.55 MW pilot power plant that uses a Kalina cycle fitted wit a novel radial outflow turbine (Euler turbine) with a titanium rotor. The turbine oper ates at 28,000 rpm and drives an induction generator through a single speed-reducin gearbox. The turbine has a reported efficiency of 82.4% [51]. The geofluid temperatur is 118°C. The unit came online late in 2009.

22.5 Proposed projects

22.5.1 Australia

Australia's southeastern sector could become a resource for EGS development. It is ar area currently devoid of any overt manifestations of natural geothermal activity, bu has deep hot rocks that have been shown by drilling to be water-saturated. The reservoirs also are highly overpressured by about a factor of two relative to hydrostatic conditions at the depths of 4000–5000 m.

The region evolved over the last 350 million years beginning with an uplift of molten magma that solidified into a granite mass roughly 20 km × 50 km in areal extent. The stages in the evolution are depicted schematically in Fig. 22.37.

Weathering, glaciation, and sedimentary deposition left a high-temperature, partially-fractured, water-saturated body of granite isolated from the surface by impermeable, highly altered sedimentary layers. Radiogenic decay of uranium, thorium, and potassium in the granite has maintained the rock at an elevated temperature. Owing to the 250°C nominal rock temperature, the thermal gradient is about 60°C/km, roughly double the average, normal value. The last important aspect relates to the regional stress field that switched from extensional to compressive, helping to created fractures in the granite.

This resource has been studied and researched through drilling by Geodynamics Ltd., a company formed in 2000 specifically to exploit this resource. Early business plans from 2003 were aggressive and envisioned the building of a commercial demonstration EGS power plant in 2006, and a large-scale base-load plant by the end of 2007. However, numerous drilling problems have disrupted development despite much excellent work in understanding the unique characteristics of the potentially valuable but exceedingly challenging resource.

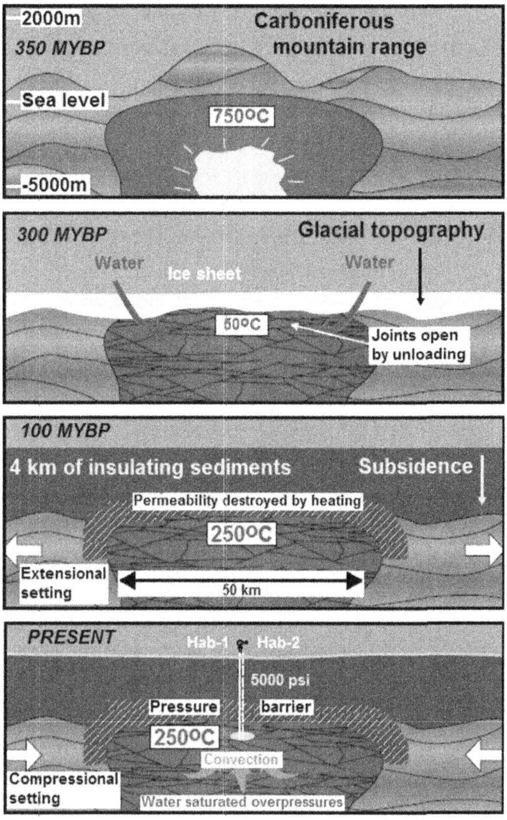

Figure 22.37 Geological evolution of Cooper Basin, Australia, after [52] [WWW].

The first well, intended as an injection well, Habenero-1 (Hab-1), was drilled and completed to a depth of 4421 m in October 2003 at a site 8 km south of the small community of Innamincka in the state of South Australia, just west of the border with Queensland. The well confirmed the high temperatures that were expected, about 250°C, but this well also discovered the overpressured nature of the formation, along with the surprising presence of water in the fractures of the granite. The geofluid was at 750 bar in the formation. The developers had anticipated a true HDR project but found that the well flowed on its own. In 2004 Hab-1 was massively stimulated to create a very large fracture zone, actually seven times greater than anticipated [52]. Further stimulation in 2005 extended the fracture volume to cover an area of about 4 km^2, based on acoustic emissions. The total fractured volume, in the shape of an elliptical pancake, was estimated at 1 km^3.

The second well, Hab-2, a production well, was drilled and completed in December 2004 to a depth of 4359 m. In May 2005, high temperature geofluid flowed to the surface in the first such demonstration in Australia. However, Hab-2 would be plagued by problems caused by well obstructions in the deep fracture zone that could not be retrieved, and which led, after several failed attempts at side-tracking, to its abandonment. The difficulty in drilling was caused by the combination of high temperatur and high pressure in the fracture zones.

A new production well, Hab-3, was drilled in 2007 with a new, more robust rig t a depth of 4221 m into the stimulated fracture zone created by Hab-1. The two well were in immediate communication as evidenced by pressure readings on Hab-1. Hab- has an open hole with a diameter of 8-1/2" to promote a higher flow rate. A circulation flow test between Hab-1 and -3 showed a mass flow of 27 kg/s, essentially th minimum required for a demonstration plant.

A 1 MW binary plant was built to operate with the Hab-1/Hab-3 doublet and supply free electricity to Innamincka. Just prior to the completion of the plant, the nev production well experienced a "well control incident," i.e., it suffered a blowout. Thi occurred on April 24, 2009 late in the evening, only days before the hot commissioning of the 1 MW demonstration plant [53]. A loud noise was heard followed by the discharge of steam and water from the area around the wellhead; see Fig. 22.38. It took 28 days to bring the well under control with cement plugs, during which time th well discharged geofluid to the surface unabated.

Analysis pointed to failed steel casings under corrosive attack by the geofluid in th form of hydrogen embrittlement from the dissolved gases in the brine, exacerbated b the high temperature and pressure. Since Hab-1 and -2 were cased with the sam materials, both of them were plugged and abandoned as a precaution, leaving th Innamincka site with no wells to serve the power plant. Further analysis showed tha the geofluid would cause corrosion in the surface equipment, and stibnite (or antimonite) deposition was also observed. This crystalline substance would cause sever problems in the heat exchanger tubes of a binary plant. One final difficulty in implementing a power plant at Innamincka is the extremely high pressures of the geofluid for which no supplier could be found for the reinjection pumps. The surface pressur

Figure 22.38 Habanero-3 discharging water and steam after blowout on April, 24, 2009; courtesy of Geodynamics Limited [53] [WWW].

f the brine is 350 bar (much higher than the steam inlet pressures at advanced fossil-ıeled power plants). The reinjection pumps need to raise the pressure to 450 bar to ›rce it back into the formation; the seals on commercially available pumps cannot ,ithstand these pressures.

Despite all these serious issues, plans call for the drilling of a new deep well, Hab-4, ometime in 2011 close to the original three wells, while other sites in the vicinity are eing drilled as possible future sites for an EGS plant [54]. In addition to pursuing the eep EGS program, Geodynamics has embarked on a "shallows" program aimed at xploiting the hot sedimentary aquifers that are only 2000–3000 m deep. These prob-bly will not require massive stimulation and fracture creation, but contain fluids at ›wer temperatures than the deep formation, about 135–150°C. Exploitation of the shallows" would be similar to what is being done in the Molasse Basin in Germany, s described in Sect. 22.4.4. As of July 2011, $400 million have been spent on this ffort.

22.5.2 Germany

'he geothermal industry in Germany is currently being stimulated by the generous eed-in tariff provided by the federal government of €0.15/kWh ($0.19/kWh) for elec-ricity generated by geothermal energy. There are three areas that are suitable for ow-temperature binary cycle applications including the North German Basin, NGB e.g., Neustadt-Glewe), the Upper Rhine Graben, URG (e.g., Landau), and the Molasse 3asin, MB, in the south (e.g., Unterhaching). Not including the 24 large hydrothermal listrict heating plants now in operation, there are several electric power projects in ›rogress or under consideration. These include, but are not limited to the following: Dürrnhaar, Mauerstetten, Taufkirchen, and Sauerlach, (MB); Bruchsal, Riedstadt, and Speyer (URG); and Groß Schönebeck (NGB). As was mentioned earlier, geofluids from he URG and NGB tend to be highly mineralized having TDS concentrations 2–7 times greater than sea water, whereas geofluids from the MB are very clean. A research acility has been installed at Groß Schönebeck to study means of coping with the ıigh-TDS, high-NCG geofluids in the NGB. It will include three small binary units 140 kW, 330 kW, and 550 kW) that can operate in various combinations. The facil-ty is expected to be in operation in 2012 [55].

22.5.3 United States

There was an aborted attempt to stimulate an old, unproductive well at The Geysers ield in California by a company called AltaRock. The project was cancelled because he formation was so porous and "mushy" that it could not be drilled, let alone stimu-ated. A few other efforts were made on the fringes of commercial geothermal fields, such as Coso in California and Desert Peak in Nevada, but these have not been out-standing successes.

Currently the main area for an EGS project is the Newberry Volcano in Oregon, about 32 km (20 mi) south of Bend. AltaRock is also the developer for Newberry, but as a partner with Davenport Newberry [56,57]. This is a basic research project aimed at demonstrating that the stimulation techniques used elsewhere can be adopted, or adapted to United States fields. The developers plan to use a method called "hydraulic shearing" instead of hydrofracking to minimize the risk of induced seismicity. There are five goals for the project:

- Demonstrate current technology and advances in EGS by
  - stimulating at least three fracture zones
  - demonstrating diverter technology for multiple zone stimulation
  - demonstrating single-well test methods to assess productivity after stimulation
- Drill two production wells into a newly created EGS reservoir
- Produce economic quantities of fluid per production well
- Establish circulation through a three-well system (a triplet)
- Develop a conceptual model of the complete EGS system [56].

Newberry Volcano is classified as a shield volcano, being made up of hundreds of small vents clustered on a raised mound landform. It is one of the largest and youngest volcanoes in the U.S. The area is a well-known geothermal resource, having been estimate to potentially yield 740 MW for 30 years by the U.S. Geological Survey. In fact in 1995, CalEnergy drilled two deep wells that produced steam.

Funding for the project includes a grant of $21.45 million from the U.S. American Reinvestment and Recovery Act (ARRA) and $22.36 million from the AltaRock-Davenport Partnership. The project is being subjected to intense scrutiny by the environmental community in light of the European experiences with induced seismicity. Extraordinary precautions will be in place before any stimulation of the formation happens. The test site lies about 15 km (9 mi) east of the small Pinecrest neighborhood and about 20 km (12 mi) from the town of La Pine. The Paulina Lake Lodge lies 3 km (1.9 mi) southeast of the injection well site. The future of EGS projects anywhere in the U.S. is probably riding on the success or failure of the Newberry Volcano project.

References

[1] Thorsteinsson, H., C. Augustine, B.J. Anderson, M.C. Moore and J.W. Tester, "The Impacts of Drilling and Reservoir Technology Advances on EGS Exploitation," *Proc. 33rd Workshop on Geothermal Reservoir Engineering*, Stanford University, Stanford, CA, Jan. 28–30, 2008, SGP-TR-18.

[2] Duchane, D.V., "Hot Dry Rock Development Program: Progress Report, Fiscal Year 1993," LA-12903-PR, Los Alamos National Laboratory, Los Alamos, NM, January 1995.

[3] Tester, J.W., D.W. Brown and R.M. Potter, "Hot Dry Rock Geothermal Energy – A New Energy Agenda for the 21st Century," LA-11514-MS, Los Alamos National Laboratory, Los Alamos, NM, July 1989.

[4] Brown, D.W., "1995 Reservoir Flow Testing at Fenton Hill, New Mexico," *Proc. 3rd International HDR Forum*, Santa Fe, NM, 1996, pp. 34–37.

[5] Batchelor, A. S., "The Stimulation of a Hot Dry Rock Geothermal Reservoir in the Cornubian Granite, England," *Proc. 8th Workshop on Geothermal Reservoir Engineering*, Stanford, CA, 14–16 Dec. 1982, pp. 237–248.

[6] Tester, J.W., B.J. Anderson, A.S. Batchelor, D.D. Blackwell, R. DiPippo, E.M. Drake, J. Garnish, B. Livesay, M.C. Moore, K. Nichols, S. Petty, M.N. Toksöz, and R.W. Veatch, Jr., *The Future of Geothermal Energy: Impact of Enhanced Geothermal Systems (EGS) on the United States in the 21st Century*, ISBN 0-615-13438-6, Massachusetts Institute of Technology, Cambridge, MA, 2006; http://geothermal.inel.gov.

[7] Batchelor, A.S., "Hot Dry Rock and Its Relationship to Existing Geothermal Systems," *Proc. Camborne School of Mines Int. Hot Dry Rock Conf.*, 27–30 Jun. 1989, Robertson Scientific Publications, ISBN 1-85365-217-2, 1989, pp. 13–29.

8] "Rosemanowes Quarry," *Wikipedia*: http://en.wikipedia.org/wiki/Rosemanowes_Quarry.
9] Matsunaga, I., H. Niitsuma and Y. Oikawa, "Review of the HDR Development at Hijiori Site, Japan," *Proc. World Geothermal Congress 2005*, Paper 1635, Antalya, Turkey, April 24–29, 2005.
0] Xu, T., H. Kaieda, A. Ueda, K. Sugiyama, A. Ozawa, Y. Wan, and K. Pruess, "Fluid-Rock Interactions in Enhanced Geothermal Systems with CO_2 as Working Fluid: Modeling of Geochemical Changes Induced by CO_2 Injection into the Ogachi (Japan) EGS Site," *Geothermal Resources Council TRANSACTIONS*, V. 34, 2010, pp. 497–502.
1] Vuataz, F.-D.,"Hijiori Hot Dry Rock Project, Northern Japan," Swiss Deep Heat Mining Project, Steinmaur, Switzerland, September 2004.
2] NEDO, "Development of a Hot Dry Rock Power Generation System," New Energy and Industrial Technology Development Organization, Kanagawa, Japan, 2004.
3] Yamaguchi, T., N. Hiwaki, T. Abe and Y. Oikawa, "90-Day Circulation Test at Hijiori HDR Test Site," *Geothermal Resources Council TRANSACTIONS*, V. 16, 1992, pp. 417–422.
4] Tenma, N., T. Yamaguchi, K. Tezuka, Y. Oikawa and G. Zyvoloski, "Comparison of Heat Extraction from Production Wells in the Shallow and the Deep Reservoirs at the Hijiori HDR Test Site Using Fehm Code," *Proc. 26th Workshop on Geothermal Reservoir Engineering*, Stanford, CA, Jan. 29–Feb. 1, 2001, SGP-TR-168.
5] Kruger, P, H. Karasawa, N. Tenma and K. Kitano, "Analysis of Heat Extraction from the Hijiori and Ogachi HDR Geothermal Resources in Japan," *Proc. World Geothermal Congress 2000*, International Geothermal Association, pp. 2677–2682.
16] Kawasaki, K., Y. Oikawa, Y. Sato, N. Tenma, and T. Tosha, "Heat Extraction Experiment at Hijiori Test Site (First Year)," *Proc. 27th Workshop on Geothermal Reservoir Engineering*, Stanford, CA, Jan. 28–30, 2002, SGP-TR-171.
17] Yanagisawa, N., "Ca and CO_2 Transport and Scaling in the Hijiori HDR System, Japan," *Proc. World Geothermal Congress 2010*, Paper 3123, Bali, Indonesia, April 25–29, 2010.
18] Hori, Y., K. Kitano and H. Kaieda, "Outline of Ogachi Project for HDR Geothermal Power in Japan," *Geothermal Resources Council TRANSACTIONS*, V. 18, 1994, pp. 439–443.
19] Kaieda, H., H. Ito, H. Suenaga, K. Kusunoki, K. Suzuki, K. Kiho and H. Li, "Review of the Ogachi HDR Project: Search for Water Flow Paths in HDR Reservoir," *Geothermal Resources Council TRANSACTIONS*, V. 26, 2002, pp. 225–228.
20] Ueda, A., Y. Kuroda, K. Sugiyama, A. Ozawa, H. Kaieda, Y. Kaji, S. Mito and H. Wakahama, "CO_2 Sequestration into Hydrothermal System at Ogachi HDR Site," *Proc. World Geothermal Congress 2010*, Paper 3703, Bali, Indonesia, April 25–29, 2010.
21] Pruess, K, "Enhanced Geothermal Systems (EGS) Using CO_2 as Working Fluid – A Novel Approach for Generating Renewable Energy with Simultaneous Sequestration of Carbon," *Geothermics*, V. 35, 2006, pp. 351–367.
22] Häring, M.O. and R.J. Hopkirk, The Swiss Deep Heat Mining Project – The Basel Exploration Drilling," *GeoHeat Center Bulletin*, Oregon Inst. of Technology, Klamath Falls, OR, March 2002, pp. 31–33.
23] Hopkirk, R.J., "Annual Report for 2002 on the Swiss Participation in the GIA – the IEA Geothermal Implementing Agreement; Includes the Tasks Undertaken by the Chairman of the Executive Committee and Progress in Annex III (Hot Dry Rock)," IEA, 2002.
24] Häring M.O., F. Ladner, U. Schanz and T. Spillmann, "Deep Heat Mining Basel, Preliminary Results," Geothermal Explorers Ltd, Schlossstrasse 3, CH-4133 Pratteln, Switzerland, 2007.
25] Hopkirk, R.J. and M.O. Häring, "The Swiss Deep Heat Mining Programme: Activities & Perspectives," ENGINE Launching Conf., Feb. 12–15, 2006, BRGM, Orléans, France.
26] Ladner, F. and M.O. Häring, "Hydraulic Characteristics of the Basel 1 Enhanced Geothermal System," *Geothermal Resources Council TRANSACTIONS*, V. 33, 2009.
27] Hopkirk, R.J. and M.O. Häring, "The Swiss Deep Heat Mining Programme: Activities & Perspectives," Presentation slides, ENGINE Launching Conf., Feb. 12–15, 2006, BRGM, Orléans, France.
28] "Induced Seismicity in Basel," *Wikipedia*, Dec. 28, 2010: http://en.wikipedia.org/wiki/Induced_seismicity_in_Basel.
29] Brutschin, C. and J. Hofer,"Geothermal Project 'Deep Heat Mining Basel' Cannot Be Resumed," Dept. of Economics, Society and Environment for Basel-Stadt, Basel, Switzerland, Dec. 10, 2009 (in German): http://www.bs.ch/mm/showmm.htm?url = 2009-12-10-wsd-001.htm.

[30] Häring, M.O., U. Schanz, F. Ladner and B.C. Dyer, "Characterisation of the Basel 1 Enhance Geothermal System," *Geothermics*, V. 37, 2008, pp. 469–495.

[31] Hettkamp, T., J. Baumgärtner, R. Baria, A. Gérard, T. Gandy, S. Michelet and D. Teza, "Electricit Production From Hot Rocks," *Proc. 29th Workshop on Geothermal Reservoir Engineering*, Stanfor University, Stanford, CA, Jan. 26–28, 2004, SGP-TR-175.

[32] Cuenot, N., M. Frogneux, C. Dorbath and M. Calo, "Induced Microseismic Activity During Recen Circulation Tests at the EGS Site of Soultz-Sous-Forêts (France)," *Proc. 36th Workshop on Geotherm Reservoir Engineering*, Stanford University, Stanford, CA, Jan. 31–Feb. 2, 2011, SGP-TR-191.

[33] "Special Issue: The Deep EGS (Enhanced Geothermal System) Project at Soultz-Sous-Forêts (Alsace France)," *Geothermics*, V. 35, 2006, pp. 473–710.

[34] Genter, A., K.F. Evans, N. Cuenot, D. Fritsch and B. Sanjuan, "Contribution of the Exploration of Dee Crystalline Fractured Reservoir of Soultz to the Knowledge of Enhanced Geothermal Systems (EGS), *Comptes Rendus Geoscience*, V. 342, 2010, pp. 502–516.

[35] Genter, A., D. Fritsch, N. Cuenot, J. Baumgärtner and J.-J. Graff, "Overview of the Current Activitie of the European EGS Soultz Project: From Exploration to Electricity Production," *Proc. 34th Worksho on Geothermal Reservoir Engineering*, Stanford University, Stanford, CA, Feb. 9–11, 2009, SGP-TR 187.

[36] Cuenot, N., J.-P. Faucher, D. Fritsch, A. Genter and D. Szablinski, "The European EGS Project a Soultz-sous-Forêts: From Extensive Exploration to Power Production," ENGINE Report, April 2008.

[37] Schneider, H., P. Seibt and H. Menzel, "Hydrogeotherrnal Energy Use - The Example of the Neustadt Glewe Geothermal Plant, Germany," *Proc. European Geothermal Conf. Basel '99*, V. 2, Basel Switzerland, Sep. 28–30, 1999.

[38] Schellschmidt, R., B. Sanner, R. Jung and R. Schulz, "Geothermal Energy Use in Germany," *Proc European Geothermal Congress 2007*, Unterhaching, Germany, May 30–Jun. 1, 2007.

[39] BINE Information Service, "Geothermal Electricity Generation in Neustadt-Glewe," Bonn, Germany March 2009.

[40] EGEC – European Geothermal Energy Council, K4RES-H, Key Issue 5: Innovative Applications "Combined Geothermal Heat And Power Plants (CHP)," c.2005.

[41] Broßmann, E., and M. Koch, "First Experiences with the Geothermal Power Plant in Neustadt-Glewe (Germany)," *Proc. World Geothermal Congress 2005*, Paper 1316, Antalya, Turkey, April 24–29 2005.

[42] Funke, T., "ORC Power Plant Neustadt-Glewe: Operational Experience Since 2004," Presentation a Electricity Generation, Combined Heat and Power, Strasbourg, September 15, 2006.

[43] Köhler, S., "Analysis of the Combined Heat and Power Plant Neustadt-Glewe," *Proc. World Geotherma Congress 2005*, Paper 1309, Antalya, Turkey, April 24–29, 2005.

[44] Schönwiesner-Bozkurt, C., "Geothermal Energy Pilot Project Unterhaching, Germany," Presentation Strasbourg, September 15, 2006.

[45] BINE Information Service, "Geothermal Electricity Generation in Landau," 2007: http://www.bine.info/en/hauptnavigation/topics/renewable-energy-sources/geothermal-energy/publikation/geothermische-stromerzeugung-in-landau/?artikel=221.

[46] Schindler, M., J. Baumgärtner, T. Gandy, P. Hauffe, T. Hettkamp, H. Menze, P. Penzkofer, D. Teza, T. Tischner and G. Wahl, "Successful Hydraulic Stimulation Techniques for Electric Power Production in the Upper Rhine Graben, Central Europe," *Proc. World Geothermal Congress 2010*, Paper 3163, Bali, Indonesia, April 25–29, 2010.

[47] "Geothermal Plant Likely Cause of Earthquakes," *The Local*, Dec. 8, 2010: http://www.thelocal.de/sci-tech/20101208-31671.html.

[48] Imolauer, K., "Geothermal Project Unterhaching & Risk Management," Presentation, Asian-Pacific Berlin Week, Oct. 9, 2009.

[49] Wolfgramm, M., J. Bartels, F. Hoffmann, G. Kittl, G. Lenz, P. Seibt, R. Schulz, R. Thomas and H.J. Unger, "Unterhaching Geothermal Well Doublet: Structural and Hydrodynamic Reservoir Characteristic; Bavaria (Germany)," *Proc. European Geothermal Congress 2007*, Unterhaching, Germany, May 30-Jun. 1, 2007.

[50] Rubner, J., "Test Facilities – Solar and Geothermal Power Plants: Power from Heaven and Earth," Siemens, Research & Development, Fall 2007: http://www.siemens.com/innovation/en/publikationen/publications_pof/pof_fall_2007/solar_and_geothermal_power_plants.htm.

[51] "Bruchsal Germany 600 kW Kalina Cycle Euler Turbine," Energent Corporation, Santa Ana, CA, 2011: http://www.energent.net/documents/bruchsal.pdf.
[52] "Annual Report 2006," ABN 55 095 006 090, Geodynamics Ltd., Milton, Queensland, Australia, pp. 18–19.
[53] "Annual Report 2009," ABN 55 095 006 090, Geodynamics Ltd., Milton, Queensland, Australia, p. 8.
[54] "Quarterly Report, Period Ending 31 March 2011," Geodynamics Ltd., Milton, Queensland, Australia, pp. 4–5.
[55] Frick, S., A. Saadat, S. Kranz, S. Regenspurg and E. Huenges, "Geothermal Research at Groß Schönebeck, Research Goals and Power Plant Concept," Tech. Report, Helmholtz-Zentrum Potsdam, Deutsches GeoForschungsZentrum GFZ, 2011.
[56] "Newberry EGS Enhanced Geothermal Systems Demonstration," Poster, AltaRock Energy, Inc., Seattle, WA: http://www.altarockenergy.com/.
[57] "AltaRock Energy and Davenport Newberry to Demonstrate Innovative Geothermal Technology," Press Release, AltaRock Energy, Inc., June 2010: http://www.newberrygeothermal.com/index.htm.

Chapter 23

Environmental Impact of Geothermal Power Plants

"While offering no detailed challenge to the proposition that beauty is in the eye of the beholder, I do insist that the Wairakei borefield ranks high in New Zealand's superb hierarchy of visual delights. If a tramper on Highway 1 were to pause at dusk 8 km north of Taupo on a moist day with a stiff breeze, he would see an eerie scene of haunting beauty. Scores of fleecy plumes are skyward only to be seized and devoured by green demons that haunt the boughs of imperial conifers; bundles of silvery bullwhips, cracked by an invisible giant who lurks behind the western hill, are caught in stop-action as they rise and fall in unison. It is an odd amalgam of technology and nature, of the Tin Woodsman of Oz and the Sorcerer's Apprentice, gently underscored by the whispering, slightly syncopated 'whuff-whuff... whuff... whuff' of the wellhead silencers."

Robert C. Axtmann – 1975

Geothermal Power Plants: Principles, Applications, Case Studies and Environmental Impact, Third Edition

© 2012 Elsevier Ltd. All rights reserved.

23.1 Overview

At various points throughout this book (e.g., in Sects. 5.7, 6.7, 7.6 and 8.6), and in the case studies, some of the environmental effects of geothermal power plant operations have been discussed in the context of the subject at hand. This concluding chapter aims to present a comprehensive picture of the environmental advantages offered by geothermal power plants as well as their possible detrimental effects.

Certain environmental impacts associated with the development of geothermal sites and the operation of plants are inevitable. However, under normal conditions they are generally confined to the immediate vicinity of the plant and are of lesser impact than those of other electric power generation technologies, particularly those using carbon based fossil fuels and nuclear fuels.

There have now been more than one hundred years of experience in developing geothermal fields, and in building, operating, upgrading, and even decommissioning geothermal plants of various types. In the earliest days, drilling of wells could be a hazardous undertaking and the behavior of geothermal reservoirs was mysterious. Early developers and operators learned by doing, and eventually a scientific understanding of the nature of the resource evolved. Along with this came the technology of how best to exploit geothermal energy and how to deal with the potential environmental impacts.

This chapter is loosely based on Chapter 8 of Tester *et al* [1], much of which was written by the author. Besides this source, the reader may wish to consult Refs [2–11] for further information.

23.2 Regulations

Most countries have laws that regulate the construction and operation of power plants with the intent of preserving the natural environment and safeguarding the health and well-being of people as well as the flora and fauna of the region. The United States has federal, state, and local regulations that cover a broad range of possible environmental impacts. Being subject to these regulations means that geothermal power plants must be built and designed so as to minimize, if not eliminate, all possible adverse environmental impacts. The following laws and regulations must be adequately addressed before any geothermal project can be completed [1, 8]:

- Clean Air Act
- National Environmental Policy Act
- National Pollutant Discharge Elimination System Permitting Program
- Safe Drinking Water Act
- Resource Conservation and Recovery Act
- Toxic Substance Control Act
- Noise Control Act
- Endangered Species Act
- Archeological Resources Protection Act
- Hazardous Waste and Materials Regulations
- Occupational Health and Safety Act, and
- Indian Religious Freedom Act.

The comprehensive spectrum of impacts covered in this list practically ensures that no geothermal power plant will be a threat to the environment anywhere in the United States. Although there are no uniform international standards regarding the environmental impact of geothermal plants, it is common for most countries to require that plants meet appropriate environmental regulations; see, e.g., Refs. [12–15].

23.3 General impacts of electricity generation

The list of possible environmental impacts from any kind of electricity-generating power plant is long. Most of the items shown below apply to all power plants, but all of them apply to geothermal plants in varying degrees:

- Gaseous emissions to the atmosphere
- Water pollution
- Solids emissions to the surface and the atmosphere
- Noise pollution
- Land usage
- Land subsidence
- Induced seismicity
- Induced landslides
- Water usage
- Disturbance of natural hydrothermal manifestations
- Disturbance of wildlife habitat and vegetation
- Alteration of natural vistas
- Catastrophic events.

Of these some are of serious concern for geothermal plants. Abatement technology is available and usually deployed to mitigate the most potentially harmful impacts. Compared with other types of power plants, geothermal plants hold significant advantages for many of these impacts. In the sections that follow, it will be seen that geothermal plants are relatively benign in the areas of atmospheric emissions, particularly with regard to "greenhouse gases," land usage, solids emissions, water usage, and water pollution. Although they are rarely encountered, land subsidence and induced seismicity can have serious consequences. Matters relating to noise abatement, assuring personnel safety, avoiding catastrophic events, and preserving natural thermal features and vistas all deserve the power plant developer's attention.

23.4 Environmental advantages of geothermal plants

There is great concern worldwide about atmospheric emissions of carbon dioxide, CO_2, owing to its heat-trapping properties and the fear of its effect on the global climate. Geothermal power plants have very low gaseous emissions, albeit most of which is CO_2, on a per MWh-generated basis, when compared with all other power generation technologies that emit CO_2 as a normal part of operation. Geothermal binary plants normally emit no gases at all. Using the same basis of comparison, geothermal plants use much less land than any other type of power plant. With one notable exception, geothermal fluids used in power plants are fairly innocuous chemically and pose little hazard in terms of solids pollution. Reinjection of waste brines from geothermal plants avoids contamination of surface and groundwater aquifers. Thus, taken in broad scope, geothermal power plants are one of the most, if not the most, environmentally benign sources of electrical power.

23.4.1 Gaseous emissions

Gaseous emissions from dry- and flash-steam geothermal plants stem from the noncondensable gases (NCG) that are carried in the geofluid in dissolved form. Unless the NCG are removed upstream of the turbine (which is currently not done in commercial plants), the NCG will accumulate in the condenser, thereby raising the backpressure on the turbine. This will cause a significant reduction in turbine power output. With reference to Fig. 5.9, as the condenser pressure rises, so does the condensing temperature, thereby elevating the line 6–7. This causes the enthalpy of state 5 to increase and reduces both the enthalpy drop across the turbine and the power output. For this reason, the NCG are removed from the condenser by some means (e.g., steam-jet ejectors, vacuum pumps, turbocompressors; see Figs. 5.6, 6.6, 7.10, 11.14, and 11.15). This comes at the expense of some steam or power, but always results in a higher net power than if the NCG were left to accumulate.

Carbon dioxide and hydrogen sulfide, H_2S, are the most common and prominent NCG in geothermal steam, but gases such as methane, hydrogen, sulfur dioxide, or ammonia can also be found, usually in very low concentrations. Currently it is not required to capture or treat CO_2, but H_2S is strictly regulated in the United States owing to its offensive odor at very low concentrations, 30 parts per billion, and to its toxicity at higher levels [7]. If necessary, the NCG can be chemically treated or scrubbed after removal from the condenser to remove the H_2S. It is also possible to compress the NCG, redissolve them into the waste brine, and inject them back into the reservoir. This latter approach, however, may lead to an increase in the NCG concentration in the geofluids coming from the production wells. In the case of binary plants, where the brine is produced in the liquid phase with the NCG still in solution, the NCG remain in solution throughout the heat transfer processes and return to the reservoir with the waste brine during reinjection.

In an attempt to control global warming, rules and regulations are being discussed around the world that would penalize plants that emit carbon into the atmosphere. One approach would place a "cap" on carbon emissions for power plants; another would institute a "carbon tax" on emissions. However, currently there are no restrictions on the discharge of CO_2 from power plants. Geothermal binary plants emit no CO_2 whereas steam and flash plants emit much less CO_2 per MWh generated than do fossil-fueled plants. Having no, or relatively low, CO_2 emissions, geothermal plants should be in a favorable position if and when some type of restriction is placed on the emissions of carbon. If a "carbon tax" were to be implemented, the cost to generate a kilowatt-hour of electricity from fossil-fueled plants would be affected far more than less polluting technologies, such as geothermal. Under a program of "carbon emission credits," geothermal power plants could gain an additional revenue stream by selling carbon credits in a trading market.

able 23.1 Gaseous emissions from various power plants [8].

lant type	CO_2 kg/MWh	SO_2 kg/MWh	NO_x kg/MWh	Particulates kg/MWh
oal-fired steam plant	994	4.71	1.955	1.012
il-fired steam plant	758	5.44	1.814	NA
as turbine	550	0.0998	1.343	0.0635
ydrothermal – flash-steam	27.2	0.1588	0	0
ydrothermal – The Geysers dry-steam	40.3	0.000098	0.000458	Negligible
ydrothermal – closed loop binary	0	0	0	Negligible
PA average, all US plants	631.6	2.734	1.343	NA

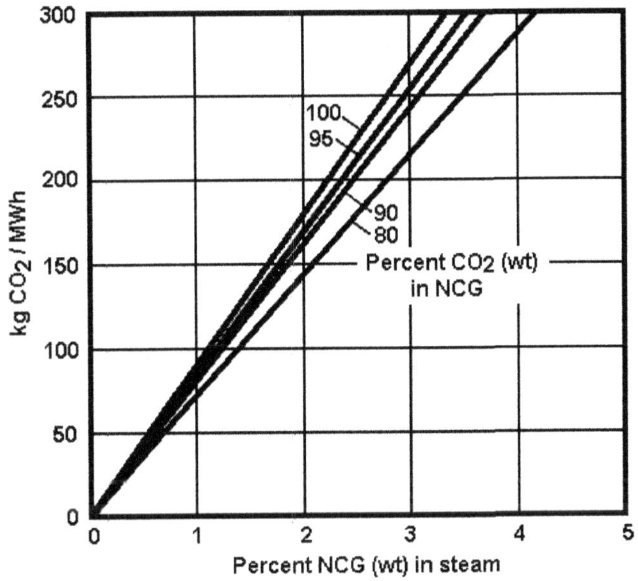

igure 23.1 Potential unabated carbon dioxide emissions (kg/MWh) for various concentrations of CO_2 in the NCG s a function of the concentration of NCG in the steam [6].

With regard to the U.S.-regulated pollutants, nitrogen oxides, NO_x, and sulfur dioxide, SO_2, geothermal steam and flash plants emit negligible amounts. Table 23.1 shows a comparison of gaseous emissions from typical geothermal plants with other types of power plants [8]. It is worth noting that the NO_x and SO_2 emissions at The Geysers only result from the method used to treat H_2S in the NCG, namely, a combustion process that oxidizes the H_2S in a few of the units. Most geothermal steam plants do not rely on combustion for H_2S abatement and therefore emit no NO_x at all.

The amounts of CO_2 and H_2S that could theoretically be emitted are shown in Figs. 23.1 and 23.2 as functions of their concentrations in the NCG and of the concentration of the NCG in the steam [6]. Since typical NCG concentrations range from 0.5–1.0% (wt) of steam, with CO_2 constituting about 95% of the NCG and H_2S generally no more than 1–2% of the NCG, the typical *unabated* emissions of these two gases range from 50–80 kg/MWh for CO_2 and 0.5–1.8 kg/MWh for H_2S. Abatement

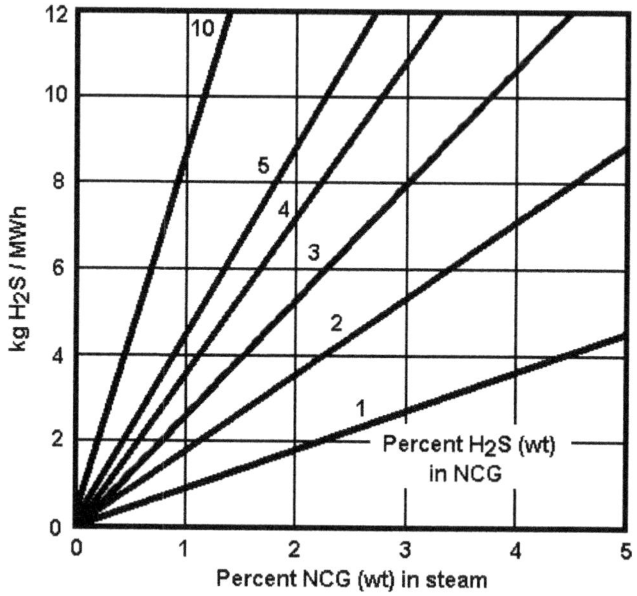

Figure 23.2 Potential unabated hydrogen sulfide emissions (kg/MWh) for various concentrations of H_2S in the NCG as a function of the concentration of NCG in the steam [6].

systems are commercially available that drastically lower the H_2S emissions. For example, one manufacturer claims that a redox chemical treatment system can be designed to achieve over 99.9% H_2S removal [16].

23.4.2 Land usage

The area required to support a geothermal power plant, including the well field, substation, access roads, and auxiliary buildings depends on the power plant rating, the type of energy conversion system, the properties of the geothermal reservoir fluid, and the piping system chosen for collecting the geofluid from the production wells and disposing of the waste brine to the injection wells. The power plant must be built close to the production wells to avoid thermodynamic losses caused by long geofluid pipelines. Although a well field for a 20–50 MW power plant can cover a considerable area, 5 to 10 km^2 or more, the well pads themselves typically cover only about 2% of the total area. Directional drilling allows multiple wells to be drilled from a single pad and minimizes the area needed for the well pads.

The pipelines used to convey the geofluids are usually mounted on stanchions, run along service roads, and incorporate vertical and horizontal expansion loops. Well fields are thus compatible with other uses such as agriculture, aquaculture, and the raising of livestock. Figure 23.3 illustrates this, showing a wellhead, two-phase pipeline, and some cattle grazing at the Miravalles plant in Costa Rica (see Chap. 16). Another example of multiple use is the prawn farm that is adjacent to the Wairakei power station in New Zealand (see Fig. 23.6). The prawns are raised in fresh-water ponds that are heated to about 24–28°C using residual brine from the plant, creating a product that cannot otherwise be grown in New Zealand.

igure 23.3 *Typical pipeline at Miravalles geothermal power plant, Costa Rica. Photo by author [WWW].*

able 23.2 Comparison of land requirements for typical power generation options [1, 6].

ower plant technology	Land usage m^2/MW	Land usage m^2/GWh
10 MW geothermal flash plant (including wells)	1,260	160
0 MW geothermal binary plant (excluding wells)	1,415	170
9 MW geothermal FCRC plant (excluding wells)$^{(1)}$	2,290	290
258 MW coal plant (including strip mining)	40,000	5,700
70 MW nuclear plant (plant site only)	10,000	1,200
5 MW hydroelectric plant (reservoir only)$^{(2)}$	1,200,000	250,000
7 MW solar thermal plant (Mojave Desert, CA)	28,000	3,200
0 MW solar PV plant (Southwestern U.S.)$^{(3)}$	66,000	7,500
5 MW wind farm (10 × 2.5 MW)$^{(4)}$	16,000	7,300

1) Typical flash-crystallizer/reactor-clarifier plant at Salton Sea, CA; see Sect. 9.9.
2) Average of 10 plants at the Cumberland River Basin, Tennessee Valley Authority system.
3) By deploying PV panels on rooftops of existing buildings, no additional land would be needed.
4) Assumes a clear area with a radius equal to the hub height.

Table 23.2, using data from [6] and elsewhere, presents a comparison of land usages for typical geothermal flash and binary plants with those of coal, nuclear, hydroelectric, solar thermal, photovoltaic, and wind plants [1]. Realistic capacity factors have been used in the calculations for each technology; furthermore, average power outputs, not rated values, were used for the solar plants.

A geothermal flash or binary plant requires (per MW) 5% of the area needed for a solar thermal plant, and 2% for a solar photovoltaic plant located in the best insolation area in the United States. The ratios are similar on a per MWh basis. The coal plant, including 30 years of strip mining, requires between 30 and 35 times the surface area for a flash or binary plant, on either a per MW or MWh basis. The nuclear plant occupies about seven times the area of a flash or binary plant. The highest land usage among geothermal plants occurs in the case of those using hypersaline brines; they require about 75% more land than either simple flash or binary, owing to the size of the chemical treatment facilities that render the brines manageable. The advantage that geothermal plants hold over the alternatives is striking.

We have chosen as one example to illustrate the land usage for a binary plant the water-cooled 53 MW Heber 2 plant complex (formerly designated and referred to elsewhere in this book as SIGC); see Sect. 17.4. The 33 MW SIGC plant was augmented by a 2 × 10 MW two-unit binary plant, part of the Gould Project [17]. Figure 23.4, an aerial view courtesy of *Google Earth* [18], shows how a power plant can operate compatibly within an agricultural setting; see Fig. 17.6. The developed area sits among irrigated crop fields and covers a total of 0.29 km^2 (385 m × 765 m); however only about 60% of this (the northern part) is used for the power station, i.e., Heber 2 Gould II, Heber South, and the main production wellpad. The unused land to the south was the location of the original Heber Binary Demonstration plant that was decommissioned and dismantled; see Sect. 17.3 and Fig. 17.3. The specific area usage for the current 53 MW Heber 2 complex is about 3,340 m^2/MW. The plant equipment by itself (14 individual units, cooling towers, and on-site wells, excluding the outlying wells), covers approximately 0.081 km^2 or about 1,530 m^2/MW.

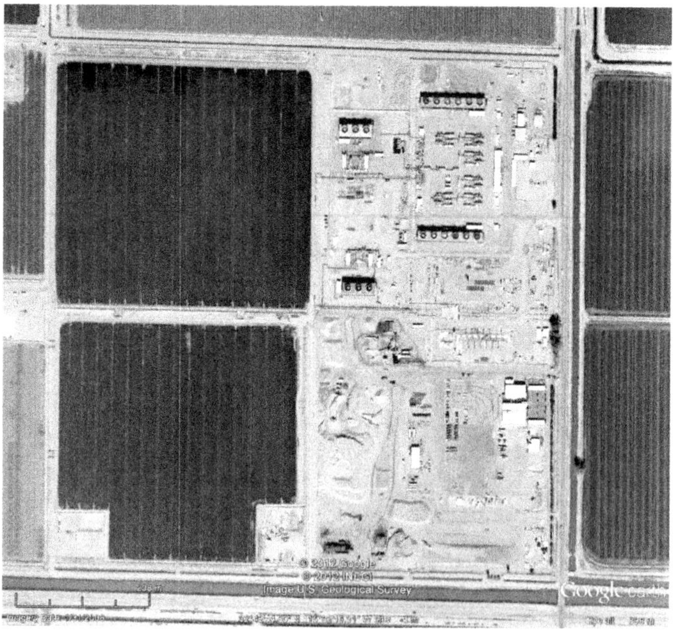

Figure 23.4 Aerial view of the Heber 2 binary plant near Heber, CA [18]. The 10 MW Gould II unit (with a 3-cell cooling tower) lies just west of the original SIGC plant. Heber South is just south of Gould II; see Fig. 17.6. Google Earth image from 1-31-2008 [WWW].

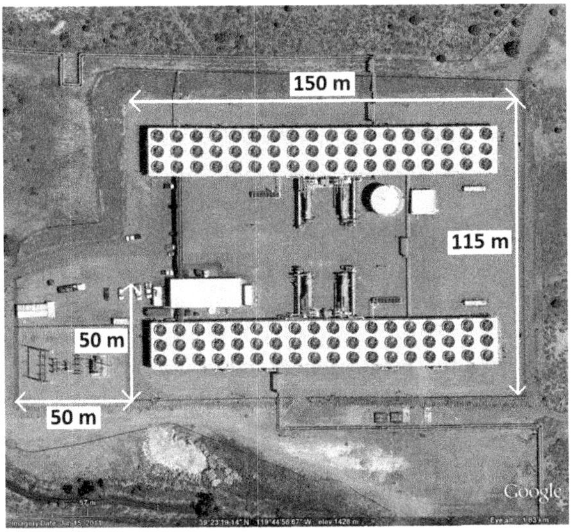

Figure 23.5 Steamboat Galena 3 binary plant, with air-cooled condensers, Nevada, [18a] [WWW].

The next example is for a relatively new air-cooled binary power plant. Figure 23.5, courtesy of *Google Earth* [18a], is an aerial view of the 2008 Steamboat Galena 3 binary plant located about 10 miles (16 km) south of Reno, Nevada. The plant has an installed power capacity of 26 MW using two 13 MW units. The area occupied by the power station, excluding wells and pipelines, but including the switchyard, is 19,750 m^2 or about 760 m^2/MW. This may be compared to the water-cooled Heber 2 plant with 1,780 m^2/MW (33 MW, 58,800 m^2).

The smaller specific area needed for the air-cooled plant versus the water-cooled plant is surprising. In general, the use of air-cooled condensers for heat rejection significantly increases the specific footprint of the plant compared to a plant of the same power rating using a wet cooling tower. For instance, the water cooling towers at Heber 2 occupy only 5% of the land area proper for the Heber 2 plant and require roughly 91 m^2/MW, whereas the air-cooled condensers at Galena occupy about 31.5% of the power station area (excluding the switchyard) and require 209 m^2/MW. In these cases, the more than 2:1 disadvantage in cooling area per MW is more than offset by improvements in binary turbine technology, both in terms of increased capacity and improved efficiency, which allows for a more compact site arrangement per installed power.

23.4.3 Solids discharge

The solids that could potentially be discharged into the environment from geothermal plants are confined to materials that are initially dissolved in the geofluid and which precipitate during the processes undergone within the power plant. Of all the plants now in operation around the world, only those at the Salton Sea field in Southern California (see Sect. 9.8) pose a threat in this regard. The concentration of dissolved solids in a sample brine is given earlier in Table 9.2. Clever engineering, evolved over years of research and development sponsored by government and private industry, has tamed these very aggressive brines to the extent that over 300 MW of generating capacity are installed and operating with negligible impact from solids pollution.

The two methods for coping with these high-salinity brines, namely, flash crystallizer/reactor-clarifier (FCRC) and pH-modification (pH-mod) systems, were discussed in Sects. 9.8.1 and 9.8.2. By controlling the precipitation of the solids, these methods allow either for the solids to remain in solution long enough to pass through the plant and be reinjected back into the reservoir (pH-mod) or for the solids to precipitate in a manner and place where they can be removed from the geofluid and collected for proper disposal (FCRC). The latter approach cleans the brine and permits it to be reinjected without the possibility of solids precipitation within the reservoir where it could adversely affect the permeability of the formation. Thus, with proper design of the treatment system, the solids naturally occurring in the brine are not allowed to escape uncontrolled into the environment.

Other common dissolved minerals such as silica and calcium occur in modest concentrations that can be managed relatively easily, either by chemical pretreatment of the brines while they are still in the production wells (see Sect. 16.5) or by post-utilization settling ponds [19].

In the case of highly mineralized geofluids, the levels of valuable elements might be sufficient to economically justify a mineral recovery system. A pilot zinc recovery facility was designed and built to accompany the Salton Sea Unit 5, but it did not perform well and was eventually dismantled. Since the recovery plant was one part of an integrated brine processing and power generation complex, any problems with the pilot plant had adverse effects on the brine handling and power generation sides of the operation.

23.4.4 Water usage

Water is needed at every stage of development of a geothermal project. This is no different from any other large power development project. However, the needs for geothermal projects are relatively easy to satisfy. Furthermore, water use can be managed in most cases to minimize environmental impacts. The two main areas of water usage are the drilling of wells and the discharge of waste heat if a water cooling tower is used.

Well drilling. Chapter 3 describes the drilling operations for geothermal wells. The water required during this phase of development cools the drill bit, removes rock chips, and provides structural integrity of the hole until casing can be set. This water, or "mud," is actually a mixture of water and chemicals. It is recirculated after being cooled in a small cooling tower and strained to remove the rock fragments at the surface. Thus a very small amount of fresh water is needed to compensate for evaporation losses during cooling. Usually this can be provided by local surface water from ponds, streams or shallow wells.

Cooling water for heat rejection. Whenever power is generated on a continuing basis, the rejection of heat into the environment is an inevitable consequence of the Second Law of thermodynamics. The customary method of discharging waste heat in eothermal steam or flash plants is the use of water cooling towers. Furthermore, such lants provide their own make-up water requirements for the cooling towers since the team condensate is sufficient to cover the evaporative loss of water from the tower. here is nevertheless a need for relatively small quantities of fresh water for blowdown) avoid a buildup of solids in the cooling tower cold well. This water, like that for rilling fluid make-up, can usually be taken from streams or other water sources.

It is not necessary to use any water for cooling purposes if a dry cooling system is dopted. Air-cooled condensers are widely used with binary plants where water may e in short supply since binary plants do not supply their own make-up water as do ash-steam and dry-steam plants (see Chap. 7). While air-cooled condensers eliminate he need for fresh make-up water, they occupy large tracts of land, as mentioned earer, owing to the poor heat transfer properties of air versus water. Additionally, there s a larger parasitic power requirement compared to water cooling towers owing to the arge number of electric motor-driven fans. For example, in the case of the 15.5 MW Jnit 5 bottoming binary plant at the Miravalles field in Costa Rica, a design comparion between a water cooling tower and an air-cooled condenser showed that the airooled condenser would cost more than three times as much, weigh more than two nd a half times as much, cover about three times as much surface area, and consume bout three times more fan power than a water cooling tower [20].

3.4.5 Water pollution

here are several places where geofluids may get into the environment during field levelopment or normal operations. Since these fluids may contain minerals and elenents harmful to humans, flora, or fauna, the onus is on the plant designers to rovide barriers to prevent these fluids from entering the biosphere. The amount of disolved solids increases significantly with temperature, making high-temperature geoluids more risky than moderate- or low-temperature ones. Some of these dissolved ninerals (e.g., boron and arsenic) could poison surface or groundwaters and also arm vegetation or animals.

The well casing is the first barrier against pollution of groundwaters. Damaged casngs may allow brines to mingle with fresh water aquifers. Therefore, particular care is aken to install and cement multiple casings at shallow depths to provide extra barriers see Fig. 3.7). Cement-bond logs (integrity tests) are performed to assure the driller hat there are no blind spots behind the casing that could rupture under thermal stress aused by repeated opening and closing of the well.

Liquid streams might endanger surface waters through run-off during well testing. Thus, fluids discharging during tests are directed to impermeable holding ponds. Also, steam pipelines are fitted with traps to remove condensate (see Fig. 7.7) and that liquid s sent by pipelines to holding ponds. Later the collected fluids are reinjected deep underground.

Despite all these design precautions, it is nevertheless prudent to have monitoring wells strategically located in the well field to rapidly detect any problems with subsurace leakage and permit prompt remediation. For those few developments where 100% reinjection of residual brines is still not practiced, it is essential to monitor all discharge streams to avoid exceeding allowable limits of contaminants.

23.5 Environmental challenges of geothermal plants

23.5.1 Land subsidence

It is difficult to draw generalizations about subsidence because of a wide range of experiences among geothermal fields. Larderello, the large dry-steam field in Italy, has been in operation for over 100 years with negligible subsidence. In New Zealand, both the Wairakei and Ohaaki fields have experienced subsidence: at Wairakei the depression or "bowl" is very large but located some 1.5 km from the center of the production field, whereas at Ohaaki the subsidence is relatively small but, being centered on the production field, has caused serious problems with wells and pipelines. The large geothermal area in the Imperial Valley of the United States supports nearly 500 MW of power generation, and Cerro Prieto in Mexico, in an extension of the same geological formation, supports over 700 MW, and neither area has experienced significant subsidence [21]. Subsidence would be extremely disruptive here since the land in the Imperial Valley is generally flat and arid, with irrigation canals in place to distribute water to vast farms (e.g., see Fig. 23.4).

Geothermal reservoir production at rates much greater than recharge can lead to surface subsidence. This was observed, for example, beginning with the first few years of operation of the power plant at Wairakei when all the residual brine was allowed to flow to the adjacent Waikato River. The production wells at Wairakei were drilled through a relatively shallow cap rock (Huka Falls Formation) containing pumice breccias and mudstones. The thickness of the cap rock varies from 150–200 m in the northern part of the field to only 30–90 m in the western part. Tests have shown that the pumice breccias and mudstones exhibit compressibilities sufficiently high to account for the subsidence [22]. It is important to note that the greatest subsidence correlates with the thickest part of the cap rock (R. Glover, personal communication, May 16, 2007).

During nearly 50 years of operation, much of it without reinjection, subsidence rates in one area of Wairakei reached nearly 500 mm/y; the maximum depression now exceeds 15 m. Although recently the rate of subsidence has decreased substantially to about 70 mm/y within the deepest depression, a projection indicates that the total subsidence should increase by an additional 20 m by the year 2050 [23,24]. Furthermore, the subsidence has now spread across the field, covering an area extending to about 5 km to the southwest and 4 km to the northwest of the current bowl [25]. Figure 23.6 is a map showing the rates of subsidence and other important features and structures in the most affected part of the field [26].

Figure 23.7 shows the Wairakei drop structure (drainage channel to the Wairakei Stream) that was impacted by the subsidence. The area of maximum subsidence lies just to the north of the Wairakei Stream some 150–200 m upstream of the drop structure. About 20 years after the first power units came online, subsidence caused a flume that fed the waste brine into the canal to fail. All plant operations were halted and it took three days to repair the broken connection. The canal is still in operation but carries only about 40% of the waste brine flow since over 50% is now being reinjected (I. Thain, personal communication, May 1, 2007).

Subsidence is more likely to occur in formations where the fluid in place is under lithostatic, rather than hydrostatic pressure. For example, this condition is present in

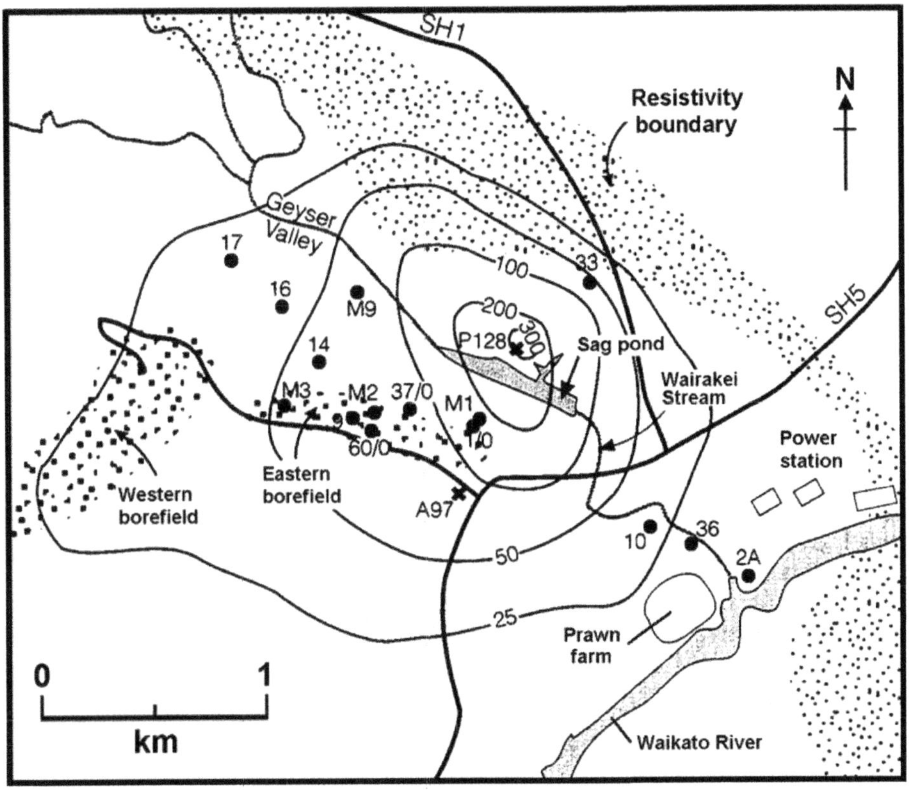

'igure 23.6 *Subsidence in the northern sector of the Wairakei field: numbered contours show average rates of subsidence in ım/y from 1986 to 1994; filled circles are selected wells; crosses are leveling benchmarks. Redrawn from [26].*

he geopressured resources along the Gulf Coast of the United States; see Sect. 9.6.3. Iere the pressure of the fluid contributes to the support of the overburden and its emoval leaves the overburden partially unsupported. Subsidence is less likely in competent formations with fracture-dominated permeability. For example, at The Geysers ield in California, which consists of fractured greywacke, a maximum subsidence rate of 47 mm/y was observed from 1977 to 1996, an order of magnitude lower than the naximum seen at Wairakei [27].

Although reinjection does not guarantee the avoidance of subsidence, it can reduce he risk, provided it is carried out so as to maintain reservoir fluid pressure. Nowadays, geothermal developers normally incorporate reinjection into reservoir management programs right from the start both to minimize this risk and to prolong the ife of the reservoir.

23.5.2 Induced seismicity

nduced seismicity is a phenomenon in which a change in fluid pressure within a stressed rock formation leads to movement of the fractured rocks. The energy released

Figure 23.7 Wairakei drop structure (hot-water drainage channel) to the Wairakei Stream not far from the area of maximum subsidence. Photo courtesy of S. Tamanyu, Geological Survey of Japan; photo first published on the cover of Chishitsu News, No. 531, 1998 [WWW].

is transmitted through the rock and may reach the surface with enough intensity to be heard or felt by persons in the area. This may happen, for example, when the reservoir for a hydroelectric station is first filled, when fossil fuels are extracted from oil and gas fields, or when fluids are injected underground at high pressure. The likelihood and the severity of the event depend on the local state of stress within the formation.

Nearly every geothermal field under exploitation has experienced induced seismicity to some degree [28]. In a normal hydrothermal setting this has not been a problem since high pressures are not needed for the reinjection of residual brines. This may be a more important problem for the emerging technology of Enhanced (or Engineered) Geothermal Systems or EGS [1]. Whereas injection of waste fluids is an ancillary, albeit important, activity at a hydrothermal power plant, injection of high-pressure fluids is one of the critical features associated with the creation and maintenance of an EGS servoir; see Chap. 22. A 2007 incident at an EGS site in Basel, Switzerland resulted a magnitude 3.3 earthquake and caused considerable alarm among residents [29]. he problem is deemed significant enough to warrant annual workshops on the subct; see, e.g., [30].

Granted the obvious problems that can be created by strong induced seismic events, ere are several positive aspects of induced microseismicity, the type more commonly ssociated with hydrothermal operations. The acoustic noise generated may be monired with sensitive, high-precision instruments to provide real-time information egarding the behavior of the reservoir. For example, the 3-dimensional acoustic noise attern can shed light on the movement of the injected water front as it moves rough the fractures in the formation. It can help delineate the permeable portions of e formation, i.e., the locations of the fractures and faults.

In EGS applications, it is being used to map the growth and extent of the fractures as hey are being created. Figure 23.8 shows one such application for the Soultz field France [31]. During the six days while the well was shut in after stimulation, he 3-dimensional swarm of acoustic noise grew to the extent shown in the figure, dicating the growth of the fracture pattern in the formation.

Since it is often difficult to discern natural from induced seismic events, it is wise to ollect baseline data prior to field development and drilling at a selected site. Also, a horough scientific study should be carried out before drilling to determine the geologic nd tectonic conditions existing at the site. This should provide the data needed to void the inadvertent lubrication of a major fault that could cause a significant seismic

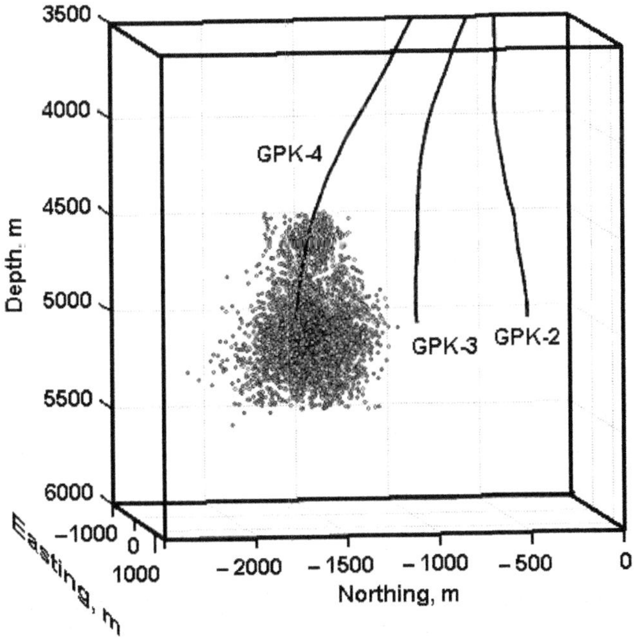

Figure 23.8 Stimulation of well GPK-4 during a 2004 test: seismic event swarm shown after the well was shut in for six days [31].

event. It is useful also to monitor the site for any unexpected natural or induce microseismic events after field work commences. Finally, if there are residents in th neighborhood of the site, an educational program should be put in place to inforn people of the possibility, unlikely though it may be, of felt seismic events, and to set u a hotline where they can report such occurrences [1].

23.5.3 Induced landslides

Many geothermal fields lie in rugged volcanic terrain prone to natural landslides Indeed, some fields have been developed atop ancient landslides. Landslides can be trig gered by earthquakes, and, as we have discussed, while it is possible that geotherma production or injection could lead to induced seismicity, it is highly improbable tha such activities could lead to an event large enough to cause a major earthquake.

Landslides have occurred at geothermal fields, but the cause is often unclear. Th worst disaster happened at the Zunil field in Guatemala in January 1991 in whic at least 23 people were killed [32]. Figure 23.9 shows the devastation when a larg portion of a steep slope above the field collapsed, spreading rock and moisture-lader debris a distance of 800–1200 m onto a relatively flat plateau [33]. Approximatel 10^6 m^3 of earth was released. The slide also buried one of the wells, ZCQ-4, that stooc at the base of the scarp, close to the Zunil fault; see Fig. 23.10. A thorough examina tion of the site done soon after the event (before the excavation of the buried well

Figure 23.9 January 5, 1991 landslide at Zunil geothermal field, Guatemala. At least six houses were destroyed and the main highway was buried [33]. Photo courtesy of Geothermal Resources Council.

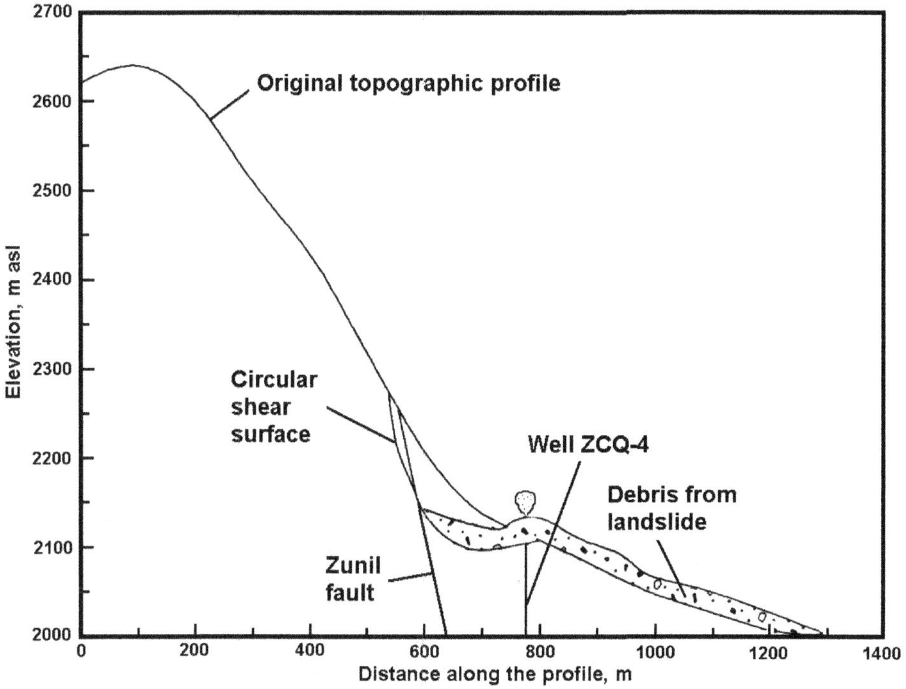

igure 23.10 *Cross-section through the Zunil landslide [33].*

oncluded that the scarp had been in a state of unstable equilibrium owing to the teepness of the slope, the abundance of hydrothermally altered rock, and the proximty to the Zunil fault. Seasonally high groundwater levels also played a role in priming he slope for failure. A later analysis [34] identified several precursor events – a minor andslide at the foot of the scarp about one week earlier, the increase in flow of spring water issuing from the toe of the scarp, followed by a change in the color of the water, and finally its abrupt cessation just before the large slide. Construction and road building activities that began in 1981 probably also set the stage for the 1991 catastrophe.

While it is difficult to pin down the precise trigger for the disaster, it seems that the lope was gradually weakening and the imperceptible movement of the slope had at irst increased the permeability around the Zunil fault allowing the spring flow to ncrease, but eventually the movement sealed off the flow paths. The subsequent uildup of piezometric pressure behind the scarp then became high enough to cause he slope first to slump near the toe and then for the weakened scarp wall to completely collapse [34].

Both studies [33] and [34] offered general recommendations for the prevention of similar catastrophes. These include: (1) development of a hazard map identifying all potential landslide areas, (2) slope monitoring instrumentation, (3) monitoring of springs for changes in flow rate, temperature, chemistry, and clarity, (4) installation of drains in slopes to remove moisture, and (5) avoidance of obvious unstable areas for wells, roads, and other construction activity.

23.5.4 Noise pollution

In the development stage of a geothermal power project, noise is generated during road construction, excavation for drilling sites, well drilling, and well testing. While these may be disturbing to nearby residents (if any), they are of limited duration. Furthermore the most objectionable sounds can be significantly reduced with appropriate mufflers and other sound deadening materials. During normal plant operations, various components are sources of noise; these include: transformers, generators, water cooling towers motors, pumps and fans for circulating water and air associated with the heat rejection system, brine and steam flowing through pipes, etc. These are generally confined to the area within the plant fence boundary. However, cooling towers or air-cooled condensers are relatively tall structures having fans on top. Thus these may be the dominant source of noise during normal operation. Since air-cooled condensers use numerous cells each fitted with a fan (e.g., see Fig. 9.12), they generate more noise than water cooling towers which are smaller and use far fewer cells for a given plant power rating.

If the turbine should trip for any reason, such as a problem in the electrical transmission system, it is usually necessary to direct the full steam flow away from the turbine for a short time using the emergency steam venting system. This allows the wells to continue to flow avoiding a sudden closure that could damage the casings or wellhead equipment. During the outage, steam is typically sent to relief valves and rock mufflers where the steam velocity is reduced drastically to lower the noise associated with the venting process; see Figs. 7.8 and 7.9. Since the noise generated by a moving gas stream is proportional to the velocity raised to the 8th power, reducing the steam velocity by a mere factor of 2 will cut the sound emitted by a factor of 256, or more than two and a half orders of magnitude.

The highest noise levels occur during the drilling and testing of wells. Noise levels can reach 114 decibels A-weighted (dB-A) near the drill site when drilling with air, as would be the case if the reservoir were characterized by tiny fractures in the production zone. Since the intensity of sound drops rapidly as one moves away from the source, geofluids issuing vertically from a wide open well would register only 71–83 dB-A at a distance of 900 m. Thus, under normal operating conditions, the noise emanating from a plant situated within the boundaries of a large geothermal field should not be objectionable to anyone living nearby.

To give some perspective to these numbers, Table 23.3 gives the noise levels for several everyday activities and various geothermal processes.

23.5.5 Disturbance of natural hydrothermal manifestations

As described in Sect. 5.7, there have been numerous cases where hydrothermal developments have compromised or totally destroyed natural hydrothermal manifestations such as geysers, hot springs, mud pots, etc. [35,36]. The drawdown, or lowering of the hydrostatic water level, from production wells disturbs the natural thermohydraulic balance that gives rise to the manifestations. In particular, geysers are delicate phenomena that are subject to nature's whims, even without humankind's interference. A study of the geysers at Yellowstone National Park in the U.S. will show that geysers are ephemeral, coming and going with the seismic events associated with the region. Commercial developments at Beowawe and Steamboat in Nevada extinguished

able 23.3 Geothermal noise relative to other sources.

oise source	Level (dB-A)
t aircraft @ 30 m	120–130
eothermal air drilling rig @ 8 m (25 kg/s steam entry, no muffler)	114
utomobile freeway or a subway train @ 6 m	90
eothermal air drilling rig @ 8 m (25 kg/s steam entry, with muffler)	84
neumatic drill @ 15 m	80
eothermal steam well, wide-open vertical discharge @ 900 m	71–83
acuum cleaner @ 3 m	70
eothermal air drilling rig @ 75 m (25 kg/s steam entry, with muffler)	65
ormal speech @ 0.3 m	65
usiness office	50
esidential area in evening	40

he geysers that used to be seen in both places, thus depriving future generations from njoying their beauty.

The Wairakei Geyser Valley, a famous tourist attraction in the late nineteenth and arly twentieth centuries, straddled the upper Wairakei Stream for a distance of some 50 m, and was home to 22 active geysers along with other thermal features until eld development began (see Fig. 23.6). By the mid-1970s, the geysers were extin-uished [37]. Instead tracts of steaming and hot ground intensified, particularly in an rea to the northeast of the Geyser Valley. The center of the subsidence bowl, dis-ussed earlier in Sect. 23.5.1, is about 1 km to the southeast of the old Geyser Valley. sag pond has since formed in the Wairakei Stream near the bottom of the subsi-ence bowl (R. Glover, personal communication, May 14, 2007).

Apart from these generally negative consequences of development, there is another iewpoint on this subject. It has been argued that there can be beneficial effects on hermal manifestations as a result of geothermal development [38]. For example, in New Zealand, besides the loss of many thermal features, several have been created or nhanced, such as the enlarged areas of steaming ground at Wairakei, and increased oiling of hot springs at other sites.

3.5.6 *Disturbance of wildlife habitat, vegetation, and scenic vistas*

onsidering the relatively small area needed for geothermal development and opera-ions, the potential impact on wildlife, vegetation, and vistas can be minimized with roper planning and engineering. Furthermore, in the U.S. and most countries, an nvironmental Impact Statement must be filed before any permits are granted for a eothermal project.

Certainly any power generation facility constructed where none existed will alter the iew of the landscape. Conventional power plants in developed, commercial, or industrial ettings, while objectionable to many for other reasons, do not stand out as sharply as a eothermal plant in a flat agricultural region or on the flank of a volcano. Even so, with are and creativity geothermal plants can be designed to blend into the surroundings.

In areas having natural foliage, it is necessary to clear some of it to make room for oads, well pads, pipe routes, separator stations, holding ponds, and the power house

Figure 23.11 Ahuachapán geothermal facility after commissioning around 1977 [39]. Owing to areas of acidic soils and hot ground, initially there was little natural vegetation around the plant [WWW].

and its associated facilities. Once the construction phase is over and operations begin the disturbed area can be replanted to regain a semblance of its original natural appearance.

The Ahuachapán plant in El Salvador is an interesting case in point. Figures 23.11 and 23.12 show the facility soon after commissioning around 1977 [39] and then after the planting of trees and vegetation in 2005.

Compared to wind turbines, solar thermal "power towers," or fossil-fueled or nuclear plants, geothermal plants generally have a low profile and are less conspicuous. The exceptions are the geothermal plants that use natural-draft cooling towers: several older plants at Larderello (Italy), and those at Matsukawa (Japan) and Ohaaki (New Zealand). Buildings and pipelines can be painted appropriate colors to help mask them from a distance (e.g., see Fig. 11.17). While it is impossible to conceal steam rising from silencers or cooling towers at geothermal plants, most people find the sight of white steam clouds unobjectionable.

It should be obvious that sites of rare natural beauty such as national parks ought to be off-limits to geothermal or any other type of industrial or commercial development.

23.5.7 Catastrophic events

Besides landslides, some other serious events that might occur at a geothermal plant include well blowouts, phreatic explosions, ruptured steam pipes, turbine failures, fires, etc. Most of these accidents are similar to what can happen at any power generation facility and have been known to cause casualties. The ones that are unique to

'igure 23.12 Ahuachapán geothermal facility showing growth of eucalyptus trees circa 2005 [40]. Eucalyptus rees are resistant to the hostile soil conditions found there. Other areas that were formerly crop lands have been lanted with various native species. Thus the plant site is now more foliated than before the plant was built and the rea has been partially restored to what it might have resembled hundreds of years ago. (J.A. Rodriguez, personal ommunication, May 2, 2007.) Photo courtesy of LaGeo S.A. de C.V., El Salvador [WWW].

geothermal power plants involve well drilling and testing. In the early days of geothermal energy exploitation, well blowouts during drilling were a fairly common occurence, but nowadays the use of fast-acting blowout preventers have practically eliminated this potentially life-threatening problem. Better understanding of the geology of the site obtained using modern geoscientific methods further reduces the danger of surprises during drilling. The monitoring of reservoir pressures, commonly done at most fields, will give an early indication of a potentially dangerous situation that might result in a phreatic explosion. Lastly, proper engineering and adherence to design and building codes should also minimize, if not completely eliminate, the chance of a mechanical or electrical failure that could cause serious injury to plant personnel or local inhabitants. Since many geothermal plants are located in earthquake-prone regions, it is worth noting that geothermal plants have many times withstood strong earthquakes, including the massive 9.0 magnitude earthquake off the coast of Japan on March 11, 2011, without suffering major damage or failures leading to environmental contamination; see Sect. 14.4.

23.5.8 Thermal pollution

Although thermal pollution is currently not a specifically regulated quantity, it does represent an environmental impact for all power plants that rely on a heat source for

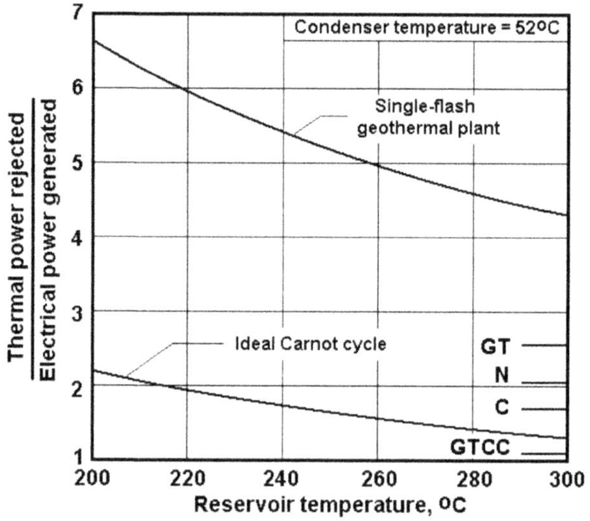

Figure 23.13 Heat rejection per unit of electrical generation for various plants: GT = simple gas turbine; N = nuclear; C = coal; GTCC = gas turbine combined cycle [5].

their motive force. As was discussed in Sect. 5.4.6, heat rejection from geothermal plants is higher per unit of electricity production than for fossil-fueled or nuclear plants. Figure 23.13 shows a comparison between a geothermal single-flash plant and several alternative cycles. For example, using a reservoir fluid temperature of 220°C the flash plant rejects about three times as much heat per unit of useful electrical generation as an ideal Carnot cycle operating between the reservoir temperature and the assumed condensing temperature of 52°C. The other practical cycles all reject far less heat per unit of generation than the flash plant.

23.6 Summary

From the long list of potential environmental impacts from geothermal power plants, we have seen that some of them warrant the serious attention of scientists and engineers. These include subsidence, induced seismicity, and landslides, all of which can have serious and possibly disastrous consequences. Several of the others are fairly easily controlled with current technology – air and water pollution, and noise abatement. Failure of equipment can be avoided through proper design, engineering, and construction. Geothermal plants are far less environmentally intrusive than alternative plants in several respects – very low carbon dioxide emissions, essentially no NOx and SO_2 emissions, extremely low H_2S emissions using appropriate NCG cleanup equipment, low land usage per installed megawatt, and very low water usage for steam plants. Geothermal binary plants are environmentally benign, essentially zero-emission plants, especially when fitted with air-cooled condensers.

eferences

1] Tester, J.W., B.J. Anderson, A.S. Batchelor, D.D. Blackwell, R. DiPippo, E.M. Drake, J. Garnish, B. Livesay, M.C. Moore, K. Nichols, S. Petty, M.N. Toksöz and R.W. Veatch, Jr., *The Future of Geothermal Energy: Impact of Enhanced Geothermal Systems (EGS) on the United States in the 21st century*, Massachusetts Institute of Technology, Cambridge, MA, 2006: http://geothermal.inel.gov.

2] Armstead, H.C.H., *Geothermal Energy, 2nd Ed.*, E. and F. N. Spon, London, 1983.

3] Armstead, H.C.H. and J.W. Tester, *Heat Mining*, E. and F. N. Spon, London, 1987.

4] Burnham, L. (Exec. Ed.), T.B. Johansson, H. Kelly, A.K.N. Reddy and R.H. Williams, *Renewable Energy: Sources for Fuels and Electricity*, Island Press, Washington DC, 1993.

5] DiPippo, R., "Geothermal Energy: Electricity Production and Environmental Impact, A Worldwide Perspective," *Energy and Environment in the 21st Century*, MIT Press, Cambridge, MA, 1991, pp. 741–754.

6] DiPippo, R., "Geothermal Energy: Electricity Generation and Environmental Impact," *Energy Policy*, V. 19, 1991, pp. 798–807.

7] Hartley, R.P., "Environmental Considerations," Chap. 9 in *Sourcebook on the Production of Electricity from Geothermal Energy*, J. Kestin, Ed. in Chief, R. DiPippo, H.E. Khalifa, and D.J. Ryley, Eds., U.S. Dept. of Energy, DOE/RA/4051-1, U.S. Gov. Printing Office, Washington, DC, 1980.

8] Kagel, A., D. Bates and K. Gawell, *A Guide to Geothermal Energy and the Environment*, Geothermal Energy Association, Washington, DC, 2007: http://www.geo-energy.org/publications/reports/Environmental%20Guide.pdf.

9] Mock, J.E., J.W. Tester and P.M. Wright, "Geothermal Energy from the Earth: Its Potential Impact as an Environmentally Sustainable Resource," *Ann. Rev. Energy Environ.*, V. 22, 1997, pp. 305–356.

10] Pasqualetti, M.J., "Geothermal Energy and the Environment – The Global Experience," *Energy (UK)*, V. 5, 1980, pp. 111–165.

11] Tester, J.W., E.M. Drake, M.W. Golay, M.J. Driscoll and W.A. Peters, *Sustainable Energy: Choosing Among Options*, The MIT Press, Cambridge, MA, 2005.

12] Pascual, R.V.J., "Impacts of Philippine Environmental Regulatory Policies on PNOC-EDC's Corporate Environmental Management Initiatives," *Proc. World Geothermal Congress 2005*, Paper No. 0305, Int'l. Geothermal Ass'n., Antalya, Turkey, 2005.

13] Dickie, B.N. and K.M. Luketina, "Sustainable Management of Geothermal Resources in the Waikato Region, New Zealand," *Proc. World Geothermal Congress 2005*, Paper No. 0303, Int'l. Geothermal Ass'n., Antalya, Turkey, 2005.

14] Mwawughanga, F.M., "Regulatory Framework for Geothermal in Kenya," *Proc. World Geothermal Congress 2005*, Paper No. 0312, Int'l. Geothermal Ass'n., Antalya, Turkey, 2005.

15] Guido, H., S., O. Vallejos R. and E. Sánchez R., "Environmental Management at the Miravalles Geothermal Field," *Proc. World Geothermal Congress 2005*, Paper No. 0248, Int'l. Geothermal Ass'n., Antalya, Turkey, 2005.

16] Merichem Chemicals & Refinery Services LLC, Gas Technology Products, 2007: http://www.gtp-merichem.com/downloads/geothermal_plants.pdf.

17] Ormat Technologies, Inc., "10K Report," p. 32, March 12, 2007.

18] *Google Earth*, 6.1.0.5001, Build date: 10/17/2011, Image date: 1/31/2008.

18a] *Google Earth*, 5.2.1.1588, Build date: 9/1/2010, Image date: 6/14/2011.

19] DiPippo, R., *Geothermal Energy as a Source of Electricity: A Worldwide Survey of the Design and Operation of Geothermal Power Plants*, U.S. Department of Energy, DOE/RA/28320-1, U.S. Government Printing Office, Washington, DC, 1980.

20] Moya, R.P. and R. DiPippo, "Miravalles Unit 5 Bottoming Binary Plant: Planning, Design, Performance and Impact," *Geothermics*, V. 36, 2007, pp. 63–96.

21] Narasimhan, T.N. and K.P. Goyal, "Subsidence Due to Geothermal Fluid Withdrawal," Earth Sciences Div., Lawrence Berkeley Laboratory, Rep. No. LBL-10967, GSRMP-14, October 1982.

22] Allis, R.G. and P. Barker, "Update on Subsidence at Wairakei," *Proc. Pacific Geothermal Conference and 4th New Zealand Geothermal Workshop*, Part 2, 1982, pp. 365–370.

23] Allis, R.G. and X. Zhan, "Predicting subsidence at Wairakei and Ohaaki Geothermal fields, New Zealand," *Geothermics*, V. 29, 2000, pp. 479–497.

[24] White, P.J., J.V. Lawless, S. Terzaghi and W. Okada, "Advances in Subsidence Modelling of Exploite Geothermal Fields," *Proc. World Geothermal Congress 2005*, Paper No. 0222, Int'l. Geothermal Ass'n Antalya, Turkey, 2005.

[25] White, P.J., "Latest Developments in Subsidence at Wairakei-Tauhara," *Proc. 27th New Zealan Geothermal Workshop*, 2005.

[26] Allis, R.G., X. Zhan and B. Carey, "Modelling of Subsidence at Wairakei and Ohaaki Fields," *Pro 19th New Zealand Geothermal Workshop*, 1997.

[27] Mossop, A. and P. Segall, "Subsidence at The Geysers Geothermal Field, N. California from Comparison of GPS and Leveling Surveys," *Geophysical Research Letters*, V. 24, 1997, pp. 1839–1842

[28] Kugaenko, Y., V. Saltykov and V. Chebrov, "Seismic Situation and Necessity of Local Seismi Monitoring in Exploited Mutnovsky Steam-Hydrothermal Field (Southern Kamchatka, Russia)," *Pro World Geothermal Congress 2005*, Paper No. 0260, Int'l. Geothermal Ass'n, Antalya, Turkey, 2005.

[29] Pancevski, B., "'Green energy' project gives Swiss the shakes," *London Sunday Telegraph*, February 19 2007: http://www.telegraph.co.uk/news/worldnews/1543048/Green-energy-project-gives-Swiss-the shakes.html.

[30] Majer E. and R. Baria, "Induced seismicity associated with Enhanced Geothermal Systems: State o knowledge and recommendations for successful mitigation," Working paper presented at the Stanfor Geothermal Workshop, Stanford University, CA., 2006.

[31] Baria, R., J. Baumgaertner, D. Teza and S. Michelet, "Reservoir Stimulaton and Testing Technique for EGS Systems (Soultz)," Presented at EGS Workshop, Massachusetts Institute of Technology Cambridge, MA, November 10, 2005.

[32] Global Volcanism Program, Smithsonian National Museum of Natural History, "Landslide in geother mal field; 23 people reported dead": http://www.volcano.si.edu/world/volcano.cfm?vnum=1402 04=&volpage=var&VErupt=Y&VSources=Y&VRep=Y&VWeekly=N.

[33] Flynn, T., F. Goff, E. Van Eeckhout, S. Goff, J. Ballinger and J. Suyama, "Catastrophic Landslide at Zuni I Geothermal Field, Guatemala, January 5, 1991," *Geothermal Resources Council TRANSACTIONS*, V. 15 1991, pp. 425–433.

[34] Voight, B., "Causes of Landslides: Conventional Factors and Special Considerations for Geothermal Sites and Volcanic Regions," *Geothermal Resources Council TRANSACTIONS*, V. 16, 1992, pp. 529–533.

[35] Jones, G.L., "Geysers/Hot Springs Damaged or Destroyed by Man," 2006: http://www.wyojones.com/ destroye.htm.

[36] Keam, R.F., K.M. Luketina and L.Z. Pipe, "Definition and Listing of Significant Geothermal Feature Types in the Waikato Region, New Zealand," *Proc. World Geothermal Congress 2005*, Paper No. 0209, Int'l. Geothermal Ass'n., Antalya, Turkey, 2005.

[37] White, P.J. and T. Hunt, "Cessation of Spring Flow and Spring Feed Depths, Geyser Valley, Wairakei," *Proc. 23th New Zealand Geothermal Workshop*, 2001, pp. 173–178.

[38] Bromley, C.J., "Advances in Environmental Management of Geothermal Developments," *Proc. World Geothermal Congress 2005*, Paper No. 0236, Int'l. Geothermal Ass'n., Antalya, Turkey, 2005.

[39] DiPippo, R., "The Geothermal Power Station at Ahuachapán, El Salvador," *Geothermal Energy Magazine*, V. 6, October 1978, pp. 11–22.

[40] LaGeo, Brochure "Energía limpia para un planeta mejor," Santa Tecla, El Salvador.

Appendices

\ppendix A: Worldwide State of Geothermal Power Plant Development as of August 2011

"If you cannot measure it, you cannot improve it." Lord Kelvin

As of the writing of this book (August 2011), there are 28 countries that either urrently produce electricity from geothermal energy or used to do so. Four countries Argentina, Democratic Republic of the Congo, Greece, and Zambia) have shut down ery small power plants.

The growth of geothermal power worldwide is shown in a semi-log plot in Fig. A.1. It an be seen that the geothermal industry has experienced several different periods of evelopment. Until World War II, Italy was the only country with commercial geothermal ower plants. From 1930 to 1944, the average annual growth was about 14%. New ealand (1959) and the United States (1960) were next to place geothermal plants into ommercial operation, followed by Japan (1967), Iceland (1969), Mexico (1973), and the hilippines (1979). From 1920 to 1977, the annual average growth rate was 8.2%.

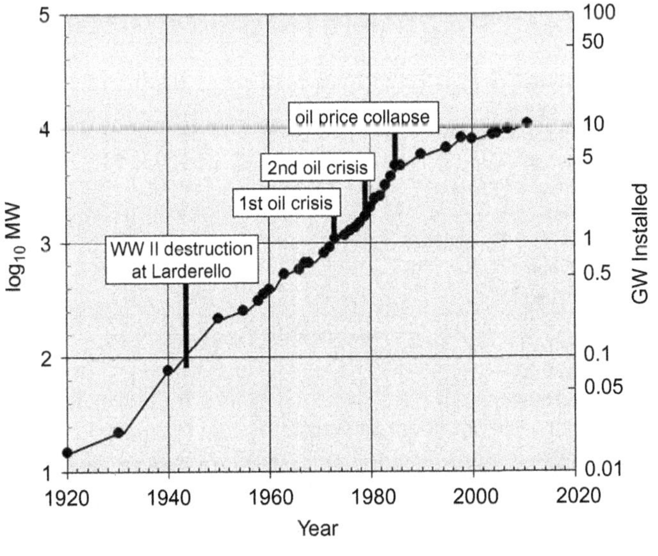

ig. A.1 Growth of worldwide geothermal power installation.

The first great oil crisis (1973) and the second one (1979) spurred many countrie to develop their geothermal resources, but it took a few years before the momentun built. From 1977 to 1985 the annual growth rate accelerated to 17.1%. Since the until 2011, the growth rate has slowed to about 3.2%.

The rest of this appendix consists of a series of tables showing the details of th installed geothermal power around the world by country, by capacity (Table A.1) by number of units (Table A.2), and by type of energy conversion systen (Tables A.3–A.5). Detailed compilations for each country that has or had operatin geothermal power plants are also included (Tables A.6–A.8). It must be borne i mind that the power capacities listed are the installed values, but that the actua running capacities may be different, usually lower, and depend on local conditions such as field and reservoir behavior and demand for electricity.

Table A.1 Status of geothermal power plant development: by MW.

Rank	Country	No. units	MWe
1	United States	253	2774.43
2	Philippines	48	1840.9
3	Indonesia	23	1134
4	Mexico	39	983.3
5	Italy	35	882.5
6	New Zealand	43	783.3
7	Iceland	31	715.4
8	Japan	21	535.26
9	Costa Rica	8	205
10	El Salvador	7	204.3
11	Kenya	13	166.2
12	Turkey	8	94.98
13	Nicaragua	5	87.5
14	Russia	12	79
15	Papua-New Guinea	6	56
16	Guatemala	9	44.6
17	Portugal – Azores	6	26
18	China	8	24
19	France – Guadeloupe	2	14.7
20	Ethiopia	1	8.5
21	Germany	4	6.75
22	Austria	3	1.45
23	Thailand	1	0.3
24	Australia	1	0.15
25	Argentina	0	0
26	Democratic Republic of the Congo	0	0
27	Greece	0	0
28	Zambia	0	0
Totals		587	10668.52
Average (MW/unit)			18.17

able A.2 Status of geothermal power plant development: by number of units.

ank	Country	No. units	MWe
1	United States	253	2774.43
2	Philippines	48	1840.9
3	New Zealand	43	783.3
4	Mexico	39	983.3
5	Italy	35	882.5
6	Iceland	31	715.4
7	Indonesia	23	1134
8	Japan	21	535.26
9	Kenya	13	166.2
0	Russia	12	79
1	Guatemala	9	44.6
2	Costa Rica	8	205
3	China	8	24
4	Turkey	8	94.98
5	El Salvador	7	204.3
6	Papua-New Guinea	6	56
7	Portugal – Azores	6	26
8	Nicaragua	5	87.5
9	Germany	4	6.75
0	Austria	3	1.45
1	France – Guadeloupe	2	14.7
2	Ethiopia	1	8.5
3	Thailand	1	0.3
4	Australia	1	0.15
5	Argentina	0	0
6	DRC – Congo	0	0
7	Greece	0	0
8	Zambia	0	0
otals		587	10668.52
verage (MW/unit)			18.17

Table A.3 Geothermal power plants: by installed MW for each type of plant.

Country	Dry steam	1-Flash	2-Flash	3-Flash	Binary	Flash-binary	Hybrid	Total
United States	1477	49	707.3	49	466.13	20	6	2774.4
Philippines	0	1648	50.9	0	0	142	0	1840.9
Indonesia	445	689	0	0	0	0	0	1134
Mexico	0	540	440	0	3.3	0	0	983.3
Italy	862.5	20	0	0	0	0	0	882.5
New Zealand	55	5	326.8	176.8	84.2	135.5	0	783.3
Iceland	30	614.7	60	0	10.7	0	0	715.4
Japan	23.5	351.75	160	0	2.49	0	0	537.74
Costa Rica	0	144	0	0	61	0	0	205
El Salvador	0	160	35	0	9.3	0	0	204.3
Kenya	0	116.4	0	0	1.8	48	0	166.2
Turkey	0	20.4	47.4	0	27.18	0	0	94.98
Nicaragua	0	80	0	0	7.5	0	0	87.5
Russia	0	79	0	0	0	0	0	79
Papua-New Guinea	0	56	0	0	0	0	0	56
Guatemala	0	0	0	0	0	44.6	0	44.6
Portugal – San Miguel	0	0	0	0	26	0	0	26
China	0	0	24	0	0	0	0	24
France – Guadeloupe	0	10	4.7	0	0	0	0	14.7
Ethiopia	0	0	0	0	0	8.5	0	8.5
Germany	0	0	0	0	6.75	0	0	6.75
Austria	0	0	0	0	1.45	0	0	1.45
Thailand	0	0	0	0	0.3	0	0	0.3
Australia	0	0	0	0	0.15	0	0	0.15
Totals	2893	4581.26	1856.1	225.8	707.76	398.6	6	10668.52
Percent of total	27.12	42.94	17.40	2.12	6.63	3.74	0.06	100

able A.4 Geothermal power plants: by number of units for each type of plant.

ountry	Dry steam	1-Flash	2-Flash	3-Flash	Binary	Flash-binary	Hybrid	Total
nited States	26	3	28	1	184	10	1	253
hilippines	0	42	1	0	0	5	0	48
ew Zealand	1	1	10	5	10	16	0	43
lexico	0	32	4	0	3	0	0	39
aly	34	1	0	0	0	0	0	35
eland	1	20	2	0	8	0	0	31
ndonesia	8	15	0	0	0	0	0	23
pan	1	16	3	0	1	0	0	21
enya	0	6	0	0	1	6	0	13
ussia	0	12	0	0	0	0	0	12
uatemala	0	0	0	0	0	9	0	9
hina	0	0	8	0	0	0	0	8
osta Rica	0	4	0	0	4	0	0	8
urkey	0	1	1	0	6	0	0	8
Salvador	0	5	1	0	1	0	0	7
apua-New Guinea	0	6	0	0	0	0	0	6
ortugal – San Miguel	0	0	0	0	6	0	0	6
icaragua	0	4	0	0	1	0	0	5
ermany	0	0	0	0	4	0	0	4
ustria	0	0	0	0	3	0	0	3
rance – Guadeloupe	0	1	1	0	0	0	0	2
ustralia	0	0	0	0	1	0	0	1
thiopia	0	0	0	0	0	1	0	1
hailand	0	0	0	0	1	0	0	1
otals	71	169	59	6	234	47	1	587
ercent of total	12.10	28.79	10.05	1.02	39.86	8.01	0.17	100

able A.5 Summary: Geothermal power plants by number of units and installed capacity.

ype:	Dry steam	1-Flash	2-Flash	3-Flash	Binary	Flash-binary	Hybrid	Total
o. units:								
otals	71	169	59	6	234	47	1	587
% of total	12.10	28.79	10.05	1.02	39.86	8.01	0.17	100
IW:								
otals	2893	4581.26	1856.1	225.8	707.76	398.6	6	10668.52
% of total	27.12	42.94	17.40	2.12	6.63	3.74	0.06	100
IW/unit	40.75	27.11	31.46	37.63	3.02	8.48	6.00	18.17

Table A.6.1 Iceland.

Plant	Year	Type	MW-rated	No. units	MW-total	Comments
Namafjall: Bjarnarflag	1969	1-Flash	3.2	1	3.2	
Krafla:						
Unit 1	1977	2-Flash	30	1	30	
Unit 2	1997	2-Flash	30	1	30	
Svartsengi:						
Unit 1	1978	1-Flash	1	1	1	
Unit 2	1979	1-Flash	1	1	1	
Unit 3	1981	1-Flash	6	1	6	
Unit 4	1989	Binary	1.3	3	3.9	
Unit 4 Extension	1993	Binary	1.2	4	4.8	
Unit 5	1999	1-Flash	30	1	30	
	2007	Dry steam	30	1	30	
Nesjavellir:						
Units 1&2	1998	1-Flash	30	2	60	
Unit 3	2001	1-Flash	30	1	30	
Unit 4	2005	1-Flash	30	1	30	
Husavik	2000	Binary	2	1	2	Kalina cycl
Reykjanes:						
Salt Plant	1983	1-Flash	0.5	1	0.5	
Units 1&2	2006	1-Flash	50	2	100	
Unit 3	2010	1-Flash	50	1	50	
Hellisheidi: Units 1&2	2006	1-Flash	45	2	90	
Unit 3	2007	1-Flash	33	1	33	
Units 4&5	2008	1-Flash	45	2	90	
Units 6&7	2011	1-Flash	45	2	90	
Totals				31	715.4	

able A.6.2 Indonesia.

ant	Year	Type	MW-rated	No. units	MW-total	Comments
VA:						
amojang:						
Wellhead unit	1978	Dry steam	0.25	1	0.25	Retired
Unit 1	1982	Dry steam	30	1	30	
Unit 2	1987	Dry steam	55	1	55	
Unit 3	1987	Dry steam	55	1	55	
Units 4&5	2007	Dry steam	30	2	60	
ieng:						
Wellhead unit	1980	1-Flash	2	1	2	Moved
Unit 1	2002	1-Flash	60	1	60	
arajat:						
Unit I	1994	Dry steam	55	1	55	
Unit II	2000	Dry steam	80	1	80	
Unit III	2008	Dry steam	110	1	110	
unung Salak:						Aka Awibengkok
Units 1&2	1994	1-Flash	55	2	110	
Units 3-6	1997	1-Flash	55	4	220	
/ayang Windu:						
Unit 1	1999	1-Flash	110	1	110	
Unit 2	2009	1-Flash	117	1	117	
ORTH SULAWESI:						
ahendong:						
Pilot unit	1992	1-Flash	2.5	1	2.5	Inactive
Unit 1	2002	1-Flash	20	1	20	
Unit 2	2008	1-Flash	20	1	20	
Unit 3	2009	1-Flash	20	1	20	
UMATRA:						
ibayak:						
Unit 1	1996	1-Flash	2	1	2	From Dieng
Units 2&3	2007	1-Flash	5	2	10	
otals				26	1138.75	
ctive				23	1134	

Table A.6.3 Italy.

Plant	Year	Type	MW-rated	No. units	MW-total	Comment
LARDERELLO:						
Larderello 2: Units 1-4	1938	Dry steam	14.5	4	58	Retired
Unit 5	1938	Dry steam	11	1	11	Retired
Larderello 3: Unit 1	1965	Dry steam	24	1	24	Retired
Unit 2	1983	Dry steam	26	1	26	Retired
Unit 3	1964	Dry steam	24	1	24	Retired
Unit 4	1965	Dry steam	24	1	24	Retired
Nuova Larderello	2005	Dry steam	20	1	20	
Castelnuovo: Unit 1	1946	Dry steam				
Unit 2	1948	Dry steam	11	1	11	Retired
Unit 5	1967	Dry steam	11	1	11	Retired
Nuova Castelnuovo	2000	Dry steam	14.5	1	14.5	
Gabbro	1969	Dry steam	15	1	15	Retired
Nuova Gabbro	2002	Dry steam	20	1	20	
Valle Secolo: Unit 1	1991	Dry steam	60	1	60	
Unit 2	1992	Dry steam	60	1	60	
Farinello	1995	Dry steam	60	1	60	
Le Prata	1996	Dry steam	20	1	20	
Monteverdi 1&2	1997	Dry steam	20	2	40	
Selva	1997	Dry steam	20	1	20	
Lago: Unit 1	1960	Dry steam	6.5	1	6.5	Retired
Unit 2	1960	Dry steam	12.5	1	12.5	Retired
Unit 3	1964	Dry steam	14.5	1	14.5	Retired
Nuova Lago	2002	Dry steam	10	1	10	
Cornia: Unit 1	1987	Dry steam	20	1	20	Retired
Unit 2	1994	Dry steam	20	1	20	
San Martino: Unit 2	1985	Dry steam	20	1	20	Retired
Unit 3	1985	Dry steam	20	1	20	Retired
Nuova San Martino	2005	Dry steam	40	1	40	
Molinetto	1982	Dry steam	8	1	8	Retired
Nuova Molinetto	2002	Dry steam	20	1	20	
La Leccia	1983	Dry steam	8	1	8	Inactive
Lagoni Rossi: Unit 1	1960	Dry steam	3.5	1	3.5	Retired
Unit 2	1969	Dry steam	3	1	3	Retired
Unit 3	1981	Dry steam	8	1	8	Retired
Nuova Lagoni Rossi	2009	Dry steam	20	1	20	
Monterotondo	1968	Dry steam	12.5	1	12.5	Retired
Nuova Monterotondo	2002	Dry steam	10	1	10	
Sasso Pisano 1: Units 1&2	1969	Dry steam	3.5	2	3.5	Retired
Sasso Pisano 2: Unit 1	1968	Dry steam	12.5	1	12.5	Retired
Unit 2	1968	Dry steam	3.2	1	3.2	Retired
Serrazzano:						
Units 1&2	1967	Dry steam	12.5	2	25	Retired
Unit 5	1975	Dry steam	15	1	15	Retired
Nuova Serrazzano	2002	Dry steam	60	1	60	
Carboli:						
Unit 1	1998	Dry steam	20	1	20	
Unit 2	1997	Dry steam	20	1	20	
Nuova Sasso	1996	Dry steam	20	1	20	
Sasso 2	2009	Dry steam	20	1	20	
Sesta	2002	Dry steam	20	1	20	

(*Continued*)

able A.6.3 (Continued).

lant	Year	Type	MW-rated	No. units	MW-total	Comments
T. AMIATA:						
agnore:						
Bagnore 1&2	1945	Dry steam	3.5	2	7	Retired
Unit 1	1959	Dry steam	0.9	1	0.9	Retired
Unit 2	1962	Dry steam	3.5	1	3.5	Retired
Unit 3	1998	1-Flash	20	1	20	
Bellavista	1987	Dry steam	20	1	20	Retired
iancastagnaio:						
Unit 1	1969	Dry steam	15	1	15	Retired
Unit 2	1969	Dry steam	8	1	8	
Unit 3	1990	Dry steam	20	1	20	
Unit 4	1991	Dry steam	20	1	20	
Unit 5	1994	Dry steam	20	1	20	
RAVALE-RADICONDOLI:						
adicondoli: Units 1&2	1979	Dry steam	15	2	30	Retired
ancia	1986	Dry steam	20	1	20	
ianacce	1987	Dry steam	20	1	20	
ancia 2	1988	Dry steam	20	1	20	
ravale 21	1991	Binary	0.7	1	0.7	Retired
ravale 3	2000	Dry steam	20	1	20	
ravale 4	2002	Dry steam	40	1	40	
uova Radicondoli	2002	Dry steam	40	1	40	
uova Radicondoli 2	2010	Dry steam	20	1	20	
hiusdino 1	2010	Dry steam	20	1	20	
otals				75	1359.3	
ctive				35	882.5	

Table A.6.4 Japan.

Plant	Year	Type	MW-rated	No. units	MW-total	Comments
KYUSHU:						
Tsurumi	1925	Dry steam	0.00112	1	0.00112	Experimenta
Hakuryu	1951	Dry steam	0.3	1	0.3	Experimenta
Otake	1967	1-Flash	12.5	1	12.5	
Otake Pilot	1978	Binary	1	1	1	Retired
Hatchobaru 1	1977	2-Flash	55	1	55	
Hatchobaru 2	1990	2-Flash	55	1	55	
Hatchobaru 3	2003	Binary	2	1	2	
Suginoi	1981	1-Flash	1.9	1	1.9	
Kirishima Kokusai	1984	1-Flash	0.22	1	0.22	
Takenoyu	1991	1-Flash	0.05	1	0.05	
Kujukannko	1998	1-Flash	0.99	1	0.99	
Takigami	1996	1-Flash	25	1	25	
Takigami Binary	1997	Binary	0.49	1	0.49	Retired
Ogiri	1996	1-Flash	30	1	30	
Yamagawa	1995	1-Flash	30	1	30	
HONSHU:						
Matsukawa	1966	Dry steam	23.5	1	23.5	
Onuma	1974	1-Flash	9.5	1	9.5	
Onikobe	1975	1-Flash	12.5	1	12.5	
Kakkonda 1	1978	1-Flash	50	1	50	
Kakkonda 2	1996	1-Flash	30	1	30	
Sumikawa	1995	1-Flash	50	1	50	
Uenotai	1994	1-Flash	28.8	1	28.8	
Yanaizu-Nishiyama	1995	1-Flash	65	1	65	
Hachijo-Jima	1999	1-Flash	3.3	1	3.3	
HOKKAIDO:						
Mori	1982	2-Flash	50	1	50	
Nigorikawa Pilot	1978	Binary	1	1	1	Retired
Totals				26	540.04	
Active				21	535.26	

Table A.6.5 Kenya.

Plant	Year	Type	MW-rated	No. units	MW-total	Comments
Olkaria I:						aka East Olkaria
Unit 1	1981	1-Flash	15	1	15	
Unit 2	1982	1-Flash	15	1	15	
Unit 3	1985	1-Flash	15	1	15	
Olkaria II: Units 1&2	2003	1-Flash	35	2	70	aka NE Olkaria
Olkaria III:						aka West Olkaria
Phase 1, Units 1&2	2000	Flash-binary	4.6	2	9.2	
Phase 1, Unit 3	2000	Flash-binary	2.8	1	2.8	
Phase 2, Units 4–6	2008	Flash-binary	12	3	36	
Oserian:						Flower company
Unit 1	2004	Binary	1.8	1	1.8	
Unit 2	2007	1-Flash	1.4	1	1.4	Back-pressure
Totals				13	166.2	

able A.6.6 Mexico.

.ant	Year	Type	MW-rated	No. units	MW-total	Comments
ATHE: Pilot plant	1959	1-Flash	3.5	1	3.5	Dismantled
ERRO PRIETO:						
erro Prieto I:						
Units 1&2	1973	1-Flash	37.5	2	75	
Units 3&4	1979	1-Flash	37.5	2	75	
Unit 5	1981	1-Flash	30	1	30	
erro Prieto II: Units 1&2	1984	2-Flash	110	2	220	
erro Prieto III: Units 1&2	1985	2-Flash	110	2	220	
erro Prieto IV: Units 1–4	2000	1-Flash	25	4	100	
OS AZUFRES:						
Unit 1	1982	1-Flash	5	1	5	Dismantled
Units 2–5	1982	1-Flash	5	4	20	Noncondensing
Unit 6	1986	1-Flash	5	1	5	Noncondensing
Unit 7	1988	1-Flash	50	1	50	
Units 9–10	1990–92	1-Flash	5	2	10	Noncondensing
Units 11&12	1993	Binary	1.5	2	3	
Units 13–16	2003	1-Flash	25	4	100	
OS HUMEROS: Units 1–7						
Units 1&2	1990	1-Flash	5	2	10	Noncondensing
Units 3&5	1991	1-Flash	5	2	10	Noncondensing
Unit 6	1992	1-Flash	5	1	5	Noncondensing
Unit 7	1994	1-Flash	5	1	5	Noncondensing
Unit 4	2004	1-Flash	5	1	5	Noncondensing
Unit 8	2008	1-Flash	5	1	5	Noncondensing
OS HUMEROS II: Unit 9	2011	1-Flash	25	1	25	
RES VIRGINES: Units 1&2	2002	1-Flash	5	2	10	
IAGUARICHIC:						
Piedras de Lumbre	2001	Binary	0.3	1	0.3	
'otals				41	991.8	
ictive				39	983.3	

Table A.6.7 New Zealand.

Plant	Year	Type	MW-rated	No. units	MW-total	Comments
WAIRAKEI:						
Unit 1	1959	2-Flash	11.2	1	11.2	
Unit 2	1958	1-Flash	6.5	1	6.5	Retired
Unit 3	1959	1-Flash	6.5	1	6.5	Retired
Unit 4	1959	2-Flash	11.2	1	11.2	
Units 5&6	1962	1-Flash	11.2	2	22.4	Moved
Units 7&8	1959	3-Flash	11.2	2	22.4	
Units 9&10	1960	3-Flash	11.2	2	22.4	
Unit 11	1962	2-Flash	30	1	30	
Units 12&13	1963	2-Flash	30	2	60	
Unit 14	na	1-Flash	5	1	5	
Poihipi	1996	Dry steam	55	1	55	
Bottoming unit	2005	Binary	5	3	15	
KAWERAU:						
Unit 1	1961	1-Flash	10	1	10	Retired
TG1	1989	Binary	1.2	2	2.4	
TG2	1993	Binary	3.5	1	3.5	
KA24	2008	Binary	8.3	1	8.3	
Mighty River Power	2008	2-Flash	100	1	100	
TAUHARA:	2010	Binary	23	1	23	aka Te Huka
OHAAKI:						
Units 1&2	1988	2-Flash	11.2	2	22.4	From Wairake
Units 3&4	1988	2-Flash	46	2	92	
MOKAI:						
Mokai I	1999	Flash-binary	25, 5	1, 6	55	
Mokai II	2005	Flash-binary	33, 6	1, 1	39	
Mokai IA	2007	Binary	17	1	17	
NGAWHA:						
Unit 1	1998	Flash-binary	4.5	2	9	
Unit 2	2008	Binary	15	1	15	
ROTOKAWA:						
Combined Cycle	1997	Flash-binary	13, 4.5	1, 3	26.5	
Extension	2003	Flash-binary	6	1	6	
Nga Awa Purua	2010	3-Flash	132	1	132	
Totals				48	828.7	
Active				43	783.3	

able A.6.8 Philippines.

ant	Year	Type	MW-rated	No. units	MW-total	Comments
JZON:						
akiling-Banahaw:						aka Mak-Ban
Plant A: Units 1&2	1979	1-Flash	63.2	2	126.4	
Plant B: Units 3&4	1980	1-Flash	63.2	2	126.4	
Plant C: Units 5&6	1984	1-Flash	54.6	2	109.2	
Binary I, II, III	1994	Binary	3	5	15	Inactive
Binary	1994	Binary	0.73	1	0.73	Inactive
Plant D: Unit 9	1996	1-Flash	20	2	40	Standby
Plant E: Unit 10	1996	1-Flash	20	2	40	
iwi:						
Plant A: Units 1&2	1979	1-Flash	55	2	110	
Plant B: Units 3&4	1980	1-Flash	55	2	110	Decommissioned
Plant C: Units 5&6	1982	1-Flash	55	2	110	
acon-Manito:						aka Bac-Man
Palayan: Units 1&2	1993	1-Flash	55	2	110	
Cawayan: Unit 3	1994	1-Flash	20	1	20	Inactive
Botong: Unit 4	1998	1-Flash	20	1	20	Inactive
Manito Lowland	1998	1-Flash	1.5	1	1.5	
EYTE:						
WH Unit	1977	1-Flash	3	1	3	Retired, 1994
Tongonan I Units 1–3	1983	1-Flash	37.5	3	112.5	
Mahanagdong Units 13–15	1997	1-Flash	60	3	180	
Tongonan Unit 16	1997	1-Flash	6.5	3	19.5	
Mahanagdong Units 18–20	1997	1-Flash	6.5	3	19.5	
Malitbog Unit 8	1996	1-Flash	77.5	1	77.5	
Malitbog Unit 17	1997	1-Flash	16.7	1	16.7	
Upper Mahiao Units 4–7	1996	Flash-binary	34.12, 5.5	5	142	
Malitbog Units 21&22	1997	1-Flash	77.5	2	155	
OUTHERN NEGROS:						
alinpinon I:						
WH Units 1&2	1980	1-Flash	1.5	2	3	Retired, 1994
WH Units 3&4	1982	1-Flash	1.5	2	3	Retired
Units 1–3	1983	1-Flash	37.5	3	112.5	
alinpinon II:						
Balas-balas Unit 5	1993	1-Flash	20	1	20	aka Okoy 5
Nasuji Unit 4	1994	1-Flash	20	1	20	
Sogongon Units 6&7	1995	1-Flash	20	2	40	
IINDANAO:						
Matingao Unit 1	1997	1-Flash	52.3	1	52.3	
Sandawa Unit 2	1999	2-Flash	50.9	1	50.9	
IORTHERN NEGROS:						
Unit 1	2007	1-Flash	49	1	49	
otals				63	2015.63	
ctive				48	1840.9	

Table A.6.9 Russia.

Plant	Year	Type	MW-rated	No. units	MW-total	Comment
Paratunka: Unit 1	1967	Binary	0.68	1	0.68	Retired
Pauzhetka:						
Units 1&2	1967	1-Flash	2.5	2	5	
Unit 3	1981	1-Flash	6	1	6	
Verkhne-Mutnovsky:						
Unit 1	1998	1-Flash	4	1	4	
Units 2&3	1999	1-Flash	4	2	8	
Mutnovsky:						
Unit 1	2002	1-Flash	25	1	25	
Unit 2	2004	1-Flash	25	1	25	
Okeansky:						
Unit 1	2000	1-Flash	1.7	1	1.7	
Unit 2	2001	1-Flash	1.7	1	1.7	
Goryachii Plyazh:	2004	1-Flash	1.3	2	2.6	
Totals				13	79.68	
Active				12	79	

able A.6.10 United States: California.

ant	Year	Type	MW-rated	No. units	MW-total	Comments
HE GEYSERS:						
Unit 1	1960	Dry steam	11	1	11	Dismantled
Unit 2	1963	Dry steam	13	1	13	Dismantled
Unit 3	1967	Dry steam	27	1	27	Dismantled
Unit 4	1968	Dry steam	27	1	27	Dismantled
McCabe	1971	Dry steam	2 × 53	2	106	
Ridge Line	1972	Dry steam	2 × 53	2	106	
Fumarole	1973	Dry steam	2 × 53	2	106	Dismantled
Eagle Rock	1975	Dry steam	106	1	106	
Cobb Creek	1979	Dry steam	106	1	106	
Big Geysers	1980	Dry steam	78	1	78	
Sulphur Springs	1980	Dry steam	65	1	65	
Unit 15	1979	Dry steam	59	1	59	Dismantled
Quicksilver	1985	Dry steam	113	1	113	
Lake View	1982	Dry steam	113	1	113	
Socrates	1983	Dry steam	113	1	113	
Calistoga	1984	Dry steam	2 × 40	2	80	
Grant	1985	Dry steam	113	1	113	
Bottlerock	1985	Dry steam	15	1	15	Reactivated
Sonoma	1983	Dry steam	72	1	72	
NCPA 1	1983	Dry steam	2 × 55	2	110	
NCPA 2	1985,86	Dry steam	2 × 55	2	110	
Coldwater Creek	1988	Dry steam	2 × 65	2	130	Dismantled
Bear Canyon	1988	Dry steam	2 × 11	2	22	
West Ford Flat	1988	Dry steam	2 × 14.5	2	29	
J.W. Aidlin	1989	Dry steam	2 × 10	2	20	
otals				35	1850	
ctive				26	1477	
MPERIAL VALLEY:						
rawley:						
Pilot Plant	1980	1-Flash	10	1	10	Dismantled 1985
North Brawley	2010	Binary	10	5	50	
otals				6	60	
ctive				5	50	
ast Mesa:						
GEM 1	1979	Binary	13.4	1	13.4	Dismantled
GEM 2	1989	2-Flash	18.5	1	18.5	
GEM 3	1989	2-Flash	18.5	1	18.5	
ORMESA I	1987	Binary	0.923	26	24	Original plant
ORMESA I	2003	Binary	10, 1	4	22	
ORMESA II	1988	Binary	0.825	20	16.5	
ORMESA IE	1988	Binary	0.8	10	8	
ORMESA IH	1989	Binary	0.542	12	6.5	
otals				75	127.4	
ctive				48	90	

(Continued)

Table A.6.10 (Continued).

Plant	Year	Type	MW-rated	No. units	MW-total	Comments
Heber:						
Binary Demo	1985	Binary	45	1	45	Dismantled
Heber 1	1985	2-Flash	47	1	47	
Heber 2	1993	Binary	2.75	12	33	Orig. SIGC
Gould I	2006	Binary	2.5	4	10	
Gould II	2006	Binary	10	1	10	
Heber South	2008	Binary	10	1	10	
Totals				20	155	
Active				19	110	
Salton Sea:						
Unit 1	1982	1-Flash	10	1	10	1st FC/RC plant
Unit 2	1990	2-Flash	10,5,4	3	19	
Unit 3	1989	2-Flash	49.8	1	49.8	
Unit 4	1989	2-Flash	39.6	1	39.6	
Unit 5	2000	3-Flash	49	1	49	
Vulcan	1985	2-Flash	29,8	2	37	
A.W. Hoch	1989	2-Flash	35.8	1	35.8	Orig. Del Ranch
J.J. Elmore	1989	2-Flash	35.8	1	35.8	
J.M. Leathers	1989	2-Flash	35.8	1	35.8	
Totals				12	311.8	
CASA DIABLO:						aka Mammoth
MP-I	1984	Binary	3.5	2	7	
MP-II	1990	Binary	5	3	15	
PLES-I	1990	Binary	3.3	3	10	
Totals				8	32	
HONEY LAKE:						
Wineagle	1985	Binary	0.35	2	0.7	
Amedee	1988	Binary	0.8	2	1.6	
Honey Lake:	1989	Hybrid	30	1	6	
Totals				5	8.3	
COSO:						
Navy I Unit 1	1987	2-Flash	30	1	30	
Navy I Unit 2	1988	2-Flash	25	1	30	
Navy I Unit 3	1988	2-Flash	25	1	30	
Navy II Unit 4	1989	2-Flash	28	1	30	
Navy II Unit 5	1989	2-Flash	28	1	30	
Navy II Unit 6	1989	2-Flash	28	1	30	
BLM I Unit 7	1988	2-Flash	24	1	30	
BLM I Unit 8	1988	2-Flash	24	1	30	
BLM I Unit 9	1989	2-Flash	28	1	30	
Totals				9	270	
Totals				170	2814.5	
Active				132	2349.1	

able A.6.11 United States: Nevada, Utah, Hawaii, Idaho, Alaska, New Mexico, Oregon, nd Wyoming.

ant	Year	Type	MW-rated	No. units	MW-total	Comments
EVADA:						
Wabuska 1	1984	Binary	0.5	1	0.5	
Wabuska 2	1987	Binary	0.7	1	0.7	
Beowawe	1985	2-Flash	16	1	16	
Desert Peak 1	1985	2-Flash	7	1	7	
Desert Peak 2	2007	Binary	6	2	12	
San Emidio	1987	Binary	0.9	4	3.6	aka Empire
Steamboat 1	1986	Binary	0.86	7	6	
Steamboat 1A	1988	Binary	0.55	2	1.1	
Steamboat 2	1992	Binary	7	2	14	
Steamboat 3	1992	Binary	7	2	14	
Steamboat Hills	1988	1-Flash	13	1	13	
Steamboat Hills 2	2007	Binary	6	1	6	
Burdette	2006	Binary	13	2	26	Orig. Galena 1
Galena 2	2007	Binary	13	1	13	
Galena 3	2008	Binary	13	2	26	
Soda Lake 1	1987	Binary	1.2	3	3.6	
Soda Lake 2	1991	Binary	2	6	12	
Stillwater I	1989	Binary	0.93	14	13	Inactive
Stillwater 2	2009	Binary	12	4	48	
Dixie Valley	1988	2-Flash	60.5	1	60.5	
Brady I	1992	2-Flash	6	3	17	
Brady II	2002	Binary	3	1	3	
Rye Patch	2001	Binary	2.5	5	12.5	Inactive
Salt Wells	2009	Binary	9	2	18	
Jersey Valley	2010	Binary	15	1	15	
otals				70	361.5	
ctive				51	336	
TAH:						
Blundell Unit 1	1984	1-Flash	26	1	26	
Cove Fort 1	1985	Binary	0.5	4	2	Inactive
Cove Fort 2	1988	Dry steam	2	1	2	Inactive
Bonnett	1990	Dry steam	7	1	7	Inactive
Blundell Unit 2	2007	Binary	11	1	11	
Thermo HS: Hatch	2008	Binary	0.2	50	10	
otals				58	58	
ctive				52	47	
IAWAII:						
Puna PGV-1	1992	Flash-binary	3	10	20	
Puna Extension	2011	Binary	8	1	8	
otals				11	28	

(*Continued*)

Table A.6.11 (Continued).

Plant	Year	Type	MW-rated	No. units	MW-total	Comments
IDAHO:						
Raft River	1981	Binary	5	1	5	Dismantled 198
USGeo Phase 1	2008	Binary	13	1	13	
Totals				2	18	
Active				1	13	
ALASKA:						
Chena H. S. Units 1&2	2006	Binary	0.2	2	0.4	
Chena H. S. Unit 3	2010	Binary	0.2	1	0.2	Inactive
Totals				3	0.6	
Active				2	0.4	
NEW MEXICO:						
Burgett Greenhouse	1995	Binary	0.35	1	0.35	Inactive
Burgett Greenhouse	1995	Binary	0.4	2	0.8	Inactive
Burgett Greenhouse	2008	Binary	0.2	2	0.4	
Totals				5	1.55	
Active				2	0.4	
OREGON:						
Oregon Inst. of Tech.	2009	Binary	0.28	1	0.28	
WYOMING:						
RMOTC	2009	Binary	0.25	1	0.25	
Totals				151	468.18	
Active				121	425.33	

able A.7 Countries with fewer than 10 units.

ountry: Plant	Year	Type	MW-rated	No. units	MW-total	Comments
USTRALIA						
ulka: Unit 1	1986	Binary	0.02	1	0.02	Inactive
rdsville: Unit 1	1992	Binary	0.15	1	0.15	
otals				1	0.15	
USTRIA						
ad Blumau	2001	Binary	0.25	1	0.25	
ltheim: Unit 1	2002	Binary	1	1	1	
mbach/Braunau	2009	Binary	0.2	1	0.2	
otals				3	1.45	
HINA						
engwu:						
Unit 1	1970	1-Flash	0.086	1	0.086	Retired
Unit 2	1977	Binary	0.2	1	0.2	Retired
Unit 3	1984	1-Flash	0.3	1	0.3	Retired; aka Fengshun
uailai	1971	Binary	0.285	1	0.285	Retired
Ventang	1971	Binary	0.05	1	0.05	Retired
uitang	1975	1-Flash	0.3	1	0.3	Retired
hingshui	1981	1-Flash	3	1	3	Inactive
iongyue	1978	Binary	0.1	1	0.1	Retired
uchang	1985	Binary	0.3	1	0.3	Inactive
angjiu:						
Unit 1	1987	2-Flash	1	1	1	Retired
Unit 2	1988	2-Flash	1	1	1	Retired
angbajing:						
Unit 1	1977	1-Flash	1	1	1	Retired
Unit 2	1984	2-Flash	3	1	3	
Unit 3	1981	2-Flash	3	1	3	
Unit 4–9	1982–92	2-Flash	3	6	18	
Nagqu: Unit 1	1993	Binary	1	1	1	Retired
Totals				8	24	
OSTA RICA						
Miravalles:						
Wellhead Unit 1	1995	1-Flash	5	1	5	
Wellhead Unit 2	1996	1-Flash	5	1	5	Dismantled 1998
Wellhead Unit 3	1997	1-Flash	5	1	5	Dismantled 1998
Unit 1	1994	1-Flash	55	1	55	
Unit 2	1998	1-Flash	55	1	55	
Unit 3	2000	1-Flash	29	1	29	
Unit 5	2004	Binary	9.5	2	19	
Las Pailas: Unit 1	2011	Binary	21	2	42	
Totals				8	205	

(Continued)

Table A.7 (Continued).

Country: Plant	Year	Type	MW-rated	No. units	MW-total	Comments
EL SALVADOR						
Ahuachapán:						
Unit 1	1975	1-Flash	30	1	30	
Unit 2	1976	1-Flash	30	1	30	
Unit 3	1981	2-Flash	35	1	35	
Berlín:						
Wellhead Unit 1	1992	1-Flash	5	1	5	Retired
Wellhead Unit 2	1992	1-Flash	5	1	5	Retired
Units 1&2	1999	1-Flash	28	2	56	
Unit 3	2007	1-Flash	44	1	44	
Bottoming unit	2007	Binary	9.3	1	9.3	
Totals				7	204.3	
ETHIOPIA						
Langano: Unit 1	1998	Flash-binary	3.9, 4.6	1	8.5	Reactivated 2007
FRANCE						
Guadeloupe:						
Bouillante:						
Unit 1	1987	2-Flash	4.7	1	4.7	
Unit 2	2004	1-Flash	10	1	10	
Totals				2	14.7	
GERMANY						
Neustadt-Glewe	2003	Binary	0.2	1	0.2	
Landau	2008	Binary	3	1	3	
Unterhaching	2008	Binary	3	1	3	Kalina cycle
Bruchsal	2009	Binary	0.55	1	0.55	Kalina cycle
Totals				4	6.75	
GUATEMALA						
Zuníl: Orzunil Unit 1	1999	Flash-binary	3.5	7	24.6	
Amatitlán:						
Wellhead Unit 1	1997	1-Flash	5	1	5	Dismantled 2001
Geotermica Calderas	2003	1-Flash	5	1	5	Inactive
Unit 1	2007	Flash-binary	10	2	20	
Totals				9	44.6	
NICARAGUA						
Momotombo:						
Unit 1	1983	1-Flash	35	1	35	
Unit 2	1989	1-Flash	35	1	35	
Unit 3	2002	Binary	7.5	1	7.5	
San Jacinto:						
Unit 1	2006	1-Flash	5	2	10	Back-pressure
Totals				5	87.5	

(Continued)

able A.7 (Continued).

ountry: Plant	Year	Type	MW-rated	No. units	MW-total	Comments
APUA NEW GUINEA						
hir:						
Unit 1	2003	1-Flash	6	1	6	Back-pressure
Unit 2	2005	1-Flash	10	3	30	Condensing
Unit 3	2006	1-Flash	10	2	20	Condensing
otals				6	56	
ORTUGAL: The Azores:						
ao Miguel:						
ico Vermelho:						
Wellhead Unit 1	1980	1-Flash	3	1	3	Dismantled 2005
Unit 2	2006	Binary	6.5	2	13	
ibeira Grande:						
Phase A	1994	Binary	2.5	2	5	
Phase B	1998	Binary	4	2	8	
otals				6	26	
HAILAND						
ang: Unit 1	1989	Binary	0.3	1	0.3	
otals				1	0.3	
URKEY						
ızıldere:						
Wellhead Unit	1974	1-Flash	0.5	1	0.5	Dismantled
Unit 1	1984	1-Flash	20.4	1	20.4	
alavatlı: Dora 1	2006	Binary	7.4	1	7.4	
araköy: Bereket	2007	Binary	6.35	1	6.35	Inactive
ermencik	2009	2-Flash	47.4	1	47.4	
uzla-Çanakkale	2009	Binary	7.5	1	7.5	
alavatlı: Dora 2	2010	Binary	11.5	1	11.5	
enizli: Jeoden	2011	Binary	0.26	3	0.78	
otals				8	94.98	

able A.8 Countries with only retired/inactive plants.

ountry: Plant	Year	Type	MW-rated	No. units	MW-total	Comments
RGENTINA						
opahue: Unit 1	1988	Binary	0.67	1	0.67	Retired 1996
DRC						
Kiabukwa	1952	Binary	0.2	1	0.2	Dismantled c. 1956
GREECE						
Milos: Unit 1	1985	1-Flash	2	1	2	Retired 1989
ZAMBIA						
Kapisya	1986	Binary	0.1	2	0.2	Inactive
Totals				5	3.07	

Appendix B: Units Conversions

he book was written in mixed units, reflecting the lack of uniformity of usage across ne world. While the United States persists in using their own version of the old British ystem of units, now referred to as the U.S. Customary System of Units, the rest of the vorld has adopted the modern version of the metric system, the Système International es Unites (International System of Units) or the S.I. System. However, even among sers of the S.I. system, there are departures from the strict S.I. units. Therefore, it is ecessary frequently to consult a table of units conversions. The one given below is ntended to allow the reader to move back and forth between the units used in this ook and the ones that the reader may be more accustomed to using.

Multiply →	by or	→ to obtain
to obtain ←	by	← Divide
ft	0.3048	m
m	3.281	ft
in	25.4	mm
km	0.6214	mi
cm/s^2	0.0328	ft/s^2
gal	0.1337	ft^3
gal/min or GPM	0.2272	m^3/h
g/cm^3	0.06242	lbm/ft^3
kg	2.2046	lbm
kg/s	7,936.56	lbm/h
t/h	0.2778	kg/s
g/cm^3	1000	kg/m^3
lbm/ft^3	16.02	kg/m^3
lbf/in^2 or psia	0.06895	bar
lbf/in^2 or psia	6.895	kPa
MPa	145	lbf/in^2 or psia
bar	14.5	lbf/in^2 or psia
kg/cm^2 or ata	0.9807	bar
in Hg	3.386	kPa
°C	1.8 and add 32	°F
°F	0.5556 and subtract 32	°C
°C/m	0.549	°F/ft
K	1.8	°R
J/g or kJ/kg	0.430	Btu/lbm
$cal/cm^2 \cdot s$	4.184	$J/cm^2 \cdot s$
$cal/cm^2 \cdot s$	41.84	$kJ/m^2 \cdot s$
W/m · °C	0.5779	Btu/h · ft · °F
kJ/kg · K	0.2388	Btu/lbm · °R
kW or kJ/s	3412	Btu/h

\ppendix C: Energy Equivalents

'hese tables give the *approximate* equivalent electrical energy, in kW · h, for various rimary sources of energy. We state the type of power plant in use; typical utilization fficiencies were assumed to arrive at the equivalent electricity. Actual plants may differ from these approximate values by +/−15%. For geothermal binary plants, the rine outlet temperature is very important and will strongly affect the output.

able C.1 U.S. Customary units.

rimary "fuel" source	Electricity kW · h	Conversion system
,000 ft^3 natural gas	100	Simple cycle gas turbine
,000 ft^3 natural gas	160	Combined steam & gas turbine
barrel fuel oil	720	Double-reheat, regenerative Rankine cycle
ton (2,000 lbm) lignite	1,260	Single-reheat Rankine cycle
ton (2,000 lbm) bituminous coal	2,820	Double-reheat, regenerative Rankine cycle
ton (2,000 lbm) garbage & trash	1,025	Single-reheat Rankine cycle
lbm natural uranium	310	Pressurized water reactor
un on 10 ft^2 of earth, 1 day	0.75	Photovoltaic cell @ 10% efficiency
,000 lbm geoliquid, 300°F	3.3	Binary plant
,000 lbm geoliquid, 350°F	5.6	Single-flash plant
,000 lbm geoliquid, 400°F	10.4	Double-flash plant
,000 lbm geoliquid, 450°F	14.5	Double-flash plant
,000 lbm geosteam, 350°F	53.5	Typical dry steam plant

'able C.2 S.I. units.

'rimary "fuel" source	Electricity kW · h	Conversion system
.00 m^3 natural gas	353	Simple cycle gas turbine
.00 m^3 natural gas	565	Combined steam & gas turbine
.00 liter fuel oil	450	Double-reheat, regenerative Rankine cycle
. tonne (1,000 kg) lignite	1,390	Single-reheat Rankine cycle
. tonne (1,000 kg) bituminous coal	3,110	Double-reheat, regenerative Rankine cycle
. tonne (1,000 kg) garbage & trash	1,130	Single-reheat Rankine cycle
. kg natural uranium	140	Pressurized water reactor
;un on 1 m^2 of earth, 1 day	0.75	Photovoltaic cell @ 10% efficiency
.,000 kg geoliquid, 150°C	7.3	Binary plant
.,000 kg geoliquid, 175°C	12.3	Single-flash plant
.,000 kg geoliquid, 200°C	22.9	Double-flash plant
.,000 kg geoliquid, 235°C	32.0	Double-flash plant
.,000 kg geosteam, 175°C	118	Typical dry steam plant

Appendix D: Elements of Thermodynamics

"Die Energie der Welt ist constant. Die Entropie der Welt strebt einem Maximum zu."
Rudolf J. Clausius, on the First and Second Laws of thermodynamics – 1865

D.1 Purpose

This appendix is intended to give the reader a thumbnail sketch of the elements of thermodynamics, sufficient to allow an understanding of the concepts used in Parts 2 and 3 of the book. It cannot replace a proper exposition of the subject, which normally takes several semesters of study in an engineering curriculum.

D.2 Systems and properties

A thermodynamic *system* is a well-defined region of space, within which resides the focus of one's attention, e.g., a turbine, heat exchanger, flash vessel, etc. A system can be: (1) *closed* (no mass crosses the system boundary), (2) *open* (mass crosses the boundary), or (3) *isolated* (the system does not interact with its surroundings in any way).

A system that is *perfectly insulated* from its surroundings is said to have *adiabatic walls* or to be *adiabatic*; one that is *perfectly coupled thermally* to its surroundings is said to have *diathermal walls* or to be *diathermal*.

The system is described by its *properties*, physical characteristics that can be measured or calculated. Properties can be: (1) *extensive* (dependent on the size or extent of the system, such as mass, volume, energy), (2) *intensive* (independent of the size or extent of the system, such as temperature, pressure), or (3) *specific* (an extensive property per unit mass, such as specific volume). When the array of properties has a fixed set of values, the system exists in a certain *state*.

The *Gibbs Phase Rule* says that the number of *independent intensive thermodynamic properties*, f, is given by

$$f = C - P + 2 \tag{D.1}$$

where C is the number of pure substances present and P is the number of distinct phases present. Thus, for a single pure substance existing in a single phase (vapor, liquid, or solid), there are two independent intensive properties. Any other property can therefore be expressed in terms of the two independent ones, giving a surface in the space defined by the three properties. Special cases occur when more than one phas is present, reducing the generality of the surface. For example, a ruled surface result when two phases are present (say, liquid and vapor).

The properties are related through the *equation of state.* Gases and some vapors ca be approximately described by the simple *perfect gas equation of state,*

$$PV = mRT \tag{D.2}$$

where P is the absolute pressure, V is the volume, m is the mass, T is the absolute tem perature (kelvins or degrees Rankine), and R is the individual gas constant given by

$$R = \mathbf{R}/M \tag{D.3}$$

where M is the molar mass and $\mathbf{R}$ is the Universal Gas Constant, 8.314 kJ/kmol · K o 1.986 Btu/lbmol · R. More complex equations of state, such as the van der Waals, th Beattie-Bridgeman, or the Benedict-Webb-Rubin, can more accurately describe othe real substances. Computer programs are available for the properties of a large numbe of substances and their mixtures, e.g., the National Institute of Standards an Technology (NIST) *Standard Reference Database 23*, known as *REFPROP*.

A system undergoes a *process* when at least some of its properties change thei values. Processes may be: (1) *irreversible* (real, with a definite direction, finite rate accompanied by losses), (2) *reversible* (ideal, in equilibrium throughout, equidirec tional, infinitely slow, no losses), or (3) *quasistatic* (sufficiently close to equilibrium internally to allow reversible equations to be used as a good approximation). Once a system has undergone an irreversible process, it is impossible to restore the system to its original state without causing the surroundings to change from their initial state Thus, every irreversible process leaves a lasting effect.

When an *open system* undergoes a process, it can be either (1) *steady* (the values o system properties may vary spatially but are constant with time) or (2) *unsteady* (values of system properties at any fixed point in space vary with time). The thermody namic equations used to describe a system depend on whether it is open or closed, and whether it operates steady or unsteady.

D.3 First Law of thermodynamics for closed systems

The most general form of the First Law for a closed system is

$$Q_{1,2} - W_{1,2} = E_2 - E_1 \tag{D.4}$$

where $Q_{1,2}$ is the heat transfer, $W_{1,2}$ is the work transfer, and $E_2 - E_1$ is the change in the system energy during the process. This form applies to all closed thermodynamic systems, no matter the type of process.

If the process is assumed to be *ideally reversible* or *quasistatic*, then the following differential form may be used:

$$\delta Q - \delta W = dE \tag{D.5}$$

where the terms are defined as follows:

δQ = heat exchanged during an infinitesimal step in the process; mathematically, this is an *imperfect differential* that integrates to $Q_{1,2}$;

W = work exchanged during an infinitesimal step in the process; another *imperfect dif-rential* that integrates to $W_{1,2}$;
E = change in total system energy during an infinitesimal step in the process; this is a *erfect differential* that integrates to $E_2 - E_1$.

).4 First Law of thermodynamics for open systems

he First Law for open systems undergoing a *steady process* was given in Sect. 10.2, nd is repeated here:

$$\dot{Q} - \dot{W}_s = -\sum_{i=1}^{n} \dot{m}_i (h_i + 0.5\, V_i^2 + gz_i) \qquad \text{(D.6)}$$

ach term is defined as follows:

... rate of heat transfer (thermal power) between the system and its surroundings + when heat enters the system)
V... rate of work transfer (mechanical power) between the system and the urroundings (+ when work is delivered to the surroundings by the system)
... an index that accounts for all inlets and outlets of the system
... total number of inlets and outlets
$_i$ ... mass flow rate crossing each inlet or outlet
$_i$ ... specific enthalpy of the fluid at each inlet or outlet
$_i$ ... velocity of the fluid at each inlet or outlet
$_i$ ... elevation of each inlet or outlet relative to an arbitrary datum
... local gravitational acceleration, 9.81 m/s^2 or 32.2 ft/s^2.

To analyze a system fully, it is necessary also to use the *principle of conservation of ıass*, which in steady state requires that

$$\sum_{i=1}^{n} \dot{m}_i = 0 \qquad \text{(D.7)}$$

It is important to note that when using eqs. (D.6) and (D.7) mass flows are taken as ositive when entering the system and negative when leaving.

The First Law for open systems undergoing an *unsteady process* is mathematically ıore complex than for steady operation. Rather than an algebraic working equation, a ifferential equation must first be integrated before numerical values can be substituted. ince the electro-mechanical systems covered in this book have all been assumed to be perating steadily, this form of the First Law is not needed for our current purposes, but s included here for the sake of completeness. The most general form is:

$$\delta Q - \delta W - P\, dV = d\,(mu) - \sum_{i=1}^{n} dm_{\xi i}\, (h_\xi + \tfrac{1}{2}V_\xi^2 + gz_\xi)_i \qquad \text{(D.8)}$$

The new terms in the equation are defined as follows:

dV = expansion or compression work exchanged during an infinitesimal step in the rocess via a reversible deformation of system boundary; generally negligible for rigid ystems;

$d(mu)$ = change in system internal energy during an infinitesimal step in the process; mathematically this is a *perfect differential* that integrates to $m_2u_2 - m_1u_1$;
ξ = subscript referring to properties of streams entering or leaving the system;
dm = increment of mass entering (positive) or leaving (negative) system during infinitesimal step in process; another *perfect differential.*

This time the *principle of conservation of mass* requires that

$$\sum_{i=1}^{n} \dot{m}_i = \frac{dm}{dt} \tag{D.9}$$

or that the net influx of mass across the boundary equals the time rate of increase of mass within the system.

D.5 Second Law of thermodynamics for closed systems

The Second Law appears in many forms, some verbal and some mathematical. Often it is expressed as a negative statement or a mathematical inequality because the nature of the Second Law is to place restrictions on what can be accomplished during energy conversion processes. Here we summarize some of the common expressions of the Second Law.

D.5.1 Clausius Statement

It is impossible for heat to flow *spontaneously* from a body of lower temperature to one of higher temperature. If this were not true, it would be possible for refrigerators to operate without any motive energy being supplied.

D.5.2 Kelvin-Planck Statement

It is impossible to operate a cycle such that the only effects are the transfer of heat from a *single heat source* and the delivery of an equal amount of work from the cycle to the surroundings. If this were not true, it would be possible to operate power plants without rejecting any heat to the surroundings, and the plants would have a thermal conversion efficiency of 100%.

D.5.3 Clausius Inequality

Clausius also gave us a mathematical expression of the Second Law, namely:

$$\oint \frac{\delta Q}{T} \leq 0 \tag{D.10}$$

where $\oint$ means a *closed line integral* or a sum taken completely around the cycle of operations. The equality sign is used when all the heat transfer takes place *reversibly*, i.e., when there is no temperature difference across the heat transfer surface, whereas the inequality applies to *irreversible* heat transfer.

The Clausius Inequality may also be expressed in the following manner:
is impossible for any system to operate on a *cycle* such that

$$\oint \frac{\delta Q}{T} > 0 \tag{D.11}$$

.5.4 Existence of entropy

he Second Law implies the existence of a property called entropy, S, which is found om the equation:

$$dS = \frac{\delta Q^{rev}}{T} \tag{D.12}$$

The term dS is a mathematical *perfect differential*, and eq. (D.12) may be integrated o obtain:

$$S_2 - S_1 = \int_1^2 \frac{\delta Q^{rev}}{T} \tag{D.13}$$

his is quite significant in light of the fact that δQ^{rev} by itself is an *imperfect differential.* he quantity $1/T$ is an *integrating factor* for heat, creating a quantity that is a perfect ifferential. To carry out the integration of eq. (D.13), it is necessary to substitute for he reversible heat transfer using eq. (D.5).

Since entropy is a *thermodynamic property* and a *mathematical potential*, we are only nterested in *differences* in entropy, not the actual value of entropy. Furthermore, once he entropy difference has been found between a given pair of states by using q. (D.13), *the result will be the same no matter what type of process connects the two tates.* This latter point is often not understood by beginners to the subject who mistaknly believe that the entropy change will differ if one process is irreversible and one is deally reversible. Only the initial and final states determine the entropy change, not he process between them.

That having been said, the *calculation* of the entropy change must be done using a *eversible process*, as shown explicitly in eq. (D.13).

D.5.5 Principle of Entropy Increase (PEI)

This statement of the Second Law is intrinsically tied to the concept of an irreversible rocess. The PEI may be stated as follows:

When an *adiabatic* system undergoes an *irreversible* process from an initial state 1 to a final state 2, the entropy of the system can only *increase*, i.e.,

$$S_2 - S_1 > 0 \tag{D.14}$$

Since all real processes are irreversible and since the selection of the system boundary is arbitrary, some people have extended the boundary to include the entire Universe, and have concluded that the entropy of the Universe is constantly increasing. This would be true provided that the Universe is bounded by an adiabatic wall, which has yet to be verified.

The PEI may be put into an alternative, negative form in keeping with the genera theme of the Second Law, namely:

When an *adiabatic* system undergoes an *irreversible* process from an initial state 1 t a final state 2, it is impossible for the entropy of the system to *decrease*.

If an *adiabatic* system were to undergo an ideally *reversible* process, then the entrop of the system would remain *constant*, i.e., $S_2 = S_1$.

D.6 Second Law of thermodynamics for open systems

We presented the general form of the Second Law in Sect. 10.3 while introducing th concept of exergy, and repeat it here for completeness. The most general form tha applies to open systems is:

$$\dot{\theta}_p = \frac{dS}{d\tau} - \sum_{i=1}^{n} \dot{m}_i s_i - \int_{\tau_1}^{\tau_2} \frac{1}{T}\frac{dQ}{d\tau} \tag{D.15}$$

Each new term is defined as follows:

$\dot{\theta}_p$ ... rate of entropy production for the system caused by irreversibilities
τ ... time
s_i ... specific entropy of the fluid at each inlet or outlet, S_i/m_i
T ... absolute temperature (in K or °R) associated with the heat transfer.

The first term on the right-hand side vanishes for steady operation, the term on the left-hand side vanishes for reversible processes, and the last term on the right vanishes for adiabatic systems.

D.7 Thermodynamic state diagrams

We will conclude this brief review of basic thermodynamics with some examples of thermodynamic diagrams. It is helpful to view processes in thermodynamic state diagrams. Such diagrams provide a "road map" for visualizing, understanding, and analyzing the changes that occur when systems undergo various changes of state. Processes are usually characterized by whatever property is remains constant during the process. Thus, a process at constant temperature is called "isothermal," one at constant pressure is called "isobaric," etc.

The framework for the diagram is an appropriate projection of the working fluid surface of equilibrium states. The fluid can be represented by a 3-dimensional surface as long as it is a pure substance, in accordance with the Gibbs Phase Rule.

Some of the more useful choices for the independent properties, i.e., the coordinates of the two independent axes of the diagram, are: (1) pressure-volume, *P-v;* (2) temperature-entropy, *T-s;* (3) enthalpy-entropy, *h-s;* and (4) pressure-enthalpy, *P-h.* The latter three are particularly helpful for the cases studied in this book.

Figure D.1 shows a skeleton *T-s* diagram for the full spectrum of possible phases, followed by Fig. D.2, which gives the liquid-vapor and superheated portions of the phase space for water.

Figure D.3 gives the high-quality liquid-vapor and superheated regions for water, ollowed by Fig. D.4, which shows a skeleton *h-s* diagram for all possible phases.

Figure D.5 shows a skeleton *log P-h* diagram for all possible phases, followed by ig. D.6, which gives the liquid, liquid-vapor, and superheated vapor regions for ropane.

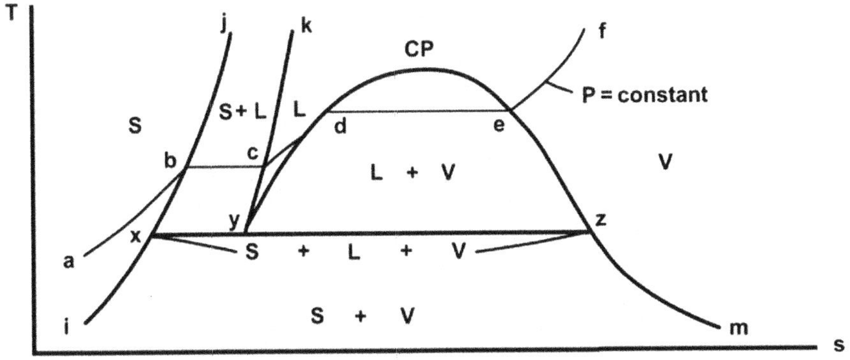

ig. D.1 Temperature-entropy diagram for a typical pure substance. Key: S = solid, L = liquid, V = vapor, P = Critical point, x-y-z = triple line, a-b-c-d-e-f = typical isobar, i-x = saturated solid in equilibrium with vapor, -j = saturated solid in equilibrium with liquid, y-k = saturated liquid in equilibrium with solid. y-CP = saturated quid in equilibrium with vapor, z-CP = saturated vapor in equilibrium with liquid, and m-z = saturated vapor in quilibrium with solid.

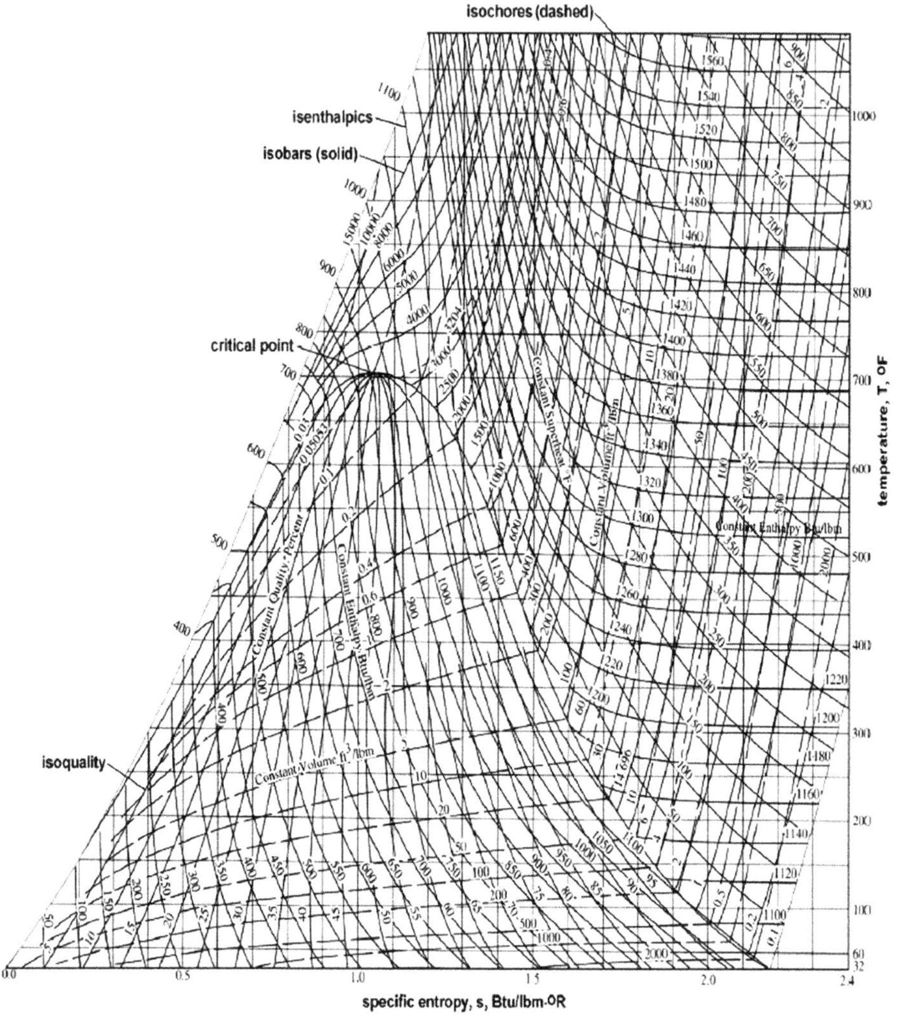

Fig. D.2 Temperature-entropy diagram for water in U.S. Customary units, showing the liquid-vapor and superheated regions.

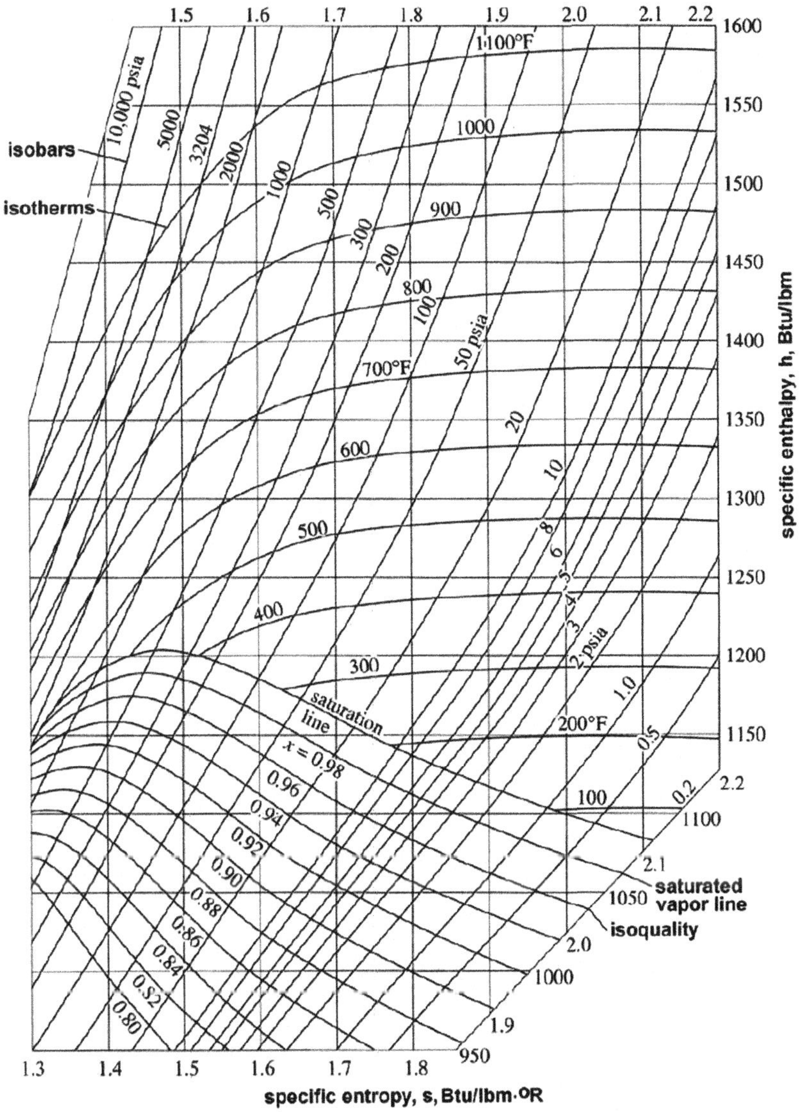

Fig. D.3 Enthalpy-entropy (Mollier) diagram for water in U.S. Customary units, showing the superheated region and the high-quality liquid-vapor region.

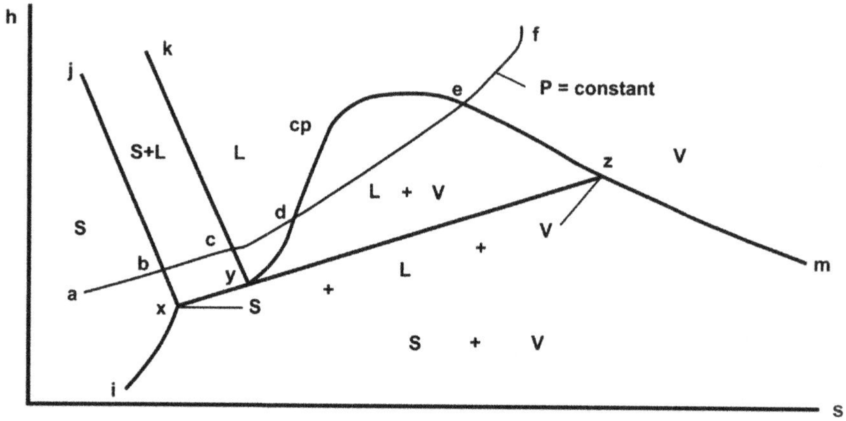

Fig. D.4 Enthalpy-entropy (Mollier) diagram for a typical pure substance. Key: S = solid, L = liquid, V = vapor, CP = critical point, x-y-z = triple line, a-b-c-d-e-f = typical isobar, i-x = saturated solid in equilibrium with vapor x-j = saturated solid in equilibrium with liquid, y-k = saturated liquid in equilibrium with solid, y-CP = saturated liquid in equilibrium with vapor, z-CP = saturated vapor in equilibrium with liquid, m-z = saturated vapor in equilibrium with solid.

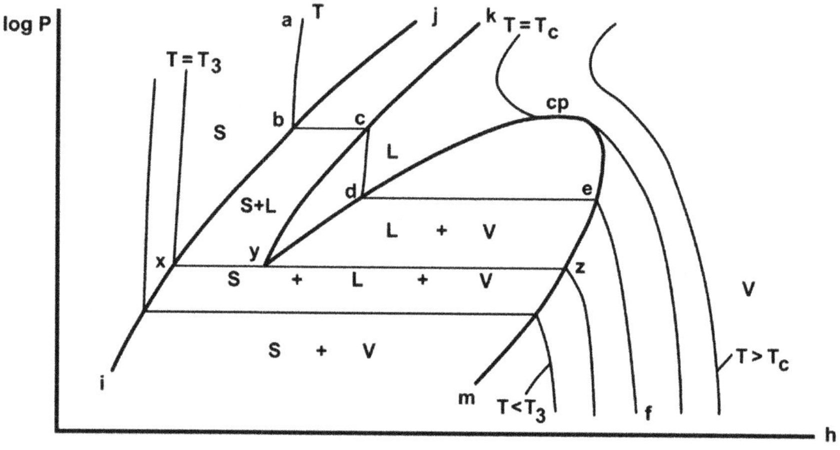

Fig. D.5 Log (pressure)-enthalpy diagram for a typical pure substance. Key: S = solid, L = liquid, V = vapor, CP = critical point, x-y-z = triple line, a-b-c-d-e-f = typical isotherm (T_3 T T_c), i-x = saturated solid in equilibrium with vapor, x-j = saturated solid in equilibrium with liquid, y-k = saturated liquid in equilibrium with solid, y-CP = saturated liquid in equilibrium with vapor, z-CP = saturated vapor in equilibrium with liquid, m-z = saturated vapor in equilibrium with solid.

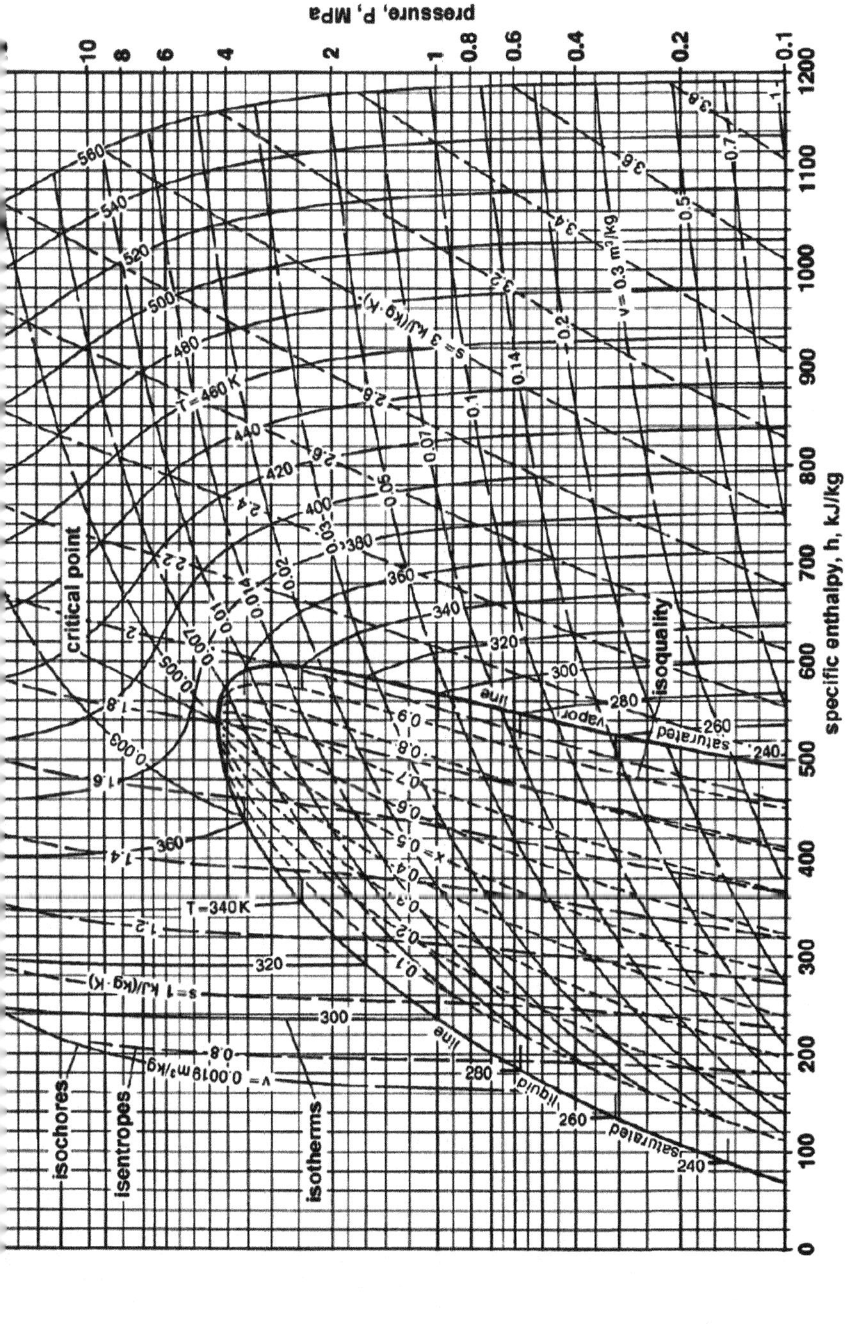

Fig. D.6 Pressure-enthalpy, semi-log diagram for propane in S.I. units, showing the liquid, liquid-vapor, and superheated regions.

Appendix E: Answers to Selected Practice Problems

"For every complex question, there is an easy answer – neat, simple and wrong."
H.L. Menchen

Chapter 1

1.1 Thickness = 19.6 km = 12.25 mi.
1.2 Temperature = 1075°C, for 25°C surface temperature.
1.3 Temperature difference = 630°C.
1.4 Depth = 6.9 km.
1.5 Depth = 3,790 m = 12,500 ft.

Chapter 2

2.1 (a) Shallow gradient = 14°F/100 ft; (b) deep gradient = 2.5°F/100 ft.
2.2 (a) SiO_2 (msl): (1) AH-l, 236°C; AH-5, 221°C; AH-26, 226°C. (2) H-1, 273°C; H-6, 241°C. (3) CP-avg, 266°C.
(c) Na/K (Fournier): (1) AH-l, 269°C; AH-5, 245°C; AH-26, 261°C. (2) H-1, 226°C; H-6, 221°C. (3) CP-avg, 292°C.
2.3 (a) Temperature = 230°C; (b) 251°C.
2.5 (a) Temperature = 195°C; (c) 229°C; (e) 254°C.

Chapter 4

4.1 Critical point: 7,688 ft (const. density); 12,370 ft (var. density); 10,500 ft (James).
4.2 Flash depth = 590 ft.
4.3 (a) At 100 m radius, ΔP = 106.7 atm; (b) At 100 l/s, ΔP = 117.8 atm.
4.4 (a) At 100 kg/s, flash depth = 373 m; at 150 kg/s, flash depth = 616 m.

Chapter 5

5.2 Turbine A: 77,933 kW; Turbine B: 67,746 kW.
5.3 (a) Specific work = 77.86 kJ/kg.
5.4 (a) Power = 8,922 kW; (b) ratio = 4.39; (c) (i) 38.3%, (ii) 44.6%.

Chapter 6

6.1 (a) Specific work = 185.4 kJ/kg.
6.2 (a) Temperature = 448.2°F; (e) plant power = 51.94 Btu/lbm; (f) 56.2%.

Chapter 7

7.1 (a) Pressure = 80 lbf/in^2; (b) about 12 wells in either case.
7.2 (b) Efficiency = 80.3%.
7.3 (a) Efficiency = 78.9%; (b) specific work = 527.1 kJ/kg; (c) 58.9%.
7.4 (a) Temperature = 185.2°C; (b) specific work = 532.0 kJ/kg.
7.5 (a) (i) Power = 7.43 MW, (ii) 12.51 MW.

Chapter 8

8.1 (a) Flow rate = 55.6 kg/s; (b) area = 6,839 ft^2; (c) functional eff. = 52.9%.
8.2 (a) Specific work = 34.11 Btu/lbm; (b) temperature = 71.6°F.
8.3 (a) Efficiency = 8.56%; (b) flow rate = 629,170 lbm/h; (d) 16.0%.
8.4 (a) (i) *LMTD* = 33.19°F; (ii) 18.20°F; (b) area ≅ 4,000 ft^2; (c) 83.3%.
8.5 (a) Specific work = 11.75 Btu/lbm R-12; (e) efficiency = 12.05%.
8.6 (a) ratio = 0.5427; (b) temperature = 163.1°F; (c) 74.8%; (d) brine flow rate = 1,733,000 lbm/h; (g) power = 7,367 kW.
8.7 (c) Efficiency = 17.9%; (d) 2. 24.97 kg/s; (e) 48.9%.
8.8 (c) Temperature ≅ 329 K; (d) (i) 22.20 kg/s, (ii) 25.77 kg/s; (e) 13.2%.

Chapter 9

9.1 (a) Specific work = 476.1 Btu/lbm; (b) 476.7 Btu/lbm; (c) 18.1 Btu/lbm (d) 1.110; (e) 1.115; (f) 3.616; (g) 4 wells.
9.2 (a) 42.46%; (b) 54.13%.
9.3 Efficiency = 52.86%.
9.4 (a) Flow rate = 34.95 kg/s; (b) (i) 5,630 kW, (ii) 1,932 kW; (c) 160.7 kW (d) 45.7%.
9.5 (a) Flow rate = 3,567,000 lbm/h; (c) 1.186; (d) 1.191; (e) 8.896.
9.6 (a) Specific work = 466 Btu/lbm; (b) 308.5 Btu/lbm; (c) 1.193.
9.7 t (SiO_2 (nsl) = 194.5°C; t (SiO, (msl) = 186.6°C; t (Na/Cl) = 320.5°C; t (Na/K/Ca) = 244.9°C.

Chapter 10

10.2 (b) Loss = 34.94 kJ/kg; (c) 90.88 kJ/kg steam; (d) (i) 9.16%.
10.3 (a) Temperature ≅ 339 K; (b) 0.555 kJ/kg; (c) *LMTD* = 17.6 K.

Appendix F: Supplementary Problems

"The fastest way to get an engineer to solve a problem is to declare that the problem is unsolvable."

Scott Adams, *The Dilbert Principle* – 2004

Chapter 1 Geology of Geothermal Regions

SP 1.1 Write a short essay describing a typical hydrothermal geothermal system. Consider a liquid-dominated reservoir; include a geological cross-section through the reservoir. Show a typical temperature–depth curve and explain its features. List the requirements for the site to be viable for long-term electric power production.

SP 1.2 Write a short essay on the origins and nature of geothermal energy from a global perspective. Give a thorough discussion of the theory of plate tectonics. Discuss the source of the normal geothermal temperature gradient and normal heat flux received at the surface of the earth. List and describe as many surface thermal manifestations of anomalous geothermal behavior as you can.

SP 1.3 Write a short essay on the origins and nature of the magma-type of geothermal energy. Include discussions of the geologic environment, typical characteristics of the resource, status of commercial development and prospects for continued development, impediments to widespread use, potential environmental impacts, etc. Include sketches and schematic diagrams wherever appropriate to illustrate your essay.

SP 1.4 Write a short essay on the origins and nature of the Hot Dry Rock (or Enhanced Geothermal Systems) type of geothermal energy. Include discussions of the geologic environment, typical characteristics of the resource, status of commercial development and prospects for continued development, impediments to widespread use, potential environmental impacts, etc. Include sketches and schematic diagrams wherever appropriate to illustrate your essay.

SP 1.5 Write a short essay on the origins and nature of the geopressured typed of geothermal energy. Include discussions of the geologic environment, typical characteristics of the resource, status of commercial development and prospects for continued development, impediments to widespread use, potential environmental impacts, etc. Include sketches and schematic diagrams wherever appropriate to illustrate your essay.

SP 1.6 (A) Describe the three major types of interaction between adjacent crusta plates according the theory of plate tectonics. Include simple sketches t illustrate your answer.

(B) Describe how each of these interactions may give rise to geotherma anomalies.

(C) Name at least one currently exploited geothermal area (i.e., generatin electrical power) associated with each of the three types.

Chapter 2 Exploration Strategies and Techniques

SP 2.1 Write a short essay on the methods of prospecting for geothermal systems. Ar certain methods more useful for high- or low-temperature systems? Includ sketches and schematic diagrams wherever appropriate to illustrate your essay.

SP 2.2 Three potential geothermal sites have been studied using temperature gradi ent surveys (typically to depths of about 500 m) and electrical resistivity sur veys (typically down to about 1000 m). The results are summarized in th table below.

Site name	Temperature gradient °C/100 m	Electrical resistivity ohm · m
Diablo Springs	75	3 (minimum)
Caliente	100	300 (minimum)
Evergreen	5	2 (minimum)

(A) Which site(s) appear to have reasonable potential as a hydrotherma resource?

(B) Which site(s) appear to be a possible candidate for a Hot Dry Rock resource?

(C) Which site(s) appear to be a possible candidate for a dry stean resource?

(D) Which site appears to be the worst candidate for a practical geotherma resource?

In all cases, fully explain the reasoning behind your answers.

SP 2.3 A 150 m deep temperature gradient hole shows a gradient that is six times greater than the normal temperature gradient. Based on this information calculate the expected temperature in the reservoir at depths of (A) 500 m and (B) 2000 m. Discuss the reliability of these two answers and the factors that affect the reliability.

SP 2.4 An ultra-simplified mixing model for a hydrothermal system is shown in the figure on the next page. The system is shown in an unexploited state in which geofluid from the reservoir rises through the formation. At some point underground, steam separates from the geofluid and continues to rise to the surface, emerging as a fumarole. The hot geothermal liquid flows laterally from the separation point until it mixes with cold ground water. The mixed fluid then rises to the surface and emerges as a hot spring.

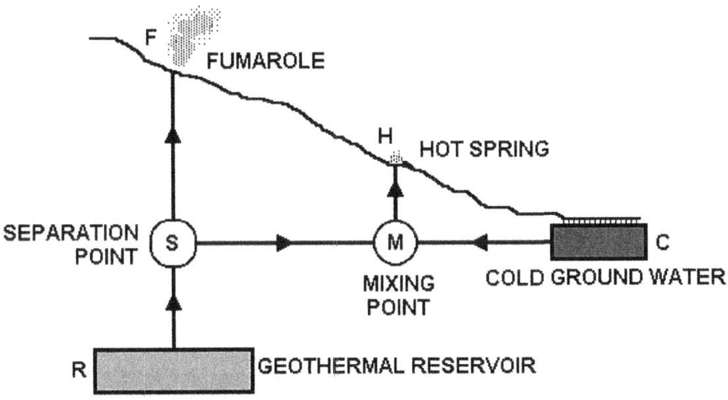

Physical quantities that are easily measured are:

T_C, temperature of the ground water
T_H, temperature of the hot spring
T_F, temperature of the fumarole
P_A, atmospheric pressure (assumed = 100 kPa).

Your task is to develop the analytical model for this hypothetical situation. Your objective is to determine the temperature of the fluid in the reservoir, T_R. Include all assumptions you make and justify them. For example, it might be reasonable to take all underground processes as adiabatic (no heat loss). Hint: The amount of silica present in the reservoir geofluid is conserved as the geofluid moves through the formation but is partitioned preferentially into the liquid phase when the steam separates from the geofluid.

Using your analytical model, estimate T_R for the following sets of conditions:
(A) $T_C = 20°C$; $T_H = 65°C$; $T_F = 105°C$
(B) $T_C = 25°C$; $T_H = 75°C$; $T_F = 110°C$
(C) $T_C = 30°C$; $T_H = 85°C$; $T_F = 115°C$.

SP 2.5 As explained in Sect. 2.3.5, the amount of silica in the liquid separated from a 2-phase geofluid mixture at the surface can be used to estimate the temperature of the liquid in the reservoir. Equation (2.7) applies to separation at one standard atmosphere of pressure. Do not use eq. (2.7) in solving this problem; use instead eqs. (2.1), (2.2), (2.3), (2.5), and (2.6).

For a local atmospheric pressure of 91 kPa (corresponding to an elevation of roughly 850 m above sea level), calculate the estimated reservoir temperature if the observed silica concentration in the separated liquid is: (A) 300 ppm; (B) 500 ppm; (C) 700 ppm; (D) 900 ppm.

SP 2.6 This problem will illustrate the correspondence (or lack thereof) between various geothermometers. Consider the following set of geochemical data obtained from the Ahuachapán field in El Salvador.

Component	Concentration, ppm
Chloride	10,430
Sodium	5,690
Potassium	950
Calcium	443
Silica, SiO_2	537

(A) Calculate the estimated reservoir temperature from: (a) the Na/K geothermometer, eq. (2.8) and (b) the Na/K/Ca geothermometer eq. (2.9).

(B) Assuming that the result obtained for the Na/K/Ca geothermometer i correct, calculate the expected concentration of silica that you woul find in the liquid separated from the 2-phase well flow if the separatio temperature is 100°C.

(C) Compare the answer for Part (B) with the value given in the data set and discuss any difference between these values.

SP 2.7 In the absence of other data, which of the following, each taken by itself would be more encouraging for the existence of a high-temperature, liquid dominated geothermal reservoir:

(A) High or low temperature gradient?
(B) High or low heat flux?
(C) High or low gravity anomaly?
(D) High or low magnetic anomaly?
(E) High or low silica concentration in springs?
(F) High or low mercury concentration in soils?

Explain the level of confidence you have in each case.

SP 2.8 Consider the following set of geochemical data obtained from the Pauzhetka field in the Kamchatka geothermal region of Russia:

Component	Concentration, ppm
Chloride	1,633
Sodium	986
Potassium	105
Calcium	52

(A) Calculate the estimated reservoir temperature from: (a) the Na/K geothermometer, eq. (2.8) and (b) the Na/K/Ca geothermometer, eq. (2.9).

(B) Assuming that the result obtained for the Na/K geothermometer is correct, calculate the expected concentration of silica that you would find in the liquid separated from the 2-phase well flow if the separation temperature is 100°C.

P 2.9 Geophysical measurements have been performed on five geothermal prospective sites. The data are shown in the table below where the gradients were obtained from shallow wells and the resistivity values are the minimum observed.

Site designation	Temperature gradient °C/100 m	Electrical resistivity ohm · m
A	25	1
B	100	500
C	5	10
D	10	200
E	2	2

Bearing in mind that either indicator can be misleading by itself, match the descriptions listed below (only five will be relevant) to the sites in the table.

Option	Description
1	Moderate-temperature reservoir
2	Deep, high-temperature reservoir
3	Gas or dry steam pockets or cap
4	Cold, igneous rocks
5	Hot dry rocks
6	Cold, high-salinity reservoir

Chapter 3 Geothermal Well Drilling

SP 3.1 An intermediate casing is about to be cemented in place in a vertical well using the displacement method (see Fig. 3.6). The previous casing has 22-in outside diameter (OD) with a 0.5-in wall thickness, and extends from the surface down to a depth of 250 ft. The new casing has a 16-in OD and a 0.438-in wall thickness and weighs 75 lbf/ft. It will be suspended and centralized in the hole a distance of 15 ft above the bottom of the 20-in diameter hole which was drilled to a total of 1450 ft. If the float collar is 10 ft above the bottom of the casing and a 5% margin of safety is allowed for the return of cement, calculate the required volume of cement (in units of cubic yards) to secure the new casing to the formation.

SP 3.2 Using the same conditions given in Problem 3.1, calculate:

(A) The net force on the rig hoist when the drill bit is at the bottom of the hole, if a 4-in OD drill pipe is used that weighs 12.93 lbf/ft. The bit assembly weighs approximately 800 lbf. The drilling fluid has a density of 80 lbm/ft^3 and is assumed to be circulating and completely filling the annulus between the drill string and the formation.

(B) The required holding strength of the hoist when the casing is being positioned for cementing in the well.

Chapter 4 Reservoir Engineering

SP 4.1 A liquid-dominated reservoir has a geofluid temperature of 230°C, a static rese voir pressure of 100 bar, and a linear drawdown coefficient of 0.50 bar/(kg/s A 1200 m well having an inside diameter of 9.625 in is drilled into the res ervoir. The Moody friction factor may be taken as constant and equal to 0.01 The local gravitational acceleration is the standard value of 9.81 m/s Determine the depth from the wellhead to the flash point if the total mass flo rate in the well is 50 kg/s, assuming the feed zone is at the bottom of the well.

SP 4.2 A well is drilled into a liquid-dominated reservoir where the fluid tempera ture is 250°C, the static reservoir pressure is 65 bar, and a linear drawdow coefficient is 0.35 bar/(kg/s). (A) Determine the mass flow rate of geoflui that will cause the fluid to flash just as it enters the well. (B) Describe i some detail qualitatively what will happen if the mass flow rate is (a) les than this value and (b) greater than this value.

SP 4.3 It is possible to find the static reservoir pressure and the assumed linea drawdown coefficient from measurements taken during flow tests. Pressur readings are taken downhole next to the feed zone and the total flow is mea sured on the surface using a separator. Data from tests at two different flo rates are given in the table below.

Test No.	Mass flow rate, kg/s	Feed zone pressure, bar
1	45	140
2	58	133

(A) Calculate the linear drawdown coefficient and (B) the static reservoi pressure.

SP 4.4 The first successful geothermal well to be drilled in Honduras was at th Platanares field. PLTG-1 was a slim exploration well but was capable of produc ing geofluid. Some data pertaining to the well are given in the following table.

Item	Value
Depth	625 m
Inside diameter	3.0 in
Well roughness	0.00006 m
Reservoir pressure	6000 kPa
Reservoir fluid temperature	160°C
Drawdown coefficient	50 kPa/(kg/s)
Mass flow rate	5.152 kg/s

(A) Calculate the depth of the flash horizon from the surface.
(B) Estimate the quality (or dryness fraction) of the geofluid at the wellhead.
(C) Assuming thermodynamic equilibrium between the liquid and vapo phases at the wellhead, calculate the average pressure gradient over th 2-phase section of the well.

P 4.5 A geothermal well is constructed with two casings of different diameter, thereby having an abrupt step increase in cross-sectional area at a certain point along the well. The lower casing has an inside diameter of 5.921 in; the upper one has an ID of 8.921 in. The lower casing runs from the bottom of the well where the feed zone is located to a height of 3,500 ft; the upper casing runs from the top of the lower casing to the surface and is 5,000 ft long. The casing material is commercial steel and has a roughness factor of 0.0018 in.

The reservoir fluid is a pressurized liquid at a temperature of 500°F. During flow tests the following data were obtained:

Test No.	Mass flow rate, lbm/s	Feed zone pressure, lbf/in^2
1	95	2056.25
2	125	1958.75

(A) Determine the linear drawdown coefficient (units = (lbf/in^2)/(lbm/s)).
(B) If the well were flowed at 100 lbm/s, what drawdown coefficient would be needed for the fluid to just be on the verge of flashing in the formation?
(C) For the actual drawdown coefficient, find the level of the flash horizon if the mass flow rate is 150 lbm/s. In doing this calculation, you may ignore the effect of the Reynolds number on the Moody friction factor in the Swamee-Jain equation.
(D) Find the minimum mass flow rate for the flash point to be located in the lower casing.
(E) Find the location of the flash horizon if the flow rate is 50 lbm/s. You may ignore the pressure drop associated with the abrupt enlargement.
(F) Find the liquid level below the surface when the well is closed in, i.e., when there is no flow.

SP 4.6 A geothermal well is completed into a liquid-dominated reservoir. The casing has an inside diameter of 9.626 in; the friction factor is 0.005; and the total length of the well is 2,500 ft, with the feed zone at the bottom. The static reservoir pressure is 1,150 lbf/in^2 and the reservoir has a linear drawdown coefficient of 1.5 (lbf/in^2)/(lbm/s). The geofluid temperature in the reservoir and up to the flash point is 400°F.
(A) Calculate the depth of the flash horizon from the surface if the wellhead valve is adjusted to give a mass flow rate of 150 lbm/s.
(B) Assuming this flow rate is achieved with a wellhead pressure of 120 lbf/in^2, abs., calculate the wellhead quality (or dryness fraction).

SP 4.7 A geothermal well is completed into a liquid-dominated reservoir. The well is of the stepped type having an abrupt step from a 7–5/8-in slotted liner to a 9–5/8-in production casing. During long-term flow testing at a constant wellhead valve setting, a very gradual decline in flow rate is observed. At some moment, a much more rapid decrease in flow occurs eventually causing the well to die. It is known that the geofluid contains about 1% of carbon dioxide (by weight of steam flow).

(A) Discuss possible reasons for the loss of the well.
(B) Suggest some ways to restore the well to production and to assure tha future wells do not suffer the same fate.

SP 4.8 During the early stages of development at the Ahuachapán field in E Salvador, flow tests were conducted on several wells in October 1976; th results are summarized in the table below showing the separator pressure the steam flow rate, and the total flow rate.

Estimate the reservoir temperature for each of the wells and for the fiel average. Other methods indicated that the reservoir temperature was abou 230°C. Discuss the results and the validity of the approximations used ir the analysis.

Well	P_{sep}, kPa	$\dot{m}_v$, kg/s	$\dot{m}_{tot}$, kg/s
AH-1	665.3	13.20	94.90
AH-4	699.4	23.66	126.63
AH-6	670.5	17.65	62.62
AH-20	626.3	10.74	55.46
AH-21	650.9	12.51	93.80
AH-26	640.9	12.37	70.71
10-well avg.	653.4	13.55	72.55

Chapter 5 Single-Flash Steam Power Plants

SP 5.1 A single-flash plant is to be designed to operate at a liquid-dominated reser voir having a temperature of 245°C. The plant condensing temperature wil be 40°C and the dead-state temperature is 25°C. The separator temperature may be found from the "equal-temperature-split" rule of thumb. The turbine efficiency should be found using the Baumann rule and a dry isentropic turbine efficiency of 85%.

The wells at the field may be characterized by the following average productivity curve:

$$\dot{m} = 100.23 - 0.02339\,P + 2.06 \times 10^{-5}\,P^2$$

where $\dot{m}$ is the total mass flow rate in kg/s and P is the wellhead pressure in kPa.

(A) Calculate the power output from one of the wells.
(B) Calculate the utilization efficiency of the plant.
These results were obtained by ignoring any pressure loss between the wellhead and the turbine inlet. Clearly this is an oversimplification. The rest of this problem will explore the effect of this assumption.
(C) Keep the same turbine inlet pressure as before, but find the required wellhead pressure if an 18-in diameter steam pipe from the wellhead separator to the turbine runs for an equivalent length of 800 m.
(D) Now, recalculate the total mass flow rate from the well, the steam flow to the turbine, and the (reduced) power output.
(E) Discuss the power loss and how it could be mitigated. Include economic considerations.

P 5.2 A single-flash plant is to be designed to operate at approximately optimum utilization efficiency. It will be located at a liquid-dominated reservoir having a temperature of 270°C and a geofluid enthalpy of 1185 kJ/kg. The turbine exhaust and the condenser are maintained at 50°C. The dead-state temperature is 25°C. The separator temperature may be found from the "equal-temperature-split" rule of thumb. The turbine efficiency should be found using the Baumann rule and a dry isentropic turbine efficiency of 85%. The wells at the field may be characterized by the following average productivity curve:

$$\dot{m} = 100.23 - 0.02339\,P + 4.028 \times 10^{-5} P^2 - 1.02 \times 10^{-7} P^3$$

where $\dot{m}$ is the total mass flow rate in kg/s and P is the wellhead pressure in kPa.

(A) Calculate the power output using one of the wells.
(B) Calculate the ratio of the heat rejected in the condenser to the power output of the turbine.
(C) From the result of Part (B), calculate the so-called "pseudo thermal efficiency," defined as the ratio of the power output to the imagined "heat added" as if the plant operated on a cycle.
(D) Calculate the utilization efficiency based on (a) wellhead conditions and (b) reservoir conditions.

Chapter 6 Double-Flash Steam Power Plants

P 6.1 Your task is to assess the potential for a silica scaling problem associated with the operation of flash plants at a 225°C, liquid-dominated resource. You may assume that the condenser will run at 50°C, and that the "equal-temperature-split" rule of thumb will be used to determine the separator (and flash) temperature for a single- (and double-) flash plant. There is a quartz concentration in the reservoir geofluid of 360 ppm. Assess the silica scaling potential for (A) a single-flash plant and (B) a double-flash plant. In case you expect a problem, suggest some ways of averting trouble (without abandoning the site).

SP 6.2 A double-flash plant is being considered for a high-temperature geothermal field. Your task is to assess the power potential and the likelihood of a silica scaling problem. The resource is liquid-dominated at 300°C. The condenser will run at 60°C. A dual-admission turbine will be employed (see Fig. 6.6). The "equal-temperature-split" rule of thumb will be used to determine the separator and flash temperatures.

(A) Calculate the specific work output of the turbine (i.e., in kJ/kg of total well flow).
(B) Assuming a typical well produces a total flow of 75 kg/s, calculate the power output per well.
(C) Assuming the reservoir fluid is saturated with respect to quartz, calculate the supersaturation of the residual brine with respect to amorphous silica.

(D) If the brine is in fact supersaturated and the plant will be designed t generate 55 MW, calculate the potential amount of silica that could pre cipitate (in kg) per day. Discuss the ramifications of this, e.g., where wi the precipitation occur and how to alleviate the problem.

SP 6.3 Two variants of a double-flash plant are being considered for a geotherma field. The resource is liquid-dominated at 400°F. The condenser will ru at 100°F. The "equal-temperature-split" rule of thumb will be used t determine the separator and flash temperatures. One variant will use a dual admission turbine (see Fig. 6.6) and the other will use two separate high and low-pressure turbines (see Fig. 6.7a). The efficiency of each turbin should be found using the Baumann rule and a dry isentropic turbine effi ciency of 85%. Determine for each plant:

(A) The specific work output (i.e., in Btu/lbm of total well flow).
(B) The utilization efficiency if the dead state is at 80°F.
(C) Assuming a typical well produces a total flow of 750,000 lbm/h, calcu late the number of wells needed for each plant to generate 50 MW Discuss the findings.

SP 6.4 Consider the double-flash power plant shown in the schematic below. Th geofluid is produced from a liquid-dominated reservoir (state 1). The separa tor, S, and the flasher, F, generate high- and low-pressure steam (saturated for the two turbines T1 and T2. Turbine T1 is equipped with a moistur removal section where moisture which forms during the expansion fron state 4 to state 9 is drained away at state 11, and flashed to the condenser This drain is located at the point where the steam temperature is exactl 50°C below that of the inlet steam.

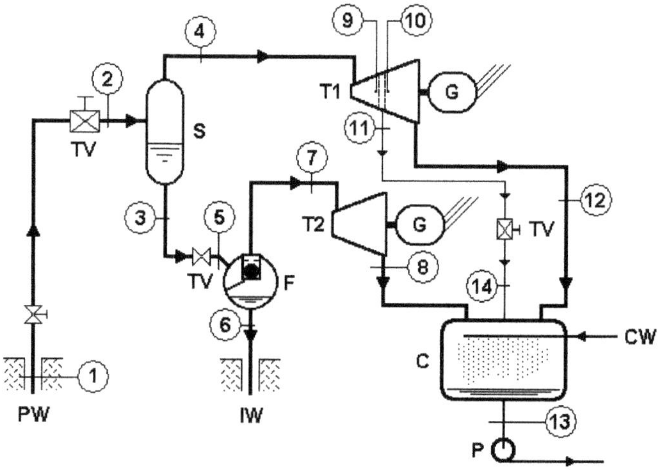

Data: $T_1 = 280°C$; $T_9 = T_{10} = T_4 - 50°C$; $T_{13} = 40°C$; dead state $T_0 = 25°C$.

Use the "equal-temperature-split" rule for setting the separator and flasher conditions.

(A) Make a careful sketch of the plant processes in a T–s diagram. Label all states in accordance with the state points shown in the schematic. Use the Baumann Rule for determining the actual efficiency of each turbine, i.e., the two sections of T1 and the whole of T2.

(B) Find the specific enthalpy h at the following state points: 1, 2, 3, 4, 5, 6, 7, 8, 9, 10, 11, and 12.

(C) Assuming that the total mass flow rate from the well is 1 kg/s, calculate the power output (in kW) of the:

(a) high-pressure section of turbine T1

(b) low-pressure section of turbine T1

(c) turbine T2.

(D) If the actual total well flow is 1000 kg/s, what is the total power of the plant in MW?

(E) Calculate the utilization efficiency, η_u, for the plant based on the condition of the geofluid in the reservoir.

(F) If the original geofluid in the reservoir has a concentration of 595 ppm of silica (as quartz), calculate the concentration of silica in the separated liquid at (a) state 3 and at (b) state 6.

(G) Using the Fournier and Marshall correlation for amorphous silica solubility, namely,

$$\log_{10} s = -6.116 + 0.01625\,T - 1.758x10^{-5}\,T^2 + 5.257x10^{-9}\,T^3,$$

where T is in kelvins and s must be multiplied by 58,400 to obtain ppm, determine whether the separated liquid at states 3 and 6 is supersaturated with respect to amorphous silica. Discuss the implications of your findings.

Chapter 7 Dry-Steam Power Plants

SP 7.1 During a performance test at The Geysers Unit 6 (now one unit at McCabe plant), the following data were taken: gross power output = 55 MW, net power = 52 MW, steam flow to the turbine = 835,000 lbm/h, turbine inlet pressure = 100 lbf/in^2, abs., turbine inlet temperature = 360°F, condensing temperature = 124°F, wet-bulb temperature = 66°F. The resource is a saturated vapor at 460°F.

(A) Find the specific enthalpy of the (a) turbine inlet steam and (b) the reservoir steam. Compare the values and discuss.

(B) Calculate the turbine isentropic efficiency.

(C) Calculate the overall net utilization efficiency of the unit based on the reservoir condition.

SP 7.2 The heat balance diagram for the historic Unit 1 at The Geysers in 1961 shows that the plant received a total steam flow of 250,820 lbm/h from several wells, of which 240,000 lbm/h was admitted to the turbine, the res being used mainly in the noncondensable gas ejectors. The generator pro duced 12,500 kW of electrical power, of which 402 kW was used for plan auxiliaries. The inlet steam pressure was 93.9 lbf/in^2, a and its temperatur was 348°F. The plant had a barometric condenser that operated at pressure of 4 in Hg. The wet bulb temperature was 66°F.

(A) Calculate the plant utilization efficiency, (a) gross and (b) net, based or the inlet steam condition.

(B) Calculate the turbine isentropic efficiency, assuming the generator is 95% efficient.

(C) Calculate the plant net utilization efficiency, based on the differenc between steam inlet condition and the steam condition at the turbin exhaust.

SP 7.3 Consider a dry steam field characterized by a closed-in wellhead pressure o 440 lbf/in^2, a and a maximum steam flow rate of 1,500,00 lbm/h when the wells are wide open, i.e., the wellhead pressure is 14 lbf/in^2, a. You may assume the productivity curve is a section of an ellipse given in the form

$$\left(\frac{\dot{m}}{\dot{m}_{max}}\right)^2 + \left(\frac{P}{P_{ci}}\right)^2 = 1$$

The steam may be taken as dry, saturated at closed-in conditions and that it is simply throttled as the wellhead valve is gradually opened. The turbine may be assumed isentropic to simplify the calculations.

(A) Determine the wellhead pressure and the steam flow rate to yield the maximum power from the resource for (a) a turbine exhausting to the atmosphere and (b) condensing at 2 lbf/in^2, a.

(B) Find the maximum power for both cases.

(C) Discuss the results, including the penalty in loss of power for off-optimum operation.

Chapter 8 Binary Cycle Power Plants

SP 8.1 A basic binary plant uses isopentane (i-C_5H_{12}) as the cycle working fluid (see Figs. 8.1 and 8.2). The brine inlet temperature is 440 K; the pinch-point temperature difference in the preheater-evaporator is 5 K. The i-C_5H_{12} pressure in the preheater-evaporator is 2.0 MPa; the condenser runs at 320 K. The isentropic efficiency for the turbine is 85% and 75% for the feed pump. The geothermal brine heat capacity is 4.19 kJ/kg · K (constant) and density is 897 kg/m^3. The dead-state temperature is 25°C. The cycle is designed to produce 1200 kW net power.

Calculate the following specific terms for the cycle, i.e., in kJ/kg of i-C_5H_{12}:

(A) Turbine work

(B) Heat rejected

(C) Feed pump work

(D) Heat added.

Calculate:

(E) Cycle thermal efficiency

(F) Mass flow rate of i-C_5H_{12}

(G) Mass flow rate of brine

(H) Brine outlet temperature

(I) Utilization efficiency (assuming pure water properties for the brine)

(J) Number of wells needed if a typical well produces 850 gallons/minute (GPM).

P 8.2 A binary plant uses isopentane (i-C_5H_{12}) as the cycle working fluid. The working fluid is superheated before entering the turbine (see figure below).

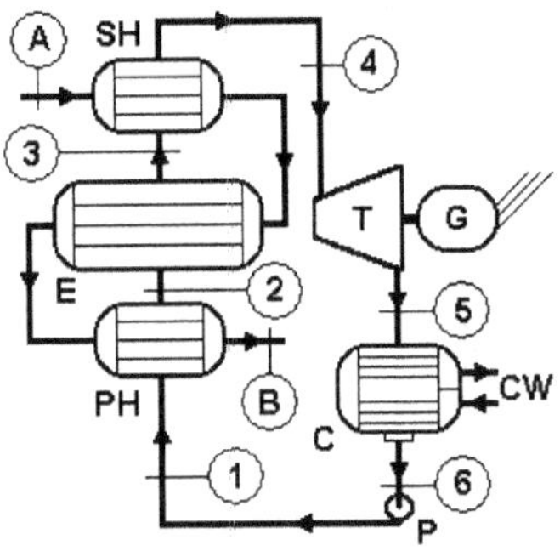

The following data are specified:

Brine inlet temperature at superheater = 425 K
i-C_5H_{12} temperature in evaporator = 400 K
Turbine inlet temperature = 420 K
Turbine isentropic efficiency = 90%
Condensing temperature = 320 K
Feed pump isentropic efficiency = 100% (ideal)
Pinch-point temperature difference in heat exchanger train = 5 K
Cycle net output = 10 MW
Dead-state temperature = 298 K.

(A) Construct the pressure–enthalpy process flow diagram (schematically, not to scale).

(B) Determine the specific enthalpy of the i-C_5H_{12} at each state point in the cycle.

(C) Calculate the brine outlet temperature from the preheater.

(D) Calculate the mass flow rate of the i-C_5H_{12}.

(E) Calculate the mass flow rate of the brine.

(F) Calculate the utilization efficiency of the binary plant based on the exergy of the brine entering the evaporator.

SP 8.3 A binary plant uses isobutane (i-C_4H_{10}) as the cycle working fluid. The working fluid is superheated before entering the turbine. A throttle valve is located between the outlet of the superheater and the turbine inlet. Normally it operates wide open (no pressure drop). The pressure leaving the superheater is kept constant at all times at a value P_{SH}. The design turbine inlet temperature is $T_1 = 460$ K. The turbine is assumed isentropic and the turbine exhaust pressure is 0.4 MPa.

(A) Find the highest value of P_{SH} such that the turbine expansion line remains entirely in the vapor region. A pressure–enthalpy process flow diagram will be very useful to solve this question.

(B) Using the fixed value for P_{SH} found above, the temperature of the i-C_4H_{10} leaving the superheater drops to $T_2 = 455$ K. The throttle valve is now activated to adjust the turbine inlet condition so that the full expansion line remains just dry. Determine the minimum pressure drop across the throttle to assure dry expansion.

(C) Calculate and compare the turbine output under design and off-design conditions, indicating the percentage loss in output.

SP 8.4 A dual-pressure binary plant depicted in the figure at the top of the next page uses isobutane (i-C_4H_{10}) as the cycle working fluid. Saturated liquid i-C_4H_{10} is pumped from the condenser C through a heater H1 from which it emerges as a saturated liquid. A proportioning valve PV allows 30% of the working fluid to flow to the evaporator E1 and 70% to flow to a feed pump FP. From there this stream is heated (H2), evaporated (E2), and superheated (SH) before entering the turbine T1. The exhaust from T1 is mixed isobarically with the saturated vapor (state 3) coming from evaporator E1 before passing into the second turbine T2. The geothermal brine passes through five heat exchangers in series before being reinjected (state b).

Data: $P_1 = P_9 = P_{10} = P_{11} = 3362$ kPa; $T_1 = 180$°C; $T_6 = 30$°C; $T_0 = 25$°C; $\eta_{T1} = \eta_{T2} = 85\%$; $\eta_{CP} = \eta_{FP} = 100\%$; $T_a = 200$°C; $T_b = 100$°C.

Also, the temperature in E1 is chosen as the average of the condenser temperature and the temperature in E2.

(A) Construct the pressure–enthalpy process flow diagram (schematically, not to scale).

(B) Determine the specific enthalpy of the i-C_4H_{10} at each state point in the cycle.

(C) Calculate the specific work output of the turbines in units of kJ/kg of i-C_4H_{10} flowing at state 6.

(D) Calculate the thermal efficiency of the cycle.

(E) Calculate the required mass flow rate of brine (assumed pure water) per unit flow rate of i-C_4H_{10}.

(F) Calculate the utilization efficiency of the plant based on the exergy of the incoming brine.

(G) If a typical well can produce 90,000 kg/h, how many wells will be needed to generate a net power output of 10 MW?

(H) Is it safe to assume that the plant as described in this problem (i.e., for the given temperatures, pressures, etc.) is actually feasible thermodynamically? To be sure of its feasibility, what should be checked?

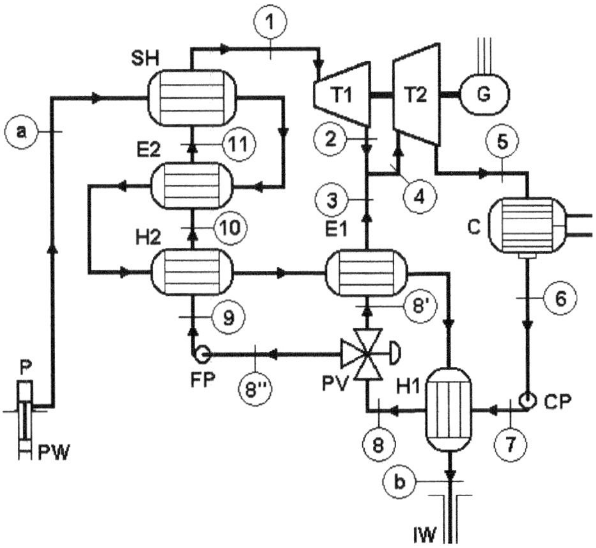

Chapter 9 Advanced Geothermal Energy Conversion Systems

P 9.1 A separating-expander incorporates a total-flow machine and a steam turbine in an integrated package as shown in the figure below. The 2-phase flow is admitted to the SE where the steam is separated and sent to the ST while the liquid generates power as it is separated centrifugally in the SE. The liquid is sent directly to the injection wells from the SE.

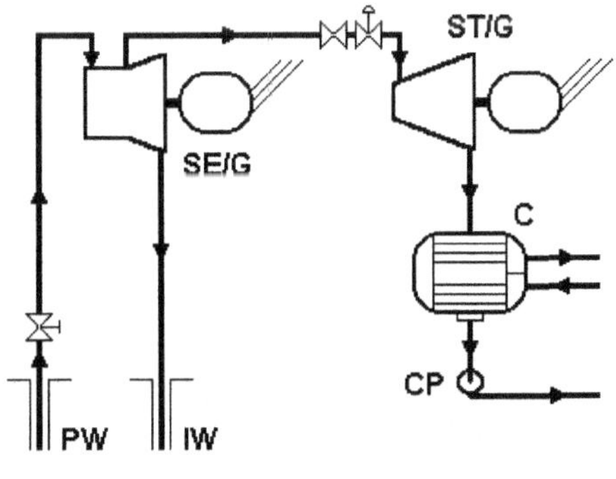

The geofluid temperature is 200°C, the steam turbine efficiency is 75%, the separating-expander efficiency is 50%, and the condenser temperature is 50°C. Calculate the dependence of the utilization factor η_u on the intermediate temperature T_1, i.e., the temperature at which the steam is sent to the ST. At what temperature does the optimum performance occur? How does this compare with the "equal-temperature-split" rule for a simple 1-stage flash plant? Use 25°C as the sink temperature.

SP 9.2 The flash-steam power plant shown in the accompanying schematic is designed to utilize highly contaminated brine. The brine is produced from the reservoir (state 0) at a temperature of 480°F. The temperature at the cyclone separator (CS) is 400°F. The brine carries 100,000 ppm total dissolved solids plus an amount of noncondensable gases (NCG). Clean steam (state 5) is generated in the primary flasher. Since the steam from the separator is too contaminated for direct use in the turbine, a fraction of it is sent to a direct-contact heater (DCH) where it reheats the essentially pure condensate from the turbine. Then a secondary flasher generates more clean steam (state 5′) which merges with the primary steam before entering the turbine. The fraction of the separated steam (state 3′) used in the DCH is chosen such that a saturated liquid is produced at the outlet of the DCH (state 2′). Additional specifications are: $T_4 = 324°F$; $T_7 = T_8 = 140°F$; dead state temperature = 80°F.

The plant is not optimized and is not very efficient, but does allow the use of a challenging geofluid. Waste occurs at two places: vent steam and NCG at state 3″ and concentrated brine at state 6″. Note that states with the same numerical label have identical thermodynamic properties, but differ in mass flow rates, as denoted by the primes. Throughout the analysis of the plant, assume that the geofluid flow from the reservoir is unity, i.e., perform a specific analysis per unit mass.

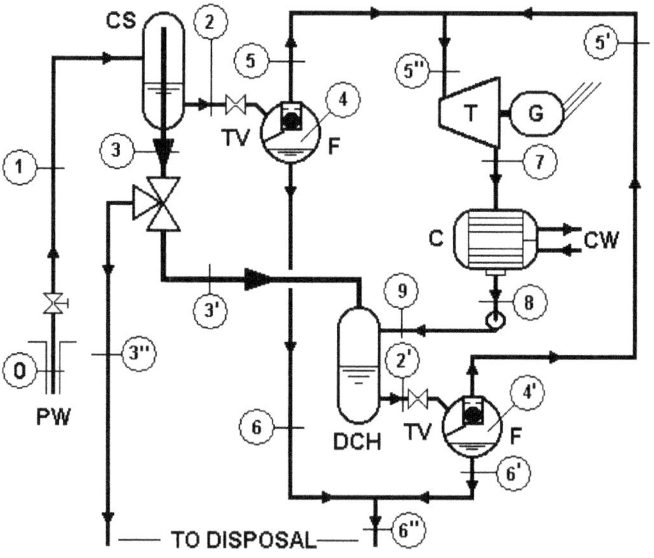

(A) Construct the temperature–entropy process flow diagram (not to scale, schematically).

(B) Calculate the following mass flows: (a) vent steam, (b) waste brine, (c) steam flow to the turbine.

(C) Calculate (a) specific work output of the turbine (i.e., in Btu/lbm of brine), using the Baumann rule and a dry efficiency of 85%, (b) specific work to run the condensate pump (assumed isentropic).

(D) Calculate the exergy of the following flows: (a) brine from the reservoir, (b) fluid entering the separator, (c) steam entering the turbine, (d) vent steam, (e) waste brine. In each case, express the result in units of Btu/lbm of geofluid at state 0. Note that the exergy of brine containing N% of dissolved solids is less than for pure water at the same conditions by a factor, (1−0.0085 N), i.e., the fluid loses 0.85% of its exergy for each 1% of TDS in the brine.

(E) Calculate the utilization efficiency of the power plant based on the reservoir conditions.

P 9.3 Your task is to design a bottoming binary plant to be used in conjunction with an existing single-flash plant. The binary plant will capture some of the waste energy (and exergy) leaving the flash plant with the residual brine. The plant schematic is shown below.

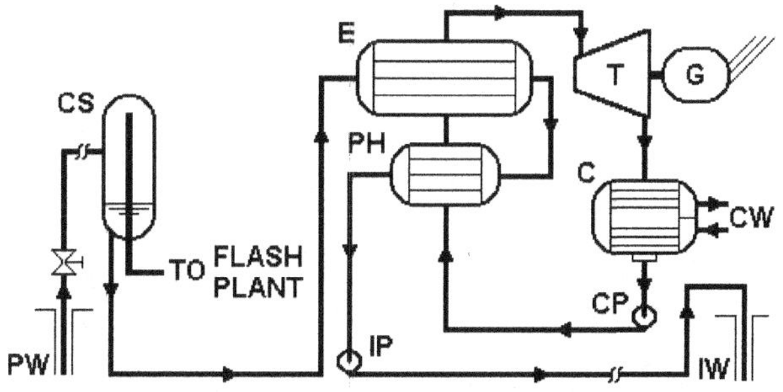

The following data are specified:

Brine inlet temperature at evaporator = 425 K
Cycle working fluid = isopentane (i-C_5H_{12})
Turbine inlet temperature = 390 K (saturated vapor)
Turbine isentropic efficiency = 85%
Condensing temperature = 310 K
Feedpump isentropic efficiency = 100% (ideal)
Pinch-point temperature difference in preheater-evaporator = 4 K
Cycle net output = 6000 kW
Dead-state temperature = 25°C.

(A) Construct the pressure–enthalpy process flow diagram (not to scale, schematically).

(B) Determine the specific enthalpy of the i-C_5H_{12} at each state point in the cycle (6 in all, including saturation states).

(C) Calculate the specific work output of the turbine (i.e., in kJ/kg of i-C_5H_{12}).

(D) Calculate the specific work to run the feed pump.
(E) Calculate the specific heat input.
(F) Calculate the net cycle thermal efficiency.
(G) Calculate the mass flow rate of the i-C_5H_{12}.
(H) Calculate the brine outlet temperature from the preheater.
(I) Calculate the mass flow rate of the brine.
(J) Calculate the utilization efficiency of the binary plant based on the exergy of the brine entering the evaporator.
(K) Calculate the functional Second Law efficiency of the combined preheater-evaporator based on the transfer of exergy from the brine to the i-C_5H_{12} (see Sect. 10.6.3).

SP 9.4 A schematic diagram of the so-called Gravity-Head Binary Power Plant is shown in the figure below. This was proposed by Matthews Geothermal, Inc. in the 1980s as a more efficient method of generating power from geothermal fluids. Brine is pumped at 360°F from the reservoir (state A) by a downhole turbo-pump unit, TPU. The TPU is powered by the binary cycle working fluid R-114 (no longer available), which is heated in a downhole, gravity-head heat exchanger (GHHX). The brine temperature may be taken as 360°F at state B; the pump in the TPU increases the brine pressure by 50 lbf/in^2. You may assume that the outer shell of the GHHX is perfectly insulated.

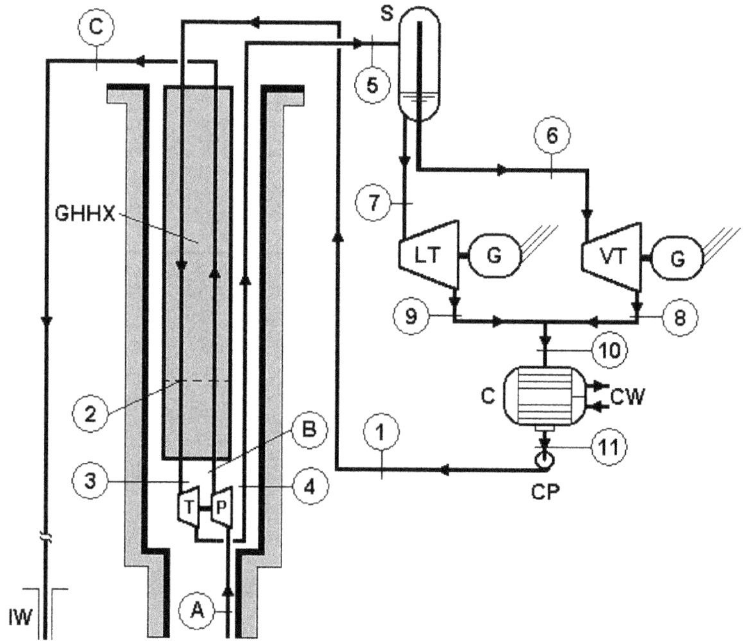

The R-114 enters the GHHX (state 1) as a saturated liquid at 100°F. The internals of the GHHX are so designed that the heating line of the R-114

coincides with the saturated liquid line from state 1 to 2. From state 2 to 3 it follows a path of constant density, $\rho = 50$ lbm/ft^3, until it reaches $P_3 = 900$ lbf/in^2. The pinch-point in the GHHX is 10°F.

The turbine in the TPU has an efficiency of 75%; the pump in the TPU is 70% efficient. From state 4 to the surface, the R-114 may be assumed to flow at constant enthalpy. At the separator, the R-114 has a temperature of 260°F. The vapor stream is used in an 85% efficient turbine, VT, while the liquid fraction is used in a 35% efficient liquid expander, LT. The exhaust streams from the two machines are combined and sent to a common condenser before being returned to the GHHX. The condensate pump, CP, has negligible work input.

(A) Sketch the processes in a pressure–enthalpy diagram for R-114.

(B) Construct the temperature–heat transfer diagram for the GHHX, making careful use of a scale *P-h* diagram for R-114.

(C) From your construction, find the brine outlet temperature, T_C.

(D) Assuming the brine has a specific heat of 1.0 Btu/lbm · °F, calculate the ratio of the flow rates, $\dot{m}_{R\text{-}114}/\dot{m}_{brine}$.

(E) Calculate (a) the isentropic and (b) the actual specific work of the pump in the TPU in units of Btu/lbm of brine.

(F) Calculate the actual change in specific enthalpy of the R-114 as it passes through the turbine in the TPU.

(G) Calculate the quality (or dryness fraction) at state 5.

(H) Calculate the specific work of (a) the vapor turbine and (b) the liquid expander in units of Btu/lbm of R-114.

(I) Calculate the gross power output of the plant if the brine flow rate is 500,000 lbm/h.

(J) Assuming that all auxiliary power requirements amount to 20% of the gross power found above, calculate the net utilization efficiency for a dead-state at 80°F.

SP 9.5 This problem investigates the efficiency of multiple-flash plants. Multiple-flash steam plants have been employed on certain resources as a means of coping with challenging conditions such as high concentrations of dissolved solids and noncondensable gases. However, they could be used at normal geothermal resources. Consider a hydrothermal resource having a saturated liquid in a reservoir at 500°F. The condensing temperature in the power plant is 120°F. The dead-state temperature is 80°F.

Calculate (A) the specific output of the turbine(s) (i.e., in units of Btu/lbm of incoming geofluid) and (B) the utilization efficiency for (a) single-flash, (b) double-flash, (c) triple-flash, and (d) quadruple-flash plants.

In each case, use the "equal-temperature-split" rule to find the separator and flash temperatures. Use the Baumann rule for wet turbine efficiency where the dry efficiency is 85%. Also use individual turbines operating between their inlet condition and the condenser (i.e., do not employ pass-in turbines).

Plot the results and draw conclusions about the use of multiple-flash steam plants.

Appendix G: Answers to Selected Supplementary Problems

"I wish I had an answer to that because I'm tired of answering that question."
Yogi Berra

Chapter 2

SP 2.3 (A) 125°C; fairly reliable. (B) 425°C; unreliable, too large an extrapolation.

SP 2.6 SP 2.6(A) (a) 250.2°C; (b) 254.2°C. (B) 696 ppm. (C) Since the measured is less than this value, the quartz concentration in the reservoir fluid is less than the 484 ppm found for the assumed temperature, and the reservoir fluid temperature is less also. The actual temperature in the reservoir was about 235°C.

Chapter 4

SP 4.1 632 m.

SP 4.2 (A) 72.2 kg/s. (B) (a) Flashing will occur inside the well; (b) flashing will occur in the formation leading to "excess" steam and enthalpy at the wellhead.

SP 4.4 (A) 92 m. (B) 0.044 (4.4% vapor). (C) 3 kPa/m.

SP 4.5 (B) 16.84 $(\text{lbf/in}^2)/(\text{lbm/s})$. (C) 2628 ft below the surface.

SP 4.6 (A) 702 ft. (B) 0.071 (7.1% vapor).

Chapter 5

SP 5.1 (A) 9.49 MW. (B) 10.8%. (C) 161 kPa. (D) 8.97 MW.

SP 5.2 (B) 5.43. (C) 15.6%.

Chapter 6

SP 6.1 (A) Undersaturated. (B) Supersaturated, expect silica precipitation.

SP 6.3 Dual-admission turbine: (A) 36.58 Btu/lbm. (B) 51.2%. (C) 6.2 (7) wells. Separate turbines: (A) 36.42 Btu/lbm. (B) 51.0%. (C) 6.2 (7) wells.

Chapter 7

SP 7.1 (B) 0.849. (C) 49.1%.

SP 7.2 (A) (a) 48.4%; 46.8%. (B) 71.2%. (C) 67.4%.

SP 7.3 (B) (a) 76.5 MW; (b) 119.4 MW.

Chapter 8

SP 8.1 (A) 76.1 kJ/kg. (B) 410.6 kJ/kg. (C) 4.07 kJ/kg. (D) 482.6 kJ/kg. (E) 14.9% (F) 16.66 kg/s. (G) 95.95 kg/s. (H) 420 K. (I) 11.3%. (J) 2 wells, minimum.

SP 8.4 (C) 77.83 kJ/kg. (D) 14.05%. (E) 1.202. (F) 37.5%. (G) 7. (H) Check th pinch-point. It is about 20°C and occurs at the hottest end of th superheater.

Chapter 9

SP 9.3 (C) 63.65 kJ/kg. (D) 1.466 kJ/kg. (E) 458.93 kJ/kg. (F) 13.6%. (G) 96.49 kg/s (H) 95.6°C. (I) 184.89 kg/s. (J) 36.0%. (K) 79.8%.

SP 9.4 (C) 120°F. (D) 3.514. (H) (a) 9.013 Btu/lbm; (b) 0.2339 Btu/lbm (I) 4.76 MW. (J) 46.5%.

Appendix H: REFPROP Tutorial with Application to Geothermal Binary Cycles

"If the automobile had followed the same development cycle as the computer, a Rolls-Royce would today cost $100, get a million miles per gallon, and explode once a year, killing everyone inside."

Robert X. Cringely, *InfoWorld* magazine

H.1 Introduction

REFPROP – Reference Fluid Thermodynamic and Transport Properties – is a product of the United States National Institute of Standards and Technology (NIST) [1]. It is a very accurate and comprehensive software program that yields all thermophysical properties of interest to scientists and engineers and covers a very large number of pure substances and predefined mixtures. Users can also define mixtures of particular interest to them. Version 8.0 was released in April 2007; Version 9.0 was released in November 2010. A free student version is also available, albeit with a limited number of substances.

This appendix will show how REFPROP can be used to analyze geothermal power plants. As such, it will presume a basic familiarity with the program. A User's Guide is available from NIST and should be consulted for the basic principles used and the general capabilities of the program.

H.2 Typical geothermal binary power cycle

To illustrate REFPROP we will examine a simple Rankine cycle with superheat. Such a cycle may be used in a binary geothermal power plant.

Figure H.1(a) shows the plant. The following nomenclature describes each component: T = turbine, G = generator, C = condenser, P = pump, PH = preheater, EV = evaporator, SH = superheater, CW = cooling water, AIR = cooling air, B = geothermal brine. We will assume that the cycle working fluid is n-pentane.

Figure H.1(b) shows the processes undergone by the n-pentane in pressure-enthalpy coordinates. State 2s is the ideal turbine exhaust state ($s_{2s} = s_1$) and state 4s is the ideal pump outlet state ($s_{4s} = s_3$).

The high pressure in the cycle is P_E and the condensing pressure is P_C. We will assume that both pressures are given along with the turbine inlet temperature, T_1. Furthermore we will assume that the turbine has an isentropic efficiency of 85% while

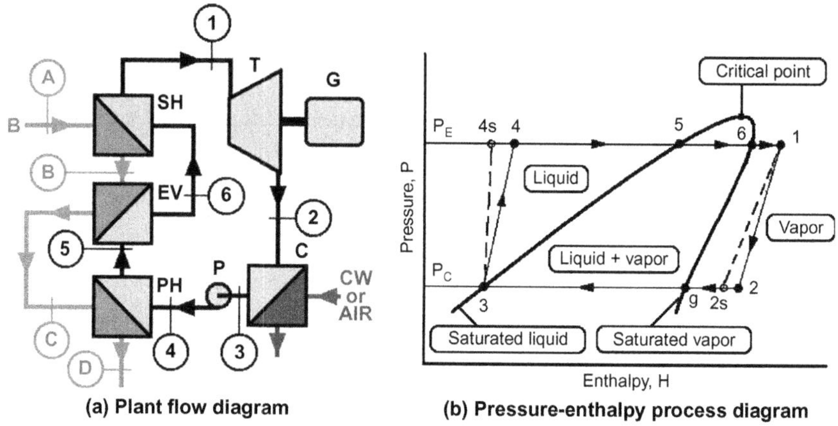

Fig. H.1 (a) and (b) Diagrams for binary plant with superheat.

the pump efficiency is 75%. These data are sufficient to allow the determination of all heat transfer and work transfer terms, as well as cycle efficiencies.

The overall plant performance depends on the properties of the geothermal brine, B which will be taken as pure water. We will assume that the inlet brine pressure, temperature, and mass flow rate are all given.

The analysis of the cycle and plant requires us to find the enthalpy, h, at all state points and entropy, s, at selected points. This is where REFPROP comes in. REFPROP can be used in either of two ways to solve the problem: (1) it can find specific properties for a given set of independent properties, or (2) it can be used as an Excel function, similar to trigonometric or statistical functions, allowing the cycle to be analyzed parametrically with Excel. We will illustrate both approaches.

H.3 REFPROP state-point properties

Let us establish the given data set first. We will present the data in a table.

State	Pressure kPa	Temperature °C	Quality —	Entropy kJ/kg · K	Enthalpy kJ/kg
1	**1,300**	**145**	**n.d.**		
2s	**250**				
2	**250**				
g	**250**		**1**		
3	**250**		**0**		
4s	**1,300**		**n.d.**		
4	**1,300**		**n.d.**		
5	**1,300**		**0**		
6	**1,300**		**1**		
A	**1,500**	**165**	**n.d.**		

Notice that states 2s and 4s are not actual states but are needed to find states 2 and 4, respectively. Similarly, state g (saturated vapor) occurs somewhere in the condenser between states 2 and 3. Note also that the quality is not defined (n.d.) outside the 2-phase liquid-vapor region. For this tutorial, we will ignore pressure losses in all piping and heat exchangers.

Now REFPROP will be used to find the enthalpy values at the required state points. First, select “pentane” from the Substance menu. Then under Options, go to Units, and select “SI with Celsius”; also select “kPa” as the pressure units since the problem uses kPa, not MPa. Under the Calculate tab, check “Specified state points”. This will allow us to find state 1; enter 145 and 1300, and get 35.846 kg/m^3 for the density, 534.80 kJ/kg for the enthalpy, and 1.3878 kJ/kg-K for the entropy. We do not need the density and could have deselected it from the Options/Properties menu. Copy the h and s values into our table.

The table below will be filled in according to our use of REFPROP, so follow the text below for the sequence of operations.

State	Pressure kPa	Temperature °C	Quality —	Entropy kJ/kg · K	Enthalpy kJ/kg
1	**1,300**	**145**	**n.d.**	1.3878	**534.80**
2s	**250**	98.419		1.3878	469.89
2	**250**				**479.627**
g	**250**	65.409	**1**	1.2024	404.06
3	**250**	65.409	**0**	0.22095	**71.799**
4s	**1,300**	65.876	**n.d.**	0.22095	73.610
4	**1,300**		**n.d.**		**74.214**
5	**1,300**	138.73	**0**	0.75865	**275.54**
6	**1,300**	138.73	**1**	1.3483	**518.40**
A	**1,500**	**165**	**n.d.**	1.9913	**697.69**

Next we realize that state 2s has the same entropy as state 2, so copy that value into 2s. Next we have the pressure and entropy at state 2s and can go to REFPROP with those to get the enthalpy at 2s. The obtained values have been pasted into the table.

Now to find the state 2 we need the definition of the turbine efficiency, namely,

$$\eta_t = \frac{w_{t,act}}{w_{t,s=const}} = \frac{h_1 - h_2}{h_1 - h_{2s}} \tag{H.1}$$

Thus we can calculate the enthalpy at state 2 from

$$h_2 = h_1 - \eta_t(h_1 - h_{2s}) \tag{H.2}$$

and add the answer to the table.

Continuing through the cycle in order, we see that state g is next. Since it is a saturation state, we need to select Calculate/Saturation points (at equilibrium). This will open a new window in REFPROP, which can be dragged on the screen to allow it to be viewed together with our first window. Enter the pressure of 250, and get the results. The vapor values belong to state g, and the liquid values to state 3.

Next we deal with the pump. The entropy at state 4s is the same as at state 3, so we can use the first REFPROP window to get the enthalpy there. Enter an entropy of 0.22095, and a pressure of 1300, and see the results.

Note: You can easily copy and paste between REFPROP windows by highlighting and using Ctrl-C and then Ctrl-V. This conveniently eliminates possible typing and round-off errors.

Now to find the state 4 we need the definition of the pump efficiency, namely,

$$\eta_p = \frac{w_{t,s=const}}{w_{t,act}} = \frac{h_{4s} - h_3}{h_4 - h_3} \tag{H.3}$$

Thus we can calculate the enthalpy at state 4 from

$$h_4 = h_3 + (h_{4s} - h_3)/\eta_p \quad \text{(H.4}$$

and add the answer to the table.

Only states 5 and 6 remain, and they are both saturation states. So we revert to th second REFPROP window using a pressure of 1300 and get the results, placing th vapor values in the table for state 6 and the liquid values for state 5. Thus the table i now complete for the n-pentane enthalpy values. If needed, property values for the sti blank entries could be found from REFPROP.

To find the brine properties, we need to change the Substance in REFPROP t "water". This opens a third window. Then enter the brine pressure, 1500, and th temperature, 165, and get the results.

The rest of the analysis of the cycle and plant can now be carried out using th standard thermodynamic equations. The results are summarized in the table below The reader should verify the results for him/herself.

Item	Value
Total heat added	460.586 kJ/kg of n-pentane
Heat rejected	407.828 kJ/kg of n-pentane
Work of turbine	55.173 kJ/kg of n-pentane
Work of pump	2.415 kJ/kg of n-pentane
Net cycle work	52.758 kJ/kg of n-pentane
Cycle thermal efficiency	0.1145 (11.45%)

In the case of compressed liquid states, a word of caution is in order. Owing to the non-unique relationship between the enthalpy h and temperature T, for certain values of these independent variables, two different physical states are possible. This is illustrated in Fig. H.2, a Mollier diagram (enthalpy plotted against entropy) for water substance, when T and h are used as independent variables. All substances behave similarly. Namely, isotherms exhibit a minimum in enthalpy for intermediate values o entropy, whereas they rise steeply at low values, and rise and taper off asymptotically at high values. Thus, for a given temperature, say 575.02°C, and a given enthalpy say 3000 kJ/kg, there are two possible values of entropy, namely, point A, 3.72 kJ/kg · K and point B, 5.46 kJ/kg · K. Under these and similar conditions, REFPROP might return a message of non-convergence. Point A corresponds to an extremely high pressure, ~1,200 MPa, and is unrealistic for geothermal applications. If this convergence problem should arise, there is a way around it. From Fig. H.2, it is clear that the enthalpy h is monotonic in pressure P. Thus, switching the independent variables to P and h should eliminate any problem.

H.4 REFPROP as an Excel function

As explained in the REFPROP 8.0 User's Guide (p. 86*ff*.), REFPROP can be linked to Excel using a .dll statement. Once this is properly set up, it allows Excel to call up thermophysical properties in the same way as any other library function, such as

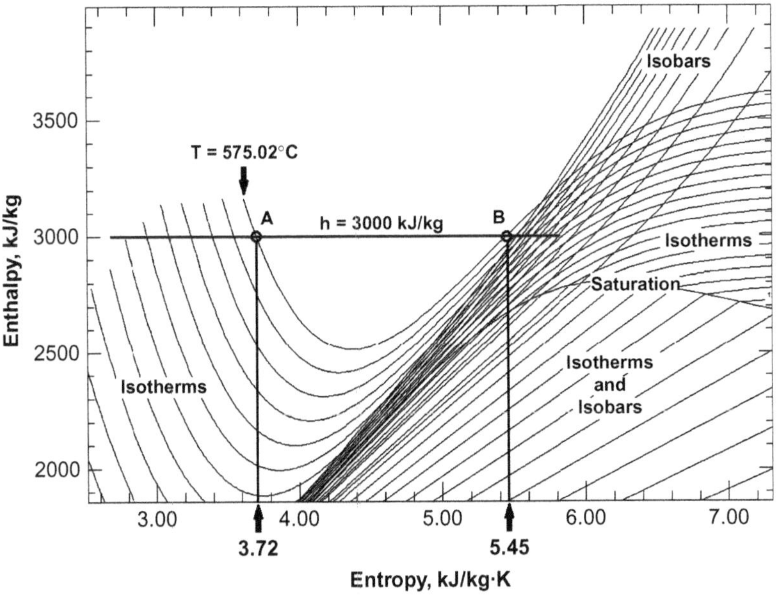

ig. H.2 *Mollier diagram for water.*

average," "sum," etc. The properties of most interest to us include the enthalpy, ntropy, pressure, and temperature, both for saturated states and single-phase liquid ınd vapor states. There are many other properties available and they can be found in he User's Guide (pp. 87–88).

For *single-phase liquid or vapor states for a pure substance*, we need two independent roperties to nail the value of any other property. Thus if we need the enthalpy, we nust supply to the calling function the values of, say, the pressure and temperature. REFPROP allows the use of many combinations of independent properties (User's Guide, p. 89). The most useful to us in our work are the following: TP, PH, and PS, vhere T = temperature, P = pressure, H = enthalpy, and S = entropy.

The general format of the calling function for the enthalpy as a function of tempera- ure and pressure is as follows:

= enthalpy(*substance name*, "TP", *units*, *T-value*, *P-value*)

For *substance name*, you can either type in the name, such as water, or enter an Excel cell in which the name of the substance has been typed. The latter way is very handy especially when the problem involves two fluids, as in the example being stud- ed. For *units*, the same options apply; it is convenient to just type the preferred units nto a cell and insert that cell into the function calling statement. Also SI with C can be abbreviated as C. English units are also available in REFPROP. The pair of indepen- dent variables (in this case "TP") must be enclosed within quotes.

The entropy can be found from a similar function:

= entropy(*substance name*, "TP", *units*, *T-value*, *P-value*)

Another useful function is:

= enthalpy(*substance name*, "PS", *units*, *P-value*, *S-value*)

This one comes in handy when we seek the ideal outlet states from turbines or pumps where we know the pressure and entropy.

If we need to know the temperature that corresponds to a given pressure and enthalpy, we can use:

= temperature(*substance name*, "PH", *units*, *P-value*, *H-value*)

For saturated liquid states, the general calling function for the enthalpy as a function of saturation temperature is:

= enthalpy(*substance name*, "Tliq", *units*, *T-value*)

For saturated vapor states, the general calling function for the enthalpy as a function of saturation temperature is:

= enthalpy(*substance name*, "Tvap", *units*, *T-value*)

For saturated liquid states, the general calling function for the enthalpy as a function of saturation pressure is:

= enthalpy(*substance name*, "Pliq", *units*, *P-value*)

For saturated vapor states, the general calling function for the enthalpy as a function of saturation pressure is:

= enthalpy(*substance name*, "Pvap", *units*, *P-value*)

The saturation pressure for a pure substance that corresponds to a given temperature can be found from:

= pressure(*substance name*," "Tliq", *units*, *T-value*)

The saturation temperature for a pure substance that corresponds to a given pressure can be found from:

= temperature(*substance name*, "Pliq", *units*, *P-value*)

In the case of azeotropic mixtures, the last two expressions suffice. For non-azeotropic mixtures, the last two expressions will give the values on the saturated liquid line, but to find the values on the saturated vapor line, one must insert "Tvap" and "Pvap" for "Tliq" and "Pliq", respectively.

Let us now revisit the Rankine cycle and use Excel to analyze the system. Using the appropriate functions, the following spreadsheet can be obtained.

WF = n-pentane: work & heat per kg/s of n-pentane flowing in the cycle

PE	PC	tC	tE	h3	s3	h4s	eta-P
MPa	MPa	C	C	kJ/kg	kJ/kgK	kJ/kg	
1.3	0.25	65.41	138.73	71.80	0.220954	73.61	0.75
						pentane	C
h4	h5	h6	T1	h1	s1	s2s	sg
kJ/kg	kJ/kg	kJ/kg	C	kJ/kg	kJ/kgK	kJ/kgK	kJ/kgK
74.22	275.54	518.40	145.00	534.80	1.3878	1.3878	1.2024
h2s	eta-T	h2	w-T	w-P	**w-NET**	q-IN	**eta-TH**
kJ/kg		kJ/kg	kW/(kg/s)	kW/(kg/s)	**kW/(kg/s)**	kW/(kg/s)	**%**
469.89	0.85	479.63	55.17	2.42	**52.75**	460.58	**11.45**

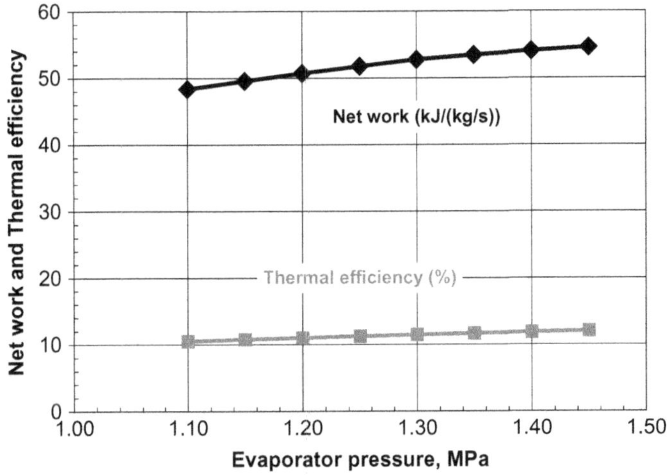

Fig. H.3 Variation in specific net work and thermal efficiency with evaporator pressure.

In the actual Excel spreadsheet, these numbers would be arrayed across the page; they have been repackaged here to fit the width of the page.

The results are of course identical with the first approach. This method facilitates parametric studies. For example, the high pressure PE can be varied over a range to see its effect on net work and efficiency. For example, the results shown in Fig. H.3 were obtained rapidly by sweeping the spreadsheet over pressure values from 1.10 to 1.45.

To complete the analysis of the plant, one must apply the First Law of thermodynamics to the heat exchanger train, knowing the brine inlet pressure, temperature, and mass flow rate. One additional parameter is needed, namely, the pinch-point temperature difference between the brine and the n-pentane as they interact thermally in the heat exchangers. REFPROP will yield all the enthalpies and temperatures needed to complete the solution, using the same methodology illustrated in these notes together with the governing equations developed in Chapter 8.

Reference

[1] National Institute of Standards and Technology (NIST), U.S. Department of Commerce http://www.nist.gov/srd/nist23.cfm.

Index

'ote: Page numbers followed by "t" indicates table and "f" indicates figure.

B

C

D

I

M

P

Q

R

T

U

Y

Z

Printed in the United States
By Bookmasters